VOLUME ONE HUNDRED AND THIRTY

Current Topics in DEVELOPMENTAL BIOLOGY

Extracellular Matrix and Egg Coats

CURRENT TOPICS IN DEVELOPMENTAL BIOLOGY

"A meeting-ground for critical review and discussion of developmental processes"
A.A. Moscona and Alberto Monroy (Volume 1, 1966)

VOLUME ONE HUNDRED AND THIRTY

CURRENT TOPICS IN DEVELOPMENTAL BIOLOGY

Extracellular Matrix and Egg Coats

Edited by

EVELINE S. LITSCHER
Department of Cell, Developmental, and Regenerative Biology, Icahn School of Medicine at Mount Sinai, New York, NY, United States

PAUL M. WASSARMAN
Department of Cell, Developmental, and Regenerative Biology, Icahn School of Medicine at Mount Sinai, New York, NY, United States

ELSEVIER

ACADEMIC PRESS
An imprint of Elsevier

Academic Press is an imprint of Elsevier
50 Hampshire Street, 5th Floor, Cambridge, MA 02139, United States
525 B Street, Suite 1650, San Diego, CA 92101, United States
The Boulevard, Langford Lane, Kidlington, Oxford OX5 1GB, United Kingdom
125 London Wall, London, EC2Y 5AS, United Kingdom

First edition 2018

ISBN: 978-0-12-809802-8
ISSN: 0070-2153

For information on all Academic Press publications
visit our website at https://www.elsevier.com/books-and-journals

Publisher: Zoe Kruze
Acquisition Editor: Zoe Kruze
Editorial Project Manager: Shellie Bryant
Production Project Manager: Denny Mansingh
Cover Designer: Greg Harris

Typeset by SPi Global, India

CONTENTS

CONTRIBUTORS

Josephine C. Adams
School of Biochemistry, University of Bristol, Bristol, United Kingdom

Marcel Bokhove
Department of Biosciences and Nutrition & Center for Innovative Medicine, Karolinska Institutet, Huddinge, Sweden

Douglas W. DeSimone
Department of Cell Biology, University of Virginia, School of Medicine, Charlottesville, VA, United States

Bette J. Dzamba
Department of Cell Biology, University of Virginia, School of Medicine, Charlottesville, VA, United States

Stephen Frankenberg
School of BioSciences, University of Melbourne, Parkville, VIC, Australia

Richard J. Goodyear
Sussex Neuroscience, School of Life Sciences, University of Sussex, Brighton, United Kingdom

Satish K. Gupta
Reproductive Cell Biology Laboratory, National Institute of Immunology, New Delhi, India

David F. Holmes
Wellcome Trust Centre for Cell-Matrix Research, Faculty of Biology, Medicine and Health, Manchester Academic Health Science Centre, University of Manchester, Manchester, United Kingdom

Luca Jovine
Department of Biosciences and Nutrition & Center for Innovative Medicine, Karolinska Institutet, Huddinge, Sweden

Karl E. Kadler
Wellcome Trust Centre for Cell-Matrix Research, Faculty of Biology, Medicine and Health, Manchester Academic Health Science Centre, University of Manchester, Manchester, United Kingdom

Emily E. Killingbeck
Department of Genome Sciences, University of Washington, Seattle, WA, United States

Eveline S. Litscher
Department of Cell, Developmental, and Regenerative Biology, Icahn School of Medicine at Mount Sinai, New York, NY, United States

Yinhui Lu
Wellcome Trust Centre for Cell-Matrix Research, Faculty of Biology, Medicine and Health, Manchester Academic Health Science Centre, University of Manchester, Manchester, United Kingdom

Tsukasa Matsuda
Graduate School of Bioagricultural Sciences, Nagoya University, Nagoya, Japan

Robert P. Mecham
Department of Cell Biology and Physiology, Washington University School of Medicine, St. Louis, MO, United States

Jonathon M. Muncie
Center for Bioengineering and Tissue Regeneration, University of California; Graduate Program in Bioengineering, University of California San Francisco and University of California Berkeley, San Francisco, CA, United States

Shunsuke Nishio
Graduate School of Bioagricultural Sciences, Nagoya University, Nagoya, Japan

Hiroki Okumura
Faculty of Agriculture, Meijo University, Nagoya, Japan

Francesco Ramirez
Department of Pharmacological Sciences, Icahn School of Medicine at Mount Sinai, New York, NY, United States

Marilyn B. Renfree
School of BioSciences, University of Melbourne, Parkville, VIC, Australia

Guy P. Richardson
Sussex Neuroscience, School of Life Sciences, University of Sussex, Brighton, United Kingdom

Rei Sekiguchi
Cell Biology Section, National Institute of Dental and Craniofacial Research, National Institutes of Health, Bethesda, MD, United States

Tobias Starborg
Wellcome Trust Centre for Cell-Matrix Research, Faculty of Biology, Medicine and Health, Manchester Academic Health Science Centre, University of Manchester, Manchester, United Kingdom

Willie J. Swanson
Department of Genome Sciences, University of Washington, Seattle, WA, United States

Paul M. Wassarman
Department of Cell, Developmental, and Regenerative Biology, Icahn School of Medicine at Mount Sinai, New York, NY, United States

Valerie M. Weaver
Center for Bioengineering and Tissue Regeneration; Eli and Edythe Broad Center of Regeneration Medicine and Stem Cell Research, The Helen Diller Family Comprehensive Cancer Center, University of California, San Francisco, CA, United States

Kenneth M. Yamada
Cell Biology Section, National Institute of Dental and Craniofacial Research, National Institutes of Health, Bethesda, MD, United States

PREFACE

This volume of CTDB consists of 14 chapters, 7 on extracellular matrix (ECM) and 7 on the egg's zona pellucida (ZP) contributed by leading investigators in Australia, India, Japan, Sweden, the United Kingdom, and the United States.

Research on ECM has a long and extensive history with nearly 100,000 primary papers and 20,000 reviews published on the subject during the past 70 years or so (PubMed). ECM is a collection of diverse molecules, including proteins (e.g., collagens, fibronectins, and laminins), proteoglycans (e.g., heparin, keratin, and chondroitin sulfates), and polysaccharides (e.g., hyaluronic acid), that reside outside animal cells. ECM serves many roles, such as providing structural support for cells, tissues, and organs, and regulating numerous events, including gene expression, differentiation, morphogenesis, cellular adhesion, cellular migration, and intercellular communication. Similarly, nearly all animal eggs are surrounded by a unique, specialized ECM, called the ZP, that provides both structural support for eggs and follicle cells surrounding them and species-restricted receptors for binding of sperm during fertilization.

In this volume of CTDB coverage of ECM includes the following seven chapters: (Chapter 1) How the ECM functions as a highly complex entity that regulates tissue organization and cell behavior through dialogue with cellular constituents of tissues (J.M. Muncie and V.M. Weaver); (Chapter 2) Research on matricellular proteins of ECM in nonmammalian animals that provide access to early stages of embryonic development and limb and organ regeneration (J.C. Adams); (Chapter 3) Mechanisms underlying the generation of a collagen fibril network of defined architecture and mechanical properties (D.F. Holmes, Y. Lu, T. Starborg, and K.E. Kadler); (Chapter 4) Developmental and disease dynamics of basement membranes of nematodes, flies, and vertebrates (R. Sekiguchi and K.M. Yamada); (Chapter 5) How ECM composition and physical properties influence cell fate decisions associated with vascular tissue development and disease (R.P. Mecham and F. Ramirez); (Chapter 6) The structure, function, and development of the tectorial membrane and major events during its evolution (R.J. Goodyear and G.P. Richardson); (Chapter 7) How biophysical features of ECM, such as stiffness and viscoelasticity, influence the sculpting of embryonic tissues and regulation of cell fates (B.J. Dzamba and D.W. DeSimone).

In this volume of CTDB coverage of the ZP includes the following seven chapters: Various aspects of the egg's ZP in fish (Chapter 8, E.S. Litscher and P.M. Wassarman), birds (Chapter 9, S. Nishio, H. Okumura, and T. Matsuda), mice (Chapter 10, P.M. Wassarman and E.S. Litscher), marsupials and monotremes (Chapter 11, S. Frankenberg and M. Renfree), and human beings (Chapter 12, S.K. Gupta); (Chapter 13) The atomic structures of ZP module proteins of mammals and nonmammals in the context of sperm–egg interactions (M. Bokhove and L. Jovine); (Chapter 14) The structure and function of metazoan egg coat proteins with emphasis on the potential role their evolution has played in creation and maintenance of species boundaries (E.E. Killingbeck and W.J. Swanson).

We are very grateful to Richard Goodyear and Guy Richardson for providing a beautiful and relevant image for the cover of the volume. Finally, we thank all of the authors for their scholarly contributions to this volume and for their considerable efforts to complete their manuscripts on time in order to meet publisher's deadlines.

Eveline S. Litscher
Paul M. Wassarman

CHAPTER ONE

The Physical and Biochemical Properties of the Extracellular Matrix Regulate Cell Fate

Jonathon M. Muncie[*,†], **Valerie M. Weaver**[*,‡,1]

[*]Center for Bioengineering and Tissue Regeneration, University of California, San Francisco, CA, United States

[†]Graduate Program in Bioengineering, University of California San Francisco and University of California Berkeley, San Francisco, CA, United States

[‡]Eli and Edythe Broad Center of Regeneration Medicine and Stem Cell Research, The Helen Diller Family Comprehensive Cancer Center, University of California, San Francisco, CA, United States

[1]Corresponding author: e-mail address: valerie.weaver@ucsf.edu

Contents

Current Topics in Developmental Biology, Volume 130
ISSN 0070-2153
https://doi.org/10.1016/bs.ctdb.2018.02.002

Abstract

The extracellular matrix is a complex network of hydrated macromolecular proteins and sugars that, in concert with bound soluble factors, comprise the acellular stromal microenvironment of tissues. Rather than merely providing structural information to cells, the extracellular matrix plays an instructive role in development and is critical for the maintenance of tissue homeostasis. In this chapter, we review the composition of the extracellular matrix and summarize data illustrating its importance in embryogenesis, tissue-specific development, and stem cell differentiation. We discuss how the biophysical and biochemical properties of the extracellular matrix ligate specific transmembrane receptors to activate intracellular signaling that alter cell shape and cytoskeletal dynamics to modulate cell growth and viability, and direct cell migration and cell fate. We present examples describing how the extracellular matrix functions as a highly complex physical and chemical entity that regulates tissue organization and cell behavior through a dynamic and reciprocal dialogue with the cellular constituents of the tissue. We suggest that the extracellular matrix not only transmits cellular and tissue-level force to shape development and tune cellular activities that are key for coordinated tissue behavior, but that it is itself remodeled such that it temporally evolves to maintain the integrated function of the tissue. Accordingly, we argue that perturbations in extracellular matrix composition and structure compromise key developmental events and tissue homeostasis, and promote disease.

1. INTRODUCTION

The extracellular matrix (ECM) is a complex network of proteins, polysaccharides, and water that comprise the acellular stromal microenvironment in all tissues and organs. Historically, the ECM was thought to provide structural information required to maintain the physical integrity of the tissue. However, it is now understood that the ECM is a biologically active component of all tissues that directs cell fate and influences tissue development and homeostasis (Fig. 1).

As organisms develop, they continuously generate and reorganize their ECM to provide the necessary structural framework to support the growth and development of emerging tissues. The ECM in turn provides critical biochemical and biophysical cues that guide cell fate, drive morphogenetic movements to sculpt the tissue, and induce tissue-specific differentiation.

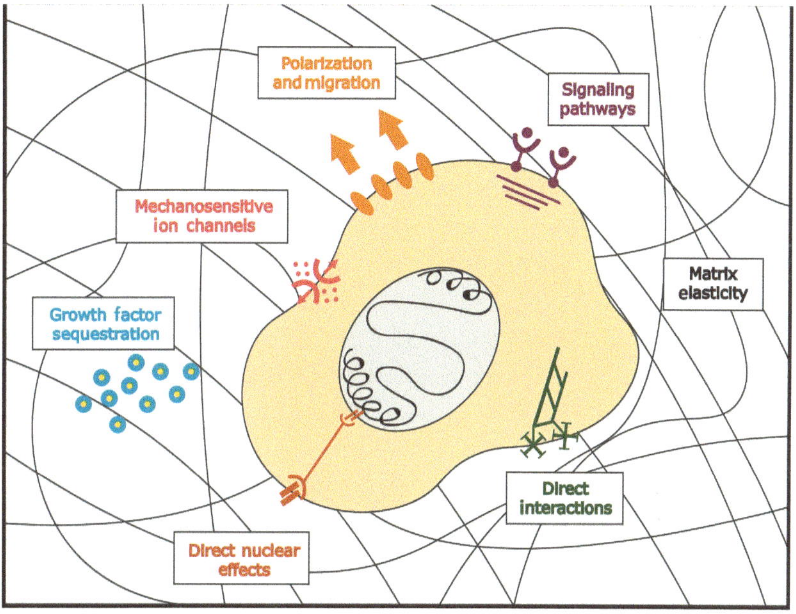

Fig. 1 Illustration of the various physical and biochemical cues integrated by the extracellular matrix, which are simultaneously sensed by cells through parallel mechanisms and are critical for determining cell fate, inducing tissue-specific differentiation, and promoting developmental morphogenesis.

The concept of "dynamic reciprocity" which maintains that the evolving ECM dictates cell and tissue fate which feedback to modulate ECM composition and organization represents a critical concept in developmental biology (Bissell, Hall, & Parry, 1982; Paszek & Weaver, 2004). "Tensional homeostasis" incorporates the viscoelasticity of the ECM and cell tension into the dynamic reciprocity paradigm thereby providing a unified working hypothesis with which to understand how the evolving biochemical and biophysical properties of the ECM direct development and maintain tissue homeostasis.

The ECM is also a key component of the adult stem cell niche and refers to the local microenvironment that sustains stem cell quiescence and facilitates the maintenance of stem cells through regulated self-replication and retention of multipotency. The ECM physically buffer stem cells that reside within the niche from differentiation cues sequesters critical growth factors and morphogens, and facilitates efficient nutrient exchange to sustain the long-term growth and survival and pluripotency of the stem cells.

2. ECM COMPOSITION

The main functional units of the ECM are cell-secreted macromolecular proteins. There are four general classes of ECM proteins: collagens, proteoglycans, glycoproteins, and elastins (Tsang, Cheung, Chan, & Cheah, 2010).

2.1 Collagens

Collagens provide the tissue with tensile strength and structural integrity (Gordon & Hahn, 2010; Lodish, Berk, Zipursky, et al., 2000). Collagens are composed of three alpha chains that assemble into heterotrimeric and homotrimeric molecules, various combinations which comprise the 28 recognized types of collagen. Fibrillar collagens are the most common type and are assembled in woven triple-helical structures that arise from long repeating stretches of Gly-X-Y residues in the alpha chains. In these repeated stretches, X is typically proline and Y is typically hydroxyproline. These triple helices then self-assemble into thin and thick fibrils. Nonfibrillar collagens arise from disruptions in the Gly-X-Y repeats of the alpha chains. Instead of forming fibrils, these nonfibrillar collagens form mesh-like networks in the ECM, such as collagen IV found in basement membranes.

The biochemical and mechanical properties of fibrillar collagens are largely dependent on posttranslational modifications that drive the cross-linking of the collagen fibrils, and thus dictate the tensile strength of the tissue. Proline and lysine residues on procollagens are hydroxylated intracellularly by specific enzymes (Yamauchi & Sricholpech, 2012). These hydroxylated residues can be further modified within the cell by enzyme-mediated glycosylation, resulting in the addition of galactose or glucose. Collagen glycosylation alters cell–collagen interactions, and thus broadly influences biological functions of collagen, such as the ability of collagens to direct angiogenesis or bone mineralization (Jürgensen et al., 2011; Palmieri et al., 2010; Tenni, Valli, Rossi, & Cetta, 1993). Once procollagens are secreted to the extracellular space and are self-assembled into fibrils, specific lysine and hydroxylysine residues are deaminated by an enzyme called lysyl oxidase (LOX) that produces reactive aldehyde groups that initiate covalent cross-links via condensation reactions. Lysyl oxidase and hydroxylase cross-linking are critical for the tensile strength and structural stability of tissues, and loss of their activity has deleterious consequences to the organism including bone fragility, cardiovascular disease, and in some instances lethality (Kagan & Li, 2003; Mäki et al., 2002). Thus, extracellular posttranslational modifications to fibrillar collagen expand its structure function and modify its biological role in tissues.

2.2 Proteoglycans

Proteoglycans consist of a core protein domain covalently linked to glycosaminoglycans (GAGs) (Mouw, Ou, & Weaver, 2014). These GAGs form long, negatively charged, linear repeats of disaccharide units, and provide proteoglycans with their unique ability to bind water, which is critical for imparting compressive resistance to tissues. Aggrecan and versican are common examples of proteoglycans that are ubiquitously expressed in many tissues. Hyaluronan, or hyaluronic acid (HA), is a common GAG that is also abundantly expressed in all tissues but has unique properties because it does not covalently attach to peptides (Laurent & Fraser, 1992). Thus, while not a true proteoglycan, HA possesses similar biological functions exhibited by proteoglycans. HA is abundantly expressed in the brain, where it forms a relatively "loose" and unorganized network. However, HA can also bind and form cross-links with other proteoglycans and glycoproteins, such as versican and tenascin, to support more condensed perineural networks (PNNs) that stiffen the neuronal microenvironment (Kwok, Dick, Wang, & Fawcett, 2011). The formation of PNNs coincides with brain stiffening, reduced neuron growth and extension, and loss of plasticity in the developing brain, further illustrating how tissue-specific biochemical properties of the ECM guide cell fate and tissue development (Kwok et al., 2011; Yamaguchi, 2000).

2.3 Glycoproteins

Glycoproteins are similar to proteoglycans in that they consist of peptide units and covalently bound carbohydrate groups. However, whereas proteoglycans are characterized by long, linear chains of repeating disaccharides, glycoproteins contain bulky, branched carbohydrates with no, or few, repeating structures. Glycoproteins serve as connector molecules within the ECM because they contain functional groups that recognize and bind other ECM molecules, as well as to cell adhesion molecules, secreted growth factors, and morphogens. Fibronectin and laminin are two predominant and important glycoproteins found in the ECM stroma.

Fibronectin is secreted into the extracellular microenvironment and then forms polymerized fibrils with other fibronectin molecules or with collagen by a cell-dependent process termed fibrillogenesis (Kadler, Hill, & Canty-Laird, 2008; Singh, Carraher, & Schwarzbauer, 2010). In fact, cross-linking of collagen by LOX depends upon fibronectin, emphasizing the importance of intimate interactions between ECM proteins in regulating its structure function (Kubow et al., 2009). The RGD sequence (Arg-Gly-Asp), which can bind to specific cellular transmembrane proteins called integrins, is a particularly important cell-binding motif within the fibronectin molecule. However, the strength of cell–integrin interactions with fibronectin is greatly enhanced through binding to the separate fibronectin synergy sequence (Schwarzbauer & DeSimone, 2011; Sechler, Corbett, & Schwarzbauer, 1997). Fibroblasts in particular are capable of unfolding fibronectin to reveal cryptic cell-binding sites, including the synergy sequence (Friedland, Lee, & Boettiger, 2009; Seong et al., 2013; Smith et al., 2007). Once revealed, the synergy sequence can then be ligated by cellular $\alpha5\beta1$ integrin to potentiate intracellular tension and alter epithelial or endothelial morphogenesis or fibroblast function (Benito-Jardón et al., 2017; Friedland et al., 2009; Miroshnikova et al., 2017). In this way, fibronectin structure function can be discretely modulated by cell-generated force to alter cell and tissue function.

Laminins are composed of α, β, and γ heterotrimer chains that connect via central domains to form cross-shaped, Y-shaped, or single-arm structures (Mouw, Ou, et al., 2014; Timpl & Brown, 1994). Laminins are localized to the basal lamina of tissues and function primarily to connect various components of the ECM. Laminins are classified according to their subunit composition, such that laminin 111 refers to laminin trimers comprised of $\alpha1$, $\beta1$, and $\gamma1$ chains (Aumailley et al., 2005). Laminin heterogeneity generates significant functional diversity. For instance, while laminin 111 polymerizes to form a meshwork that contributes to the tissue basement membrane, laminin 332, that is secreted during wound healing and hair follicle development, is subject to extensive proteolytic cleavage, which is critical for its ability to facilitate coordinated keratinocyte migration (Sugawara, Tsuruta, Ishii, Jones, & Kobayashi, 2008).

2.4 Elastin

Elastin is a unique structural protein that imparts, as its name suggests, elastic properties to tissues such as the skin and vasculature. Whereas fibrillar collagen provides tensile strength to resist deformation, elastin provides tissues with resilience so that when they are stretched they can return to their original conformation. Elastins are typically more hydrated and flexible in their relaxed, as compared to their stretched, state. The precise structure function relationship that attributes this unique property of elastins is still a matter of debate, but is frequently described as the relaxed state being entropically favorable toward stretched configurations (Debelle & Tamburro, 1999).

Here, we reviewed the different types of ECM proteins as discrete units according to their structure and functional properties. Nevertheless, it is important to appreciate that the tissue-specific biochemical and physical properties of the ECM arise from its complex organization, posttranslational modifications, and its ability to be dynamically remodeled; and these are absolutely critical for the ECM's ability to influence cell and tissue fate.

2.5 The Dynamic Nature of the ECM

The ECM is a fibrous network or scaffold within which tissues reside, prompting the perspective that the ECM is a static physical structure. In fact, the ECM is quite dynamic (Lu, Takai, Weaver, & Werb, 2011; Page-McCaw, Ewald, & Werb, 2007; Streuli, 1999). The protein components of the ECM are turned over regularly, and the cells within the tissue are actively engaged in remodeling their local ECM (Bonnans, Chou, & Werb, 2014). Virtually all cells produce and secrete ECM proteins into their microenvironment together with matrix metalloproteinases (MMPs), which degrade ECM proteins (Bonnans et al., 2014; Mecham, 2012; Page-McCaw et al., 2007). Cells also assemble and disassemble their local ECM by binding specific residues on matrix proteins and exerting physical forces generated by actomyosin contractility to modify the three-dimensional (3D) conformation and organization of these proteins (Ohashi, Kiehart, & Erickson, 2002; Pankov et al., 2000; Zhong et al., 1998). Cellular remodeling can promote interactions between different ECM molecules or reveal cryptic binding sites within the ECM molecule that promote cell adhesion and proliferation (Baneyx, Baugh, & Vogel, 2001, 2002; Zhong et al., 1998). The ability of the ECM to be constantly remodeled and reorganized is critical for development, tissue-specific differentiation, and stem cell fate specification.

Indeed, the ECM is a complex network of macromolecular structures whose physical and biochemical properties play a key role in cell and tissue biology. Moreover, the ECM is assembled and remodeled via cell-mediated processes, and its organization and reorganization are important in development and stem cell niche regulation. Rather than merely being a passive component of the stromal microenvironment within which cells and tissues exist, it is now appreciated that the ECM can directly regulate cell and tissue fate.

3. ECM REGULATES DEVELOPMENT AND MODULATES STEM CELL FATE

3.1 The ECM Regulates Development

Beginning in the 1980s and early 1990s, the ECM was recognized as a critical regulator of embryogenesis and tissue-specific development. Gene analysis and manipulation studies in model organisms revealed that several developmental defects could be mapped back to ECM proteins (Adams & Watt, 1993). In *Caenorhabditis elegans*, *dpy-13* (dumpy) was identified as a member of the collagen gene family, and mutations to *dpy-13* were shown to cause a short and chunky body shape (von Mende, Bird, Albert, & Riddle, 1988). Mutations to *Sqt-1* and *clb-2* were also identified as sequences for collagen proteins that caused significant developmental defects in body shape, with the *clb-2* mutation causing embryonic lethality (Guo, Johnson, & Kramer, 1991; Kramer, Johnson, Edgar, Basch, & Roberts, 1988). Mutations to *unc-6*, a sequence for a laminin-related protein, caused defects in neural development that resulted from misguiding of axon extensions (Ishii, Wadsworth, Stern, Culotti, & Hedgecock, 1992). Similar studies in *Drosophila* identified mutations to the gene-coding sequences of laminin A and the *Drosophila* analog for fibrinogen, which both led to significant developmental defects (Baker, Mlodzik, & Rubin, 1990; Hortsch & Goodman, 1991). Furthermore, mutations to *Drosophila* integrins, the cellular receptors for ECM proteins, generated a host of defects ranging from wing, to eye, to muscle development (Brower & Jaffe, 1989; Leptin, Bogaert, Lehmann, & Wilcox, 1989; Volk, Fessler, & Fessler, 1990; Wilcox, DiAntonio, & Leptin, 1989; Zusman, Patel-King, Ffrench-Constant, & Hynes, 1990).

In addition to mapping genetic mutations to ECM proteins and receptors, function-blocking manipulation of cell–ECM interactions has served to illustrate the importance of cell–matrix adhesion to tissue development. Decoupling interactions between mesodermal progenitors and fibronectin using function-blocking antibodies in the blastocoel roof of amphibian embryos prevented development from proceeding beyond gastrulation (Boucaut et al., 1984, 1985). Later in development, injection of the integrin-binding motif of fibronectin, the RGD peptide, prevented the proper establishment of left/right asymmetry in *Xenopus* embryos (Yost, 1992). Similarly, injection of RGD-containing peptides in *Gallus gallus domesticus* (chick) embryos prevented neural crest cell migration (Boucaut et al., 1984).

3.2 ECM-Directed Cell Migration in Development

ECM-directed cell migration is a critical determinant of cell fate specification during development. This concept was demonstrated by studies exploring the impact of cell–ECM interactions in heart development. The heart originates as a bilayered tube formed by the ECM-dependent migration of myocardial precursors from the left and right lateral sides toward the midline of the organism. Zebrafish hearts exhibit serious developmental defects when mutations are induced in the fibronectin-encoding *natter* gene that impede myocardial precursor migration (Trinh & Stainier, 2004). Similarly, morpholino-mediated knockdown of zebrafish fibronectin in early embryogenesis compromised cardiac development that could be rescued by injection of exogenous fibronectin (Matsui et al., 2007). Coordinated and directed ECM deposition is also a critical aspect of neural crest cell migration during embryogenesis. Neural crest cells migrate from the dorsal region of the developing neural tube and contribute to multiple cell types and tissues including neurons and osteoblasts (Knecht & Bronner-Fraser, 2002). Conditional knockout of β1 integrin in mouse neural crest precursors compromised fibronectin, laminin, tenascin, and collagen IV deposition, prevented neural crest cell migration, and led to lethal neuronal defects, emphasizing the link between ECM deposition and tissue development (Pietri et al., 2004). Indeed, ECM deposition is important for guiding and directing cell migration during development and by so doing plays a key role in shaping the embryo, as has been illustrated by the attractive and repellent effects of versican and aggrecan on neural crest cell migration (Perissinotto et al., 2000). Moreover, laminin α5 appears to restrict neural crest cell migration along confined pathways, as was illustrated by the unfettered broad migration of neural crest cells and compromised development of the organism when this laminin chain was knocked out (Coles, Gammill, Miner, & Bronner-Fraser, 2006). Not surprisingly, distinct ECM components direct cell migration in different tissues during development. For example, fibronectin facilitates migration of cranial neural crest cells, whereas vitronectin, collagen I, and laminin do not (Alfandari, Cousin, Gaultier, Hoffstrom, & DeSimone, 2003).

While these and other experiments emphasize the importance of the ECM in development, the main inference from these pioneering studies is that the ECM provides structural integrity to maintain cell adhesion-dependent tissue organization and that it facilitates directed cell migration. Over the past several decades this perspective has expanded to encompass a role for the ECM and its receptors as an instructive developmental regulator and important cue that directs stem cell differentiation.

3.3 ECM as a Critical Component of the Stem Cell Niche

The ECM not only regulates stem cell fate and tissue-specific differentiation but also modulates stem cell behavior in adult organisms. Stem cells are characterized by two unique properties: (i) they are capable of self-renewal, dividing to produce additional stem cells in a process termed "symmetric" division; (ii) they are also capable of giving rise to differentiated cells with specialized functions, termed "asymmetric" division (Fuchs & Chen, 2012). Despite the potential for both self-renewal and differentiation, adult stem cells often remain quiescent for long periods of time (Cheung & Rando, 2013; Li & Clevers, 2010). It is the specialized ECM microenvironment of these adult stem cells, the stem cell niche, which has been credited with regulating stem cell behavior, including the maintenance of their prolonged quiescence. Indeed, the ECM is a critical part of this niche, providing a microenvironment within which stem cells can maintain quiescence until they receive stimuli that promote their expansion and differentiation (Gattazzo, Urciuolo, & Bonaldo, 2014). Often, stem cell expansion and differentiation are preceded or accompanied by changes to the niche. Thus, the ECM both maintains stem cell populations in the adult tissue and regulates their differentiation.

4. CELL–ECM SIGNALING

The ECM influences cell fate by binding to cell surface receptors that recognize specific ECM moieties. Once bound to the ECM, these cellular receptors alter cell growth, survival, motility, and differentiation by initiating intracellular signaling and cytoskeletal reorganization. The cellular receptors that bind to the ECM include integrins as well as discoidin domain receptors (DDRs), syndecans, CD44, and receptor for hyaluronic acid-mediated motility, as well as Robo receptors.

4.1 ECM–Integrin Binding

Integrins, which are the best-studied ECM receptors, are transmembrane heterodimer molecules that bind to specific ECM residues and interact with adhesion plaque proteins and cytoskeletal linker proteins via their cytoplasmic tail. There are 18 alpha subunits and 8 beta subunits, which combine to form 24 distinct integrin heterodimers (Hynes, 2002). The combination of the alpha and beta subunits that recognize and bind different ECM molecules imparts functional specificity to integrins. Not only do integrins facilitate the physical interaction of cells with the ECM, but they also induce intracellular signaling, alter cytoskeletal organization, and stimulate cellular tension to regulate cell growth and survival, promote invasion and migration, and direct cell differentiation and stem cell fate.

Integrin adhesion and signaling are highly context dependent. For instance, α5β1, αvβ5, α6β1, and α9β1 integrins are highly expressed in mouse embryonic stem cells (mESCs) (Lee et al., 2010), and their combined ligation to peptide-conjugated polyethylene glycol (PEG) hydrogels with fibronectin-derived peptide sequence RGDSP, CCN1-derived sequence TTSWSQ, and the tenascin-C-derived sequence AEIDGIE can maintain their pluripotency in culture. By contrast, preventing the initial, but not the late, ligation of the collagen-binding integrins α1β1 and α2β1, using function-blocking antibodies, inhibited osteoblast differentiation of both multipotent 2T3 cells and freshly harvested bone marrow cells, as was indicated by reduced alkaline phosphatase production and cellular mineralization (Jikko, Harris, Chen, Mendrick, & Damsky, 1999; Mizuno, Fujisawa, & Kuboki, 2000). Indeed, blocking integrin binding to collagen I reduced osteoblast differentiation even in the presence of a constitutively active bone morphogenetic protein 2 (BMP-2) receptor, presumably by blocking BMP-2 receptor activation. Similarly, knocking down the fibronectin receptor α5β1 integrin prevented osteoblast differentiation of human mesenchymal stem cells (MSCs), whereas its activation and ligation by collagen promoted osteoblast differentiation (Hamidouche et al., 2009). These findings underscore the critical context-dependent role of integrins in ECM-dependent regulation of stem cell fate.

Integrins are critical regulators of cell fate during tissue-specific development and differentiation. For instance, Fässler and colleagues used β1 integrin-deficient embryonic stem cells to demonstrate the importance of β1 integrin in early cardiogenesis (Fässler et al., 1996). Similarly, although conditional β1 integrin knockdown in mammary epithelial cells (MECs) had no effect on branching morphogenesis of the adolescent mammary gland, lack of β1 integrin prevented pregnancy-associated MEC alveolar differentiation (Klinowska et al., 1999; Naylor et al., 2005). Thus, in the absence of β1 integrin MECs failed to lactate, presumably because the cells could not respond to prolactin to activate transcription 5 (STAT5) signaling that is critically required for beta casein expression (Xu et al., 2009). Indeed, early studies identified β1 integrin signaling as necessary for MEC growth and survival (Boudreau, Sympson, Werb, & Bissell, 1995; Klinowska et al., 1999; Naylor et al., 2005).

4.2 Nonintegrin Receptors

Analogous to integrin receptors, the DDR subfamily of receptor tyrosine kinases that bind to extracellular collagens also modulate cell growth, survival, migration, and tissue morphogenesis by activating intracellular signaling and regulating gene transcription. For instance, DDR1 binding to collagen promotes the pluripotency of mESCs via Ras-mediated activation of phosphoinositide 3-kinase (PI3K)/Akt, and extracellular signal-regulated kinase (ERK) that regulate BMI-1 (Suh & Han, 2011). Similarly, DDR2 ligation by collagen and ERK activation is critical for runt-related transcription factor 2 (Runx2) induction of osteogenic differentiation that likely explains the dwarfism and aberrant bone formation observed in the DDR2-null mouse (Labrador et al., 2001; Zhang et al., 2011).

CD44 receptors, which bind to hyaluronan (HA), elicit profound effects on cell behavior by activating intracellular signaling and altering cytoskeletal organization. Thus, inhibiting CD44 binding to HA significantly reduced hematopoietic stem/progenitor cell (HSC/HPC) homing and engraftment into the bone marrow, presumably by impeding association with their endothelium and endosteum niche (Avigdor et al., 2004). Similarly, a role for CD44 in cell differentiation was underscored in CD44-null mice which exhibited severely compromised keratinocyte differentiation and epidermal barrier formation, presumably because HA ligation is critical for lamellar body formation and secretion and epidermal barrier development (Bourguignon et al., 2006). Interestingly, the CD44-null mice also had significantly reduced levels of extracellular HA, implicating cellular ligation in HA homeostasis.

Roundabout (Robo) proteins are another group of transmembrane receptors that bind to ECM glycoproteins termed Slits. Robo protein adhesion is critical for branching morphogenesis in the mammary gland (Macias et al., 2011), and Slit binding to Robo4 mediates HSC homing to the bone marrow niche and efficient bone marrow engraftment (Smith-Berdan et al., 2011). In *Drosophila*, Robo–Slit interactions play an integral role in mediating intestinal tissue homeostasis. *Drosophila* intestinal epithelial cells secrete Slit, which suppresses endocrine lineage differentiation by binding and activating Robo2 receptors on adjacent stem cell progenitors (Biteau & Jasper, 2014).

5. ECM-DEPENDENT MODULATION OF GROWTH FACTOR AND MORPHOGEN FUNCTION

The ECM controls the availability and presentation of growth factors and morphogens. This ECM-soluble factor-binding function is mediated primarily by the extracellular proteoglycans and heparin sulfates that contain binding sites for a multitude of growth factors and secreted morphogens and enzymes. The binding of soluble secreted factors to the ECM creates a signaling "reservoir" that can be rapidly accessed by the cells within the tissue. ECM binding of growth factors also modulates and sometimes potentiates the cellular signaling elicited by these molecules and can in some instances also bind inhibitors to temper signaling.

5.1 ECM-Transforming Growth Factor Beta Superfamily Interactions

Members of the transforming growth factor beta (TGF-β) superfamily, including BMPs and TGF-β, play a major role in regulating embryogenesis, tissue development, and stem cell fate. The BMPs and TGF-β bind to specific cellular receptors and activate intracellular signaling to alter cell and tissue behavior by regulating gene expression (ten Dijke & Hill, 2004). Through their receptor activation, BMPs and TGF-β regulate a multitude of developmental programs including embryonic gastrulation and tissue-specific differentiation such as cardiogenesis, chondrogenesis, and osteoblast differentiation (Abdel-Latif et al., 2008; Akhurst, Lehnert, Faissner, & Duffie, 1990; Chen, Deng, & Li, 2012; Dabovic et al., 2002; Dickson, Slager, Duffie, Mummery, & Akhurst, 1993; Erlebacher & Derynck, 1996; Goumans et al., 2008; van der Kraan, Blaney Davidson, Blom, & van den Berg, 2009). Given their critical role during embryogenesis, development, and tissue homeostasis, it is not surprising that the storage, release, and activation of TGF-β and BMPs are both tightly regulated by ECM binding (Robertson et al., 2015; Wipff, Rifkin, Meister, & Hinz, 2007). To begin with, cells secrete TGF-β in its inactive form where it associates with a large macromolecular protein aggregate termed the latent TGF-β complex. Proteins within the latent complex contain the RGD peptide, which can then be recognized by cellular transmembrane receptors, including integrin αvβ6. Integrin-mediated binding of the latent protein complex can then release the active TGF-β ligand from the latent complex in an actomyosin tension-dependent manner (Hinz, 2015). BMPs also bind to the ECM and this binding influences tissue development by generating BMP gradients that critically direct cell migration and differentiation. For example, BMPs directly bind to specific collagen 11α and collagen IV domains (Wang, Harris, Bayston, & Ashe, 2008), and this bound BMP directs body axis formation, heart development, and sensory neuron development in the inner ear (Chang et al., 2008; Khetarpal, Robertson, Yoo, & Morton, 1994; Larraín et al., 2000; Li et al., 2005). Bone morphogenetic proteins also associate with fibrillins, which are ECM glycoproteins (Sengle et al., 2008), and this interaction is critical for limb digit formation patterning (Arteaga-Solis et al., 2001). Similarly, the simultaneous dynamic ECM remodeling and collagen fibrillogenesis that are absolutely critical for chondrogenesis and the cartilage-to-bone transition are tightly choreographed by ECM-bound BMP (Yoon et al., 2005; Zhu, Oganesian, Keene, & Sandell, 1999). The ECM-bound BMP not only stimulates cell migration and differentiation, but its ECM-bound inhibitors may also repress signaling and differentiation, as has been proposed by studies in a simplified embryogenesis model (Warmflash, Sorre, Etoc, Siggia, & Brivanlou, 2014).

5.2 ECM–Growth Factor Collaboration

ECM-bound growth factors can elicit profound effects on tissue development and differentiation. This concept was elegantly illustrated by studies conducted in *Xenopus*, in which platelet-derived growth factor subunit A (PDGFA) bound to ectodermal-generated fibronectin was shown to be absolutely critical for the directional migration of mesodermal cells during gastrulation (Keller, 2005). The importance of ECM binding of PDGFA was illustrated by showing that expression of either a dominant-negative PDGFA receptor or reducing the level of matrix-bound PDGFA severely compromised mesoderm migration toward the animal pole (Nagel, Tahinci, Symes, & Winklbauer, 2004). Similarly, IGF-I and IGF-II are often highly abundant in tissues where they associate with the ECM even when no insulin-like growth factor (IGF) mRNA can be easily detected (Han,

D'Ercole, & Lund, 1987; Hill et al., 1989). This ECM-associated IGF has been shown to play an instructive role in directed neural development and was found to be critical for inducing oligodendrocyte differentiation from neural progenitors (Hsieh et al., 2004), and for the cooperative BMP-mediated induction of chondrogenesis from adipose-derived MSCs (An, Cheng, Yuan, & Li, 2010). The ECM also modulates IGF signaling through competitive binding of their carrier proteins the insulin growth factor-binding proteins (IGFBPs). The association of IGFBPs to the ECM not only retards their MMP-mediated degradation but additionally alters the cellular-binding affinity for IGF-1, as was illustrated by a sevenfold decrease in IGF-1 binding affinity induced by ECM-bound IGFBP-5 (Jones, Gockerman, Busby, Camacho-Hubner, & Clemmons, 1993; Martin, Fowlkes, Babic, & Khokha, 1999). Yet another example of how matrix-bound growth factors can influence tissue development is demonstrated by fibroblast growth factor (FGF), which associates strongly with ECM-localized or membrane-tethered proteoglycans that contain heparin sulfate side chains (Jia et al., 2009). In fact there are several different FGFs, each with different affinities for particular heparin sulfate side chains, so that variations in ECM composition can demonstrate unique specificity for sequestering particular FGFs (Harada et al., 2009). The proteoglycan association of FGFs can also stabilize interactions between FGFs and FGF receptors to influence cellular FGF signaling and cell fate (Ornitz, 2000). More frequently, however, proteolytic release of FGFs from the heparin sulfate-containing proteoglycans permits binding of the FGF ligand to cellular FGF receptors to stimulate cellular signaling and regulate cell growth, survival, and migration. Released FGF is key for tissue-specific differentiation, including lung and mammary gland branching morphogenesis (Cardoso, 2006; Patel et al., 2007; Pond et al., 2013; Tholozan et al., 2007).

6. ECM-DEPENDENT SPATIAL–MECHANICAL REGULATION OF CELL PHENOTYPE

6.1 Control of Cell Shape

ECM ligation-dependent changes in cell shape can significantly alter cell growth and survival and modulate cell fate. Initial evidence for a relationship between cell shape and ECM ligation and cell differentiation was provided using cultured explanted chick embryonic vertebrate chondrocytes. Over 50 years ago, Abbott and colleagues showed that when they plated isolated chondrocytes as clusters of "rounded" cells they retained their chondrocyte fate, as indicated by chondroitin sulfate and collagen production. However, if they plated the chondrocytes as single "spread" cells, sustained differentiation was inhibited (Abbott & Holtzer, 1966). Solursh and colleagues also reported that only when rounded but not spread MSCs were plated on a type I collagen matrix could the cells continue to express collagen II and stain positively for alcian blue (Solursh, Linsenmayer, & Jensen, 1982). Similarly, embryonic mesenchyme cells could only be efficiently differentiated into α-actin muscle myosin and desmin-expressing smooth muscle cells when they were plated as elongated cells on laminin basement membranes (Yang, Palmer, Relan, Diglio, & Schuger, 1998). Indeed, links between cell shape and mesenchymal cell smooth muscle cell differentiation were also indicated in another study in which the investigators used a microporous culture system that permitted precise control of ECM ligation, cell shape, and spreading to interrogate the impact of ECM context on cell fate specification (Yang, Relan, Przywara, & Schuger, 1999).

The advent of microcontact printing, in which ECM ligands can be precisely patterned, permitted unprecedented assessment of the role of cell shape, ECM ligation, and cell differentiation, and facilitated delineation of molecular mechanism. In a seminal study by McBeath and colleagues, microcontact-printed fibronectin deconstructed the impact of cell shape on MSC fate, independent of ECM composition and density. McBeath and colleagues showed that plating MSCs on small fibronectin islands, in which the cells retained a "rounded" morphology, promoted their differentiation into adipocytes. By contrast, when the same cells were plated on large fibronectin islands that permitted their spreading, they activated more Ras homolog gene family member A (RhoA) and RhoA kinase (ROCK) and had higher cytoskeletal tension that fostered their osteogenic differentiation (McBeath, Pirone, Nelson, Bhadriraju, & Chen, 2004). More provocatively, when mesenchymal cells grown in mixed osteo/adipogenic media were plated on fibronectin patterns with concave edges and sharp vertices that enhanced cytoskeletal tension, they showed a bias toward osteogenesis. By contrast, mesenchymal cells grown on patterns with convex, rounded edges, which do not enhance cellular tension, preferentially underwent adipogenesis. These findings imply that cell shape may function as a cell fate "switch" by regulating cell tension (Kilian, Bugarija, Lahn, & Mrksich, 2010). Nevertheless, cell shape does not always unambiguously induce cell differentiation. For instance, keratinocytes plated on patterned ECM rectangles that restricted their spreading fostered their epidermal differentiation, whereas when they were plated on substrata that enhanced their spreading they maintained their multipotency (Connelly et al., 2010).

In recent years there has been a growing focus on understanding how the 3D microenvironment that cells experience in vivo results in different cell behaviors than those observed in studies of cell behavior in vitro using traditional two-dimensional (2D) cultures (Lund, Yener, Stegemann, & Plopper, 2009; Rubashkin, Ou, & Weaver, 2014; Sasai, 2013). In this regard, natural collagen and fibrin gels have been used for many years but are complex, lack reproducibility, and fail to control for cell shape and ligand concentration (Janmey, Winer, & Weisel, 2009; Lee, Kasper, & Mikos, 2014; Parenteau-Bareil, Gauvin, & Berthod, 2010). Decellularized tissues have also enjoyed some degree of success, but again are not well defined and cannot be easily controlled, despite showing exciting success in directing stem differentiation for cardiac tissue (Song & Ott, 2011). More recent developments including PEG, alginate, and HA hydrogels have been used extensively for encapsulation to improve cell survival and function both in vitro and in vivo, due to the ability to tightly control cross-linking, cell attachment, permeability, biodegradability, and numerous other physical properties through chemical modifications (Augst, Kong, & Mooney, 2006; Burdick & Prestwich, 2011; Phelps et al., 2012). However, these hydrogels, while highly defined, do not always faithfully recapitulate the architecture and pore size of native ECMs (Griffith, 2002; Griffith & Swartz, 2006; Rubashkin et al., 2014). Toward this objective, microprinting technologies that were originally developed for DNA and protein microarrays have been adapted to build 3D cellular environments with microscale control of the scaffold design (Barbulovic-Nad et al., 2006; MacBeath & Schreiber, 2000). These may prove instrumental in delineating the effects of microenvironment composition and architecture in guiding cell fate and function (Flaim, Chien, & Bhatia, 2005; Flaim, Teng, Chien, & Bhatia, 2008; LaBarge et al., 2009; Montanez-Sauri, Beebe, & Sung, 2015).

6.2 ECM Stiffness and Cell Fate

All tissues within the body exhibit distinct mechanical properties that reflect their structure function and that are imparted predominantly by the composition and organization of their ECM. For example, the brain is quite soft and this compliant tissue primarily reflects the abundant hydrated HA found in the ECM. By contrast, bone is very rigid and this stiffness is mainly due to high amounts of densely packed, mineralized, cross-linked collagens.

The stiffness of the ECM in specific differentiated tissues is not only critical for the structure and function of the tissue but also regulates the growth, survival, and motility of the cellular constituents and may even direct their lineage-specific differentiation. For instance, human MSCs may be directed to differentiate down multiple independent lineages, apparently simply by varying substrate elasticity (Engler, Sen, Sweeney, & Discher, 2006). Thus, MSCs could be induced to display neurogenic markers when cultured on soft substrates (100–1000 Pa) analogous to the measured compliance of brain tissue, myogenic markers on substrates of intermediate stiffness (8000–17,000 Pa) reflecting the stiffness of skeletal muscle, and osteogenic markers on the most rigid substrates (25,000–40,000 Pa) representing the values measured in cortical bone tissue. Inhibition of nonmuscle myosin abolished elasticity-dependent differentiation of these MSCs, implying acto-myosin contractility is likely necessary to transmit information about the elasticity of the ECM into the cell to regulate its cell fate.

Neural stem cell differentiation was similarly found to be modulated by substrate stiffness (Saha et al., 2008) so that adult neural stem cells cultured in neurogenic media differentiated into neurons with the highest efficiency on substrates with an elasticity that is most similar to the brain (500 Pa). Even in mixed media that enables differentiation to multiple lineages, adult neural stem cells apparently only committed to a neuronal fate on softer substrates (100–500 Pa) and to a glial fate on stiffer substrates (1000–10,000 Pa). Importantly, only on the softest substrates (10 Pa) did the adult neural stem cells stop proliferating and develop prominent neural processes, recapitulating the distinctive behavior documented previously by neural vs glial cells on soft and stiff matrices (Flanagan, Ju, Marg, Osterfield, & Janmey, 2002).

Matrix elasticity can apparently directly regulate cell proliferation by modulating cyclin D1-dependent G1 cell cycle progression (Klein et al., 2009). The ability of ECM stiffness to modulate cell cycle progression could explain why MSC quiescence is induced by plating the cells on soft polyacrylamide gels with a compliance that is similar to that measured in the central bone marrow, whereas they proliferate and differentiate into adipocytes and osteoblasts when plated on stiffer substrates (Winer, Janmey, McCormick, & Funaki, 2009). Regardless, recognizing the impact of ECM compliance on cell cycle transit has permitted the development of culture conditions that maintain the pluripotency of both embryonic and induced pluripotent stem cells and may help to design new strategies to potentiate iPS cell generation (Dixon et al., 2014; Li et al., 2017).

Although it is entirely plausible that ECM compliance can directly specify stem cell fate, it is more likely these effects are mediated through modulation of cellular responses to growth factors and morphogens. This concept was recently illustrated by studies showing enhanced human embryonic stem cell (hESC) mesoderm progenitor differentiation mediated by culturing the cells on highly compliant substrates (Przybyla, Lakins, & Weaver, 2016). In this study, hESCs plated on soft (400 Pa) and stiff (60,000 Pa) substrates failed to undergo spontaneous differentiation. However, upon morphogen-induced differentiation, the hESCs plated on the softest substrates, recapitulating the elasticity of the early embryo, underwent highly efficient mesoderm progenitor differentiation as compared to those plated on the stiffer substrates. These findings suggest matrix elasticity may serve to "prime" cells for tissue-specific differentiation. Indeed, hESCs interacting with the softest matrix showed reduced cell–ECM integrin adhesions and developed strong cell–cell adhesions that promoted apical–basal polarity, sequestered β-catenin, P120 catenin, and Kaiso at the adherens junctions and reduced levels of secreted Wnt inhibitors. Upon exposure to BMPs and differentiation stimuli, the "primed" hESC colonies synchronously degraded their E-cadherin junctions, releasing large quantities of β-catenin, which was able to rapidly stimulate mesoderm gene expression upon translocation to the nucleus to induce mesoderm progenitor differentiation.

6.3 ECM Stiffness Modulates Morphogen Activity

Given that cells increase their actomyosin tension in response to the stiffness of the ECM, it is perhaps not surprising that the ability of cells to activate TGF-β can be potentiated by growing cells on or within a mechanically strained or stiffened ECM (Wipff et al., 2007). Moreover, once released, any activated TGF-β can then stimulate ECM synthesis and deposition and elevate the expression of enzymes such as LOX that further stiffen the ECM through a positive feedforward circuit that facilitates normal wound healing but can also drive pathological fibrosis (Cox & Erler, 2011; Frantz, Stewart, & Weaver, 2010). A more direct role for ECM stiffness in tension-mediated morphogen signaling was illustrated by its impact on Notch signaling. Notch activation is enhanced by mechanical force-mediated exposure of its cleavage site on the Notch receptor through ligand-receptor binding that is potentiated by cellular actomyosin contractility stimulated by integrin–ECM anchorage (D'Souza, Miyamoto, & Weinmaster, 2008).

6.4 ECM Stiffness Modulates Cell Behavior by Regulating Mechanosensitive Ion Channels

A stiff ECM can activate mechanosensitive ion channels to induce cell growth and survival, and stimulate cell migration and differentiation.

For example, Piezo channels are ubiquitously expressed calcium ion channels that are exquisitely sensitive to mechanical activation (Coste et al., 2010). ECM ligation can activate Piezo transmembrane cation channels to regulate cell differentiation, provided the stiffness of the ECM is high enough that cellular actomyosin tension is substantially increased. This concept was illustrated by Pathak and colleagues who showed that ECM tension-mediated activation of Piezo1 could promote the calcium-dependent neurogenesis of human neural stem cells and that its ablation or inhibition promoted astrogenesis (Pathak et al., 2014). Similarly, transient receptor potential (TRP) channels are calcium-permeable and voltage-independent ion channels that also influence stem cell growth, survival, migration, and differentiation and are thus important in tissue development and homeostasis (Muramatsu et al., 2007; Pla et al., 2005; Ramsey, Delling, & Clapham, 2006). TRP channels are highly sensitive to mechanical cues, and not surprisingly, their activity is also strongly induced by Rho GTPases (Mehta et al., 2003; Vriens et al., 2004).

6.5 ECM Stiffness Modulates Yes-Associated Protein and Transcriptional Coactivator With PDZ-Binding Motif

In recent years, signaling events that converge on the coactivators Yes-associated protein and transcriptional coactivator with PDZ-binding motif (YAP/TAZ) have become a major area of interest as evidence has emerged that their activity can be directly regulated by ECM stiffness (Hao et al., 2014). Further, YAP/TAZ are key components of the Hippo signaling pathway, which regulates organ size during development (Zhao, Tumaneng, & Guan, 2011). The relationship between the Hippo pathway and YAP/TAZ was first uncovered in *Drosophila*, where overexpression of the YAP homologue resulted in dramatic tissue overgrowth, phenocopying loss of Hippo signaling (Huang, Wu, Barrera, Matthews, & Pan, 2005). Conversely, mutation of the YAP homologue prevented normal growth and expansion of developing tissues. Later mechanistic studies revealed that activation of the Hippo signaling cascade ultimately results in phosphorylation of YAP and TAZ, preventing their nuclear localization and sequestering them to the cytoplasm where they undergo proteasomal degradation (Hong & Guan, 2012). In their unphosphorylated form, YAP/TAZ are shuttled to the nucleus, where they associate with other transcription factors and activate gene expression. Direct evidence for the role of mechanotransduction in regulating YAP/TAZ signaling emerged when it was found that YAP nuclear translocation was induced by culturing cells on stiff substrates or by fostering cell spreading using large patterned ECM ligands (Dupont et al., 2011). By contrast, nuclear YAP translocation was prevented when cells were plated on soft substrates or when cell spreading was prevented (Dupont et al., 2011). Consistently, YAP/TAZ nuclear translocation was found to be necessary for the induction of MSC differentiation in a RhoA GTPase-dependent manner that required actomyosin-mediated cytoskeletal tension. Nevertheless, it is likely ECM-mediated regulation of Hippo signaling during development is mediated by both biochemical and biomechanical cues.

7. SUMMARY AND FUTURE DIRECTIONS

The ECM regulates embryogenesis, tissue-specific development, and stem cell fate by binding cellular receptors and activating ion channels to initiate intracellular signaling and gene transcription, by controlling the availability, activation, and presentation of soluble ligands, and by altering cellular tension. Nevertheless, tissue development and cell differentiation rely on tightly regulated epigenetics to induce sustained changes in gene transcription and cell behavior. Indeed, every cell within an organism is endowed with the same genetic code, so that it is the coordinated spatial and temporal regulation of gene expression that is responsible for directing tissue development, maintaining tissue homeostasis, and inducing tissue-specific differentiation.

Cell–ECM adhesion-dependent, tissue-specific differentiation may depend upon alterations in nuclear architecture and chromatin organization. For example, early studies that examined links between the ECM and MEC differentiation showed beta casein gene expression depended upon cell rounding that associated with alterations in higher order chromatin structure and epigenetic changes (Boudreau, Myers, & Bissell, 1995; Myers et al., 1998; Schmidhauser, Bissell, Myerst, & Caspersont, 1990). Consistently, lactogenic hormonal stimulation of the beta casein transcriptional enhancer could only be induced in MECs interacting with a laminin 111 ECM when the reporter cassette was stably expressed in the MECs, implying the chromatin architecture surrounding the promoter was necessary for proper regulation of beta casein expression (Boudreau, Myers, et al., 1995; Boudreau, Sympson, et al., 1995; Xu et al., 2009). ECM stiffness-dependent, tissue-specific gene expression also implies a causal link between cell shape, cytoskeletal organization, and chromatin organization (Engler et al., 2006; Flanagan et al., 2002; Gilbert et al., 2010; Xu et al., 2009). For instance, Chen and colleagues showed that cell shape regulates RhoA/Rac activity to modulate the osteogenic vs adipogenic vs chondrogenic, and myogenic fate of hMSCs

(Gao, McBeath, & Chen, 2010; McBeath et al., 2004). Indeed, matrix metalloproteinase 14 (MMP14)-mediated ECM remodeling induces cell rounding to enhance histone acetylation marks of adipogenic cell fate, presumably by reducing epigenetic constraints imposed by type I collagen (Sato-Kusubata, Jiang, Ueno, & Chun, 2011). These studies raise the intriguing possibility that the ECM and its receptors regulate cell and tissue fate by modulating nuclear organization and chromatin. The question is how?

Ingber and colleagues have suggested that cells are "hard-wired," such that force applied at cell–integrin adhesions is directly transduced to the nucleus through the cytoskeleton, resulting in rearrangements of nuclei and nuclear components (Maniotis, Chen, & Ingber, 1997). This idea was supported by the use of microsurgical techniques to pull on individual chromosomes and nucleoli, and the subsequent observation that pulling a single chromosome out of the nucleus led to sequential removal of the remaining chromosomes (Maniotis, Bojanowski, & Ingber, 1997). Further support for this model was provided by the identification of the linker of nucleoskeleton and cytoskeleton (LINC) complex, suggesting a mechanism through which physical interactions received through cellular integrin–ECM interactions might be directly transmitted to the nucleus to alter gene transcription (Crisp et al., 2006; Haque et al., 2006; Padmakumar et al., 2005; Wang, Tytell, & Ingber, 2009). Since nuclear chromatin is thought to be organized into dense heterochromatin and more open euchromatin, this paradigm suggests that alterations in the LINC complex stimulated by biomechanical cues could rapidly change nuclear shape and increase levels of transcriptionally accessible euchromatin to influence cell fate (Skinner & Johnson, 2017). The discovery of a set of nuclear receptor coactivators that were aptly named tension-induced proteins (TIPs) has illuminated another putative pathway through which biomechanical cues transmitted through cell–ECM receptors could regulate chromatin organization and alter nuclear processes to influence gene expression (Jakkaraju, Zhe, Pan, Choudhury, & Schuger, 2005). MSCs can be induced to differentiate toward a myogenic fate by exposing the cells to chronic stretch and toward an adipogenic fate by preventing their spreading (McBeath et al., 2004; Yang, Beqaj, Kemp, Ariel, & Schuger, 2000). Jakkaraju and colleagues found that myogenic differentiation of these MSCs depends upon histone acetylation and the recruitment of specific coactivators that bind to TIP proteins through a specific nuclear receptor box domain that is exposed in response to cell stretch, and that adipogenesis is the default pathway when these TIP proteins are not recruited (Jakkaraju et al., 2005). Thus, physical forces transmitted through the ECM could directly regulate the activity of nuclear transcriptional regulators to direct cell fate.

Importantly, chromatin organization is complex and dynamic, with distal sequences of DNA capable of interacting through topologically associated domains (TADs) and long sequences of DNA sequestered near the nuclear membrane in lamin-associated domains (LADs) (Gonzalez-Sandoval & Gasser, 2016; Nicodemi & Pombo, 2014). The precise organization of these TADs and LADs is cell- and tissue-type specific. Thus, while there may be physical connections between cell–ECM adhesions and the nucleus, chromatin remodeling also depends on dynamic interactions between DNA and its binding partners. For example, osteoblast-specific transcription factors such as Runx2 can physically associate with nuclear matrix components during osteoblast differentiation, whereas more ubiquitous transcription factors remain bound to nonmatrix nuclear compartments (Lindenmuth et al., 1997). These tissue-specific transcription factors apparently bind to the nuclear matrix through specific nuclear matrix-targeting sequences such that introducing mutations in these sequences compromise cell proliferation and differentiation of myeloid progenitor cells (Stein et al., 2007). Similarly, recent studies suggest cell density, cell migration, and fluid flow can all dramatically change nuclear shape to induce sustained changes in tissue-specific gene expression (Dahl, Ribeiro, & Lammerding, 2008; Denais et al., 2016; McBride & Knothe Tate, 2008).

Beyond DNA–protein interactions that regulate chromatin structure, it is becoming increasing clear that RNAs also play important roles in regulating chromatin organization and epigenetics. Studies of the human genome have revealed that while nearly all of the genome is capable of being transcribed, only about 2% of this genetic information consists of protein-encoding exons (Consortium, 2001; Djebali et al., 2012; ENCODE Consortium, 2007; Venter et al., 2001). The remainder of transcribed RNA is referred to as heteronuclear RNA and consists of numerous subclasses of RNA whose functions are still being uncovered (Guttman & Rinn, 2012; Salditt-Georgieff, Harpold, Wilson, & Darnell, 1981; Warner, Soeiro, Birnboim, Girard, & Darnell, 1966). Among these subclasses, long intergenic noncoding RNAs (lincRNAs) have emerged as a subclass of particular interest. These lincRNAs have been shown to act as important scaffolds for chromatin-modifying proteins that are responsible for establishing and maintaining the epigenetic state of the cell, and as might be expected, the expression of lincRNAs changes with differentiation and disease (Cesana et al., 2011; Kogo et al., 2011; Tsai et al., 2010; Wang et al., 2011). Although a clear demonstration of lincRNA regulation by cell–ECM interactions has yet to be reported, it seems likely that future studies will reveal such a mechanism given the number of ways the ECM is capable of regulating the comparatively few number of protein-encoding genes. Similarly, noncoding RNAs also regulate gene expression at the level of pre-RNA processing. For instance, small nucleolar RNA (snoRNA) base pair with pre-RNA to guide folding and splicing of transcripts, and some evidence even suggests snoRNAs can affect translation of mRNA similar to microRNAs (miRNAs) (Ender et al., 2008; Gerbi & Borovjagin, 2004). We found ECM stiffness can regulate miRNAs, which modulate gene expression posttranscriptionally, downstream of activated integrins and through signaling cascades initiated by Robo–Slit interactions (Le et al., 2016; Mouw, Yui, et al., 2014). However, these were not broad regulatory effects, rather, we noted that a subgroup of miRNAs is highly sensitive to ECM stiffness, implying there may be a level of specificity to which sets of genes can be epigenetically regulated by miRNAs in response to stiffness. This raises the intriguing possibility of identifying novel tension-regulated pathways that modulate RNA biogenesis and modify chromatin and gene expression. Such interplay between the ECM and the epigenetic state of the cell could play a critical role in the spatial and temporal regulation of cell fate during embryogenesis, tissue-specific differentiation, wound healing, and homeostasis.

ACKNOWLEDGMENTS

The authors would like to acknowledge support from the California Institute for Regenerative Medicine (CIRM awards TR3-05542 and RB5-07409), the National Institutes of Health and National Cancer Institute (awards 1U01CA202241-01, R01CA192914, R01CA174929, and R01CA222508-01), and the Department of Defense Breast Cancer Research Program (USAMRAA DOD-BCRP award BC122990). J.M.M. is also grateful for support from NIH Training Grant: BioE T32 GM008155 and the UCSF Discovery Fellowship.

REFERENCES

Abbott, J., & Holtzer, H. (1966). The loss of phenotypic traits by differentiated cells. 3. The reversible behavior of chondrocytes in primary cultures. *Journal of Cell Biology*, *28*(3), 473–487. https://doi.org/10.1083/jcb.28.3.473.

Abdel-Latif, A., Zuba-Surma, E. K., Case, J., Tiwari, S., Hunt, G., Ranjan, S., et al. (2008). TGF-β1 enhances cardiomyogenic differentiation of skeletal muscle-derived adult primitive cells. *Basic Research in Cardiology*, *103*(6), 514–524. https://doi.org/10.1007/s00395-008-0729-9.

Adams, J. C., & Watt, F. M. (1993). Regulation of development and differentiation by the extracellular matrix. *Development (Cambridge, England)*, *117*(4), 1183–1198.

Akhurst, R. J., Lehnert, S. A., Faissner, A., & Duffie, E. (1990). TGF beta in murine morphogenetic processes: The early embryo and cardiogenesis. *Development (Cambridge, England)*, *108*(4), 645–656. Retrieved from http://www.ncbi.nlm.nih.gov/pubmed/1696875.

Alfandari, D., Cousin, H., Gaultier, A., Hoffstrom, B. G., & DeSimone, D. W. (2003). Integrin α5β1 supports the migration of Xenopus cranial neural crest on fibronectin. Developmental Biology, 260(2), 449–464. https://doi.org/10.1016/S0012-1606(03)00277-X

An, C., Cheng, Y., Yuan, Q., & Li, J. (2010). IGF-1 and BMP-2 induces differentiation of adipose-derived mesenchymal stem cells into chondrocytes-like cells. *Annals of Biomedical Engineering*, *38*(4), 1647–1654. https://doi.org/10.1007/s10439-009-9892-x.

Arteaga-Solis, E., Gayraud, B., Lee, S. Y., Shum, L., Sakai, L., & Ramirez, F. (2001). Regulation of limb patterning by extracellular microfibrils. *Journal of Cell Biology*, *154*(2), 275–281. https://doi.org/10.1083/jcb.200105046.

Augst, A. D., Kong, H. J., & Mooney, D. J. (2006). Alginate hydrogels as biomaterials. *Macromolecular Bioscience*, *6*(8), 623–633. https://doi.org/10.1002/mabi.200600069.

Aumailley, M., Bruckner-Tuderman, L., Carter, W. G., Deutzmann, R., Edgar, D., Ekblom, P., et al. (2005). A simplified laminin nomenclature. *Matrix Biology*, *24*(5), 326–332. https://doi.org/10.1016/j.matbio.2005.05.006.

Avigdor, A., Goichberg, P., Shivtiel, S., Dar, A., Peled, A., Samira, S., et al. (2004). CD44 and hyaluronic acid cooperate with SDF-1 in the trafficking of human CD34+ stem/progenitor cells to bone marrow. *Blood*, *103*(8), 2981–2989. https://doi.org/10.1182/blood-2003-10-3611.

Baker, N. E., Mlodzik, M., & Rubin, G. M. (1990). Spacing differentiation in the developing Drosophila eye: A fibrinogen-related lateral inhibitor encoded by scabrous. *Science (New York, N.Y.)*, *250*(4986), 1370–1377. https://doi.org/10.1126/science.2175046.

Baneyx, G., Baugh, L., & Vogel, V. (2001). Coexisting conformations of fibronectin in cell culture imaged using fluorescence resonance energy transfer. *Proceedings of the National Academy of Sciences of the United States of America*, *98*(25), 14464–14468. https://doi.org/10.1073/pnas.251422998.

Baneyx, G., Baugh, L., & Vogel, V. (2002). Fibronectin extension and unfolding within cell matrix fibrils controlled by cytoskeletal tension. *Proceedings of the National Academy of Sciences of the United States of America*, *99*(8), 5139–5143. https://doi.org/10.1073/pnas.072650799.

Barbulovic-Nad, I., Lucente, M., Sun, Y., Zhang, M., Wheeler, A. R., & Bussmann, M. (2006). Bio-microarray fabrication techniques—A review. *Critical Reviews in Biotechnology*, *26*(4), 237–259. https://doi.org/10.1080/07388550600978358.

Benito-Jardón, M., Klapproth, S., Gimeno-Lluch, I., Petzold, T., Bharadwaj, M., Müller, D. J., et al. (2017). The fibronectin synergy site re-enforces cell adhesion and mediates a crosstalk between integrin classes. *eLife*, *6*, e22264. https://doi.org/10.7554/eLife.22264.001.

Bissell, M. J., Hall, H. G., & Parry, G. (1982). How does the extracellular matrix direct gene expression? *Journal of Theoretical Biology*, *99*(1), 31–68. https://doi.org/10.1016/0022-5193(82)90388-5.

Biteau, B., & Jasper, H. (2014). Slit/Robo signaling regulates cell fate decisions in the intestinal stem cell lineage of Drosophila. *Cell Reports*, 7(6), 1867–1875. https://doi.org/10.1016/j.celrep.2014.05.024.

Bonnans, C., Chou, J., & Werb, Z. (2014). Remodelling the extracellular matrix in development and disease. *Nature Reviews Molecular Cell Biology*, *15*(12), 786–801. https://doi.org/10.1038/nrm3904.

Boucaut, J. C., Darribere, T., Li, S. D., Boulekbache, H., Yamada, K. M., & Thiery, J. P. (1985). Evidence for the role of fibronectin in amphibian gastrulation. *Journal of Embryology and Experimental Morphology*, *89*(Suppl), 211–227.

Boucaut, J. C., Darribere, T., Poole, T. J., Aoyama, H., Yamada, K. M., & Thiery, J. P. (1984). Biologically active synthetic peptides as probes of embryonic development: A competitive peptide inhibitor of fibronectin function inhibits gastrulation in amphibian embryos and neural cell migration in avian embryos. *Journal of Cell Biology*, *99*(5), 1822–1830. https://doi.org/10.1083/jcb.99.5.1822.

Boudreau, N., Myers, C., & Bissell, M. J. (1995). From laminin to lamin: Regulation of tissue-specific gene expression by the ECM. *Trends in Cell Biology*, *5*(1), 1–4. https://doi.org/10.1016/S0962-8924(00)88924-2.

Boudreau, N., Sympson, C. J., Werb, Z., & Bissell, M. J. (1995). Suppression of ICE and apoptosis in mammary epithelial cells by extracellular matrix. *Science*, *267*(5199), 891–893. https://doi.org/10.1126/science.7531366.

Bourguignon, L. Y. W., Ramez, M., Gilad, E., Singleton, P. A., Man, M.-Q., Crumrine, D. A., et al. (2006). Hyaluronan-CD44 interaction stimulates keratinocyte differentiation, lamellar body formation/secretion, and permeability barrier homeostasis. *The Journal of Investigative Dermatology*, *126*(6), 1356–1365. https://doi.org/10.1038/sj.jid.5700260.

Brower, D. L., & Jaffe, S. M. (1989). Requirement for integrins during Drosophila wing development. *Nature*, *342*(6247), 285–287. https://doi.org/10.1038/342285a0.

Burdick, J. A., & Prestwich, G. D. (2011). Hyaluronic acid hydrogels for biomedical applications. *Advanced Materials*, *23*(12), H41–56. https://doi.org/10.1002/adma.201003963.

Cardoso, W. V. (2006). Regulation of early lung morphogenesis: Questions, facts and controversies. *Development*, *133*(9), 1611–1624. https://doi.org/10.1242/dev.02310.

Cesana, M., Cacchiarelli, D., Legnini, I., Santini, T., Sthandier, O., Chinappi, M., et al. (2011). A long noncoding RNA controls muscle differentiation by functioning as a competing endogenous RNA. *Cell*, *147*(2), 358–369. https://doi.org/10.1016/j.cell.2011.09.028.

Chang, W., Lin, Z., Kulessa, H., Hebert, J., Hogan, B. L. M., & Wu, D. K. (2008). Bmp4 is essential for the formation of the vestibular apparatus that detects angular head movements. *PLoS Genetics*, *4*(4), e1000050. https://doi.org/10.1371/journal.pgen.1000050.

Chen, G., Deng, C., & Li, Y. P. (2012). TGF-β and BMP signaling in osteoblast differentiation and bone formation. *International Journal of Biological Sciences*, *8*(2), 272–288. https://doi.org/10.7150/ijbs.2929.

Cheung, T. H., & Rando, T. A. (2013). Molecular regulation of stem cell quiescence. *Nature Reviews Molecular Cell Biology*, *14*(6), 329–340. https://doi.org/10.1038/nrm3591.

Coles, E. G., Gammill, L. S., Miner, J. H., & Bronner-Fraser, M. (2006). Abnormalities in neural crest cell migration in laminin alpha5 mutant mice. *Developmental Biology*, *289*(1), 218–228. https://doi.org/S0012-1606(05)00759-1.

Connelly, J. T., Gautrot, J. E., Trappmann, B., Tan, D. W.-M., Donati, G., Huck, W. T. S., et al. (2010). Actin and serum response factor transduce physical cues from the microenvironment to regulate epidermal stem cell fate decisions. *Nature Cell Biology*, *12*(7), 711–718. https://doi.org/10.1038/ncb2074.

Coste, B., Mathur, J., Schmidt, M., Earley, T. J., Ranade, S., Petrus, M. J., et al. (2010). Piezo1 and Piezo2 are essential components of distinct mechanically activated cation channels. *Science*, *330*(6000), 55–60. https://doi.org/10.1126/science.1193270.

Cox, T. R., & Erler, J. T. (2011). Remodeling and homeostasis of the extracellular matrix: Implications for fibrotic diseases and cancer. *Disease Models & Mechanisms*, *4*(2), 165–178. https://doi.org/10.1242/dmm.004077.

Crisp, M., Liu, Q., Roux, K., Rattner, J. B., Shanahan, C., Burke, B., et al. (2006). Coupling of the nucleus and cytoplasm: Role of the LINC complex. *Journal of Cell Biology*, *172*(1), 41–53. https://doi.org/10.1083/jcb.200509124.

Dabovic, B., Chen, Y., Colarossi, C., Obata, H., Zambuto, L., Perle, M. A., et al. (2002). Bone abnormalities in latent TGF-β binding protein (Ltbp)-3-null mice indicate a role for Ltbp-3 in modulating TGF-β bioavailability. *Journal of Cell Biology*, *156*(2), 227–232. https://doi.org/10.1083/jcb.200111080.

Dahl, K. N., Ribeiro, A. J. S., & Lammerding, J. (2008). Nuclear shape, mechanics, and mechanotransduction. *Circulation Research*, *102*(11), 1307–1318. https://doi.org/10.1161/CIRCRESAHA.108.173989.

Debelle, L., & Tamburro, A. M. (1999). Elastin: Molecular description and function. *International Journal of Biochemistry and Cell Biology*, *31*(2), 261–272. https://doi.org/10.1016/S1357-2725(98)00098-3.

Denais, C. M., Gilbert, R. M., Isermann, P., McGregor, A. L., te Lindert, M., Weigelin, B., et al. (2016). Nuclear envelope rupture and repair during cancer cell migration. *Science*, *352*(6283), 353–358. https://doi.org/10.1126/science.aad7297.

Dickson, M. C., Slager, H. G., Duffie, E., Mummery, C. L., & Akhurst, R. J. (1993). RNA and protein localisations of TGF beta 2 in the early mouse embryo suggest an involvement in cardiac development. Development (Cambridge, England), 117(2), 625–39. Retrieved from http://www.ncbi.nlm.nih.gov/pubmed/7687212

Dixon, J. E., Shah, D. A., Rogers, C., Hall, S., Weston, N., Parmenter, C. D. J., et al. (2014). Combined hydrogels that switch human pluripotent stem cells from self-renewal to differentiation. *Proceedings of the National Academy of Sciences of the United States of America*, *111*(15), 5580–5585. https://doi.org/10.1073/pnas.1319685111.

Djebali, S., Davis, C. A., Merkel, A., Dobin, A., Lassmann, T., Mortazavi, A., et al. (2012). Landscape of transcription in human cells. *Nature*, *489*(7414), 101–108. https://doi.org/10.1038/nature11233.

D'Souza, B., Miyamoto, A., & Weinmaster, G. (2008). The many facets of Notch ligands. *Oncogene*, *27*(38), 5148–5167. https://doi.org/10.1038/onc.2008.229.

Dupont, S., Morsut, L., Aragona, M., Enzo, E., Giulitti, S., Cordenonsi, M., et al. (2011). Role of YAP/TAZ in mechanotransduction. *Nature*, *474*(7350), 179–183. https://doi.org/10.1038/nature10137.

ENCODE Consortium. (2007). Identification and analysis of functional elements in 1% of the human genome by the ENCODE pilot project. *Nature*, *447*(7146), 799–816. https://doi.org/10.1038/nature05874.

Ender, C., Krek, A., Friedländer, M. R., Beitzinger, M., Weinmann, L., Chen, W., et al. (2008). A human snoRNA with microRNA-like functions. *Molecular Cell*, *32*(4), 519–528. https://doi.org/10.1016/j.molcel.2008.10.017.

Engler, A. J., Sen, S., Sweeney, H. L., & Discher, D. E. (2006). Matrix elasticity directs stem cell lineage specification. *Cell*, *126*(4), 677–689. https://doi.org/10.1016/j.cell.2006.06.044.

Erlebacher, A., & Derynck, R. (1996). Increased expression of TGF-β2 in osteoblasts results in an osteoporosis-like phenotype. *Journal of Cell Biology*, *132*(1–2), 195–210. https://doi.org/10.1083/jcb.132.1.195.

Fässler, R., Rohwedel, J., Maltsev, V., Bloch, W., Lentini, S., Guan, K., et al. (1996). Differentiation and integrity of cardiac muscle cells are impaired in the absence of beta 1 integrin. *Journal of Cell Science*, *109*(Pt. 1), 2989–2999.

Flaim, C. J., Chien, S., & Bhatia, S. N. (2005). An extracellular matrix microarray for probing cellular differentiation. *Nature Methods*, *2*(2), 119–125. https://doi.org/10.1038/nmeth736.

Flaim, C. J., Teng, D., Chien, S., & Bhatia, S. N. (2008). Combinatorial signaling microenvironments for studying stem cell fate. *Stem Cells and Development*, *17*(1), 29–40. https://doi.org/10.1089/scd.2007.0085.

Flanagan, L. A., Ju, Y. E., Marg, B., Osterfield, M., & Janmey, P. A. (2002). Neurite branching on deformable substrates. *Neuro Report*, *13*(18), 2411–2415. https://doi.org/10.1097/00001756-200212200-00007.

Frantz, C., Stewart, K. M., & Weaver, V. M. (2010). The extracellular matrix at a glance. *Journal of Cell Science*, *123*(24), 4195–4200. https://doi.org/10.1242/jcs.023820.

Friedland, J. C., Lee, M. H., & Boettiger, D. (2009). Mechanically activated integrin switch controls 5 1 function. *Science, 323*(5914), 642–644. https://doi.org/10.1126/science.1168441.

Fuchs, E., & Chen, T. (2012). A matter of life and death: Self-renewal in stem cells. *EMBO Reports, 14*(1), 39–48. https://doi.org/10.1038/embor.2012.197.

Gao, L., McBeath, R., & Chen, C. S. (2010). Stem cell shape regulates a chondrogenic versus myogenic fate through rac 1 and N-cadherin. *Stem Cells, 28*(3), 564–572. https://doi.org/10.1002/stem.308.

Gattazzo, F., Urciuolo, A., & Bonaldo, P. (2014). Extracellular matrix: A dynamic microenvironment for stem cell niche. *Biochimica et Biophysica Acta—General Subjects, 1840*(8), 2506–2519. https://doi.org/10.1016/j.bbagen.2014.01.010.

Gerbi, S., & Borovjagin, A. (2004). Pre-ribosomal RNA processing in multicellular organisms. *The Nucleolus*, 170–198.

Gilbert, P. M., Havenstrite, K. L., Magnusson, K. E. G., Sacco, A., Leonardi, N. A., Kraft, P., et al. (2010). Substrate elasticity regulates skeletal muscle stem cell self-renewal in culture. *Science, 329*(5995), 1078–1081. https://doi.org/10.1126/science.1191035.

Gonzalez-Sandoval, A., & Gasser, S. M. (2016). On TADs and LADs: Spatial control over gene expression. *Trends in Genetics, 32*(8), 485–495. https://doi.org/10.1016/j.tig.2016.05.004.

Gordon, M. K., & Hahn, R. A. (2010). Collagens. *Cell and Tissue Research, 339*(1), 247–257. https://doi.org/10.1007/s00441-009-0844-4.

Goumans, M. J., de Boer, T. P., Smits, A. M., van Laake, L. W., van Vliet, P., Metz, C. H. G., et al. (2008). TGF-β1 induces efficient differentiation of human cardiomyocyte progenitor cells into functional cardiomyocytes in vitro. *Stem Cell Research, 1*(2), 138–149. https://doi.org/10.1016/j.scr.2008.02.003.

Griffith, L. G. (2002). Tissue engineering—Current challenges and expanding opportunities. *Science, 295*(5557), 1009–1014. https://doi.org/10.1126/science.1069210.

Griffith, L. G., & Swartz, M. A. (2006). Capturing complex 3D tissue physiology in vitro. *Nature Reviews Molecular Cell Biology*, 7(3), 211–224. https://doi.org/10.1038/nrm1858.

Guo, X., Johnson, J. J., & Kramer, J. M. (1991). Embryonic lethality caused by mutations in basement membrane collagen of C. elegans. *Nature, 349*(6311), 707–709. https://doi.org/10.1038/349707a0.

Guttman, M., & Rinn, J. L. (2012). Modular regulatory principles of large non-coding RNAs. *Nature, 482*(7385), 339–346. https://doi.org/10.1038/nature10887.

Hamidouche, Z., Fromigué, O., Ringe, J., Häupl, T., Vaudin, P., Pagès, J.-C., et al. (2009). Priming integrin alpha5 promotes human mesenchymal stromal cell osteoblast differentiation and osteogenesis. *Proceedings of the National Academy of Sciences of the United States of America, 106*, 18587–18591. https://doi.org/10.1016/j.bone.2009.03.058.

Han, V. K., D'Ercole, A. J., & Lund, P. K. (1987). Cellular localization of somatomedin (insulin-like growth factor) messenger RNA in the human fetus. *Science, 236*(4798), 193–197.

Hao, J., Zhang, Y., Wang, Y., Ye, R., Qiu, J., Zhao, Z., et al. (2014). Role of extracellular matrix and YAP/TAZ in cell fate determination. *Cellular Signalling, 26*(2), 186–191. https://doi.org/10.1016/j.cellsig.2013.11.006.

Haque, F., Lloyd, D. J., Smallwood, D. T., Dent, C. L., Shanahan, C. M., Fry, A. M., et al. (2006). SUN1 interacts with nuclear lamin A and cytoplasmic nesprins to provide a physical connection between the nuclear lamina and the cytoskeleton. *Molecular and Cellular Biology, 26*(10), 3738–3751. https://doi.org/10.1128/MCB.26.10.3738-3751.2006.

Harada, M., Murakami, H., Okawa, A., Okimoto, N., Hiraoka, S., Nakahara, T., et al. (2009). FGF9 monomer-dimer equilibrium regulates extracellular matrix affinity and tissue diffusion. *Nature Genetics, 41*(3), 289–298. https://doi.org/10.1038/ng.316.

Hill, D. J., Clemmons, D. R., Wilson, S., Han, V. K. M., Strain, A. J., & Milner, R. D. G. (1989). Immunological distribution of one form of insulin-like growth factor (IGF)-binding protein and IGF peptides in human fetal tissues. *Journal of Molecular Endocrinology, 2*(1), 31–38.

Hinz, B. (2015). The extracellular matrix and transforming growth factor-β1: Tale of a strained relationship. *Matrix Biology, 47*, 54–65. https://doi.org/10.1016/j.matbio.2015.05.006.

Hong, W., & Guan, K. L. (2012). The YAP and TAZ transcription co-activators: Key downstream effectors of the mammalian hippo pathway. *Seminars in Cell and Developmental Biology, 23*(7), 785–793. https://doi.org/10.1016/j.semcdb.2012.05.004.

Hortsch, M., & Goodman, C. S. (1991). Cell and substrate adhesion molecules in Drosophila. *Annual Review of Cell Biology*, 7(1), 505–557.

Hsieh, J., Aimone, J. B., Kaspar, B. K., Kuwabara, T., Nakashima, K., & Gage, F. H. (2004). IGF-I instructs multipotent adult neural progenitor cells to become oligodendrocytes. *Journal of Cell Biology, 164*(1), 111–122. https://doi.org/10.1083/jcb.200308101.

Huang, J., Wu, S., Barrera, J., Matthews, K., & Pan, D. (2005). The hippo signaling pathway coordinately regulates cell proliferation and apoptosis by inactivating yorkie, the Drosophila homolog of YAP. *Cell, 122*(3), 421–434. https://doi.org/10.1016/j.cell.2005.06.007.

Hynes, R. O. (2002). Integrins: Bidirectional, allosteric signaling machines. *Cell, 110*(6), 673–687. https://doi.org/10.1016/S0092-8674(02)00971-6.

International Human Genome Sequencing Consortium. (2001). Initial sequencing and analysis of the human genome. *Nature, 409*, 860–921. https://doi.org/10.1038/35057062.

Ishii, N., Wadsworth, W. G., Stern, B. D., Culotti, J. G., & Hedgecock, E. M. (1992). UNC-6, a laminin-related protein, guides cell and pioneer axon migrations in C. elegans. *Neuron, 9*(5), 873–881. https://doi.org/10.1016/0896-6273(92)90240-E.

Jakkaraju, S., Zhe, X., Pan, D., Choudhury, R., & Schuger, L. (2005). TIPs are tension-responsive proteins involved in myogenic versus adipogenic differentiation. *Developmental Cell, 9*(1), 39–49. https://doi.org/10.1016/j.devcel.2005.04.015.

Janmey, P. A., Winer, J. P., & Weisel, J. W. (2009). Fibrin gels and their clinical and bioengineering applications. *Journal of the Royal Society Interface, 6*(30), 1–10. https://doi.org/10.1098/rsif.2008.0327.

Jia, J., Maccarana, M., Zhang, X., Bespalov, M., Lindahl, U., & Li, J. P. (2009). Lack of L-iduronic acid in heparan sulfate affects interaction with growth factors and cell signaling. *Journal of Biological Chemistry, 284*(23), 15942–15950. https://doi.org/10.1074/jbc.M809577200.

Jikko, A., Harris, S. E., Chen, D., Mendrick, D. L., & Damsky, C. H. (1999). Collagen integrin receptors regulate early osteoblast differentiation induced by BMP-2. *Journal of Bone and Mineral Research, 14*(7), 1075–1083. https://doi.org/10.1359/jbmr.1999.14.7.1075.

Jones, J. I., Gockerman, A., Busby, W. H., Camacho-Hubner, C., & Clemmons, D. R. (1993). Extracellular matrix contains insulin-like growth factor binding protein-5: Potentiation of the effects of IGF-I. *Journal of Cell Biology, 121*(3), 679–687. https://doi.org/10.1083/jcb.121.3.679.

Jürgensen, H. J., Madsen, D. H., Ingvarsen, S., Melander, M. C., Gårdsvoll, H., Patthy, L., et al. (2011). A novel functional role of collagen glycosylation: Interaction with the endocytic collagen receptor uPARAP/ENDO180. *Journal of Biological Chemistry, 286*(37), 32736–32748. https://doi.org/10.1074/jbc.M111.266692.

Kadler, K. E., Hill, A., & Canty-Laird, E. G. (2008). Collagen fibrillogenesis: Fibronectin, integrins, and minor collagens as organizers and nucleators. *Current Opinion in Cell Biology, 20*(5), 495–501. https://doi.org/10.1016/j.ceb.2008.06.008.

Kagan, H. M., & Li, W. (2003). Lysyl oxidase: Properties, specificity, and biological roles inside and outside of the cell. *Journal of Cellular Biochemistry*, *88*(4), 660–672. https://doi.org/10.1002/jcb.10413.

Keller, R. (2005). Cell migration during gastrulation. *Current Opinion in Cell Biology*, *17*, 533–541. https://doi.org/10.1016/j.ceb.2005.08.006.

Khetarpal, U., Robertson, N. G., Yoo, T. J., & Morton, C. C. (1994). Expression and localization of COL2A1 mRNA and type II collagen in human fetal cochlea. *Hearing Research*, *79*(1–2), 59–73. https://doi.org/10.1016/0378-5955(94)90127-9.

Kilian, K. A., Bugarija, B., Lahn, B. T., & Mrksich, M. (2010). Geometric cues for directing the differentiation of mesenchymal stem cells. *Proceedings of the National Academy of Sciences of the United States of America*, *107*(11), 4872–4877. https://doi.org/10.1073/pnas.0903269107.

Klein, E. A., Yin, L., Kothapalli, D., Castagnino, P., Byfield, F. J., Xu, T., et al. (2009). Cell-cycle control by physiological matrix elasticity and in vivo tissue stiffening. *Current Biology*, *19*(18), 1511–1518. https://doi.org/10.1016/j.cub.2009.07.069.

Klinowska, T. C., Soriano, J. V., Edwards, G. M., Oliver, J. M., Valentijn, A. J., Montesano, R., et al. (1999). Laminin and beta 1 integrins are crucial for normal mammary gland development in the mouse. *Developmental Biology*, *215*(1), 13–32. https://doi.org/10.1006/dbio.1999.9435.

Knecht, A. K., & Bronner-Fraser, M. (2002). Induction of the neural crest: a multigene process. *Nature Reviews Genetics*, *3*(6), 453–461. https://doi.org/10.1038/nrg819.

Kogo, R., Shimamura, T., Mimori, K., Kawahara, K., Imoto, S., Sudo, T., et al. (2011). Long noncoding RNA HOTAIR regulates polycomb-dependent chromatin modification and is associated with poor prognosis in colorectal cancers. *Cancer Research*, *71*(20), 6320–6326. https://doi.org/10.1158/0008-5472.CAN-11-1021.

Kramer, J. M., Johnson, J. J., Edgar, R. S., Basch, C., & Roberts, S. (1988). The sqt-1 gene of C. elegans encodes a collagen critical for organismal morphogenesis. *Cell*, *55*(4), 555–565. https://doi.org/10.1016/0092-8674(88)90214-0.

Kubow, K. E., Klotzsch, E., Smith, M. L., Gourdon, D., Little, W. C., & Vogel, V. (2009). Crosslinking of cell-derived 3D scaffolds up-regulates the stretching and unfolding of new extracellular matrix assembled by reseeded cells. *Integrative Biology*, *1*(11–12), 635. https://doi.org/10.1039/b914996a.

Kwok, J. C. F., Dick, G., Wang, D., & Fawcett, J. W. (2011). Extracellular matrix and perineuronal nets in CNS repair. *Developmental Neurobiology*, *71*(11), 1073–1089. https://doi.org/10.1002/dneu.20974.

LaBarge, M. A., Nelson, C. M., Villadsen, R., Fridriksdottir, A., Ruth, J. R., Stampfer, M. R., et al. (2009). Human mammary progenitor cell fate decisions are products of interactions with combinatorial microenvironments. *Integrative Biology*, *1*(1), 70–79. https://doi.org/10.1039/B816472J.

Labrador, J. P., Azcoitia, V., Tuckermann, J., Lin, C., Olaso, E., Mañes, S., et al. (2001). The collagen receptor DDR2 regulates proliferation and its elimination leads to dwarfism. *EMBO Reports*, *2*(5), 446–452. https://doi.org/10.1093/embo-reports/kve094.

Larraín, J., Bachiller, D., Lu, B., Agius, E., Piccolo, S., & De Robertis, E. M. (2000). BMP-binding modules in chordin: A model for signalling regulation in the extracellular space. *Development (Cambridge, England)*, *127*(4), 821–830. https://doi.org/10.1016/j.drugalcdep.2008.02.002.A.

Laurent, T. C., & Fraser, J. R. (1992). Hyaluronan. *FASEB Journal: Official Publication of the Federation of American Societies for Experimental Biology*, *6*(7), 2397–2404. https://doi.org/10.1016/S0140-6736(01)35637-4.

Le, L. T. N., Cazares, O., Mouw, J. K., Chatterjee, S., Macias, H., Moran, A., et al. (2016). Loss of miR-203 regulates cell shape and matrix adhesion through ROBO1/Rac/FAK in response to stiffness. *Journal of Cell Biology*, *212*(6), 707–719. https://doi.org/10.1083/jcb.201507054.

Lee, E. J., Kasper, F. K., & Mikos, A. G. (2014). Biomaterials for tissue engineering. *Annals of Biomedical Engineering*, *42*, 323–337. https://doi.org/10.1007/s10439-013-0859-6.

Lee, S. T., Yun, J. I., Jo, Y. S., Mochizuki, M., van der Vlies, A. J., Kontos, S., et al. (2010). Engineering integrin signaling for promoting embryonic stem cell self-renewal in a precisely defined niche. *Biomaterials*, *31*(6), 1219–1226. https://doi.org/10.1016/j.biomaterials.2009.10.054.

Leptin, M., Bogaert, T., Lehmann, R., & Wilcox, M. (1989). The function of PS integrins during Drosophila embryogenesis. *Cell*, *56*(3), 401–408. https://doi.org/10.1016/0092-8674(89)90243-2.

Li, L., & Clevers, H. (2010). Coexistence of quiescent and active adult stem cells in mammals. *Science*, *327*(5965), 542–545. https://doi.org/10.1126/science.1180794.

Li, H., Corrales, C. E., Wang, Z., Zhao, Y., Wang, Y., Liu, H., et al. (2005). BMP4 signaling is involved in the generation of inner ear sensory epithelia. *BMC Developmental Biology*, *5*, 16. https://doi.org/10.1186/1471-213x-5-16.

Li, Y., Li, L., Chen, Z. N., Gao, G., Yao, R., & Sun, W. (2017). Engineering-derived approaches for iPSC preparation, expansion, differentiation and applications. *Biofabrication*, *9*(3), 032001. https://doi.org/10.1088/1758-5090/aa7e9a.

Lindenmuth, D. M., Van Wijnen, A. J., Hiebert, S., Stein, J. L., Lian, J. B., & Stein, G. S. (1997). Subcellular partitioning of transcription factors during osteoblast differentiation: Developmental association of the AML/CBFα/PEBP2α-related transcription factor-NMP-2 with the nuclear matrix. *Journal of Cellular Biochemistry*, *66*(1), 123–132. https://doi.org/10.1002/(SICI)1097-4644(19970701)66:1<123::AID-JCB13>3.0.CO;2-K.

Lodish, H., Berk, A., Zipursky, S. L., et al. (2000). *Molecular cell biology* (4th ed.). (Section 22.3, Collagen: The fibrous proteins of the matrix). New York: W. H. Freeman.

Lu, P., Takai, K., Weaver, V. M., & Werb, Z. (2011). Extracellular matrix degradation and remodeling in development and disease. *Cold Spring Harbor Perspectives in Biology*, *3*(12), 1–24. https://doi.org/10.1101/cshperspect.a005058.

Lund, A. W., Yener, B., Stegemann, J. P., & Plopper, G. E. (2009). The natural and engineered 3D microenvironment as a regulatory cue during stem cell fate determination. *Tissue EngineeringPart B, Reviews*, *15*(3), 371–380. https://doi.org/10.1089/ten.teb.2009.0270.

MacBeath, G., & Schreiber, S. L. (2000). Printing proteins as microarrays for high-throughput function determination. *Science*, *289*(September), 1760–1763. https://doi.org/10.1126/science.289.5485.1760.

Macias, H., Moran, A., Samara, Y., Moreno, M., Compton, J. E., Harburg, G., et al. (2011). SLIT/ROBO1 signaling suppresses mammary branching morphogenesis by limiting basal cell number. *Developmental Cell*, *20*(6), 827–840. https://doi.org/10.1016/j.devcel.2011.05.012.

Mäki, J. M., Räsänen, J., Tikkanen, H., Sormunen, R., Mäkikallio, K., Kivirikko, K. I., et al. (2002). Inactivation of the lysyl oxidase gene Lox leads to aortic aneurysms, cardiovascular dysfunction, and perinatal death in mice. *Circulation*, *106*(19), 2503–2509. https://doi.org/10.1161/01.CIR.0000038109.84500.1E.

Maniotis, A. J., Bojanowski, K., & Ingber, D. E. (1997). Mechanical continuity and reversible chromosome disassembly within intact genomes removed from living cells. *Journal of Cellular Biochemistry*, *65*(1), 114–130. https://doi.org/10.1002/(SICI)1097-4644(199704)65:1<114::AID-JCB12>3.0.CO;2-K.

Maniotis, A. J., Chen, C. S., & Ingber, D. E. (1997). Demonstration of mechanical connections between integrins, cytoskeletal filaments, and nucleoplasm that stabilize nuclear structure. *Proceedings of the National Academy of Sciences of the United States of America*, *94*(3), 849–854. https://doi.org/10.1073/pnas.94.3.849.

Martin, D. C., Fowlkes, J. L., Babic, B., & Khokha, R. (1999). Insulin-like growth factor II signaling in neoplastic proliferation is blocked by transgenic expression of the metalloproteinase inhibitor TIMP-1. *Journal of Cell Biology*, *146*(4), 881–892. https://doi.org/10.1083/jcb.146.4.881.

Matsui, T., Raya, A., Callol-Massot, C., Kawakami, Y., Oishi, I., Rodriguez-Esteban, C., et al. (2007). miles-apart-Mediated regulation of cell-fibronectin interaction and myocardial migration in zebrafish. *Nature Clinical Practice. Cardiovascular Medicine*, *4*(Suppl. 1), S77–82. https://doi.org/10.1038/ncpcardio0764.

McBeath, R., Pirone, D. M., Nelson, C. M., Bhadriraju, K., & Chen, C. S. (2004). Cell shape, cytoskeletal tension, and RhoA regulate stem cell lineage commitment. *Developmental Cell*, *6*(4), 483–495. https://doi.org/10.1016/S1534-5807(04)00075-9.

McBride, S. H., & Knothe Tate, M. L. (2008). Modulation of stem cell shape and fate A: The role of density and seeding protocol on nucleus shape and gene expression. *Tissue Engineering. Part A*, *14*(9), 1561–1572. https://doi.org/10.1089/ten.tea.2008.0112.

Mecham, R. P. (2012). Overview of extracellular matrix. *Current Protocols in Cell Biology*, *57* 10.1.1–10.1.16. https://doi.org/10.1002/0471143030.cb1001s57.

Mehta, D., Ahmmed, G. U., Paria, B. C., Holinstat, M., Voyno-Yasenetskaya, T., Tiruppathi, C., et al. (2003). RhoA interaction with inositol 1,4,5-trisphosphate receptor and transient receptor potential channel-1 regulates Ca2+ entry: Role in signaling increased endothelial permeability. *Journal of Biological Chemistry*, *278*(35), 33492–33500. https://doi.org/10.1074/jbc.M302401200.

Miroshnikova, Y. A., Rozenberg, G. I., Cassereau, L., Pickup, M., Mouw, J. K., Ou, G., et al. (2017). α5B1-Integrinpromotes tension-dependent mammary epithelial cell invasion by engaging the fibronectin synergy site. *Molecular Biology of the Cell*, *28*(22), 2958–2977. mbc. E17-02-0126. https://doi.org/10.1091/mbc.E17-02-0126.

Mizuno, M., Fujisawa, R., & Kuboki, Y. (2000). Type I collagen-induced osteoblastic differentiation of bone-marrow cells mediated by collagen-α2β1 integrin interaction. *Journal of Cellular Physiology*, *184*(2), 207–213. https://doi.org/10.1002/1097-4652(200008)184:2<207::AID-JCP8>3.0.CO;2-U.

Montanez-Sauri, S. I., Beebe, D. J., & Sung, K. E. (2015). Microscale screening systems for 3D cellular microenvironments: Platforms, advances, and challenges. *Cellular and Molecular Life Sciences*, *72*(2), 237–249. https://doi.org/10.1007/s00018-014-1738-5.

Mouw, J. K., Ou, G., & Weaver, V. M. (2014). Extracellular matrix assembly: A multiscale deconstruction. *Nature Reviews Molecular Cell Biology*, *15*(12), 771–785. https://doi.org/10.1038/nrm3902.

Mouw, J. K., Yui, Y., Damiano, L., Bainer, R. O., Lakins, J. N., Acerbi, I., et al. (2014). Tissue mechanics modulate micro RNA-dependent PTEN expression to regulate malignant progression. *Nature Medicine*, *20*(4), 360–367. https://doi.org/10.1038/nm.3497.

Muramatsu, S., Wakabayashi, M., Ohno, T., Amano, K., Ooishi, R., Sugahara, T., et al. (2007). Functional gene screening system identified TRPV4 as a regulator of chondrogenic differentiation. *Journal of Biological Chemistry*, *282*(44), 32158–32167. https://doi.org/10.1074/jbc.M706158200.

Myers, C. A., Schmidhauser, C., Mellentin-Michelotti, J., Fragoso, G., Roskelley, C. D., Casperson, G., et al. (1998). Characterization of BCE-1, a transcriptional enhancer regulated by prolactin and extracellular matrix and modulated by the state of histone acetylation. *Molecular and Cellular Biology*, *18*(4), 2184–2195. Retrieved from http://www.pubmedcentral.nih.gov/articlerender.fcgi?artid=121460&tool=pmcentrez&rendertype=abstract.

Nagel, M., Tahinci, E., Symes, K., & Winklbauer, R. (2004). Guidance of mesoderm cell migration in the Xenopus gastrula requires PDGF signaling. *Development (Cambridge, England)*, *131*, 2727–2736. https://doi.org/10.1242/dev.01141.

Naylor, M. J., Li, N., Cheung, J., Lowe, E. T., Lambert, E., Marlow, R., et al. (2005). Ablation of β1 integrin in mammary epithelium reveals a key role for integrin in glandular morphogenesis and differentiation. *Journal of Cell Biology*, *171*(4), 717–728. https://doi.org/10.1083/jcb.200503144.

Nicodemi, M., & Pombo, A. (2014). Models of chromosome structure. *Current Opinion in Cell Biology*, *28*(1), 90–95. https://doi.org/10.1016/j.ceb.2014.04.004.
Ohashi, T., Kiehart, D. P., & Erickson, H. P. (2002). Dual labeling of the fibronectin matrix and actin cytoskeleton with green fluorescent protein variants. *Journal of Cell Science*, *115*(Pt. 6), 1221–1229.
Ornitz, D. M. (2000). FGFs, heparan sulfate and FGFRs: Complex interactions essential for development. *BioEssays*, *22*(2), 108–112. https://doi.org/10.1002/(SICI)1521-1878(200002)22:2<108::AID-BIES2>3.0.CO;2-M.
Padmakumar, V. C., Libotte, T., Lu, W. L., Zaim, H., Abraham, S., Noegel, A. A., et al. (2005). The inner nuclear membrane protein Sun1 mediates the anchorage of Nesprin-2 to the nuclear envelope. *Journal of Cell Science*, *118*(15), 3419–3430. https://doi.org/10.1242/jcs.02471.
Page-McCaw, A., Ewald, A. J., & Werb, Z. (2007). Matrix metalloproteinases and the regulation of tissue remodelling. *Nature Reviews Molecular Cell Biology*, *8*(3), 221–233. https://doi.org/10.1038/nrm2125.
Palmieri, D., Valli, M., Viglio, S., Ferrari, N., Ledda, B., Volta, C., et al. (2010). Osteoblasts extracellular matrix induces vessel like structures through glycosylated collagen I. *Experimental Cell Research*, *316*(5), 789–799. https://doi.org/10.1016/j.yexcr.2009.12.006.
Pankov, R., Cukierman, E., Katz, B. Z., Matsumoto, K., Lin, D. C., Lin, S., et al. (2000). Integrin dynamics and matrix assembly: Tensin-dependent translocation of $\alpha5\beta1$ integrins promotes early fibronectin fibrillogenesis. *Journal of Cell Biology*, *148*(5), 1075–1090. https://doi.org/10.1083/jcb.148.5.1075.
Parenteau-Bareil, R., Gauvin, R., & Berthod, F. (2010). Collagen-based biomaterials for tissue engineering applications. *Materials*, *3*(3), 1863–1887. https://doi.org/10.3390/ma3031863.
Paszek, M. J., & Weaver, V. M. (2004). The tension mounts: Mechanics meets morphogenesis and malignancy. *Journal of Mammary Gland Biology and Neoplasia*, *9*(4), 325–342. https://doi.org/10.1007/s10911-004-1404-x.
Patel, V. N., Knox, S. M., Likar, K. M., Lathrop, C. A., Hossain, R., Eftekhari, S., et al. (2007). Heparanase cleavage of perlecan heparan sulfate modulates FGF10 activity during ex vivo submandibular gland branching morphogenesis. *Development*, *134*(23), 4177–4186. https://doi.org/10.1242/dev.011171.
Pathak, M. M., Nourse, J. L., Tran, T., Hwe, J., Arulmoli, J., Le, D. T. T., et al. (2014). Stretch-activated ion channel Piezo1 directs lineage choice in human neural stem cells. *Proceedings of the National Academy of Sciences of the United States of America*, *111*(45), 16148–16153. https://doi.org/10.1073/pnas.1409802111.
Perissinotto, D., Iacopetti, P., Bellina, I., Doliana, R., Colombatti, A., Pettway, Z., et al. (2000). Avian neural crest cell migration is diversely regulated by the two major hyaluronan-binding proteoglycans PG-M/versican and aggrecan. *Development*, *127*(13), 2823–2842. Retrieved from http://www.ncbi.nlm.nih.gov/entrez/query.fcgi?cmd=Retrieve&db=PubMed&dopt=Citation&list_uids=10851128.
Phelps, E. A., Enemchukwu, N. O., Fiore, V. F., Sy, J. C., Murthy, N., Sulchek, T. A., et al. (2012). Maleimide cross-linked bioactive PEG hydrogel exhibits improved reaction kinetics and cross-linking for cell encapsulation and in situ delivery. *Advanced Materials*, *24*(1), 64–70. https://doi.org/10.1002/adma.201103574.
Pietri, T., Eder, O., Breau, M. A., Topilko, P., Blanche, M., Brakebusch, C., et al. (2004). Conditional beta1-integrin gene deletion in neural crest cells causes severe developmental alterations of the peripheral nervous system. *Development*, *131*(16), 3871–3883. https://doi.org/10.1242/dev.01264.
Pla, A. F., Maric, D., Brazer, S.-C., Giacobini, P., Liu, X., Chang, Y. H., et al. (2005). Canonical transient receptor potential 1 plays a role in basic fibroblast growth factor (bFGF)/FGF receptor-1-induced $Ca2+$ entry and embryonic rat neural stem cell proliferation. *The Journal of Neuroscience*, *25*(10), 2687–2701. https://doi.org/10.1523/JNEUROSCI.0951-04.2005.

Pond, A. C., Bin, X., Batts, T., Roarty, K., Hilsenbeck, S., & Rosen, J. M. (2013). Fibroblast growth factor receptor signaling is essential for normal mammary gland development and stem cell function. *Stem Cells*, *31*(1), 178–189. https://doi.org/10.1002/stem.1266.

Przybyla, L., Lakins, J. N., & Weaver, V. M. (2016). Tissue mechanics orchestrate wnt-dependent human embryonic stem cell differentiation. *Cell Stem Cell*, *19*(4), 462–475. https://doi.org/10.1016/j.stem.2016.06.018.

Ramsey, I. S., Delling, M., & Clapham, D. E. (2006). An introduction to TRP channels. doi:10.1146/annurev.physiol.68.040204.100431. *Annual Review of Physiology*, *68*(1), 619–647. https://doi.org/10.1146/annurev.physiol.68.040204.100431.

Robertson, I. B., Horiguchi, M., Zilberberg, L., Dabovic, B., Hadjiolova, K., & Rifkin, D. B. (2015). Latent TGF-β-binding proteins. *Matrix Biology*, *47*, 44–53. https://doi.org/10.1016/j.matbio.2015.05.005.

Rubashkin, M. G., Ou, G., & Weaver, V. M. (2014). Deconstructing signaling in three dimensions. *Biochemistry*, *53*(13), 2078–2090. https://doi.org/10.1021/bi401710d.

Saha, K., Keung, A. J., Irwin, E. F., Li, Y., Little, L., Schaffer, D. V., et al. (2008). Substrate modulus directs neural stem cell behavior. *Biophysical Journal*, *95*(9), 4426–4438. https://doi.org/10.1529/biophysj.108.132217.

Salditt-Georgieff, M., Harpold, M. M., Wilson, M. C., & Darnell, J. E. (1981). Large heterogeneous nuclear ribonucleic acid has three times as many 5' caps as polyadenylic acid segments, and most caps do not enter polyribosomes. *Molecular and Cellular Biology*, *1*(2), 179–187. https://doi.org/10.1128/MCB.1.2.179.

Sasai, Y. (2013). Next-generation regenerative medicine: Organogenesis from stem cells in 3D culture. *Cell Stem Cell*, *12*(5), 520–530. https://doi.org/10.1016/j.stem.2013.04.009.

Sato-Kusubata, K., Jiang, Y., Ueno, Y., & Chun, T.-H. (2011). Adipogenic histone mark regulation by matrix metalloproteinase 14 in collagen-rich microenvironments. *Molecular Endocrinology*, *25*(5), 745–753. https://doi.org/10.1210/me.2010-0429.

Schmidhauser, C., Bissell, M. J., Myerst, C. A., & Caspersont, G. F. (1990). Extracellular matrix and hormones transcriptionally regulate bovine β-casein 5 sequences in stably transfected mouse mammary cells. *Proceedings of the National Academy of Sciences of the United States of America*, *87*(December), 9118–9122. https://doi.org/10.1073/pnas.87.23.9118.

Schwarzbauer, J. E., & DeSimone, D. W. (2011). Fibronectins, their fibrillogenesis, and in vivo functions. *Cold Spring Harbor Perspectives in Biology*, *3*, a005041. https://doi.org/10.1101/cshperspect.a005041.

Sechler, J. L., Corbett, S. A., & Schwarzbauer, J. E. (1997). Modulatory roles for integrin activation and the synergy site of fibronectin during matrix assembly. *Molecular Biology of the Cell*, *8*(12), 2563–2573. https://doi.org/10.1091/mbc.8.12.2563.

Sengle, G., Charbonneau, N. L., Ono, R. N., Sasaki, T., Alvarez, J., Keene, D. R., et al. (2008). Targeting of bone morphogenetic protein growth factor complexes to fibrillin. *The Journal of Biological Chemistry*, *283*(20), 13874–13888. https://doi.org/10.1074/jbc.M707820200.

Seong, J., Tajik, A., Sun, J., Guan, J.-L., Humphries, M. J., Craig, S. E., et al. (2013). Distinct biophysical mechanisms of focal adhesion kinase mechanoactivation by different extracellular matrix proteins. *Proceedings of the National Academy of Sciences of the United States of America*, *110*(48), 19372–19377. https://doi.org/10.1073/pnas.1307405110.

Singh, P., Carraher, C., & Schwarzbauer, J. E. (2010). Assembly of fibronectin extracellular matrix. *Annual Review of Cell and Developmental Biology*, *26*, 397–419. https://doi.org/10.1146/annurev-cellbio-100109-104020.

Skinner, B. M., & Johnson, E. E. P. (2017). Nuclear morphologies: Their diversity and functional relevance. *Chromosoma*, *126*(2), 195–212. https://doi.org/10.1007/s00412-016-0614-5.

Smith, M. L., Gourdon, D., Little, W. C., Kubow, K. E., Eguiluz, R. A., Luna-Morris, S., et al. (2007). Force-induced unfolding of fibronectin in the extracellular matrix of living cells. *PLoS Biology*, *5*(10), 2243–2254. https://doi.org/10.1371/journal.pbio.0050268.

Smith-Berdan, S., Nguyen, A., Hassanein, D., Zimmer, M., Ugarte, F., Ciriza, J., et al. (2011). Robo4 cooperates with Cxcr4 to specify hematopoietic stem cell localization to bone marrow niches. *Cell Stem Cell*, *8*(1), 72–83. https://doi.org/10.1016/j.stem.2010.11.030.

Solursh, M., Linsenmayer, T. F., & Jensen, K. L. (1982). Chondrogenesis from single limb mesenchyme cells. *Developmental Biology*, *94*(1), 259–264. https://doi.org/10.1016/0012-1606(82)90090-2.

Song, J. J., & Ott, H. C. (2011). Organ engineering based on decellularized matrix scaffolds. *Trends in Molecular Medicine*, *17*, 424–432. https://doi.org/10.1016/j.molmed.2011.03.005.

Stein, G. S., Lian, J. B., van Wijnen, A. J., Stein, J. L., Javed, A., Montecino, M., et al. (2007). Organization of transcriptional regulatory machinery in nuclear microenvironments: Implications for biological control and cancer. *Advances in Enzyme Regulation*, *47*, 242–250. https://doi.org/10.1016/j.advenzreg.2006.12.014.

Streuli, C. (1999). Extracellular matrix remodelling and cellular differentiation. *Current Opinion in Cell Biology*, *11*, 634–640. https://doi.org/10.1016/S0955-0674(99)00026-5.

Sugawara, K., Tsuruta, D., Ishii, M., Jones, J. C. R., & Kobayashi, H. (2008). Laminin-332 and -511 in skin. *Experimental Dermatology*, *17*(6), 473–480. https://doi.org/10.1111/j.1600-0625.2008.00721.x.

Suh, H. N., & Han, H. J. (2011). Collagen I regulates the self-renewal of mouse embryonic stem cells through α2β1 integrin- and DDR1-dependent Bmi-1. *Journal of Cellular Physiology*, *226*(12), 3422–3432. https://doi.org/10.1002/jcp.22697.

ten Dijke, P., & Hill, C. S. (2004). New insights into TGF-beta-Smad signalling. *Trends in Biochemical Sciences*, *29*(5), 265–273. https://doi.org/10.1016/j.tibs.2004.03.008.

Tenni, R., Valli, M., Rossi, A., & Cetta, G. (1993). Possible role of overglycosylation in the type I collagen triple helical domain in the molecular pathogenesis of osteogenesis imperfecta. *American Journal of Medical Genetics*, *45*, 252–256. https://doi.org/10.1002/ajmg.1320450219.

Tholozan, F. M. D., Gribbon, C., Li, Z., Goldberg, M. W., Prescott, A. R., McKie, N., et al. (2007). FGF-2 release from the lens capsule by MMP-2 maintains lens epithelial cell viability. *Molecular Biology of the Cell*, *18*(11), 4222–4231. https://doi.org/10.1091/mbc.E06-05-0416.

Timpl, R., & Brown, J. C. (1994). The laminins. *Matrix Biology*, *14*(4), 275–281. https://doi.org/10.1016/0945-053X(94)90192-9.

Trinh, L. A., & Stainier, D. Y. R. (2004). Fibronectin regulates epithelial organization during myocardial migration in zebrafish. *Developmental Cell*, *6*(3), 371–382. https://doi.org/10.1016/S1534-5807(04)00063-2.

Tsai, M. C., Manor, O., Wan, Y., Mosammaparast, N., Wang, J. K., Lan, F., et al. (2010). Long noncoding RNA as modular scaffold of histone modification complexes. *Science*, *329*(5992), 689–693. https://doi.org/10.1126/science.1192002.

Tsang, K. Y., Cheung, M. C. H., Chan, D., & Cheah, K. S. E. (2010). The developmental roles of the extracellular matrix: Beyond structure to regulation. *Cell and Tissue Research*, *339*(1), 93–110. https://doi.org/10.1007/s00441-009-0893-8.

van der Kraan, P. M., Blaney Davidson, E. N., Blom, A., & van den Berg, W. B. (2009). TGF-beta signaling in chondrocyte terminal differentiation and osteoarthritis. Modulation and integration of signaling pathways through receptor-Smads. *Osteoarthritis and Cartilage*, *17*(12), 1539–1545. https://doi.org/10.1016/j.joca.2009.06.008.

Venter, J. C., Adams, M. D., Myers, E. W., Li, P. W., Mural, R. J., Sutton, G. G., et al. (2001). The sequence of the human genome. *Science (New York, N.Y.)*, *291*(5507), 1304–1351. https://doi.org/10.1126/science.1058040.

Volk, T., Fessler, L. I., & Fessler, J. H. (1990). A role for integrin in the formation of sarcomeric cytoarchitecture. *Cell*, *63*(3), 525–536. https://doi.org/10.1016/0092-8674(90)90449-O.

von Mende, N., Bird, D. M., Albert, P. S., & Riddle, D. L. (1988). dpy-13: A nematode collagen gene that affects body shape. *Cell*, *55*(4), 567–576. https://doi.org/10.1016/0092-8674(88)90215-2.

Vriens, J., Watanabe, H., Janssens, A., Droogmans, G., Voets, T., & Nilius, B. (2004). Cell swelling, heat, and chemical agonists use distinct pathways for the activation of the cation channel TRPV4. *Proceedings of the National Academy of Sciences of the United States of America*, *101*(1), 396–401. https://doi.org/10.1073/pnas.0303329101.

Wang, X., Harris, R. E., Bayston, L. J., & Ashe, H. L. (2008). Type IV collagens regulate BMP signalling in Drosophila. *Nature*, *455*(7209), 72–77. https://doi.org/nature07214.

Wang, N., Tytell, J. D., & Ingber, D. E. (2009). Mechanotransduction at a distance: Mechanically coupling the extracellular matrix with the nucleus. *Nature Reviews Molecular Cell Biology*, *10*(1), 75–82. https://doi.org/10.1038/nrm2594.

Wang, K. C., Yang, Y. W., Liu, B., Sanyal, A., Corces-Zimmerman, R., Chen, Y., et al. (2011). A long noncoding RNA maintains active chromatin to coordinate homeotic gene expression. *Nature*, *472*(7341), 120–126. https://doi.org/10.1038/nature09819.

Warmflash, A., Sorre, B., Etoc, F., Siggia, E. D., & Brivanlou, A. H. (2014). A method to recapitulate early embryonic spatial patterning in human embryonic stem cells. *Nature Methods*, *11*(8), 847–854. https://doi.org/10.1038/nmeth.3016.

Warner, J. R., Soeiro, R., Birnboim, H. C., Girard, M., & Darnell, J. E. (1966). Rapidly labeled HeLa cell nuclear RNA: I. Identification by zone sedimentation of a heterogeneous fraction separate from ribosomal precursor RNA. *Journal of Molecular Biology*, *19*(2), 349–361. https://doi.org/10.1016/S0022-2836(66)80009-8.

Wilcox, M., DiAntonio, A., & Leptin, M. (1989). The function of PS integrins in Drosophila wing morphogenesis. *Development*, *107*(4), 891–897.

Winer, J. P., Janmey, P. A., McCormick, M. E., & Funaki, M. (2009). Bone marrow-derived human mesenchymal stem cells become quiescent on soft substrates but remain responsive to chemical or mechanical stimuli. *Tissue Engineering Part A*, *15*(1), 147–154. https://doi.org/10.1089/ten.tea.2007.0388.

Wipff, P. J., Rifkin, D. B., Meister, J. J., & Hinz, B. (2007). Myofibroblast contraction activates latent TGF-β1 from the extracellular matrix. *Journal of Cell Biology*, *179*(6), 1311–1323. https://doi.org/10.1083/jcb.200704042.

Xu, R., Nelson, C. M., Muschler, J. L., Veiseh, M., Vonderhaar, B. K., & Bissell, M. J. (2009). Sustained activation of STAT5 is essential for chromatin remodeling and maintenance of mammary-specific function. *Journal of Cell Biology*, *184*(1), 57–66. https://doi.org/10.1083/jcb.200807021.

Yamaguchi, Y. (2000). Lecticans: Organizers of the brain extracellular matrix. *Cellular and Molecular Life Sciences*, *57*(2), 276–289. https://doi.org/10.1007/PL00000690.

Yamauchi, M., & Sricholpech, M. (2012). Lysine post-translational modifications of collagen. *Essays In Biochemistry*, *52*, 113–133. https://doi.org/10.1042/bse0520113.

Yang, Y., Beqaj, S., Kemp, P., Ariel, I., & Schuger, L. (2000). Stretch-induced alternative splicing of serum response factor promotes bronchial myogenesis and is defective in lung hypoplasia. *Journal of Clinical Investigation*, *106*(11), 1321–1330. https://doi.org/10.1172/JCI8893.

Yang, Y., Palmer, K. C., Relan, N., Diglio, C., & Schuger, L. (1998). Role of laminin polymerization at the epithelial mesenchymal interface in bronchial myogenesis. *Development (Cambridge, England)*, *125*(14), 2621–2629.

Yang, Y., Relan, N. K., Przywara, D. A., & Schuger, L. (1999). Embryonic mesenchymal cells share the potential for smooth muscle differentiation: Myogenesis is controlled by the cell's shape. *Development (Cambridge, England)*, *126*(13), 3027–3033.

Yoon, B. S., Ovchinnikov, D. A., Yoshii, I., Mishina, Y., Behringer, R. R., & Lyons, K. M. (2005). Bmpr1a and Bmpr1b have overlapping functions and are essential for chondrogenesis in vivo. *Proceedings of the National Academy of Sciences of the United States of America, 102*, 5062–5067. https://doi.org/10.1073/pnas.0500031102.

Yost, H. J. (1992). Regulation of vertebrate left-right asymmetries by extracellular matrix. *Nature, 357*(6374), 158–161. https://doi.org/10.1038/357158a0.

Zhang, Y., Su, J., Yu, J., Bu, X., Ren, T., Liu, X., et al. (2011). An essential role of discoidin domain receptor 2 (DDR2) in osteoblast differentiation and chondrocyte maturation via modulation of Runx2 activation. *Journal of Bone and Mineral Research, 26*(3), 604–617. https://doi.org/10.1002/jbmr.225.

Zhao, B., Tumaneng, K., & Guan, K.-L. (2011). The Hippo pathway in organ size control, tissue regeneration and stem cell self-renewal. *Nature Cell Biology, 13*(8), 877–883. https://doi.org/10.1038/ncb2303.

Zhong, C., Chrzanowska-Wodnicka, M., Brown, J., Shaub, A., Belkin, A. M., & Burridge, K. (1998). Rho-mediated contractility exposes a cryptic site in fibronectin and induces fibronectin matrix assembly. *Journal of Cell Biology, 141*(2), 539–551. https://doi.org/10.1083/jcb.141.2.539.

Zhu, Y., Oganesian, A., Keene, D. R., & Sandell, L. J. (1999). Type IIA procollagen containing the cysteine-rich amino propeptide is deposited in the extracellular matrix of prechondrogenic tissue and binds to TGF-β1 and BMP-2. *Journal of Cell Biology, 144*(5), 1069–1080. https://doi.org/10.1083/jcb.144.5.1069.

Zusman, S., Patel-King, R. S., Ffrench-Constant, C., & Hynes, R. O. (1990). Requirements for integrins during Drosophila development. *Development (Cambridge, England), 108*(3), 391–402. https://doi.org/10.1038/342285a0.

CHAPTER TWO

Matricellular Proteins: Functional Insights From Non-mammalian Animal Models

Josephine C. Adams[1]

School of Biochemistry, University of Bristol, Bristol, United Kingdom
[1]Corresponding author: e-mail address: jo.adams@bristol.ac.uk

Contents

Abstract

The extracellular matrix (ECM) has central roles in tissue integrity and remodeling throughout the life span of animals. While collagens are the most abundant structural components of ECM in most tissues, tissue-specific molecular complexity is contributed by ECM glycoproteins. The matricellular glycoproteins are categorized primarily according to functional criteria and represented predominantly by the thrombospondin, tenascin, SPARC/osteonectin, and CCN families. These proteins do not self-assemble into ECM fibrils; nevertheless, they shape ECM properties through interactions with structural ECM proteins, growth factors, and cells. Matricellular proteins also

Current Topics in Developmental Biology, Volume 130
ISSN 0070-2153
https://doi.org/10.1016/bs.ctdb.2018.02.003

promote cell migration or morphological changes through adhesion-modulating or counter-adhesive actions on cell–ECM adhesions, intracellular signaling, and the actin cytoskeleton. Typically, matricellular proteins are most highly expressed during embryonic development. In adult tissues, expression is more limited unless activated by cues for dynamic tissue remodeling and cell motility, such as occur during inflammatory response and wound repair. Many insights in the complex roles of matricellular proteins have been obtained from studies of gene knockout mice. However, with the exception of chordate-specific tenascins, these are highly conserved proteins that are encoded in many animal phyla. This review will consider the increasing body of research on matricellular proteins in nonmammalian animal models. These models provide better access to the very earliest stages of embryonic development and opportunities to study biological processes such as limb and organ regeneration. In aggregate, this research is expanding concepts of the functions and mechanisms of action of matricellular proteins.

ABBREVIATIONS

BMP bone morphogenetic protein
CCN CYR61/CTGF/NOV
ECM extracellular matrix
dpp decapentaplegic
EGF epidermal growth factor
FN fibronectin
GFP green fluorescent protein
IGFBP insulin-like growth factor-binding protein
LDL low-density lipoprotein
RNAi RNA interference
siRNA short interfering RNA
SNP single-nucleotide polymorphism
SPARC small protein acidic and rich in cysteine
SMOC SPARC-related modular calcium-binding protein
SPOCK sparc/osteonectin, cwcv, and kazal-like domains
TN tenascin
TSP thrombospondin
TSR thrombospondin repeats
VEGF vascular endothelial growth factor

1. GENERAL INTRODUCTION

The extracellular matrix (ECM) of animals is a complex, micron-scale, proteinaceous structure that is formed as a fibrous network between cells and has a tissue-specific composition of collagens, proteoglycans, glycoproteins, and glycosaminoglycans. The evolution of the ECM is viewed as one of the driving factors for the complex multicellularity of animals (Brunet & King, 2017; Sebé-Pedrós et al., 2013; Williams, Tew, Paul, & Adams, 2014). Two crucial activities of ECM components are to undergo extracellular interactions to assemble the ECM and to bind plasma membrane receptors that secure the association of ECM with cell surfaces. Integrins, dystroglycan, glypican, and syndecan proteoglycans are prime examples of cell–ECM receptors. In relation to the fibrils and networks of the ECM, these cell-surface interactions maintain transmission of mechanical forces between ECM and cells and also regulate intracellular chemical signaling pathways that control cell survival, cytoskeletal organization, motility, differentiation, and gene expression (Geiger & Yamada, 2011; Hynes, 2014; Jansen, Atherton, & Ballestrem, 2017). Interactions of ECM proteins with low-density lipoprotein (LDL)-related receptors (and others) serve in ECM remodeling by dynamic uptake of ECM proteins for intracellular degradation (Lillis, Van Duyn, Murphy-Ullrich, & Strickland, 2008). Proteomic studies of decellularized tissues demonstrate that tissue ECMs typically contain over 150 protein components; these include nonstructural proteins, cytokines, and growth factors as well as canonical ECM proteins (Beachley et al., 2015; Naba et al., 2012, 2017).

From the 1970s onward, functional studies of the cell adhesive properties of purified or recombinant ECM proteins led to an appreciation that some ECM proteins are far more adhesive for cells than others. For example, the glycoprotein fibronectin (FN) that is expressed widely in vertebrates, supports spreading and focal adhesion assembly by many different cell types (reviewed by Otey & Burridge, 1990), whereas thrombospondin-1 (TSP1) supports cell attachment, and, in certain cell types only, limited spreading and formation of lamellipodia (Adams, 1995; Adams & Lawler, 1993; Lawler, Weinstein, & Hynes, 1988). Tenascin-C (TN-C) and small protein acidic and rich in cysteine (SPARC, also known as osteonectin) became noted for their antiadhesive properties, in the case of TN-C involving actual repulsion of cell attachment or axon outgrowth in vitro (Chiquet-Ehrismann, Kalla, Pearson, Beck, & Chiquet, 1988; Sage, 1992; Sage, Vernon, Funk, Everitt, & Angello, 1989). In addition, when presented to cells in soluble form, TN-C and SPARC destabilized the focal adhesions formed by cells adherent on FN or vitronectin (Fischer, Tucker, Chiquet-Ehrismann, & Adams, 1997; Murphy-Ullrich & Höök, 1989; Murphy-Ullrich, Lane, Pallero, & Sage, 1995; Murphy-Ullrich et al., 1991). It was also noted that SPARC, tenascin-C (TN-C), and TSP1 (known originally as thrombospondin) do not self-assemble into ECM fibril structures, although they associated with the fibrils and networks formed by collagens, FN, or other structural ECM proteins (Faissner, Kruse, Kühn, & Schachner, 1990; Giudici et al., 2008; Rentz, Poobalarahi, Bornstein, Sage, & Bradshaw, 2007; Tan & Lawler, 2009). In broad terms, the adhesive ECM proteins are present throughout life and form large, stable molecular networks, fibrils, and fibers within the ECM: prime examples are fibrillar collagens, FN, and laminins. Analyses of life-course expression profiles brought to light that TSP1, TN-C, and their respective later-identified gene family paralogues were expressed in many mammalian tissues during development yet had relatively low and restricted expression in adult tissues, unless upregulated by inflammatory cues or tissue injury (for recent reviews, see Chiovaro, Chiquet-Ehrismann, & Chiquet, 2015; Midwood, Chiquet, Tucker, & Orend, 2016; Rienks & Papageorgiou, 2016). Correspondingly, in culture, these proteins were produced at highest levels by actively proliferating cells. Collectively, these findings led to the concept that the ECM includes proteins with "antiadhesive" and "adhesion-modulating" activities as well as the cell-adhesive proteins (Bornstein, 1995, 2000; Brekken & Sage, 2001; Chiquet-Ehrismann, 1991; Sage & Bornstein, 1991). The observed distinct functional properties of the less adhesive, nonstructural proteins led first to the articulation of a grouping of "antiadhesive" or "adhesion-modulatory" ECM proteins and then to the origin of the term "matricellular" (Bornstein, 1995; Sage & Bornstein, 1991; for a discussion of origins, see Murphy-Ullrich & Sage, 2014). Explicitly from the outset, this grouping related to definitions based on tissue distributions and experimentally defined functional properties, and was not based on protein structural homologies.

Continuing cell biological and biochemical studies led to evidence that adhesion-modulating proteins undergo many types of receptor and non-receptor interactions at cell surfaces and often act in coordination with cytokines/growth factors as well as structural ECM proteins (e.g., SPARC and platelet-derived growth factor; Raines, Lane, Iruela-Arispe, Ross, & Sage, 1992). From the 1990s onward, the development of gene knockout mice for many ECM proteins established that null mutations for major structural ECM proteins (e.g., single laminin or fibrillar collagen chains or FN) had profound and lethal consequences during embryonic development (Gustafsson & Fässler, 2000; Hynes, 1994). In contrast, gene deletion of SPARC, TSP1, TSP-2, or TN-C did not affect viability and had limited overt (or in the case of TN-C, initially nonapparent) phenotypic consequences (Forsberg et al., 1996; Gilmour et al., 1998; Kyriakides et al., 1998; Lawler et al., 1998; Saga, Yagi, Ikawa, Sakakura, & Aizawa, 1992). The research developments led to further discussion and extension of the "Matricellular Hypothesis" (Bornstein, 2000, 2001).

In fact, in-depth studies of the viable knockout mice for single matricellular family members, especially with regard to TSP1, TSP2, and TN-C, have revealed a multitude of subtle and tissue-specific phenotypes and altered pathological susceptibilities (reviewed by Adams & Lawler, 2011; Brekken & Sage, 2001; Chiquet-Ehrismann, Orend, Chiquet, Tucker, & Midwood, 2014; Chiquet-Ehrismann & Tucker, 2011; Midwood et al., 2016). In addition, mutations or polymorphisms in many matricellular-encoding genes are now recognized to have causal or contributory roles in human genetic diseases (Table 1). Additional ECM-affiliated molecules have been identified, by in vitro and in vivo experimentation, to have matricellular properties. The most well established of these are the CCN family of proteins (Klenotic, Zhang, & Lin, 2016; Lau, 2016). Members of all four families of matricellular proteins have now been implicated in the pathogenesis of human diseases that present major quality of life and economic burdens around the world: cancer, diabetes, cardiovascular diseases, and chronic fibrotic disorders. Thus, there is growing interest in matricellular proteins as possible therapeutic targets, sources of biomaterials, or constituents of artificial organs (Chen & Simmons, 2011; McNeill, Vulesevic, Ostojic, Ruel, & Suuronen, 2015; Morris & Kyriakides, 2014). Recent years have also brought an emerging appreciation that the expression profiles and functional activities of several other ECM-associated molecules: fibulin-4, galectins, osteopontin, periostin, pigment epithelium-derived growth factor, and certain small leucine-rich proteoglycans (e.g., lumican) have overlapping matricellular characteristics (Murphy-Ullrich & Sage, 2014).

In parallel with extensive research into properties of matricellular proteins at cellular, molecular, and structural levels, many in vivo investigations into the functional roles of matricellular proteins have been made in mice. Recent excellent reviews have discussed individual matricellular proteins and their binding partners, or tissue-specific aspects of their functions with most emphasis on mammalian biology (Bradshaw, 2012; Giblin & Midwood, 2015; Lau, 2016; Leask, 2013; Lowy & Oskarsson, 2015; Midwood et al., 2016; Stenina, Topol, & Plow, 2007; Sweetwyne & Murphy-Ullrich, 2012). Readers are encouraged to consult these reviews to gain more detailed perspectives in particular areas of interest. However,

Table 1 Roles of Mutations and Polymorphisms in Matricellular-Encoding Genes in Human Genetic Diseases

Protein	Human Gene Locus	OMIM Accession Number	Human Genetic Disease Associations	References
Thrombospondin-1	15q14	188060	N700S SNP and premature myocardial infarction (USA and Italian populations) SNPs and familial pulmonary hypertension	Topol et al. (2001) and Zwicker et al. (2006)
			5′UTR SNP and postrefractive surgery chronic dry eye	Maloney et al. (2012) and Contreras-Ruiz et al. (2014)
Thrombospondin-2	6q27	188061	Intron SNP and lumbar disc herniation	Hirose et al. (2008)
Thrombospondin-3	1q22	188062	—	
Thrombospondin-4	5q14.1	600715	A387P SNP and premature myocardial infarction	McCarthy et al. (2004)
			A387P SNP and cardiovascular risk in postinfarct patients	Corsetti et al. (2011)
Thrombospondin-5/COMP	19p13.11	600310	Multiple point and deletion mutations causal for PSACH and EDM1	Hecht et al. (1995), Briggs et al. (1995), and Briggs, Brock, Ramsden, and Bell (2014)
Tenascin-C	9q33.1	187380	Missense mutations and autosomal dominant deafness 56	Zhao et al. (2013)
			SNPs and coronary artery disease	Minear et al. (2011)

Tenascin-R	1q25.1	601995	Possible association of point mutants with susceptibility to Parkinson's disease	Farlow et al. (2016)
Tenascin-X	6p21.33	600985	Deficiency and Ehlers–Danlos syndrome	Burch et al. (1997) and Zweers et al. (2011)
			G1113R mutation and vesicoureteral reflux 8	Gbadegesin et al. (2013)
Tenascin-W (N)	1q25.1	—	—	
CCN1(CYR61)	1p22.3	602369	—	
CCN2 (CTGF)	6q23.2	121009	Promoter region SNP and systemic sclerosis	Fonseca et al. (2007)
CCN3 (NOV)	8q24.12	164958	—	
CCN4 (WISP1)	8q24.22	603398	—	
CCN5 (WISP2)	20q13.12	603399	—	
CCN6 (WISP3)	6q21	603400	Point mutations and progressive pseudorheumatoid arthropathy of childhood	Hurvitz et al. (1999)
			Spondyloepiphyseal dysplasia tarda with progressive arthropathy	Liao et al. (2004a, 2004b)
SPARC (Osteonectin)	5q33.1	182120	Missense mutations and OI type XVII	Mendoza-Londono et al. (2015)
			3′UTR SNP and bone volume regulation	Dole et al. (2015)
Hevin (SPARCL1)	4q22.1	606041	—	

Continued

Table 1 Roles of Mutations and Polymorphisms in Matricellular-Encoding Genes in Human Genetic Diseases—cont'd

Protein	Human Gene Locus	OMIM Accession Number	Human Genetic Disease Associations	References
SMOC1	14q24.2	608488	Point mutations and microphthalmia and limb anomalies (Waardenburg anophthalmia syndrome)	Abouzeid et al. (2011), Okada et al. (2011), Rainger et al. (2011)
SMOC2	6q27	607223	Point mutations and dentin dysplasia, type i, with extreme microdontia and misshapen teeth	Bloch-Zupan et al. (2011)
			Loss-of-function mutation and recessive oligodontia	Alfawaz et al. (2013)
			SNP associated with primary angle closure glaucoma	Al-Dabbagh et al. (2017)
SPOCK1 (Testican1)	5q31.2	602264	D80V mutant possibly associated with a neurodevelopmental phenotype	Dhamija, Graham, Smaoui, Thorland, and Kirmani (2014)
SPOCK2 (Testican2)	10q22.1	607988	—	
SPOCK3	4q32.3	607989	—	

SNP, single-nucleotide polymorphism.

nonmammalian animal models also provide important insights into conserved pathophysiological functions and can offer opportunities to study biological processes such as limb regeneration that are not a property of mammals. Amphibian and fish models have provided greater access to the very earliest stages of embryonic development. This review will introduce the most well-established families of matricellular ECM proteins and then discuss current understanding of their roles in invertebrate, fish, amphibian, and reptile species. Avian models have been adopted for research on tenascins and SPARC (e.g., Chiquet, Vrucinić-Filipi, Schenk, Beck, & Chiquet-Ehrismann, 1991; Halfter, Chiquet-Ehrismann, & Tucker, 1989; Kim et al., 1997; Pacifici et al., 1990), but are not discussed here for reasons of space.

2. AN INTRODUCTION TO MATRICELLULAR PROTEINS

This section will introduce briefly the protein families that are currently viewed as quintessential matricellular proteins. The properties of SPARC, TNs, and TSPs in cell culture and tissues, and the limited overt phenotypic consequences of gene knockouts in mice were instrumental to the origin of the matricellular concept.

2.1 CCN Proteins

CCN (for CYR61/CTGF/NOV) proteins were identified originally from diverse cellular contexts and studied for their growth-promoting activities. It is now known that there are six CCN proteins in mammals. All are short (281–380aa) polypeptides that include four major domains, except for CCN5 that lacks a cysteine knot domain (Fig. 1). CCNs have several "domain relatives," i.e., proteins that have one or two domains in common with the CCNs. Through the insulin-like growth factor-binding protein (IGFBP) N-terminal-like domain, CCNs are related to the family of IGFBPs (Vilmos, Gaudenz, Hegedus, & Marsh, 2001). The domain organization of a von Willebrand factor_C (vWF_C) domain followed by a TSR domain also occurs in some TSPs (Fig. 1). Although recognized initially only in vertebrates, CCN-like proteins are also encoded in the amphioxus *Branchiostoma floridae* and the urochordate *Ciona intestinalis* (Mosher & Adams, 2012). In fact, like TSPs, CCN are "CIBLIN" genes: conserved in bilaterians and lost in nematodes (Erives, 2015). A single CCN is encoded in annelids and many arthropod and mollusk species (e.g., *Drosophila* CCN,

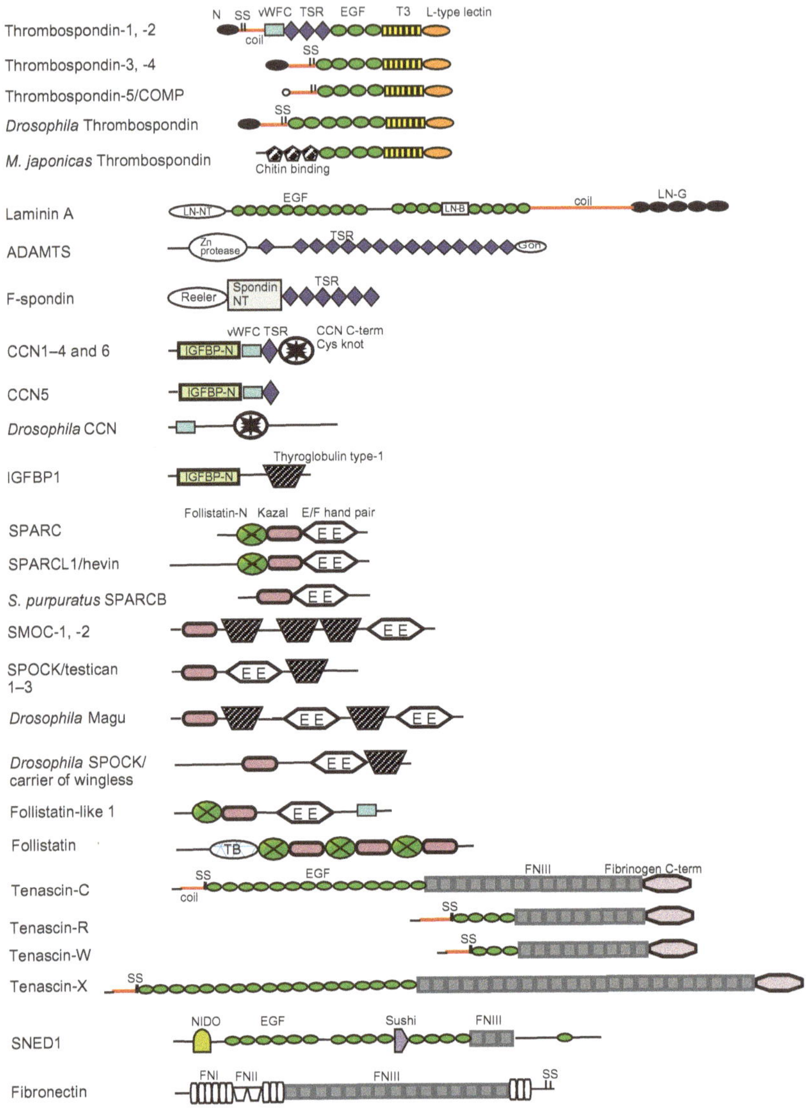

Fig. 1 Domain architectures of matricellular proteins and some prominent domain relatives from the ECM. Diagrams are based on the human proteins unless otherwise stated. Domains were identified by CCD and InterProScan, and coils were identified by MARCOIL. Not to scale: Tenascin-X includes 32 FNIII domains. *E*, E/F hand; *EGF*, epidermal growth factor-like domain; *TB*, transforming growth factor β-binding domain; *TSR*, thrombospondin type 1 domain.

GenBank NP_730294). However, CCN-like proteins have not been identified in cnidarians or sponges, suggesting that CCNs debuted in the last bilaterian common ancestor.

Collectively, mammalian CCNs have many developmental roles in cartilage, bone, and other organs; promote angiogenesis, and are elevated in many disease processes including wound healing, inflammation, fibrosis, and cancer (Hutchenreuther, Leask, & Thompson, 2017; Krupska, Bruford, & Chaqour, 2015). CCN2 in particular has profibrotic activity as a driver of myofibroblast differentiation in various tissues (Hall-Glenn & Lyons, 2011; Leask, 2013; Mason, 2013). Unlike the other matricellular gene families, several *ccn* gene knockout mice have profound developmental phenotypes, although *ccn3* and *ccn4* gene knockout mice are viable. Thus, *ccn1*$^{-/-}$ mice die during embryogeneis because of a failure of vascular development that results in undervascularization of the placenta (Mo et al., 2002). *Ctfg/ccn2*$^{-/-}$ mice die at birth with defective organ development of the pancreatic islets, skeleton, and palate and respiratory failure (Crawford et al., 2009; Ivkovic et al., 2003; Lambi et al., 2012; Tarr et al., 2017). *Cnn5*$^{-/-}$ mice die during embryogenesis at blastula implantation (Myers, Rwayitare, Richey, Lem, & Castellot, 2012). Mammary gland-specific knockout of *ccn6* results in an increased frequency of invasive mammary carcinomas (Martin et al., 2017). In humans, point mutations of *CCN6* are associated with progressive pseudorheumatoid dysplasia (Table 1).

CCNs associate with other ECM proteins and bind growth factors, cytokines, and an array of cell-surfaces receptors that include integrins $\alpha v\beta 3$ and $\alpha 5\beta 1$, plus heparan sulfate proteoglycans as integrin coreceptors (Lau, 2016). As discussed in depth by Lau (2016), there are numerous additional receptors for CCN2 on different cells including insulin-like growth factor (IGF)-2 receptor, LDL receptor-related protein-1 (LRP-1), TrkA, fibroblast growth factor receptor-2, receptor activator of nuclear factor-κB, and dendritic cell-specific transmembrane protein. Notch-1 has been identified as a receptor for CCN3 (Wagener, Yang, Kazuschke, Winterhager, & Gellhaus, 2013).

2.2 SPARC/Osteonectin and Relatives

SPARC, also known as osteonectin or BM-40, is a small (c.340aa) monomeric protein with a distinctive domain architecture. Mammalian SPARC includes an N-terminal signal peptide followed by an uncharacterized sequence region and several major domains: acidic domain I that has low affinity for Ca^{2+} ions, a follistain-like N-terminal domain with an epidermal growth factor (EGF)-like structure, a Kazal domain (as found in serine protease inhibitors) and a E/F hand domain pair that coordinates Ca^{2+} ions with high affinity (Fig. 1). A proteolytic fragment containing the peptide motif KGHR has copper-binding activity (Lane, Iruela-Arispe, Johnson, & Sage, 1994). SPARL1/hevin is a chordate-specific paralogue of SPARC, characterized by a longer N-terminal region. SPARC and SPARCL1 are also related in domain composition to follistatin-like 1, and through the kazal and follistatin-N domains to follistatin. The SPARC-related modular calcium-binding proteins (SMOC) and the sparc/osteonectin, cwcv, and kazal-like domains (SPOCK)/testican proteoglycans are related to SPARC and SPARCL1 by Kazal and E/F hand domains, and are also related to IGFBPs by their thyroglobulin type-1 domains (Fig. 1).

SPARC is a major noncollagenous component of bone with conserved collagen-binding activities (Rosset & Bradshaw, 2016) (Table 2). It also acts in various cell types to modulate growth factor signaling and angiogenesis (Bradshaw, 2012). Although mutations in *SPARC* cause osteogenesis imperfect type XVII (Table 1), SPARC is not essential for bone development in mice; instead, a prominent phenotype of *sparc*-knockout mice is disorganized collagen ECM in the lens capsule and early-onset cataracts (Gilmour et al., 1998; Norose et al., 1998). However, collagen ECM is perturbed in multiple tissues of *Sparc*-null mice and SPARC is now appreciated to bind fibrillar collagens at the same site as the discoidin domain collagen receptor (Bradshaw, 2012; Giudici et al., 2008). Effects of SPARC on cell proliferation, shape, and migration involve its binding to several integrins, endoglin, and the macrophage scavenger receptor stabilin-1 (Bradshaw, 2012; Kzhyshkowska et al., 2006). SPARCL1/hevin is expressed in the developing nervous system where it contributes to synapse assembly by acting as a connector between neurexin1α and neuroligin-1 receptors (Singh et al., 2016). *Sparcl1* null mice undergo accelerated wound repair (Sullivan, Puolakkainen, Barker, Funk, & Sage, 2008). SMOC-1 and -2 are also implicated in cell proliferation and angiogenesis and act as inhibitors of bone morphogenetic protein (BMP) signaling (discussed in later sections). Expression patterns of *Smoc-1* and *-2* in mouse embryos implicate possible roles in differentiation of the gonads and mesonephros (Pazin & Albrecht, 2009). However, reduction of *Smoc-1* level by gene trap resulted in eye and limb abnormalities similar to human Waardenburg anophthalmia syndrome (Table 1). Knockout of *Spock1* has limited overt phenotypes (Röll, Seul, Paulsson, & Hartmann, 2006).

Table 2 Expression Profiles of Matricellular Proteins in Lamprey, Fish, Amphibia, and Reptiles

Class/*Species*	Protein	Developmental Stage/Tissue	References
Cyclostomata			
Petromyzon marinus	TN-C	Developing and repairing olfactory pathway	Zaidi, Kafitz, Greer, and Zielinski (1998)
	SPARC	Adult liver	Ringuette, Damjanovski, and Wheeler (1991)
Cartilagenous Fish			
Raja radiata	SPARC	Adult liver	Ringuette et al. (1991)
Bony Fish			
Danio rerio	CCN	Expression profiles of nine CCNs throughout development	Fernando et al. (2010)
	CCN6/Wisp3	Early embryos, brain, otic vesicle, and swim bladder	Nakamura et al. (2007)
Carassius auratus	SPARC	Regenerating scales	Iimura, Tohse, Ura, and Takagi (2012)
Carassius auratus gibelio	SPARC	Regenerating pectoral fin	Stavri and Zarnescu (2013)
Danio rerio	SMOC2	Craniofacial morphogenesis	Melvin et al. (2013)
	SPARC	Skin	Hong, Choi, Myung, and Choi (2011)
	SPARC	Inner ear otoliths	Kang et al. (2008)
	SPARC	Otic vesicle and developing pharyngeal cartilage	Rotllant et al. (2008)
Oreochromis mossambicus	SPARC	Developing skeleton	Weigele, Franz-Odendaal, and Hilbig (2015)
	SPARC	Inner ear otoliths	Weigele, Franz-Odendaal, and Hilbig (2016)
Oryzias latipes	SPARC	Skeletal development	Renn et al. (2006)
	SPARC	Developing otoliths	Nemoto, Chatani, Inohaya, Hiraki, and Kudo (2008)

Continued

Table 2 Expression Profiles of Matricellular Proteins in Lamprey, Fish, Amphibia, and Reptiles—cont'd

Class/*Species*	Protein	Developmental Stage/Tissue	References
Scophthalmus maximus	SPARC	Developmental stages	Torres-Núñez et al. (2015)
Sparus auratus	SPARC	Embryonic and larval development	Estêvão, Redruello, Canario, and Power (2005)
Takifugu rubripes	SPARC	Vertebral column development	Kaneko et al. (2016)
Carassius auratus	TN-C	Macrophages and microglia after optic nerve injury	Battisti, Wang, Bozek, and Murray (1995)
Danio rerio	TN-C	Developing brain, mesoderm, and neural crest	Tongiorgi, Bernhardt, Zinn, and Schachner (1995)
	TN-C	Embryonic somites	Bernhardt, Goerlinger, Roos, and Schachner (1998)
	TN-C	Trunk structures of embryo	Tongiorgi (1999)
	TN-C	Developing myoseptum	Schweitzer et al. (2005)
	TN-C	Adult spinal cord injury	Yu et al. (2011)
	TN-C	Developing heart	Lin et al. (2013)
		Adult heart regeneration	Chablais, Veit, Rainer, and Jaźwińska (2011)
		Adult heart regeneration	Sallin, de Preux Charles, Duruz, Pfefferli, and Jaźwińska (2015)
		Caudal fin regeneration	Govindan and Iovine (2015)
	TN-R	Growing and regenerating optic nerve	Becker, Schweitzer, Feldner, Schachner, and Becker (2004)
	TN-W	Embryos (sclerotome and neural crest), larvae, and juvenile fish (myosepta and DGR)	Weber, Montag, Schachner, and Bernhardt (1998)

Table 2 Expression Profiles of Matricellular Proteins in Lamprey, Fish, Amphibia, and Reptiles—cont'd

Class/*Species*	Protein	Developmental Stage/Tissue	References
Oncorhynchus mykiss	TN-C	Muscle wound healing	Schmidt, Andersen, Ersbøll, and Nielsen (2016)
Carassius auratus	TSP1	CNS	Hoffman, Dixit, and O'Shea (1994)
Carassius auratus	TSP1	Adult optic nerve, and elevated after crush injury	Hoffman and O'Shea (1999)
Danio rerio	TSP4	Tendons of extraocular muscles	Kasprick et al. (2011)
	TSP4a	Myotendinous junctions	Subramanian and Schilling (2014)
Oreochromis niloticus	TSP1a, 1b	Developmental expression in gonads, skeletal tissue, heart, spleen, brain, intestine	Wu, Zhou, Nagahama, and Wang (2009)
Solea senegalensis	TSP4	Ovarian development	Tingaud-Sequeira et al. (2009)
Amphibians			
Notophthalmus viridescens	CCN	Regenerating heart	Looso et al. (2012)
Xenopus laevis	CCN1	Embryogenesis	Latinkic et al. (2003)
	CCN2	Embryogenesis	Abreu et al. (2002)
Rana catesbeiana	SPARC	Larval to preadult skin metamorphosis	Ishida, Suzuki, Utoh, Obara, and Yoshizato (2003)
Xenopus laevis	SMOC1	Postgastrulation development	Thomas, Canelos, Luyten, and Moos (2009)
	SPARC	Dorsal axis of early embryos	Damjanovski, Malaval, and Ringuette (1994)
Xenopus tropicalis	SPARC	Embryonic epidermis	Huynh, Hong, Delovitch, Desser, and Ringuette (2000)
		Developing skeleton	Espinoza et al. (2010)
		Adult tissues	Damjanovski, Liu, and Ringuette (1992)

Continued

Table 2 Expression Profiles of Matricellular Proteins in Lamprey, Fish, Amphibia, and Reptiles—cont'd

Class/*Species*	Protein	Developmental Stage/Tissue	References
Ambystoma mexicanum	TN-C	Healing skin wounds of adults	Seifert, Monaghan, Voss, and Maden (2012)
	TN-C	Limb buds and regenerating limb blastema	Onda, Goldhamer, and Tassava (1990)
Ambystoma tigrinum tigrinum	TN-C	Melanophore patterning	Parichy (1996)
Notophthalmus viridescens	TN-C	Skin wounds of adult	Donaldson, Mahan, Yang, and Crossin (1991)
	TN-C	Tendons, myotendonous junctions, and periosteum of intact limbs; in blastema after limb amputation	Calve et al. (2010)
	TN-C	Limb buds and regenerating limb blastema	Onda et al. (1990) and Onda, Poulin, Tassava, and Chiu (1991)
Pleurodeles waltl	TN-C	Adult heart regeneration	Mercer et al. (2013)
	TN-C	Caudal spinal cord; regenerating spinal cord	Caubit et al. (1994)
	TN-C	Retinal regeneration	Mitashov, Arsanto, Markitantova, and Thouveny (1995)
	TN-C	Optic nerve development and regeneration	Becker et al. (1995)
Rana temporaria	TN-R	Optic nerve—development and adult	Becker, Becker, Meyer, and Schachner (1999)
	TN-C	Neuromuscular junction—intact and after injury	Mège, Nicolet, Pinçon-Raymond, Murawsky, and Rieger (1992)
Taricha torosa	TN-C	Melanophore patterning	Parichy (2001)
Xenopus laevis	TN-C	Embryonic neural crest migration pathways	Epperlein, Halfter, and Tucker (1988) and Mackie, Tucker, Halfter, Chiquet-Ehrismann, and Epperlein (1988)

Table 2 Expression Profiles of Matricellular Proteins in Lamprey, Fish, Amphibia, and Reptiles—cont'd

Class/*Species*	Protein	Developmental Stage/Tissue	References
	TN-C	Embryonic mesoderm	Umbhauer, Riou, Spring, Smith, and Boucaut (1992)
	TN-C	Embryonic somites	Umbhauer, Riou, Smith, and Boucaut (1994)
	TN-C	Embryonic development	Riou, Shi, Chiquet, and Boucaut (1988) and Williamson, Parrish, and Edelman (1991a)
	TN-C	Metamorphosis	Williamson, Parrish, and Edelman (1991b)
	TN-C	Assembly of neuromuscular junctions at myotomes	Somasekhar and Nordlander (1995)
Ambystoma mexicanum	TSP1	Chondrocytes of long bones; in wound epidermis and mesenchymal cells after limb amputation	Whited, Lehoczky, Austin, and Tabin (2011)
	TSP4	Tendons, myotendonous junctions, and perichondrium of intact limbs; also in blastema after limb amputation	Whited et al. (2011)
Xenopus laevis	TSP4	Embryonic development of CNS, notochord, and skeletal muscle	Lawler et al. (1993) and Urry, Whittaker, Duquette, Lawler, and DeSimone (1998)
Reptiles			
Pseudemys scripta elegans	CCN2	Stage 25 and adult, dorsal, and medial cortex	Wang et al. (2011)
Gallotia galloti	TN-R	Regenerating visual pathway	Lang et al. (2008)
Eublepharis macularius	TSP1	Skin wound healing	Peacock, Gilbert, and Vickaryous (2015)
	TSP1	Tail regeneration blastema	Payne, Peacock, and Vickaryous (2017)

SPARC is an ancient ECM protein present in sponges (e.g., *Leucosolenia complicata* lcpid74779 and *Oscarella carmela* ORF g.306897 from the Compagen database; Hemmrich & Bosch, 2008) and cnidarians (Koehler et al., 2009) as well as bilaterians. In-depth study of the evolution of SPARC-encoding genes revealed an unexpected evolutionary history with the identification of an ancient *SPARC* paralogue, the little-studied *SPARCB,* that has a distinct phylogenetic distribution (Bertrand et al., 2013). The SPARCB protein is characterized by a shorter region between the signal peptide and domain I (Fig. 1). Duplication of an ancestral gene in the last common ancestor of cnidarians and bilaterians is proposed to have given rise to *SPARC* and *SPARCB.* Whereas both genes were retained in the bilaterian ancestor, *SPARC* was lost in the cnidarian lineage, leaving only *SPARCB.* In the cnidarian *Nematostella vectensis*, *SPARCB* gene duplications have resulted in four paralogous genes (Koehler et al., 2009). In contrast, *SPARCB* was lost in the protostome ancestor (Bertrand et al., 2013). SMOC, SPOCK, and follistatin-like 1 are also highly conserved and can be identified in cnidarians as well as bilaterians.

2.3 Tenascins

In vertebrates, TN comprises a family of four large, multidomain glycoproteins, tenascin-C, -R, -W, and -X (Fig. 1). TNs oligomerize as trimers through the action of a trimerizing coiled-coil domain, and, in the case of TN-C and TN-W, hexamers are assembled by disulfide bonds between two trimers (Kammerer et al., 1998; Scherberich et al., 2004). C-terminal to the coiled-coil domain, all TNs are characterized by multiple, tandem EGF-like domains, repeated fibronectin-type III (FNIII) domains, and a C-terminal domain related to the C-terminal globular domain of fibrinogen. The numbers of EGF and FNIII domains vary between TN family members (Fig. 1), and in TN-C the number of FNIII domains is varied by alternative splicing; these variants have distinct cell attachment and antiadhesive activities (Chiquet et al., 1991; Fischer et al., 1997).

TN-C is the prototypic and most heavily studied member of the family. TN-C is prominent during embryogenesis of various vertebrates in the developing nervous system, in cartilage, muscle–tendon attachment sites and at sites of epithelial–mesenchymal transition, and neural crest cell migration (Table 2). It is also located in tendons and stem cell niches in adults and, in common with other matricellular proteins, is upregulated during inflammation, wound repair, fibrosis, and in tumor stroma. At these sites TN-C is thought to act to facilitate cell motility, tissue remodeling, innate immune response, and possibly to protect against pathogens (Chiquet-Ehrismann & Tucker, 2011; Midwood et al., 2016). Recent data also implicate TN-C in hearing and cardiovascular disease in humans (Table 1). As discussed by Midwood et al. (2016), TN-C interacts with multiple integrins through its RGD motif, as well as heparan sulfate proteoglycans, FN, collagen, and several growth factors including transforming growth factor β (TGFβ) and Wnt3a. TN-R is expressed in the developing nervous system and located in perineural nets; in common with TN-C, TN-R is upregulated in several neurodegenerative conditions (Reinhard, Roll, & Faissner, 2017). TN-W is a target of TGFβ signaling that is expressed in the same tissues as TN-C with nonoverlapping cellular expression and is upregulated in tumor stroma (Chiovaro et al., 2015; Chiquet-Ehrismann & Tucker, 2011). TN-X is a collagen-binding protein, and its deficiency leads to human Ehlers–Danlos syndrome, in which joint hypermobility and fragile skin are associated with disorganization of collagen and elastin fibrils (Table 1).

TNs originated much later than the other matricellular proteins, within the chordate lineage (Tucker & Chiquet-Ehrismann, 2009; Tucker et al., 2006). The cephalochordate, *B. floridae* and urochordates such as *C. intestinalis* and *Ciona savignyi* each encode a single TN. The evolution of vertebrates involved several rounds of large-scale genome duplication, and these events are considered to have given rise to the family of TN paralogues, tenascin-C, -R, -W, and -X that are encoded in bony fish and tetrapods (Tucker et al., 2006). Lampreys and the shark, *Callorhinchus milii*, encode only two forms of TN; it remains unclear if this is because of lineage-specific gene losses or alternatively, if these are the products of early, independent, lineage-specific gene duplication (Adams, Chiquet-Ehrismann, & Tucker, 2015).

TNs are a unique category of proteins, yet they have a few domain relatives of high interest within the ECM. Sushi, nidogen, and EGF-like domain-containing protein 1 precursor (SNED1; HsSNED1, GenBank NP_001073906.1) is a 1413aa ECM protein with roles in breast cancer metastasis (Naba, Clauser, Lamar, Carr, & Hynes, 2014). SNED1 contains repeated EGF-like domains, three FNIII domains, and a final EGF domain and thus has partial similarity of domain organization to the central region of TNs (Fig. 1). The most striking sequence identity is that between the FNIII domains of TNs and the FNIII domains of FN. FN also debuted in the chordate lineage (Segade, Cota, Famiglietti, Cha, & Davidson, 2016;

Tucker & Chiquet-Ehrismann, 2009) but appears to have originated after TN, because a FN-like protein is encoded in urochordates but not in *B. floridae* (Adams et al., 2015; Segade et al., 2016). By BLAST homology, the tandem FNIII domains that form the central region of human FN have 24% identity with the FNIII domains of human TN-C with 99% coverage and only 4% of gaps (e-value 2×10^{-67}). This is remarkable in view that, in the TNs of humans, the same region of TN-C has 34% identity to the FNIII domains of TN-R and 27% identity to those of TN-X. This observation raises a question whether both TN and FN might have a common origin from an ancestral FNIII domain-containing precursor.

2.4 Thrombospondins

TSPs are among the most ancient of ECM proteins, being present in modern animals from sponges to humans, although apparently lost in the nematode and planarian lineages (Bentley & Adams, 2010; Özbek, Balasubramainian, Chiquet-Ehrismann, Tucker, & Adams, 2010). All TSPs are large (850–1100aa) multidomain, calcium-binding glycoproteins and most oligomerize through the action of a coiled-coil domain near the N-terminus (Fig. 1). The N-terminal region may vary in domain composition, but the majority of TSPs contain a globular N-terminal domain prior to the coiled-coil; this domain contains a high-affinity heparin-binding site that binds secreted and membrane-associated proteoglycans. The C-terminal region has a highly conserved domain architecture, consisting of repeated EGF-like domains, tandem calcium-binding TSP type 3 repeats, and a globular C-terminal domain. The TSPs of prawns and shrimps are unusual as monomeric proteins that contain repeated type 2 chitin-binding domains in their N-terminal region (Fig. 1).

The genomes of many early diverging metazoans and protostomes contain a single thrombospondin-encoding gene, but in some species lineage-specific gene duplication and diversification have resulted in a set of unique paralogues. For example, the cnidarian *N. vectensis* encodes four thrombospondins, all of which are transcribed (Tucker et al., 2013). Other gene duplication events are thought to have taken place in the last deuterostome common ancestor, because echinoderms, urochordates, cephalochordates, and chordates all encode multiple TSPs per genome (Adams & Lawler, 2011). Thus, the complement of TSPs in bony fish and tetrapods includes pentameric TSPs with domain architectures that are most similar to the TSPs of protostomes (designed TSP3, TSP4, and

TSP5 and known as subgroup B) and trimeric TSPs that include a vWF_C domain and repeated thrombospondin type 1 (TSR) domains within the N-terminal half (designated TSP1 and TSP2; subgroup A) (Fig. 1). Distinct coiled-coil sequences assemble trimeric and pentameric forms of TSPs (Vincent, Woolfson, & Adams, 2013).

In vertebrates, the different TSP gene products have distinct tissue and cell-type-specific patterns of expression (summarized for invertebrate chordates, fish, amphibia, and reptiles in Table 2; Iruela-Arispe, Liska, Sage, & Bornstein, 1993). In the ECM, TSPs bind collagens, laminin, and FN. TSPs also interact with multiple integrins and integrin-associated receptors such as CD36 and CD47 to activate cell proliferation, motility, or other signaling responses. The trimeric TSP1 and TSP2 also modulate vascular endothelial growth factor (VEGF) and Notch signaling and TSP1 uniquely activates latent TGFβ (reported mostly for TGFβ1 but also applicable to TGFβ2) (Sweetwyne & Murphy-Ullrich, 2012). In common with the other matricellular proteins, TSPs are highly multifunctional with roles in connective tissue organization (in particular TSP5 which is the causal gene for pseudoachondroplasia; Table 1), synaptogenesis, and the cardiovascular system (Adams & Lawler, 2011; Bornstein, 2001; Risher & Eroglu, 2012; Table 1). The trimeric TSP1 and TSP2 also have antiangiogenic and immunomodulatory activities.

TSPs have an enormous number of domain relatives. Many metazoan and nonmetazoan proteins contain repeated TSR domains; these include prominent proteins with ECM-related roles such as the ADAMTS family of metalloproteinases and the spondin and minden proteins that contribute to ECM structure and function in particular tissues or developmental stages (Fig. 1) (Tucker, 2004). EGF-like domains are present in many ECM proteins including TNs and laminin subunits, and are also found widely in secreted proteins of eukaryotes, for example, in brown algae (Terauchi, Yamagishi, Hanyuda, & Kawai, 2017). The follistatin-N domain, present in SPARC and follistatins, is related in structure to the EGF domain (IPR003645). The vWF_C domain is also found in CCNs (Fig. 1) as well as in von Willebrand factor and various collagens. The N-terminal domain of many TSPs (termed TSPN domain in the conserved domain database (CDD)) has a beta-jellyroll fold highly related to the laminin_G domain that is present as repeated domains in the laminin alpha subunit (Fig. 1). The TSP C-terminal domain (termed TSPC in CDD) has a beta-sandwich fold related to that of legume (L-type) lectins. All these beta-sandwich domains are characterized by 12–14 beta strands arranged in two sheets.

Both the laminin_G domain and the L-type lectin domain are members of the same fold family, the Concanavalin A legume/glucanase fold superfamily (InterPro IPR013320; SCOP 49899).

3. MATRICELLULAR PROTEINS IN EARLY DIVERGING METAZOANS AND PROTOSTOMES

3.1 CCN

Drosophila melanogaster encodes a single CCN (GenBank NP_730294, flybase CG32183) that is a larger protein than the mammalian CCNs (470aa) and lacks clear TSR or IGFBP domains; only the vWF_C and cysteine knot domains are well conserved (Fig. 1). *ccn* has not been studied specifically as a "named" gene in *Drosophila*, but indications of significant functional roles have emerged from several genetic screens.

A screen of 167 inbred *Drosophila* strains from the Drosophila Genetic Reference Panel for susceptibility to paraquat and menadione sodium bisulfite that induce chronic oxidative stress, discovered natural variation and gender-specific differences in withstanding the toxic effects of these compounds (Weber et al., 2012). The genetic basis of this variation was investigated by a genome-wide association study to correlate single-nucleotide polymorphisms (SNPs) with survival time. Many of the correlated SNPs related to gene ontology terms "DNA metabolism" and "Neuronal development," and gender differences were identified in the profile of correlated SNPs. SNPs in *ccn* stood out for their correlation with reduced survival upon treatment with either agent across both genders. Indeed, only two SNPs had predictive power in both females and males (Weber et al., 2012).

Also in *Drosophila*, several large-scale RNA interference (RNAi) screens have reported lethality upon *ccn* knockdown (Neumuller et al., 2011; Zeng et al., 2015). siRNA knockdown of *ccn* in either the airway epithelium or skeletal muscle resulted in lethality at the pupal stage (Hosono, Matsuda, Adryan, & Samakovlis, 2015; Schnorrer et al., 2010). It is striking that results from several RNAi screens implicated functions of *ccn* in maintenance of different stem cell populations. A genome-wide RNAi screen for gene products affecting neuroblast self-renewal investigated 89% of coding genes to identify 620 "hits" with effects on neuroblast shape and number. Among these, *ccn* knockdown was categorized as leading to neuroblast overproliferation and the presence of ectopic neuroblasts (Neumuller et al., 2011). A RNAi screen for gene products influencing self-renewal of female germline stem cells screened 25% of coding genes to yield 366 "hits" with effects on germ cell function. The knockdown of *ccn* resulted in defective differentiation of germline stem cells and was associated with polyploidy of the supporting nurse cells in the egg chamber (Yan et al., 2014). In a RNAi screen to identify gene products necessary for regulation and maintenance of intestinal stem cells in adult *Drosophila*, knockdown of *ccn* correlated with death of intestinal stem cells (Zeng et al., 2015). Although some studies examined only a single *ccn* knockdown line, it is interesting that *ccn* emerged as a member of a small group of nine transcripts that were "hits" in all three screens of different types of stem cells. In genome-wide RNAi screens for mediators of heat nociception (Neely, Hess, et al., 2010) or for genes conserved with mammals that could potentially function in heart development and heart function (Neely, Kuba, et al., 2010), knockdown of *ccn* was tested but did not have distinct phenotypic effects.

3.2 SPARC and Relatives

In *C. elegans* the basement membranes that ensheath the body organs are the major form of ECM. CeSPARC (encoded by *ost1*) transcript is expressed throughout life in body wall and sex muscle cells and the gonads but not in pharyngeal muscle. Worms with transgenic overexpression of SPARC developed to adults but were impaired for movement, due to loss of coordination or paralysis. Worms were also infertile with vulval protrusions including extrusion of internal organs (Schwarzbauer & Spencer, 1993). RNAi of *ost1* to deplete CeSPARC protein in adults caused lethality of 54%–93% of progeny at the embryonic or larval stages. Surviving animals had reduced fertility or were sterile and were markedly smaller as adults (Fig. 2A) (Fitzgerald & Schwarzbauer, 1998). CeSPARC protein is present in most basement membranes of adult worms, including the pharyngeal basement membrane: this implies mechanisms for transfer of CeSPARC after its secretion. Collagen-binding properties are conserved in CeSPARC (Maurer et al., 1997), yet the retention of basement membranes after either depletion or overexpression of CeSPARC implies a nonstructural mechanism of action (Fitzgerald & Schwarzbauer, 1998). A developmental context in which CeSPARC has been found to have roles in basement membrane structure is the invasion of anchor cells through the closely apposed gonadal and ventral basement membranes. This process involves remodeling of the ventral basement membrane, initially through the action of a matrix metalloproteinase. Experimental elevation of CeSPARC levels (by two- to

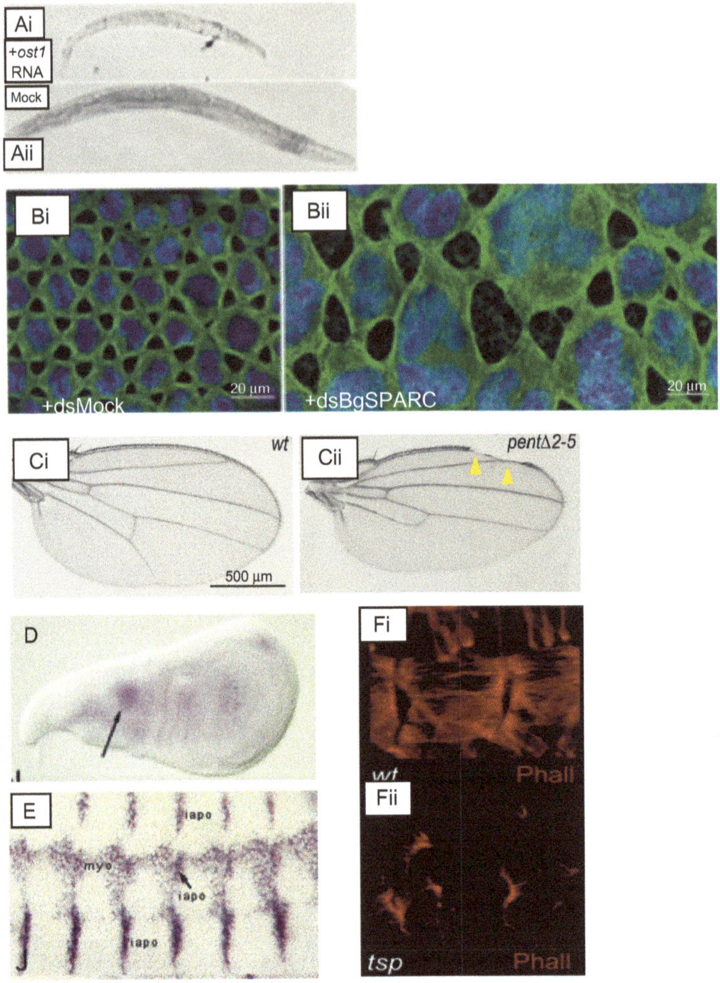

Fig. 2 Examples of activities of matricellular proteins in *C. elegans* and insects. (Ai and Aii) Reduced SPARC levels impair *C. elegans* growth. RNA was injected into young adults. Progeny that survived to adulthood in the absence of SPARC is small and lacks gut granules (Ai), compared to control injected adults (Aii). (Bi and Bii) Role of SPARC in the integrity of the egg chamber follicular epithelium. SPARC was knocked down by RNA interference in female *Blattella germanica* cockroaches at the start of the sixth instar and egg chambers examined in 5-day-old adults. Samples from mock-treated (Bi) or dsBgSPARC-treated (Bii) females were stained for tubulin (*green*) and chromatin (DAPI; *blue*) and show enlargement and altered nuclear morphology of the follicular epithelial cells after SPARC knockdown. (Ci and Cii) *Drosophila* SMOC/magu/pentagone and wing development. Compared to wild type (Ci), in the absence of SMOC, (Cii) 9% of adult wings are smaller with disrupted patterning. *Arrows* indicate region lacking chemosensory bristles. (D and E) Localization of *dtsp* transcripts by in situ hybridization

fivefold overexpression) increased anchor cell invasion in the context of mutants in which invasion is impaired, for example, *c-fos*-null worms (Morrissey et al., 2016). This process depended on the collagen-binding activity of SPARC and involved decreased incorporation of collagen IV into the gonadal basement membrane, resulting in a weakened basement membrane. Transcriptional silencing of *ost-1* in late-stage larvae resulted in decreased collagen IV within the pharyngeal basement membrane, yet led to increased collagen IV in the vicinity of sites of CeSPARC expression on body wall muscles. These results further implicate a role of CeSPARC in extracellular trafficking of collagen IV from its sites of production to remote basement membranes (Morrissey et al., 2016).

In *D. melanogaster*, *dSPARC* begins to be expressed at gastrulation onward, with hemocytes and the fat body as major sites of expression (Martinek, Shahab, Saathoff, & Ringuette, 2008; Martinek, Zou, Berg, Sodek, & Ringuette, 2002). Although initial studies indicated embryonic lethality upon gene deletion of *dSPARC* (Martinek et al., 2008), aspects of this phenotype turned out to be due to imprecise P-element excision.

during *Drosophila* embryonic development. (D) Wing imaginal disc (*arrow* indicates the notum); (E) segmental expression at muscle attachment sites. *lapo*, intersegmental apodemes; *myo*, myoblast. (Fi and Fii) Role of *Drosophila* TSP in assembly of muscle–tendon attachment sites. *Panels* show late-stage *Drosophila* embryos stained with phalloidin to visualize F-actin organization in muscles. The regular wild-type pattern of the ventrolateral muscles (Fi) is lost in *tsp* mutants due to muscle detachment and rounding (Fii). *Panels (Ai) and (Aii) reproduced from Fitzgerald, M. C., & Schwarzbauer, J. E. (1998). Importance of the basement membrane protein SPARC for viability and fertility in Caenorhabditis elegans.* Current Biology, *8(23), 285–288, with permission (Elsevier license 4231370432902). Panels (Bi) and (Bii) reproduced from Norman, M., Vuilleumier, R., Springhorn, A., Gawlik, J., Pyrowolakis, G. (2016). Pentagone internalises glypicans to fine-tune multiple signalling pathways.* eLife, *5, pii: e13301, doi: https://doi.org/10.7554/eLife.13301 (Published by eLife with open access under CC-BY 4.0). Panels (Bi) and (Bii) reproduced from Irles, P., Ramos, S., Piulachs, M. D. (2017). SPARC preserves follicular epithelium integrity in insect ovaries.* Developmental Biology, *422(2), 105–114, with permission (Elsevier license 4234270600114). Panels (D) and (E) Reproduced from Adams, J. C. Monk, R., Taylor, A. L., Ozbek, S., Fascetti, N., Baumgartner, S., Engel, J. (2003). Characterisation of* Drosophila *thrombospondin defines an early origin of pentameric thrombospondins.* Journal of Molecular Biology, *328, 479–494, with permission (Elsevier license 4231370262654). Panels (Fi) and (Fii) reproduced from Chanana, B., Graf, R., Koledachkina, T., Pflanz, R., Vorbrüggen, G. (2007). AlphaPS2 integrin-mediated muscle attachment in Drosophila requires the ECM protein thrombospondin.* Mechanisms of Development, *124(6), 463–475, with permission (Elsevier license 4231000696562).*

Targeted disruption of *dSPARC* is lethal at the larval stage, (mostly during the first instar), due to altered feeding behavior as a result of aberrant fat body development (Shahab et al., 2015). In addition to supplying nutrient energy during development, the fat body, along with the hemocytes, is a major producer of basement membrane components. Adipocytes in *dSPARC*-null or SPARC-depleted fat bodies became retracted and blebbed, with altered actin cytoskeleton and excessive surface accumulation of collagen IV and other basement membrane proteins. Interestingly, *SPARC*-null hemocytes undergo a buildup of intracellular collagen (Martinek et al., 2008). Reciprocally, depletion of collagen IV resulted in decreased levels of SPARC in basement membranes. These observations suggest homeostatic feedback mechanisms between SPARC and collagen IV (Martinek et al., 2002). During embryonic development, SPARC and collagen IV normally become deposited on the distal surface of the midgut during basement membrane assembly. Because the deposition of both proteins is diminished in *LanB2* mutants (encoding for the γ chain of laminin and present in both laminin heterotrimers of *Drosophila*), laminin deposition may be the primary driver of basement membrane assembly (Wolfstetter & Holz, 2012).

In *D. melanogaster* and other insects, oogenesis takes place in egg chambers surrounded by a follicular epithelium associated with a basement membrane. As the oocytes mature, the egg chamber elongates while remaining attached to the basement membrane in a process that involves migration of the follicle cells (Pocha & Montell, 2014). This extension involves dynamic remodeling of the surrounding basement membrane: collagen IV levels increase sharply and its fibrils become more aligned, while SPARC mRNA and protein decline rapidly between embryonic stages 3 and 7. In experiments in which SPARC expression was artificially prolonged, egg chamber elongation was inhibited and collagen IV in the basement membrane was specifically decreased (Isabella & Horne-Badovinac, 2015). Mechanistically, SPARC and collagen IV coassociated during secretory trafficking in follicle cells. Silencing of prolyl-4-hydroxylase (to block molecular assembly of collagen IV) resulted in coretention of SPARC with collagen IV within the endoplasmic reticulum. The authors proposed that the normal SPARC/collagen IV association at secretion acts to restrain the incorporation of collagen IV into the follicular basement membrane to ensure correct basement membrane assembly (Isabella & Horne-Badovinac, 2015). A similar regulatory process could be relevant to the excessive ECM deposition observed in SPARC-depleted fat bodies.

High expression of *sparc* in ovaries and fat body is conserved in the cockroach *Blattella germanica,* with elevated levels associated with the molting transition of the last larval instar to adult under the control of juvenile hormone (Irles, Ramos, & Piulachs, 2017). Transcriptional silencing of *sparc* at the start of the last larval instar affected the completion of oogenesis and prevented oviposition. However, the cellular mechanisms appear distinct to those identified in *D. melanogaster.* Defective oogenesis correlated with impaired mitoses in the follicular epithelium and reduced cell number. Silencing of *sparc* in 5-day-old adult females led to enlargement of follicle cells, with disorganized actin cytoskeletal structures, giant, lobed nuclei, and a sharp decrease in apoptosis of the follicle cells, a process normally important for egg release (Irles et al., 2017) (Fig. 2Bi and Bii).

SPARC is also functionally important in the *D. melanogaster* heart. In the insect heart, cardiomyocytes are associated with pericardial nephrocytes that act to endocytose materials from the hemolymph. Nephrocyte differentiation depends on the transcription factor dKlf15, thus loss of function of *dKlf15* has enabled study of flies lacking nephrocytes (Ivy et al., 2015). Either developmental or adult-onset loss of *dKlf15* function resulted in several heart arrhythmia phenotypes in adult flies and alterations in hemolymph composition, including elevated SPARC levels. The significance of elevated SPARC in promoting cardiomyopathy was supported by the observation of normal heart function in *dKLF15* mutants on a genetic background in which SPARC levels were reduced by 60%. However, SPARC knockdown in wild-type cardiomyocytes also caused severe cardiomyopathy, emphasizing the necessity of homeostatic levels of SPARC for normal tissue function (Hartley, Motamedchaboki, Bodmer, & Ocorr, 2016).

Drosophila SPARC has an interesting developmental role in imaginal discs, where epithelial cells proliferate prior to differentiation and a process of cell competition ensures that viable cells of reduced fitness within a wild-type environment (for example, with limited response to morphogens such as decapentaplegic (dpp, the *Drosophila* homologue of mammalian bone morphogenetic protein-2 and -4 (BMP2/4)) are eliminated by apoptosis (Claveria & Torres, 2016). Mosaic overexpression of *Drosophila* c-myc (Dmyc) in imaginal discs has been used to set up "supercompetitor" cells that out-compete the wild-type cells. In this context, *dSPARC* expression was found to be induced in the "loser" cells where it exerted a short-term protective effect against the onset of apoptosis by inhibition of caspase activation. Through experiments with cultured cells, secreted SPARC was shown to confer a survival signal against nutrient (fetal calf serum) deprivation (Portela et al., 2010). Whether this activity of SPARC depends on its interaction with collagen remains unclear.

Almost nothing is known about SPARC in mollusks. However, initial intriguing observations suggest that SPARC may contribute to shell biomineralization. In mollusks, organic components secreted by epithelia cells at the edge of the mantle (the dorsal body wall that surrounds the inner organs) are important for the layered deposition of $CaCO_3$ and contribute thereby to the mechanical properties of the shell (Kocot, Aguilera, McDougall, Jackson, & Degnan, 2016; Nudelman, 2015). Several mollusk species encode a SPARC (e.g., Clark et al., 2010; Joubert et al., 2010; Miyamoto et al., 2013), and the polypeptide sequences are well conserved with SPARC in other phyla (Irles et al., 2017; Werner, Gemmell, Grosser, Hamer, & Shimeld, 2013). An examination of SPARC transcript in *Patella vulgata* (limpet) identified its expression in mantle edge cells of larvae and adults, consistent with a role in early shell development (Werner et al., 2013). Functional studies are called for.

With regard to the SPARC-related proteins, *D. melanogaster* encodes a single Smoc orthologue, designated *magu* and also named as *pentagone* (Flybase CG2264; Li & Tower, 2009; Vuilleumier et al., 2010). *Magu* was identified in a screen for transposon-directed mutations that would extend the fecundicity and life span of female flies. When overexpressed from cDNA in all tissues of adult flies, magu extended life span in both male and female flies (Li & Tower, 2009). Since magu is expressed in the ovary and in the stem cell area near the tip of the testes (Li & Tower, 2009; Terry, Tulina, Matunis, & DiNardo, 2006), it was hypothesized to form part of the germline stem cell niches. Indeed, in the testes, *magu* is transcribed in hub cells (somatic cells that support the stem cells), leading to localization of magu protein between hub cells and germline stem cells. Adult *magu*-null mutant flies (generated by transposon insertions) were found to have smaller wings with vein patterning phenotypes (an example from Norman et al., 2016 is shown in (Fig. 2Ci and Cii) and testes that were thinner with fewer germline stem cells. Analysis of target gene expression indicated that Dpp/BMP signaling (schematized in Fig. 3A) was compromised. Indeed, the wing phenotypes were rescued by expression of an activated form of BMP receptor-1, indicating that magu acts upstream to promote BMP signaling (Zheng, Wang, Vargas, & DiNardo, 2011).

Under the name pentagone, magu/CG2264 was also identified as a regulator of BMP signaling in wing imaginal discs (Vuilleumier et al., 2010). In the wing disc, the major anterior/posterior patterning morphogen

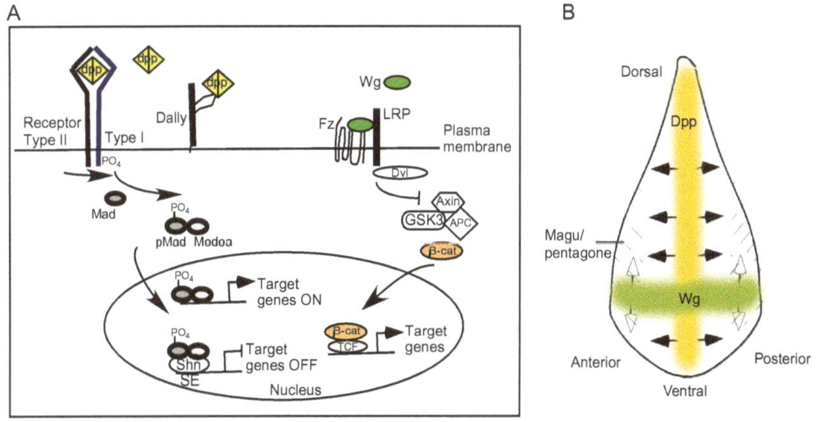

Fig. 3 Dpp and Wg signaling pathways in the wing imaginal disc of *Drosophila*. (A) Schematic diagrams of the Dpp/BMP and Wg/canonical Wnt signaling pathways. (B) Schematic of the wing imaginal disc, showing the zones of expression of Dpp (*yellow*), Wg (*green*), and magu/pent (*dashed lines*). *Arrows* indicate directions of movement of Dpp and Wg.

gradient is formed by decapentaplegic (Fig. 3B), and in addition to binding its signaling receptors, Dpp can be localized at cell-surfaces by coreceptors corresponding to glypican proteoglycans, namely, dally, and dally-like protein (Dlp). *CG2264* is expressed at the lateral margins of wing discs in a similar domain to the Dpp-inhibited target *brinker* (Fig. 3B). *CG2264* contains in its promoter region "silencer elements" that are targets for transcriptional repression by phosphorylated Mad and Medea in complex with schnurri, downstream of Dpp (Fig. 3A) (Bier & De Robertis, 2015). Silencing of *CG2264* resulted in small-sized wings (as a result of decreased cell number) with truncated longitudinal veins. The Dpp gradient was also disrupted, due to loss of the normal activity of extracellular magu/pent protein as an inhibitor of Dpp signaling. In wild-type wings, magu/pent binds to dally and dlp to extend the range of Dpp morphogenetic signaling (Vuilleumier et al., 2010). Further studies on the binding of magu/pent to dally and Dlp demonstrated that these receptors are degraded via endocytosis by a dynamin and Rab5-dependent process. By reducing cell-surface levels of dally and Dlp, magu/pent decreases the local pool of Dpp at cell surfaces and also increases the diffusability of Dpp. By a similar mechanism, magu/pent also downmodulates another major developmental signaling pathway, the wingless (Wg) pathway through reduction of extracellular Wg (Norman et al., 2016). Wg is the *Drosophila* homologue of Wnt proteins that signal canonically via frizzled receptors to regulate β-catenin-dependent transcription of target genes (Fig. 3A) (Swarup & Verheyen, 2012). Thus, magu/pent acts in wing morphogenesis by controlling the extent of two major morphogen gradients within the imaginal disc.

Expression of *CG2264* itself is susceptible to crosspathway regulation by TGFβ/activin signaling (Perrimon, Pitsouli, & Shilo, 2012). Thus, null mutants for the key SMAD in the TGFβ pathway, *smad-2,* have wider wings with altered wing vein patterning, equivalent to the phenotype of gain-of-function mutants in BMP signaling. At the molecular level, these phenotypes were explained by loss of expression of *CG2264* and other target genes via supranormal transcriptional repression through the Dpp-dependent silencer elements (Peterson & O'Connor, 2013).

During the growth of the imaginal disc and wing differentiation it is crucial that the extent of morphogen gradients remain matched to the size of the tissue. Modeling based on an "expansion–repression" circuit has predicted how magu/pent could mediate expansion of the Dpp gradient for scaling of tissue size. The model was examined by overexpression of magu/pent, which led to a mismatch between the Dpp gradient and the growth of the wing disc (Ben-Zvi, Pyrowolakis, Barkai, & Shilo, 2011). In the absence of magu/pent the Dpp gradient was sharper and did not adjust proportionately with wing tissue size, indicating a key role for magu/pent in scaling of the Dpp gradient during wing growth (Hamaratoglu, de Lachapelle, Pyrowolakis, Bergmann, & Affolter, 2011).

The single *D. melanogaster* SPOCK/testican proteoglycan, named carrier of wingless (Cow), also acts to regulate the Wg gradient of the wing imaginal disc (Chang & Sun, 2014). Cow binds via its heparan sulfate side chains to extracellular Wg. This interaction increases the rate of movement of Wg away from its zone of production, and consequentially decreases the local Wg concentration and increases the range of its gradient. During normal development, Cow has complex, biphasic effects on the expression of Wg target genes that are responsive to different Wg concentrations (Chang & Sun, 2014).

3.3 Thrombospondin

Early diverging metazoans and protostomes typically encode a single TSP (Bentley & Adams, 2010). The majority of these TSPs have domain architectures similar to TSP4, yet may include larger numbers of tandem EGF-like domains, as seen in *D. melanogaster* TSP (Fig. 1).

To date, investigation of TSPs in invertebrates has been restricted to a few species and phyla. Cnidarians are of interest for their simple tissue organization and high regenerative capacity (Tucker & Adams, 2014). In the cnidarian *N. vectensis*, the localization of TSP-168100 has been studied in intact and transected polyps. In the mesoglea, the ECM layer that separates the ectoderm from the endoderm in cnidarians, TSP-168100 is present specifically in the retractor muscles and pharynx, in neurite processes that extend within the mesoglea. After transection, the transcript is strongly upregulated, and the protein is newly detectable in the outer glycocalyx of the body wall ectoderm. These results implicate TSP-168100 in neuronal outgrowth and patterning and potentially in the tissue and ECM reorganizations associated with regeneration (Tucker et al., 2013).

There is strong commercial interest in aquaculture of shrimps and prawns for food. Research on members of this order of crustaceans has uncovered interesting roles of TSPs, which, in this lineage, uniquely include repeated chitin-binding domains at the N-terminus (Sun, Zhao, Kang, & Wang, 2006; Yamano, Qiu, & Unuma, 2004; Zhou et al., 2011) (Fig. 1). *Marsupenaeus japonicas* TSP (MjTSP) is highly expressed in oocytes throughout their development and becomes localized to the cortical rods that extrude the protective jelly coat upon fertilization (Yamano et al., 2004). Ovarian development in shrimps is negatively regulated by a neuroendocrine organ of the eyestalks, the X-organ sinus gland complex. Transection of the optic stalks removes this control and resulted in accelerated oocyte development, in correlation with a strong elevation of ovarian MjTSP protein between 2 and 7 days after transection (Okumura et al., 2006). The ovary is a conserved site of TSP expression: TSP transcripts are present at high level in the ovaries of other prawns including *Penaeus monodon* (giant tiger prawn) (Preechaphol et al., 2007). In *Fennerpenaeus chinensis*, the TSP transcript (FcTSP) is detected in hemocytes, heart, intestine, and stomach as well as the ovary (Sun et al., 2006). *F. chinensis* shrimp with a triploid genome are impaired for ovarian development, and triploid shrimp have reduced levels of FcTSP transcript in the ovary (Xie et al., 2010). All these findings imply that shrimp TSP may have roles in cell interactions that contribute to drive oocyte terminal differentiation and fertilization. Further studies on *P. monodon* oocytes have shown that PmTSPII becomes concentrated at the outer edges of mature cortical rods from whence it is secreted upon activation. Immunodepletion studies on the water-soluble proteins of the cortical rods indicated that PmTSP has a major role in inducing the acrosome reaction of sperm to permit fertilization (Magerd, Asuvapongpatana, Vanichviriyakit, Chotwiwatthanakun, & Weerachatyanukul, 2013).

Correlative evidence indicates that shrimp TSP may function in innate immune responses with a possible protective role. FcTSP transcripts were increased in *F. chinensis* injected with *Staphylococcus aureus* and *Vibrio anguillarum* as an immune challenge, including switch-on/upregulation of the transcript in the hepatopancreas (Sun et al., 2006). The shrimp lymphoid organ is a major site for phagocytosis of foreign particles and, in *P. monodon*, TSP transcripts were strongly upregulated in this organ after infection with *Vibrio harveyi* (Pongsomboon, Wongpanya, Tang, Chalorsrikul, & Tassanakajon, 2008). In contrast, infection of *F. chinensis* shrimp with white spot syndrome virus (a major economic issue in shrimp aquaculture) resulted in strong downregulation of FcTSP protein in hepatopancreas (Chai, Yu, Zhao, Zhu, & Wang, 2010). In shrimps as in other animals, activation of innate immunity depends on TOLL-like receptors and the NF-κB signaling pathway (O'Neill, Golenbock, & Bowie, 2013). Upregulation of FcTSP transcripts depends on an NF-κB family member, Relish, that is itself upregulated upon viral infection (Wang, Li, & Li, 2013).

The most well-studied TSP of invertebrates is *D. melanogaster* TSP (DmTSP). Since initial characterization of transcript expression in embryos (Adams et al., 2003; Chanana, Graf, Koledachkina, Pflanz, & Vorbrüggen, 2007; Subramanian, Wayburn, Bunch, & Volk, 2007) (Fig. 2D and E), DmTSP has been studied as a PS2 integrin ligand, for interactions with other ECM proteins, and for its functional roles at muscle–tendon attachment sites. *dtsp* deficiency is associated with embryonic lethality because the loss of tenocyte-produced DmTSP as a ligand for PS2 integrin on muscle cells results in the failure of muscle–tendon attachment sites during initial muscle contraction, and thus loss of muscle functionality (Chanana et al., 2007; Subramanian et al., 2007) (Fig. 2F). In *dtsp*-deficient larvae, other ECM components are also disorganized; notably the PS2 ligand, tiggrin, is found at muscle–muscle attachment sites but is delocalized from muscle–tendon sites (Subramanian et al., 2007). Functional cooperation between DmTSP and laminin has been indicated by a genetic analysis of the roles of laminin γ subunit (*LanB2;* this subunit is common to both of the laminin heterotrimers of *Drosophila*) in organ development. A mild *LanB2* mutant muscle phenotype at dorsal muscle–tendon attachment sites was enhanced strongly in the *dtsp* mutant background, indicating a potential role of laminin/TSP cooperation in the ECM at these sites (Wolfstetter &

Holz, 2012). The timing of localization of DmTSP at muscle ends is regulated by a secreted binding partner, slowdown. Binding of slowdown to DmTSP depends on the KGD, integrin-binding motif of DmTSP. This suggests that the normal function of slowdown may involve modulation of PS2 integrin binding by DmTSP (Gilsohn & Volk, 2010).

At the muscle–tendon attachment sites of *Drosophila*, the normal organization of ECM, PS integrin localization, intracellular cytoskeletal association, and cell contraction are all needed to balance correctly the pulling forces exerted on the muscles and for mechanical stimulation of myofibrillogenesis. The importance of correct actin cytoskeletal organization is emphasized by phenotypes of Rho kinase (DRok) loss-of-function embryos. Rho kinase is a conserved central regulator of actomyosin contractility and in its absence, DmTSP accumulated aberrantly at tendon cell tips, extensions of tendon cell membranes were deformed, and muscle–tendon cell attachments were not assembled properly (Vega-Macaya, Manieu, Valdivia, Mlodzik, & Olguín, 2016). Mutants of *kon* (the *Drosophila* orthologue of the transmembrane proteoglycan NG2/CSPG4 that is present on muscle cell surfaces) led to similar muscle–tendon detachment phenotypes as in *tsp* mutants. In vitro, Kon forms a multiprotein complex with PS2 integrins that confers strong cell–cell aggregation properties on transfected S2 cells. Kon also regulates PS2 integrin activation, as demonstrated by the presence of focal adhesion kinase (FAK, an important signaling mediator downstream of activated integrins), in Kon immunoprecipitates and reduced FAK activity (studied as the localization of phospho-FAK), in *kon* mutant muscle cells. DmTSP is also reduced at myotendinous junctions in *kon* mutants, and to a greater extent in *kon;inflated* double mutants (*inflated* encodes the αPS2 integrin subunit). These phenotypes depended on the extracellular and transmembrane domains of Kon, as reexpression restored TSP and pFAK levels and partially rescued muscle detachment (Pérez-Moreno, Espina-Zambrano, García-Calderón, & Estrada, 2017). Collectively, these studies identify DmTSP as an important PS2 ligand involved in tendon ECM organization and in the muscle–tendon cell–cell interactions that are essential for proper development and contractility of the body wall muscles.

Recently, DmTSP has been implicated through a RNAi screen in another type of cell–cell interaction, the attachments of the chordotonal organs to the cuticle. These organs contain mechanosensory neurons and function in the coordination of movement of different body parts through stretch-sensing. As at the muscle–tendon attachments, DmTSP becomes localized in the ECM of the attachment sites during larval development. Silencing of *dtsp* expression resulted in destabilization and loss of the chordotonal organ cuticle attachments (Greenblatt, Ben-El, Hassan, & Salzberg, 2017).

4. MATRICELLULAR PROTEINS IN INVERTEBRATE DEUTEROSTOMES

Data on matricellular proteins in invertebrate deuterostomes are very limited. For the most part, studies have yet to progress beyond gene and transcript identification. CCN sequences have been recognized in *B. floridae* and *C. intestinalis* (Mosher & Adams, 2012) yet transcript expression patterns or protein functions remain to be explored. Similarly, although predicted TSP proteins have been identified, laboratory studies of TSPs in invertebrate deuterostomes are missing. It is inferred that one or more TSP gene duplications took place in the last deuterostome common ancestor, as all early diverging deuterostomes examined to date have at least two B subgroup TSPs and one A-type (Adams & Lawler, 2011; Bentley & Adams, 2010).

SPARC has been identified in colocalization with collagen in oral cirri of amphioxus (*Branchiostoma belcheri*); these are rigid, sensory structures around the mouth that filter out large particles before food intake. This localization suggests a similar function of SPARC in collagen binding and fibril organization as in vertebrates (Kaneto & Wada, 2011).

B. floridae encodes a single TN with an unusually large number (38) of FNIII domains. A single TN gene and its transcript have also been identified in the urochordates *C. intestinalis* and *C. savignyi* (Tucker & Chiquet-Ehrismann, 2009; Tucker et al., 2006). FN appears to have originated in the last urochordate/vertebrate common ancestor because a FN-like protein is encoded in *Ciona* species but not in *B. floridae* (José-Edwards, Oda-Ishii, Nibu, & Di Gregorio, 2013; Segade et al., 2016; Tucker & Chiquet-Ehrismann, 2009). *C. intestinalis* FN starts to be expressed during gastrulation, and the transcript levels increase to a maximum in swimming larvae. By use of a promoter-GFP reporter, *Fn* transcripts were identified to be localized principally in the notochord. CRISPR/CAS9-targeted knockout of *Fn* in the notocord resulted in short tailed-larvae with a defective and widened notochord morphology. This is due to impaired formation of cell protrusions that would normally mediate the rearrangement of cells into a single column through the developmental processes of convergent extension and intercalation (Segade et al., 2016).

5. MATRICELLULAR PROTEINS IN CYCLOSTOMA (JAWLESS VERTEBRATES)

Lampreys are jawless fish with a cartilaginous skeleton that are members of the Cyclostoma, a sister group to the jawed vertebrates (Green & Bronner, 2014). The Japanese lamprey *Lethenteron japonicum* encodes two TNs, that both cluster with TN-R in phylogenetic analyses, and a partial sequences has been identified in the sea lamprey *Petromyzon marinus* (Adams et al., 2015). In *P. marinus* larvae, TN-C immunoreactivity is specifically associated with neuronal fascicles in the olfactory pathway, suggestive of a boundary or compartmentalization function (Zaidi et al., 1998). *Sparc* is transcribed in adult *P. marinus* liver (Ringuette et al., 1991) (Table 2).

6. MATRICELLULAR PROTEINS IN NONMAMMALIAN VERTEBRATE MODEL ANIMALS

6.1 Bony Fish

The evolution of bony fish involved a genome-wide duplication event, termed the teleost-specific genome duplication (Glasauer & Neuhauss, 2014). As a result, many gene families of bony fish include additional gene paralogues in comparison to birds and mammals. Evolutionary processes of gene loss, additional gene duplications, and/or rapid sequence divergence after the teleost-specific genome duplication add complexity to the identification of orthologous genes between bony fish and land vertebrates. In many cases, identification of syntenic chromosomal loci is needed in addition to phylogenetic analysis of protein-coding sequences (McKenzie, Chadalavada, Bohrer, & Adams, 2006; Tucker et al., 2006). For example, *Danio rerio* encodes nine CCN family members (that include multiple *ccn1*, *ccn2*, and *WISP1* paralogues but no *nov* orthologue), and at least seven TSPs (Fernando et al., 2010; McKenzie et al., 2006). The puffer fish *Takifugu rubripes* and *Tetraodon nigroviridis* each encode six TSPs, in each case corresponding to a distinct set of paralogues and with the absence of any *thbs3* orthologue from *T. nigroviridis* (Adolph, 2002; McKenzie et al., 2006). Several fish species have been identified to encode two *Thbs1*-related paralogues (McKenzie et al., 2006; Wu et al., 2009).

6.1.1 CCNs

Studies of CCNs in fish are beginning to progress after initial limited identification of CCNs.

Tracking of *ccn2* expression in *D. rerio* embryos with a promoter-GFP reporter showed that expression is concentrated in the notochord. Correspondingly, knockdown of CCN2 (by antisense morpholinos) disrupted notochord development (Chiou, Chao, Wu, Kuo, & Chen, 2006). More recently, expression of all nine CCN paralogues of *D. rerio* was examined systematically by in situ hybridization of embryos. With the exception of several *ccn3* paralogues for which expression was absent, each transcript was expressed with distinct temporal and spatial profiles, with the two *ccn2* paralogues (aka *ctgfa* and *ctgfb)* and *Cyr61-c23* having the most widespread expression (Fernando et al., 2010) (Table 2). Other studies have begun to examine pathological contexts. Expression of *ctgfa* and *ctgfb* is upregulated during embryogenesis in a *D. rerio* model of Ewing's sarcoma. In humans, this cancer of bone and cartilage occurs most commonly in teenagers and involves aberrant transcription regulation by the product of a fusion gene *EWS/FLI1* (Theisen, Pishas, Saund, & Lessnick, 2016). In the zebrafish model, CTGF upregulation in Meckel's cartilage led to failure of differentiation of hypertrophic chondrocytes. Mechanistically, the *EWS/FLI1* gene product was found to associate with Sox9 and to bind to the *ctgfa* and *ctgfb* promoter regions, suggesting that aberrant regulation of Sox9 target genes may be a central mechanism in Ewing's sarcoma (Merkes et al., 2015).

With regard to studies of spinal cord regeneration in *D. rerio, ctgfa/ccn2* has a key role in bridging of the injury by glial cells and their secreted products. Thus, *ctgfa* expression is upregulated in glial cells in the vicinity of the injury and loss-of-function mutations or overexpression of *ctgfa* inhibit or accelerate spinal cord regeneration, respectively (Mokalled et al., 2016). In contrast, ctgfa/CCN2 mediates developmental regression of blood vessels, as identified by studies of regression in the caudal vein plexus of *D. rerio*. *Ctgfa* was identified to be a target of the Yap/Taz transcriptional regulator complex in vascular endothelial cells, and its expression was necessary for normal caudal vein plexus regression, a process also dependent on ctgfa-mediated F-actin reorganization (Nagasawa-Masuda & Terai, 2017).

The *ccn6/Wisp3* orthologue of *D. rerio* is expressed in embryos from 6 h after fertilization onwards and later localizes to the brain, otic vesicles, and swim bladder (Nakamura et al., 2007) (Table 2). Knockdown (by antisense morpholinos) led to impaired development of pharyngeal cartilage structures, whereas overexpression (by mRNA injection) led to dorsalization of the embryos through inhibition of both BMP and Wnt signaling. In the case of BMP, the mechanism of inhibition involved binding of CCN6 to BMP. In the case of Wnt, CCN6 bound to the coreceptor Frizzled-8 and also to LDL receptor-related protein 6 (LRP6), to block Wnt/receptor binding and canonical Wnt signaling (Nakamura et al., 2007).

With regard to effects of temperature shifts on gene expression profiles (an area of general research interest for fish developmental biology), temperature-dependent expression of *ccn1/cry61* has been identified in an *Oryzias latipes* cell line, with levels increasing with decreasing temperature (5°C and 15°C vs 25°C). It was suggested that ECM remodeling might be important in physiological adaption to lower temperatures (Hirayama, Ahsan, Mitani, & Watabe, 2008).

6.1.2 SPARC and Relatives

In mammals, the family of secretory calcium-binding phosphoproteins have major roles in biomineralization of bones and teeth (Kawasaki, Suzuki, & Weiss, 2004), and SPARC is not essential for bone morphogenesis (Section 2.2). Similarly, knockdown studies in *D. rerio* indicate localized tissue requirements for SPARC and not a general role in skeletal development. The single *sparc* of *D. rerio* is expressed during embryogenesis in the otic vesicle, notochord, floor plate, somites, and developing caudal fin and, at later times, in skeletal elements (Rotllant et al., 2008) (Table 2). Knockdown of *sparc* (by antisense morpholinos) in 1–2 cell embryos resulted in smaller embryos by 30 h after fertilization, which had small, deformed pectoral fins and lower jaw and an altered swimming behavior suggestive of impairments to the inner ear. The SPARC-depleted embryos died within 7 days. Further studies of the lower jaw established requirements for SPARC for proper development of the pharyngeal cartilage and expression of *col2a1a;* however, neural crest morphogenesis was normal. SPARC-knockdown also led to defects in development of the otic vesicle resulting smaller ears with improperly formed semicircular canals. All these phenotypes could be rescued by injection of a synthetic *sparc* mRNA (Rotllant et al., 2008). Although this study reported normal development of inner ear otoliths after *sparc*-knockdown, another study of adult sturgeon and catfish identified SPARC as a constituent of otoliths by proteomics and confirmed its presence in *D. rerio* otoliths from embryos and adults by immunoblotting and immunofluorescence (Kang, Stevenson, Yau, & Kollmar, 2008). In this study, reduction of SPARC protein levels (by antisense morpholinos injected at nonlethal doses) led to reductions in otolith size and number up to 72 h postfertilization, along with aberrations of otolith shape or location within the tissue. Impaired swimming and development of fins and the otic vesicle were also described, as observed by Rotllant et al. (2008) and Kang et al. (2008). Effects of *sparc* knockdown by morpholinos in early *D. rerio* embryos have also been examined with regard to embryonic hematopoiesis. These studies identified impaired differentiation of erythroid progenitor cells and placed sparc downstream of the fibroblast growth factor 21 regulatory pathway for hematopoiesis (Ceinos et al., 2013).

Other studies have focused on the regulation of SPARC in the skeleton. Particularly in relation to commercial fish farming, there is interest in the effects of hormones and nutrients on fish skeletal development; in this context *sparc* is utilized as a marker of bone ECM (Table 2). By transcriptomics, *sparc* transcripts were identified in both bone and chondroid bone (gill arches) of *Sparus auratus* (sea bream). By RT-qPCR, *Sparc* transcripts were abundant in bone and were sharply downregulated after fasting (Vieira et al., 2013). Estrogen is present in the developing fish brain, and in *D. rerio* and goldfish *Carassius auratus, sparc* transcripts were shown to be downregulated by estrogen (Lehane, McKie, Russell, & Henderson, 1999; Pashay Ahi, Walker, Lassiter, & Jónsson, 2016). *Sparc* transcripts in *S. auratus* bone were also downregulated by the selective estrogen-receptor modulator raloxifene (an estrogen mimic that binds estrogen receptors to activate estrogen-dependent pathways) (Vieira, Pinto, Guerreiro, & Power, 2012). An investigation of the effects of dietary vitamin A on development of *S. auratus* found that high vitamin A resulted in smaller fish (27% decrease in dry weight after a diet with 1.5 × normal vitamin A level). This was accompanied by an altered ratio of mineralized tissue/cartilage in skeletal structures and decreased expression of *sparc* and *col1a1* by 60 h after hatching (Fernández et al., 2011). In contrast, excess leucine, an amino acid that activates the intracellular PI3k/Akt/mTor pathway (that has known roles in bone development and growth), caused a rapid transient increase in *sparc* transcripts in *S. auratus* bone (Garcia de la Serrana, Mareco, La Vieira, Power, & Johnston, 2016).

SPARC has also featured as a bone marker in studies of effects of environmental conditions on the fish skeleton. Elevated water temperatures are used in commercial fish production, yet correlate with skeletal deformities (Wargelius, Fjelldal, & Hansen, 2005). In a study of *Salmo salar* (Atlantic salmon) kept from birth at 6°C or 10°C, around 28% of fish maintained at the higher temperature had skeletal deformities by the time they reached a weight of 60 g (Ytteborg, Baeverfjord, Torgersen, Hjelde, & Takle, 2010). Several ECM transcripts, including *col1a1*, *decorin*, and *sparc*, were reduced from the earliest stages examined, indicating possible impaired terminal differentiation of osteoblasts. In addition, the hypertrophic chondrocyte zone of cartilage was enlarged, indicative of impaired chondrocyte differentiation. Both these changes would result in altered skeletal structure and mechanical properties (Ytteborg et al., 2010).

Because of ongoing interactions between ocean warming and ultraviolet (u.v.) radiation, the effects of u.v. exposure on fish development are also of research interest. A study of the effects of u.v. radiation on *D. rerio* development showed strong upregulation of *sparc* transcripts in 24 h postfertilization embryos after full spectrum u.v. exposure on 4 h embryos; this response depended on wavelengths <350 nm (Torres-Nuñez et al., 2012). The sites of increased transcript levels involved all the normal embryonic sites of *sparc* expression. The u.v. exposed fish were more prone to developmental abnormalities and had reduced (53% vs 94% of controls) survival. Overexpression (by mRNA injection) of *sparc* in 1–2 cell embryos resulted in a similar range of developmental defects as u.v. irradiation (Torres-Nuñez et al., 2012) and was reminiscent of effects of SPARC overexpression on *Xenopus* development (Section 6.2.2).

Mutations in *SMOC-1* cause Waardenburg anophthalmia syndrome in humans (Table 1). In *D. rerio*, *smoc-1* is expressed from the 10-somite stage in the brain, otic vesicle, retina, and pharyngeal arches (Table 2). Knockdown by antisense morpholino at the 1–2 cell stage resulted in microphthalmia, retinal defects, and abnormalities of the forebrain at 5 days postfertilization, thus mimicking key aspects of the human disease (Abouzeid et al., 2011).

In humans, point mutations in *SMOC-2* are associated with defective tooth development (Table 1) (Alfawaz et al., 2013; Bloch-Zupan et al., 2011). Similar phenotypes have been reported from antisense morpholino knockdown studies in *D. rerio* (Bloch-Zupan et al., 2011). A screen by morpholino knockdown for genes with roles in craniofacial morphogenesis identified that *smoc-2* also contributes to development of the viscerocranial cartilage, resulting in smaller heads and small or absent eyes (Melvin, Feng, Hernandez-Lagunas, Artinger, & Williams, 2013). In a study focused on *smoc-2* functions in early *D. rerio* embryogenesis (Mommaerts, Esguerra, Hartmann, Luyten, & Tylzanowski, 2014), *smoc-2* knockdown (by antisense morpholinos and with rescue by coinjection of *smoc-2* mRNA) resulted in impaired somitogenesis and delayed onset of blood cell circulation. The latter was associated with altered morphologies of sites of hematopoiesis, (i.e., the posterior blood islands and yolk sac) and downregulation of the marker *mpx* in all blood islands. Further analysis of anterior blood islands revealed lower levels of several markers of hemangioblast cells and reduced numbers of myeloid progenitor cells (into which hemangioblasts differentiate). At later stages, numbers of macrophages and neutrophils (the terminally differentiation derivatives of myeloid progenitor cells) were also reduced. Although the *smoc-2* knockdown did not affect BMP signaling in these tissues, expression of several BMP target genes was reduced, suggesting that, analogous to SMOC-1 in *X. laevis* (Section 6.2.2), smoc-2 acts in *D. rerio* to modulate signaling downstream of BMP receptors (Mommaerts et al., 2014).

6.1.3 Tenascins

The study of TNs in fish has emphasized developmental processes. *Tn-W* is expressed in the lateral plate mesoderm of *D. rerio* embryos and in the somatic mesoderm and migrating neural crest cells of larvae (Table 2). In young fish, *Tn-W* expression is concentrated in myosepta and dorsal root ganglia. These domains of expression overlap partially with *Tn-C*, which is also expressed strongly in the developing central nervous system (especially hindbrain, certain rhombomeres, telencephalic–diencephalic border, the border of the diencephalon with the midbrain, hypothalamus, and the basal plate of the caudal midbrain tegmentum) and axial mesoderm (Tongiorgi et al., 1995; Weber et al., 1998) (Table 2). In the developing heart TN-C and TN-W may act in coordination to provide ECM properties necessary for proper formation of the heart tube (Lin et al., 2013). TN-C protein is present in many structures of the developing trunk region, with *Tn-C* mRNA expression by the somites dependent on the presence of a notochord (Tongiorgi, 1999). TN-C is also expressed at the myoseptum, a connective tissue structure to which motor axons grow out from the spinal cord. Functional perturbations by injection or overexpression of TN-C fragments disrupted axon extension and patterning (Schweitzer et al., 2005).

Unlike adult mammals and birds, adult bony fish have capacity for tissue regeneration, as evident by full repair of spinal cord injury and regeneration capacity of the appendages without scarring. In the fish heart, cardiomyocytes retain phenotype plasticity to dedifferentiate, proliferate, and regenerate the myocardium, throughout life. TN-C may have important roles in spinal repair, as its inhibition or knockdown impaired regeneration (Yu et al., 2011). TN-R is expressed in discrete brain regions that border the pathway of the optic nerves and may have similar activities to

TN-C in promoting optic nerve growth and repair in adult zebrafish (Becker et al., 2004; Becker, Schweitzer, Feldner, Becker, & Schachner, 2003). TN-C is highly elevated at the border zone of myocardial infarcts where it may contribute in the transient "regeneration ECM" to scarless regeneration of the myocardium (Chablais et al., 2011; Mercer, Odelberg, & Simon, 2013). Close study of spatial and temporal aspects of a myocardial infact model in *D. rerio* demonstrated that the cycling cardiomyocytes form a blastema-like zone embedded in an ECM enriched for tenascin-C and FN (Sallin et al., 2015). Similarly, a transient ECM with elevated TN-C, hyaluronan, and FN is formed during caudal fin regeneration. Accumulated fibroblasts at the edge of the wound area are thought to be an important source of TN-C (Govindan & Iovine, 2015; Mateus et al., 2012). In contrast, in the rainbow trout *Oncorhynchus mykiss*, TN-C is upregulated from 14 to 100 days after deep punch biopsy muscle wounding, yet these wounds healed with fibrotic scarring (Schmidt et al., 2016).

6.1.4 Thrombospondins

The roles of TSPs in fish have barely been addressed. Several fish species encode two paralogues of TSP1, TSP1a and TSP1b. Expression patterns of these were investigated during developmental stage of tilapia, *Oreochromis niloticus*. Both genes were transcribed, but with different temporal and spatial patterns, even within the same tissue, as demonstrated with regard to skeletal muscle or ovaries during oogenesis (Wu et al., 2009) (Table 2). A study of oogenesis in the Senegalese sole, *Solea senegalensis*, identified the paralogue *Thbs4b* to be transcriptionally upregulated during vitellogenesis (the stage of oocyte growth), also that *Thbs1* transcripts are present throughout the ovarian cycle and elevated in ovaries undergoing follicular atresia, with predominant expression in the follicular cells (Tingaud-Sequeira et al., 2009) (Table 2).

In *D. rerio*, *Tsp4b* transcripts localize at muscle–tendon attachment sites. *Tsp4b* is expressed by muscle cells during development of the somites and myotendinous junctions and later becomes restricted to tenocytes (Subramanian & Schilling, 2014). Knockdown (by antisense morpholinos) resulted in weakening of muscle–tendon attachments, especially after electrical stimulation. This effect correlated with disruptions to the basement membrane (including disrupted localization of laminin), ECM disorganization at the myotendinous junctions and reduced and mislocalized active FAK and paxillin. The phenotypes depended on secreted TSP4b (i.e., a noncell autonomous role). These findings indicate that TSP4b has similar developmental roles in ECM assembly and integrin signaling regulation at zebrafish myotendinous junctions as DmTSP in *Drosophila* embryos (Subramanian & Schilling, 2014) (Section 3.3).

6.2 Amphibia

In recent decades, amphibian species have been popular models for studies of expression patterns of matricellular proteins, especially TN-C (Table 2). In addition to their suitability for studies of the earliest stages of embryogenesis, amphibians are useful models to study wound repair and tissue regeneration. In frogs, regeneration activity is limited to the larval stages, but newts, salamanders, and axolotl can regenerate the tail, limbs, and organs such as the heart throughout life (McCusker, Bryant, & Gardiner, 2015). Many studies by in situ hybridization or immunohistochemistry provide correlative evidence for upregulation or dynamic expression profiles of individual matricellular proteins at sites of wound repair, or in the early stages of regeneration of amputated limbs or heart (Table 2).

In general, amphibian limb regeneration begins with formation of blastema tissue below the wound epithelium at the amputation site. Blastema assembly depends on several forms of intercellular communication and local nerve signaling and involves the accumulation of dermal fibroblasts underneath the wound area. In the blastema, proliferative limb progenitor cells accumulate along with associated macrophages and a transitional ECM of altered composition is formed. At later stages of the growth of a replacement limb, the same morphogenetic programs that govern embryonic limb development are active (McCusker et al., 2015). Studies that have addressed matricellular proteins in amphibians are discussed below.

6.2.1 CCNs

In *X. laevis* embryos, zygotic expression of *ccn1/Cyr61* begins at the neurula stage. In early tadpoles the transcript is localized to the somites, heart, notochord, and blood vessels (Table 2). Overexpression (by injection of RNA or protein) or depletion (by antisense morpholinos) of *ccn1* leads to impairment of gastrulation (Latinkic et al., 2003). Detailed analysis of morpholino-treated embryos identified defects in epiboly and disorganization/reduction of ECM on the roof of the blastocoel cavity. In vitro cell adhesion experiments with mouse CCN1 implicated effects on cell–ECM and cell–cell adhesion and modulation of Wnt signaling as possible mechanisms (Latinkic et al., 2003). *Ccn2* has a different expression pattern in embryonic floor plate, notochord, and somites (Table 2). Its overexpression led to secondary body axes as a result of inhibition of BMP4 signaling and activation of TGFβ signaling by direct binding of the cysteine-rich domain of CCN2 to these proteins (Abreu, Ketpura, Reversade, & De Robertis, 2002).

6.2.2 SPARC and Relatives

Expression of *sparc* by *X. laevis* embryos begins after gastrulation, continues in the early stages of development of the epidermis, somites, neural tube, and notochord, and decreases after these tissues have formed (Damjanovski et al., 1994; Lane & Sage, 1994; Mason et al., 1986) (Table 2). *Sparc* overexpression (by RNA or protein injection) at the blastula stage led to arrest of postgastrula development, in association with missing somites, deformed or absent eyes, and impaired notochord formation (Damjanovski, Karp, Funk, Sage, & Ringuette, 1997; Purcell, Gruia-Gray, Scanga, & Ringuette, 1993). Axis defects were also caused specifically by injection of a peptide (peptide 4.2) based on the second E/F hand domain of SPARC (Damjanovski et al., 1997). This activity was associated with impaired cell migration and loss of lamellipodia in involuting head mesodermal cells during gastrulation (Huynh, Sage, & Ringuette, 1999). Knockdown of *sparc* transcripts by morpholinos led to developmental arrest at the neurula and tailbud stages as a result of weakened cell–cell associations (as indicated by electron microscopy view of separated somites) and loss of the epidermis. Although SPARC is a collagen-binding protein, basement membranes and connective tissue collagen fibrils were not affected by *sparc* knockdown and the phenotype was not associated with elevated apoptosis (Huynh et al., 2011).

The level of SPARC protein in the tails of swimming tadpoles increases and peaks just prior to metamorphosis when resorption of the tail begins (López-Murcia, Terni, & Llobet, 2015). At this developmental stage, neurons controlling tail movement are eliminated and glial cells are a source of SPARC. In microcultures of superior cervical ganglion cells in vitro, SPARC inhibited cholinergic synaptogenesis, and the E/F hand peptide 4.2 was sufficient for this activity. With regard to elimination of preformed synapses by the peptide, F-actin remodeling was found to be required in conjunction with increased endocytosis and enlargement of lysosomes. The role of SPARC in normal synaptic pruning in the tail was confirmed by injection of peptide 4.2 into swimming tadpoles, which resulted in reduced swimming activity, loss of neuromuscular junctions, and alterations to axon patterning of motor neurons in the tail (López-Murcia et al., 2015).

Of the other SPARC family members, SMOC-1 has been studied during early development of *X. laevis* in relation to prior knowledge of SMOC1/2 as binding partners of BMP in bovine cartilage. *X. laevis SMOC-1 (XlSMOC-1)* is expressed from the late gastrulation stage onward in the anteriorventral region and later in pharyngeal arches, neural crest cells, midbrain, and hindbrain (Thomas et al., 2009) (Table 2). Overexpression (from mRNA) led to enlargement of heads and anteriordorsal structures and an enlarged neural plate; these phenotypes are similar to effects of known BMP antagonists. In vitro experiments demonstrated that secreted XlSMOC can mediate these activities. Knockdown of *XlSMOC* transcripts (by antisense morpholinos) in early embryos resulted in developmental arrest at neurulation, indicating a unique role of XlSMOC in developmental patterning. XlSMOC acted to inhibit BMP2 signaling by increasing downstream Erk activity: active Erk phosphorylated Smad1, -5, and -8 which caused them to be targeted for proteosomal degradation (Thomas et al., 2009).

In studies with *X. laevis* blastula animal caps in vitro (Thomas et al., 2017), purified recombinant XlSMOC2 induced neural structures via inhibition of BMP2 signaling. Further analysis of the SMOC2 domain requirements for BMP inhibition identified an important role of the thyroglobulin-1-like domains, whereas the isolated E/F hand domain acted to elevate BMP signaling. The latter activity is congruent with the role of *Drosophila* SMOC in extending BMP gradients in the wing imaginal disc (Section 3.2). The BMP signal-enhancement activity of SMOC was considered to be mediated by competition, via the E/F hand domain, for BMP binding to heparan sulfate proteoglycans. As demonstrated in grafted *X. laevis* embryos this had the effect of increasing the signaling range of BMP (Thomas et al., 2017).

6.2.3 Tenascins

Studies of TNs in amphibians have focused mostly on TN-C. TN-C protein has been identified in developing and regenerating limbs of the newt *Notophthalmus viridescens* and the axolotl *Ambystoma mexicanum* (Onda, Goldhamer, & Tassava, 1990) (Table 2). In the newt, *TnC* is expressed in the wound epithelium and cells of the blastema during growth and redifferentiation (Onda, Poulin, Tassava, & Chiu, 1991). Transcript profiling also indicated major changes in ECM composition during this period (Calve, Odelberg, & Simon, 2010). In *X. laevis* embryos, *Tn-C* has limited expression at the neurula stage and is strongly expressed in mesodermal tissues of tailbud embryos. Correspondingly, expression was activated in vitro by the mesodermal inducers activin A and basic fibroblast growth factor (Umbhauer et al., 1992).

Functional properties of TN-C in amphibian embryos were explored in experiments utilizing purified chicken TN-C and mesodermal cells isolated from the dorsal marginal zone (DMZ) of gastrula stage embryos of *Pleurodeles waltl* (Riou, Shi, Chiquet, & Boucaut, 1990). As in classic experiments with mammalian cells (Chiquet-Ehrismann et al., 1988), attachment and spreading of DMZ cells on FN-coated surfaces was inhibited by addition of TN-C in solution. Notably, the concentration of TN-C needed to block attachment to mesodermal cell-derived, fibrillar ECM was 10-fold higher than for attachment to purified FN. Soluble TN-C also blocked migration of cells from DMZ explants on FN, and the cells were also repelled by coated stripes of mixed FN/TN-C. In blastula stage embryos, where TN-C is normally absent from the ECM of the blastocoel roof, injection of TN-C into the blastocoel led to arrest of gastrulation and aberrant embryonic morphology (Fig. 4A and B), with an absence of mesodermal cells on the blastocoel roof. The injected TN-C bound rapidly to the ECM of the blastocoel roof, that includes an extensive FN network, indicating an in vivo mechanism for the antiadhesive activity of TN-C (Riou et al., 1990).

Studies of axolotl tail regeneration have shown that muscle-derived cells also contribute to the blastema (Echeverri & Tanaka, 2002). TN-C is prominent in the transitional interstitial ECM that surrounds regenerating myofibers and its location correlates with regions of active cell proliferation. In contrast to the results with early embryonic cells discussed above, TN-C supported myoblast migration (Calve et al., 2010). The elastic modulus (a measure of tissue stiffness) of blastema tissue is lower than that of adult muscle. The reduced stiffness as well as high TN-C content may provide cooperating signals to active myoblast proliferation and migration (Calve & Simon, 2012).

6.2.4 Thrombospondins

Research on TSPs has made limited use of amphibian models. The presence of additional gene paralogues adds to the complexity of studying TSPs in *X. laevis* (Lawler et al., 1993). Nevertheless, the developmental expression patterns of *Thbs4* have been studied in detail (Table 2). More recently, TSP1 and TSP4 were studied in the axolotl *A. mexicanum* in relation to the process of limb regeneration. In intact adult limbs, *tsp1* and *tsp4* transcripts had distinct locations (Whited et al., 2011) (Table 2). Within

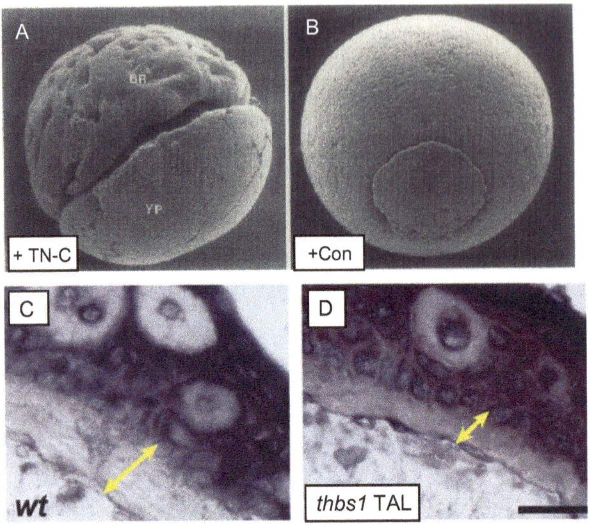

Fig. 4 Examples of effects of altered levels of TN-C or TSP1 in amphibians. (A and B) Tenascin-C overexpression blocks gastrulation in *X. laevis*. Embryos were injected with TN-C (*left panel*) or control saline (*right panel*) at stage 7 and examined 18 h later. The TN-C-injected embryo has a deformed animal pole and enlarged yolk sac and has failed to gastrulate. The control is normal for a stage 11 gastrula. *BR*, blastocoel roof, *YP*, yolk plug. (C and D) Mosaic TSP1 knockout by genome editing impairs regenerative remodeling in *A. mexicanum*. *Yellow arrows* indicate reduced depth of the subepidermal collagenous connective tissue at the regeneration site in *thbs-1*-edited axolotl, 6 days after limb amputation. *Panels (A) and (B) reproduced from Riou, J. F., Shi, D. L., Chiquet, M., Boucaut, J. C. (1990). Exogenous tenascin inhibits mesodermal cell migration during amphibian gastrulation.* Developmental Biology, 137*(2), 305–317, with permission (Elsevier license 4231400552139). Panels (C) and (D) reproduced from Kuo, T. H., Kowalko, J. E., DiTommaso, T., Nyambi, M., Montoro, D.T., Essner, J.J., et al. (2015). TALEN-mediated gene editing of the thrombospondin-1 locus in axolotl.* Regeneration (Oxford, England), *2(1), 37–43., https://doi.org/10.1002/reg2.29 (Published with open access by John Wiley and Sons under CC BY 4.0).*

1 day of limb amputation *Tsp1* transcripts were upregulated in the basal layer of the wound epidermis and remained elevated over several weeks. Isolated cells within the blastema also expressed *tsp1* as did, after 3 weeks, chondrocytes in the regenerating skeletal structures. In contrast, *tsp4* was expressed during the first week of blastema formation by mesenchymal cells that clustered under the wound epithelium, and this expression continued over several weeks. Later, *tsp4* was expressed in the chondrocytes and perichondrium of the developing digits. Because mesenchymal expression of *tsp4* does not occur during normal embryonic limb development, it was suggested that TSP4 may have a regeneration-specific role within the

blastema (Whited et al., 2011). In an exciting development for the field, gene editing of *tsp1* by TALEN technology was used to generate embryos mosaic for *tsp1* (Kuo et al., 2015). Limb development proceeded normally (as in *Thbs1*−/− mice, Lawler et al., 1998), as did limb regeneration. However, increased numbers of monocytes/macrophages were recruited to the blastema (reminiscent of *Thbs1*−/− mice), and (by histology) the thickness of the collagen-rich layer beneath the wound epidermis was decreased by about 50% (Kuo et al., 2015) (Fig. 4C and D). Whether this layer contains thicker collagen fibrils, a phenotype reported in several TSP knockout mice (e.g., Kim, Christofidou, Keene, Milde, & Adams, 2015) is of interest for further investigations.

6.3 Reptiles

Publications on matricellular proteins in reptiles are extremely limited and to date have focused on tissue localizations (Table 2). Some lizard species have capacity for adult tail regeneration and scarless wound repair; one example is the leopard gecko, *Eublepharis macularius*. Punch-biopsy cutaneous wounds on the back or tail of adult *E. macularius* healed without scarring and with complete reformation of scales between 14 and 45 days after wounding. The epithelial layer reformed rapidly over 5 days with accumulation of cells in a loose ECM low in collagen underneath the epithelium; this feature is reminiscent of the amphibian regeneration blastema. Unlike mammalian wound healing, vascularization of the wound bed was delayed until >5 days after wounding. Cells in the wound bed expressed high levels of both TSP1 and VEGF at 5 and 14 days after wounding. By analogy to mammals, TSP1 might have a role in controlling the extent or rate of VEGF-driven angiogenesis (Peacock et al., 2015). During tail regeneration, VEGF and TSP1 were also present at elevated levels in association with endothelial cells and fibroblast-like cells of the blastema, with TSP1 particularly associated with larger vessels as redifferentiation proceeded (Payne, Peacock, & Vickaryous, 2017).

7. PERSPECTIVES

Over the last several decades, studies of matricellular proteins have gone forward in a variety of nonmammalian animal models. As the above synthesis makes clear, certain animals have been preferred for study of particular proteins—for example, studies in amphibian have concentrated mostly on SPARC and TN-C. Overall, the view of matricellular proteins

Table 3 Properties of Matricellular Proteins

Properties identified from mammalian models:
Developmental expression patterns; transient or tissue-specific expression in adults
Not forming structural fibrils within extracellular matrix
Counter-adhesive to cell–ECM adhesion and/or not supporting focal adhesions
Interact with growth factors as well as cell-surface receptors
Not essential for viability in mice (SPARC, TNs, TSPs)
Properties emerging from studies of nonmammalian animals:
Mediators of cell-to-cell interactions
Regulatory interactions with morphogenetic signaling pathways
Roles in postembryonic regeneration and stem cell niches

in nonmammalian animals is incomplete and patchy. Nevertheless, the overview of current research both confirms the conservation of matricellular attributes and brings attention to additional properties that may also be fundamental and would merit further investigation in mammalian models (Table 3).

The studies emphasize the conservation of expression of matricellular family members during early embryonic development. In this context, the most in-depth research has been on the functional roles of SPARC and CCN family members. The research demonstrates important roles of these proteins during early body patterning and also an importance of maintaining the correct levels of these proteins (i.e., overexpression as well as knockdown/knockout resulted in developmental perturbations or arrest). With the advent of effective genome editing technologies, the feasibility of precise gene knockout models for advanced study of early developmental roles in amphibian species is an exciting prospect for future research.

Considering the known tissue-specific sites of function of matricellular proteins, several additional sites of major interest have emerged which relate to cell–cell bridging functions. All categories of matricellular proteins are represented in the developing nervous system and functional studies have demonstrated roles of tenascins and SPARC in synapse formation. These studies complement data on the roles of thrombospondins in synaptogenesis in mice (Risher & Eroglu, 2012). Functions in oogenesis have been identified for SPARC and TSPs and, in the case of TSPs, this appears to be a conserved property, relevant to prawns and fish (Sections 3.2 and 3.3) as well as mammals (Bagavandoss, Sage, & Vernon, 1998; Greenaway, Gentry, Feige, LaMarre, & Petrik, 2005; Thomas, Wilson, Silvestri, & Fraser, 2008). TSPs also have conserved roles in the bridging ECM at muscle–tendon attachment sites, as demonstrated in *Drosophila* and zebrafish. All these findings across diverse tissues and species emphasize that matricellular proteins contribute to cell–cell interactions as well as to ECM properties.

The use of nonmammalian models makes it possible to investigate additional biological processes that are unrepresented or inaccessible in mammals. A prime example is tissue regeneration in amphibians and fish. By demonstrating a context for matricellular proteins within the transitional ECM of the blastema, these studies reinforce and extend the concept that matricellular proteins are important players in tissue remodeling. Functions of tenascin-C in stem cell niches have been investigated in mammals and birds (e.g., Anstrom & Tucker, 1996; Hendaoui et al., 2014); the general association of matricellular proteins with the regeneration blastema suggests that all matricellular proteins should be investigated more widely for roles in the microenvironments of stem cells. New lines of research can be guided by the implication of *Drosophila* CCN in maintenance of tissue-specific stem cell populations, and the identified roles of *Drosophila* SMOC (magu/pentagone) and SPOCK (Cow) as regulators of essential morphogen gradients (dpp and Wg, respectively).

With regard to molecular mechanisms, whereas extracellular associations with growth factors are a central matricellular property, it is striking that nonmammalian models (principally *Drosophila* and amphibian embryos) have provided new insights into regulatory interactions between several matricellular proteins and key embryonic signaling pathways (BMP, Wnt) for tissue patterning and the body plan (Latinkic et al., 2003; Thomas et al., 2009). These findings can enrich studies of matricellular proteins in mammalian development and are also likely to be relevant to human cancers, where many developmental signaling pathways are recapitulated. The concept that matricellular proteins may have intracellular interactions and roles as well as their accepted extracellular functions are emerging from both mammalian and nonmammalian models (Hellewell & Adams, 2016); the intracellular association of SPARC and collagen IV that impacts on basement membrane assembly in *D. melanogaster* is one example.

To date, the majority of research on matricellular proteins in nonmammalian models has had a biomedical impetus, yet there are other important research drivers that can promote distinct new directions and insights. In the world of aquaculture for food supply, there is strong commercial interest in biological factors that affect the growth, health, immunity, and reproduction of prawns, shrimps, and edible fish species. Better understanding of environmental factors that affect the quality of land and aquatic ecosystems is a pressing issue with consequences for the health of the planet. In view of their pathophysiological expression profiles and modulatory biological roles, matricellular proteins should continue to be widely relevant in many fields of research related to animal and human health.

ACKNOWLEDGMENTS

This chapter is dedicated in memory of Ruth Chiquet-Ehrismann. I thank Jack Lawler and Deane Mosher for introducing me to the thrombospondins, and colleagues and collaborators for interesting discussions and continuing research projects. Apologies to those whose work is not cited due to space limitations.

REFERENCES

Abouzeid, H., Boisset, G., Favez, T., Youssef, M., Marzouk, I., Shakankiry, N., et al. (2011). Mutations in the SPARC-related modular calcium-binding protein 1 gene, SMOC1, cause Waardenburg anophthalmia syndrome. *American Journal of Human Genetics*, *88*(1), 92–98.

Abreu, J. G., Ketpura, N. I., Reversade, B., & De Robertis, E. M. (2002). Connective-tissue growth factor (CTGF) modulates cell signalling by BMP and TGF-beta. *Nature Cell Biology*, *4*(8), 599–604.

Adams, J. C. (1995). Formation of stable microspikes containing actin and the 55kDa actin bundling protein, fascin, is a consequence of cell adhesion to thrombospondin-1: Implications for the anti-adhesive activities of thrombospondin-1. *Journal of Cell Science*, *108*, 1977–1990.

Adams, J. C., Chiquet-Ehrismann, R., & Tucker, R. P. (2015). The evolution of tenascin and fibronectin. *Cell Adhesion and Migration*, *9*, 1–12.

Adams, J. C., & Lawler, J. (1993). Diverse mechanisms for cell attachment to platelet thrombospondin. *Journal of Cell Science*, *104*, 1061–1071.

Adams, J. C., & Lawler, J. (2011). The thrombospondins. In R. O. Hynes & K. Yamada (Eds.), *Cold spring harbor perspectives in biology*, *Vol. 3*(10). *The extracellular matrix*. (a009712). Cold Spring Harbor Laboratory Press.

Adams, J. C., Monk, R., Taylor, A. L., Ozbek, S., Fascetti, N., Baumgartner, S., et al. (2003). Characterisation of *Drosophila* thrombospondin defines an early origin of pentameric thrombospondins. *Journal of Molecular Biology*, *328*, 479–494.

Adolph, K. W. (2002). The zebrafish thrombospondin 3 and 4 genes (thbs3 and thbs4): cDNA and protein structure. *DNA Sequence*, *13*(5), 277–285.

Al-Dabbagh, N., Al-Shahrani, H., Al-Dohayan, N., Mustafa, M., Arfin, M., & Al-Asmari, A. K. (2017). The SPARC-related modular calcium binding protein 2 (SMOC2) gene polymorphism in primary glaucoma: A case-control study. *Clinical Ophthalmology*, *11*, 549–555.

Alfawaz, S., Fong, F., Plagnol, V., Wong, F. S., Fearne, J., & Kelsell, D. P. (2013). Recessive oligodontia linked to a homozygous loss-of-function mutation in the SMOC2 gene. *Archives of Oral Biology*, *58*, 462–466.

Anstrom, K. K., & Tucker, R. P. (1996). Tenascin-C lines the migratory pathways of avian primordial germ cells and hematopoietic progenitor cells. *Developmental Dynamics*, *206*(4), 437–446.

Bagavandoss, P., Sage, E. H., & Vernon, R. B. (1998). Secreted protein, acidic and rich in cysteine (SPARC) and thrombospondin in the developing follicle and corpus luteum of the rat. *The Journal of Histochemistry and Cytochemistry*, *46*(9), 1043–1049.

Battisti, W. P., Wang, J., Bozek, K., & Murray, M. (1995). Macrophages, microglia, and astrocytes are rapidly activated after crush injury of the goldfish optic nerve: A light and electron microscopic analysis. *The Journal of Comparative Neurology*, *354*(2), 306–320.

Beachley, V. Z., Wolf, M. T., Sadtler, K., Manda, S. S., Jacobs, H., Blatchley, M. R., et al. (2015). Tissue matrix arrays for high-throughput screening and systems analysis of cell function. *Nature Methods*, *12*(12), 1197–1204.

Becker, C. G., Becker, T., Meyer, R. L., & Schachner, M. (1999). Tenascin-R inhibits the growth of optic fibers in vitro but is rapidly eliminated during nerve regeneration in the salamander *Pleurodeles waltl*. *The Journal of Neuroscience*, *19*(2), 813–827.

Becker, T., Becker, C. G., Niemann, U., Naujoks-Manteuffel, C., Bartsch, U., Schachner, M., et al. (1995). Immunohistological localization of tenascin-C in the developing and regenerating retinotectal system of two amphibian species. *The Journal of Comparative Neurology*, *360*(4), 643–657.

Becker, C. G., Schweitzer, J., Feldner, J., Becker, T., & Schachner, M. (2003). Tenascin-R as a repellent guidance molecule for developing optic axons in zebrafish. *The Journal of Neuroscience*, *23*(15), 6232–6237.

Becker, C. G., Schweitzer, J., Feldner, J., Schachner, M., & Becker, T. (2004). Tenascin-R as a repellent guidance molecule for newly growing and regenerating optic axons in adult zebrafish. *Molecular and Cellular Neurosciences*, *26*(3), 376–389.

Bentley, A. A., & Adams, J. C. (2010). The evolution of thrombospondins and their ligand-binding activities. *Molecular Biology and Evolution*, *27*, 2187–2197.

Ben-Zvi, D., Pyrowolakis, G., Barkai, N., & Shilo, B. Z. (2011). Expansion-repression mechanism for scaling the Dpp activation gradient in Drosophila wing imaginal discs. *Current Biology*, *21*(16), 1391–1396.

Bernhardt, R. R., Goerlinger, S., Roos, M., & Schachner, M. (1998). Anterior-posterior subdivision of the somite in embryonic zebrafish: Implications for motor axon guidance. *Developmental Dynamics*, *213*(3), 334–347.

Bertrand, S., Fuentealba, J., Aze, A., Hudson, C., Yasuo, H., Torrejon, M., et al. (2013). A dynamic history of gene duplications and losses characterizes the evolution of the SPARC family in eumetazoans. *Proceedings of the Biological Sciences*, *280*. 2012.2963.

Bier, E., & De Robertis, E. M. (2015). BMP gradients: A paradigm for morphogen-mediated developmental patterning. *Science*, *348*(6242). aaa5838.

Bloch-Zupan, A., Jamet, X., Etard, C., Laugel, V., Muller, J., Geoffroy, V., et al. (2011). Homozygosity mapping and candidate prioritization identify mutations, missed by whole-exome sequencing, in SMOC2, causing major dental developmental defects. *American Journal of Human Genetics*, *89*(6), 773–781.

Bornstein, P. (1995). Diversity of function is inherent in matricellular proteins: An appraisal of thrombospondin 1. *The Journal of Cell Biology*, *130*(3), 503–506.

Bornstein, P. (2000). Matricellular proteins: An overview. *Matrix Biology*, *19*(7), 555–556.

Bornstein, P. (2001). Thrombospondins as matricellular modulators of cell function. *The Journal of Clinical Investigation*, *107*(8), 929–934.

Bradshaw, A. D. (2012). Diverse biological functions of the SPARC family of proteins. *The International Journal of Biochemistry & Cell Biology*, *44*, 480–488.

Brekken, R. A., & Sage, E. H. (2001). SPARC, a matricellular protein: At the crossroads of cell-matrix communication. *Matrix Biology*, *19*(8), 816–827.

Briggs, M. D., Brock, J., Ramsden, S. C., & Bell, P. A. (2014). Genotype to phenotype correlations in cartilage oligomeric matrix protein associated chondrodysplasias. *European Journal of Human Genetics*, *22*(11), 1278–1282.

Briggs, M. D., Hoffman, S. M., King, L. M., Olsen, A. S., Mohrenweiser, H., Leroy, J. G., et al. (1995). Pseudoachondroplasia and multiple epiphyseal dysplasia due to mutations in the cartilage oligomeric matrix protein gene. *Nature Genetics*, *10*(3), 330–336.

Brunet, T., & King, N. (2017). The origin of animal multicellularity and cell differentiation. *Developmental Cell*, *43*(2), 124–140.
Burch, G. H., Gong, Y., Liu, W., Dettman, R. W., Curry, C. J., Smith, L., et al. (1997). Tenascin-X deficiency is associated with Ehlers-Danlos syndrome. *Nature Genetics*, *17*(1), 104–108.
Calve, S., Odelberg, S. J., & Simon, H. G. (2010). A transitional extracellular matrix instructs cell behavior during muscle regeneration. *Developmental Biology*, *344*(1), 259–271.
Calve, S., & Simon, H. G. (2012). Biochemical and mechanical environment cooperatively regulate skeletal muscle regeneration. *The FASEB Journal*, *26*(6), 2538–2545.
Caubit, X., Riou, J. F., Coulon, J., Arsanto, J. P., Benraiss, A., Boucaut, J. C., et al. (1994). Tenascin expression in developing, adult and regenerating caudal spinal cord in the urodele amphibians. *The International Journal of Developmental Biology*, *38*(4), 661–672.
Ceinos, R. M., Torres-Nuñez, E., Chamorro, R., Novoa, B., Figueras, A., Ruane, N. M., et al. (2013). Critical role of the matricellular protein SPARC in mediating erythroid progenitor cell development in zebrafish. *Cells, Tissues, Organs*, *197*(3), 196–208.
Chablais, F., Veit, J., Rainer, G., & Jaźwińska, A. (2011). The zebrafish heart regenerates after cryoinjury-induced myocardial infarction. *BMC Developmental Biology*, *11*, 21.
Chai, Y. M., Yu, S. S., Zhao, X. F., Zhu, Q., & Wang, J. X. (2010). Comparative proteomic profiles of the hepatopancreas in Fenneropenaeus chinensis response to white spot syndrome virus. *Fish & Shellfish Immunology*, *29*(3), 480–486.
Chanana, B., Graf, R., Koledachkina, T., Pflanz, R., & Vorbrüggen, G. (2007). AlphaPS2 integrin-mediated muscle attachment in Drosophila requires the ECM protein thrombospondin. *Mechanisms of Development*, *124*(6), 463–475.
Chang, Y.-H., & Sun, Y. H. (2014). Carrier of Wingless (Cow), a secreted heparan sulfate proteoglycan, promotes extracellular transport of Wingless. *PLoS One*, *9*(10), e111573.
Chen, J. H., & Simmons, C. A. (2011). Cell-matrix interactions in the pathobiology of calcific aortic valve disease: Critical roles for matricellular, matricrine, and matrix mechanics cues. *Circulation Research*, *108*(12), 1510–1524.
Chiou, M. J., Chao, T. T., Wu, J. L., Kuo, C. M., & Chen, J. Y. (2006). The physiological role of CTGF/CCN2 in zebrafish notochond development and biological analysis of the proximal promoter region. *Biochemical and Biophysical Research Communications*, *349*(2), 750–758.
Chiovaro, F., Chiquet-Ehrismann, R., & Chiquet, M. (2015). Transcriptional regulation of tenascin genes. *Cell Adhesion & Migration*, *9*(1–2), 34–47.
Chiquet, M., Vrucinić-Filipi, N., Schenk, S., Beck, K., & Chiquet-Ehrismann, R. (1991). Isolation of chick tenascin variants and fragments. A C-terminal heparin-binding fragment produced by cleavage of the extra domain from the largest subunit splicing variant. *European Journal of Biochemistry*, *199*(2), 379–388.
Chiquet-Ehrismann, R. (1991). Anti-adhesive molecules of the extracellular matrix. *Current Opinion in Cell Biology*, *3*(5), 800–804.
Chiquet-Ehrismann, R., Kalla, P., Pearson, C. A., Beck, K., & Chiquet, M. (1988). Tenascin interferes with fibronectin action. *Cell*, *53*(3), 383–390.
Chiquet-Ehrismann, R., Orend, G., Chiquet, M., Tucker, R. P., & Midwood, K. S. (2014). Tenascins in stem cell niches. *Matrix Biology*, *37*, 112–123.
Chiquet-Ehrismann, R., & Tucker, R. P. (2011). Tenascins and the importance of adhesion modulation. *Cold Spring Harbor Perspectives in Biology*, *3*(5), pii: a004960.
Clark, M. S., Thorne, M. A., Vieira, F. A., Cardoso, J. C., Power, D. M., & Peck, L. S. (2010). Insights into shell deposition in the Antarctic bivalve Laternula elliptica: Gene discovery in the mantle transcriptome using 454 pyrosequencing. *BMC Genomics*, *11*, 362.
Claveria, C., & Torres, M. (2016). Cell competition: Mechanisms and physiological roles. *Annual Review of Cell and Developmental Biology*, *32*, 411–439.

Contreras-Ruiz, L., Ryan, D. S., Sia, R. K., Bower, K. S., Dartt, D. A., & Masli, S. (2014). Polymorphism in THBS1 gene is associated with post-refractive surgery chronic ocular surface inflammation. *Ophthalmology, 121*(7), 1389–1397.

Corsetti, J. P., Ryan, D., Moss, A. J., McCarthy, J., Goldenberg, I., Zareba, W., et al. (2011). Thrombospondin-4 polymorphism (A387P) predicts cardiovascular risk in post-infarction patients with high HDL cholesterol and C-reactive protein levels. *Thrombosis and Haemostasis, 106*(6), 1170–1178.

Crawford, L. A., Guney, M. A., Oh, Y. A., Deyoung, R. A., Valenzuela, D. M., Murphy, A. J., et al. (2009). Connective tissue growth factor (CTGF) inactivation leads to defects in islet cell lineage allocation and beta-cell proliferation during embryogenesis. *Molecular Endocrinology, 23*, 324–336.

Damjanovski, S., Karp, X., Funk, S., Sage, E. H., & Ringuette, M. J. (1997). Ectopic expression of SPARC in Xenopus embryos interferes with tissue morphogenesis: Identification of a bioactive sequence in the C-terminal EF hand. *The Journal of Histochemistry and Cytochemistry, 45*(5), 643–655.

Damjanovski, S., Liu, F., & Ringuette, M. (1992). Molecular analysis of Xenopus laevis SPARC (secreted protein, acidic, rich in cysteine). A highly conserved acidic calcium-binding extracellular-matrix protein. *The Biochemical Journal, 281*(Pt. 2), 513–517.

Damjanovski, S., Malaval, L., & Ringuette, M. J. (1994). Transient expression of SPARC in the dorsal axis of early Xenopus embryos: Correlation with calcium-dependent adhesion and electrical coupling. *The International Journal of Developmental Biology, 38*(3), 439–446.

Dhamija, R., Graham, J. M. J., Smaoui, N., Thorland, E., & Kirmani, S. (2014). Novel de novo SPOCK1 mutation in a proband with developmental delay, microcephaly and agenesis of corpus callosum. *European Journal of Medical Genetics, 57*(4), 181–184.

Dole, N. S., Kapinas, K., Kessler, C. B., Yee, S. P., Adams, D. J., Pereira, R. C., et al. (2015). A single nucleotide polymorphism in osteonectin 3′ untranslated region regulates bone volume and is targeted by miR-433. *Journal of Bone and Mineral Research, 30*(4), 723–732.

Donaldson, D. J., Mahan, J. T., Yang, H., & Crossin, K. L. (1991). Tenascin localization in skin wounds of the adult newt Notophthalmus viridescens. *The Anatomical Record, 230*(4), 451–459.

Echeverri, K., & Tanaka, E. M. (2002). Ectoderm to mesoderm lineage switching during axolotl tail regeneration. *Science, 298*, 1993–1996.

Epperlein, H. H., Halfter, W., & Tucker, R. P. (1988). The distribution of fibronectin and tenascin along migratory pathways of the neural crest in the trunk of amphibian embryos. *Development, 103*(4), 743–756.

Erives, A. J. (2015). Genes conserved in bilaterians but jointly lost with Myc during nematode evolution are enriched in cell proliferation and cell migration functions. *Development Genes and Evolution, 225*(5), 259–273.

Espinoza, J., Sanchez, M., Sanchez, A., Hanna, P., Torrejon, M., Buisine, N., et al. (2010). Two families of Xenopus tropicalis skeletal genes display well-conserved expression patterns with mammals in spite of their highly divergent regulatory regions. *Evolution & Development, 12*(6), 541–551.

Estêvão, M. D., Redruello, B., Canario, A. V., & Power, D. M. (2005). Ontogeny of osteonectin expression in embryos and larvae of sea bream (Sparus auratus). *General and Comparative Endocrinology, 142*(1–2), 155–162.

Faissner, A., Kruse, J., Kühn, K., & Schachner, M. (1990). Binding of the J1 adhesion molecules to extracellular matrix constituents. *Journal of Neurochemistry, 54*(3), 1004–1015.

Farlow, J. L., Robak, L. A., Hetrick, K., Bowling, K., Boerwinkle, E., Coban-Akdemir,- Z. H., et al. (2016). Whole-exome sequencing in familial Parkinson disease. *JAMA Neurology, 73*(1), 68–75.

Fernández, I., Darias, M., Andree, K. B., Mazurais, D., Zambonino-Infante, J. L., & Gisbert, E. (2011). Coordinated gene expression during gilthead sea bream skeletogenesis and its disruption by nutritional hypervitaminosis A. *BMC Developmental Biology*, *11*, 7.

Fernando, C. A., Conrad, P. A., Bartels, C. F., Marques, T., To, M., Balow, S. A., et al. (2010). Temporal and spatial expression of CCN genes in zebrafish. *Developmental Dynamics*, *239*(6), 1755–1767.

Fischer, D., Tucker, R. P., Chiquet-Ehrismann, R., & Adams, J. C. (1997). Cell-adhesive responses to tenascin-C splice variants involve formation of fascin microspikes. *Molecular Biology of the Cell*, *8*(10), 2055–2075.

Fitzgerald, M. C., & Schwarzbauer, J. E. (1998). Importance of the basement membrane protein SPARC for viability and fertility in Caenorhabditis elegans. *Current Biology*, *8*(23), 285–288.

Fonseca, C., Lindahl, G. E., Ponticos, M., Sestini, P., Renzoni, E. A., Holmes, A. M., et al. (2007). A polymorphism in the CTGF promoter region associated with systemic sclerosis. *The New England Journal of Medicine*, *357*(12), 1210–1220.

Forsberg, E., Hirsch, E., Fröhlich, L., Meyer, M., Ekblom, P., Aszodi, A., et al. (1996). Skin wounds and severed nerves heal normally in mice lacking tenascin-C. *Proceedings of the National Academy of Sciences of the United States of America*, *93*(13), 6594–6599.

Garcia de la Serrana, D., Mareco, E. A., La Vieira, V., Power, D. M., & Johnston, I. A. (2016). Comparison of the transcriptional responses of skeletal muscle and bone to a flooding dose of leucine in the gilthead sea bream (Sparus aurata). *Comparative Biochemistry and Physiology Part B, Biochemistry & Molecular Biology*, *199*, 50–57.

Gbadegesin, R. A., Brophy, P. D., Adeyemo, A., Hall, G., Gupta, I. R., Hains, D., et al. (2013). TNXB mutations can cause vesicoureteral reflux. *Journal of the American Society of Nephrology*, *24*(8), 1313–1322.

Geiger, B., & Yamada, K. M. (2011). Molecular architecture and function of matrix adhesions. *Cold Spring Harbor Perspectives in Biology*, *3*(5), pii: a005033.

Giblin, S. P., & Midwood, K. S. (2015). Tenascin-C: Form versus function. *Cell Adhesion & Migration*, *9*(1–2), 48–82.

Gilmour, D. T., Lyon, G. J., Carlton, M. B., Sanes, J. R., Cunningham, J. M., Anderson, J. R., et al. (1998). Mice deficient for the secreted glycoprotein SPARC/osteonectin/BM40 develop normally but show severe age-onset cataract formation and disruption of the lens. *The EMBO Journal*, *17*(7), 1860–1870.

Gilsohn, E., & Volk, T. (2010). Slowdown promotes muscle integrity by modulating integrin-mediated adhesion at the myotendinous junction. *Development*, *137*(5), 785–794.

Giudici, C., Raynal, N., Wiedemann, H., Cabral, W. A., Marini, J. C., Timpl, R., et al. (2008). Mapping of SPARC/BM-40/osteonectin-binding sites on fibrillar collagens. *The Journal of Biological Chemistry*, *283*(28), 19551–19560.

Glasauer, S. M., & Neuhauss, S. C. (2014). Whole-genome duplication in teleost fishes and its evolutionary consequences. *Molecular Genetics and Genomics*, *289*(6), 1045–1060.

Govindan, J., & Iovine, M. K. (2015). Dynamic remodeling of the extracellular matrix during zebrafish fin regeneration. *Gene Expression Patterns*, *19*(1–2), 21–29.

Green, S. A., & Bronner, M. E. (2014). The lamprey: A jawless vertebrate model system for examining origin of the neural crest and other vertebrate traits. *Differentiation*, *87*(1–2), 44–51.

Greenaway, J., Gentry, P. A., Feige, J. J., LaMarre, J., & Petrik, J. J. (2005). Thrombospondin and vascular endothelial growth factor are cyclically expressed in an inverse pattern during bovine ovarian follicle development. *Biology of Reproduction*, *72*(5), 1071–1078.

Greenblatt, Ben-El, R. T., Hassan, A., & Salzberg, A. (2017). Loss of thrombospondin reveals a possible role for the extracellular matrix in chordotonal cap cell elongation. *The International Journal of Developmental Biology*, *61*(3-4-5), 311–318.

Gustafsson, E., & Fässler, R. (2000). Insights into extracellular matrix functions from mutant mouse models. *Experimental Cell Research, 261*(1), 52–68.
Halfter, W., Chiquet-Ehrismann, R., & Tucker, R. P. (1989). The effect of tenascin and embryonic basal lamina on the behavior and morphology of neural crest cells in vitro. *Developmental Biology, 132*(1), 14–25.
Hall-Glenn, F., & Lyons, K. M. (2011). Roles for CCN2 in normal physiological processes. *Cellular and Molecular Life Sciences, 68*(19), 3209–3217.
Hamaratoglu, F., de Lachapelle, A. M., Pyrowolakis, G., Bergmann, S., & Affolter, M. (2011). Dpp signaling activity requires Pentagone to scale with tissue size in the growing Drosophila wing imaginal disc. *PLoS Biology, 9*(10), e1001182.
Hartley, P. S., Motamedchaboki, K., Bodmer, R., & Ocorr, K. (2016). SPARC-dependent cardiomyopathy in Drosophila. *Circulation. Cardiovascular Genetics, 9*(2), 119–129.
Hecht, J. T., Nelson, L. D., Crowder, E., Wang, Y., Elder, F. F., Harrison, W. R., et al. (1995). Mutations in exon 17B of cartilage oligomeric matrix protein (COMP) cause pseudoachondroplasia. *Nature Genetics, 10*(3), 325–329.
Hellewell, A. L., & Adams, J. C. (2016). Insider trading: Extracellular matrix proteins and their non-canonical intracellular roles. *BioEssays, 38*, 77–88.
Hemmrich, G., & Bosch, T. C. G. (2008). Compagen, a comparative genomics platform for early branching metazoan animals, reveals early origins of genes regulating stem-cell differentiation. *BioEssays, 30*, 1010–1018.
Hendaoui, I., Tucker, R. P., Zingg, D., Bichet, S., Schittny, J., & Chiquet-Ehrismann, R. (2014). Tenascin-C is required for normal Wnt/β-catenin signaling in the whisker follicle stem cell niche. *Matrix Biology, 40*, 46–53.
Hirayama, M., Ahsan, M. N., Mitani, H., & Watabe, S. (2008). CYR61 is a novel gene associated with temperature-dependent changes in fish metabolism as revealed by cDNA microarray analysis on a medaka Oryzias latipes cell line. *Journal of Cellular Biochemistry, 104*(4), 1297–1310.
Hirose, Y., Chiba, K., Karasugi, T., Nakajima, M., Kawaguchi, Y., Mikami, Y., et al. (2008). A functional polymorphism in THBS2 that affects alternative splicing and MMP binding is associated with lumbar-disc herniation. *American Journal of Human Genetics, 82*(5), 1122–1129.
Hoffman, J. R., Dixit, V. M., & O'Shea, K. S. (1994). Expression of thrombospondin in the adult nervous system. *The Journal of Comparative Neurology, 340*(1), 126–139.
Hoffman, J. R., & O'Shea, K. S. (1999). Thrombospondin expression in nerve regeneration II. Comparison of optic nerve crush in the mouse and goldfish. *Brain Research Bulletin, 48*(4), 421–427.
Hong, Y. J., Choi, Y. W., Myung, K. B., & Choi, H. Y. (2011). The immunohistochemical patterns of calcification-related molecules in the epidermis and dermis of the zebrafish (Danio rerio). *Annals of Dermatology, 23*(3), 299–303.
Hosono, C., Matsuda, R., Adryan, B., & Samakovlis, C. (2015). Transient junction anisotropies orient annular cell polarization in the Drosophila airway tubes. *Nature Cell Biology, 17*(12), 1569–1576.
Hurvitz, J. R., Suwairi, W. M., Van Hul, W., El-Shanti, H., Superti-Furga, A., Roudier, J., et al. (1999). Mutations in the CCN gene family member WISP3 cause progressive pseudorheumatoid dysplasia. *Nature Genetics, 23*(1), 94–98.
Hutchenreuther, J., Leask, A., & Thompson, K. (2017). Studying the CCN proteins in fibrosis. *Methods in Molecular Biology, 1489*, 423–429.
Huynh, M. H., Hong, H., Delovitch, S., Desser, S., & Ringuette, M. (2000). Association of SPARC (osteonectin, BM-40) with extracellular and intracellular components of the ciliated surface ectoderm of Xenopus embryos. *Cell Motility and the Cytoskeleton, 47*(2), 154–162.
Huynh, M. H., Sage, E. H., & Ringuette, M. (1999). A calcium-binding motif in SPARC/osteonectin inhibits chordomesoderm cell migration during Xenopus laevis gastrulation:

Evidence of counter-adhesive activity in vivo. *Development, Growth & Differentiation, 41*(4), 407–418.
Huynh, M. H., Zhu, S. J., Kollara, A., Brown, T., Winklbauer, R., & Ringuette, M. (2011). Knockdown of SPARC leads to decreased cell-cell adhesion and lens cataracts during post-gastrula development in Xenopus laevis. *Development Genes and Evolution, 220*(11–12), 315–327.
Hynes, R. O. (1994). Genetic analyses of cell-matrix interactions in development. *Current Opinion in Genetics & Development, 4*(4), 569–574.
Hynes, R. O. (2014). Stretching the boundaries of extracellular matrix research. *Nature Reviews. Molecular Cell Biology, 15*(12), 761–763.
Iimura, K., Tohse, H., Ura, K., & Takagi, Y. (2012). Expression patterns of runx2, sparc, and bgp during scale regeneration in the goldfish Carassius auratus. *Journal of Experimental Zoology. Part B, Molecular and Developmental Evolution, 318*(3), 190–198.
Irles, P., Ramos, S., & Piulachs, M. D. (2017). SPARC preserves follicular epithelium integrity in insect ovaries. *Developmental Biology, 422*(2), 105–114.
Iruela-Arispe, M. L., Liska, D. J., Sage, E. H., & Bornstein, P. (1993). Differential expression of thrombospondin 1, 2, and 3 during murine development. *Developmental Dynamics, 197*(1), 40–56.
Isabella, A. J., & Horne-Badovinac, S. (2015). Dynamic regulation of basement membrane protein levels promotes egg chamber elongation in Drosophila. *Developmental Biology, 406*(2), 212–221.
Ishida, Y., Suzuki, K., Utoh, R., Obara, M., & Yoshizato, K. (2003). Molecular identification of the skin transformation center of anuran larval skin using genes of Rana adult keratin (RAK) and SPARC as probes. *Development, Growth & Differentiation, 45*(5–6), 515–526.
Ivkovic, S., Yoon, B. S., Popoff, S. N., Safadi, F. F., Libuda, D. E., Stephenson, R. C., et al. (2003). Connective tissue growth factor coordinates chondrogenesis and angiogenesis during skeletal development. *Development, 130*, 2779–2791.
Ivy, J. R., Drechsler, M., Catterson, J. H., Bodmer, R., Ocorr, K., Paululat, A., et al. (2015). Klf15 is critical for the development and differentiation of Drosophila nephrocytes. *PLoS One, 10*, e0134620.
Jansen, K. A., Atherton, P., & Ballestrem, C. (2017). Mechanotransduction at the cell-matrix interface. *Seminars in Cell & Developmental Biology, 71*, 75–83.
José-Edwards, D. S., Oda-Ishii, I., Nibu, Y., & Di Gregorio, A. (2013). Tbx2/3 is an essential mediator within the Brachyury gene network during Ciona notochord development. *Development, 140*, 2422–2433.
Joubert, C., Piquemal, D., Marie, B., Manchon, L., Pierrat, F., Zanella-Cléon, I., et al. (2010). Transcriptome and proteome analysis of Pinctada margaritifera calcifying mantle and shell: Focus on biomineralization. *BMC Genomics, 11*, 613.
Kammerer, R. A., Schulthess, T., Landwehr, R., Lustig, A., Fischer, D., & Engel, J. (1998). Tenascin-C hexabrachion assembly is a sequential two-step process initiated by coiled-coil alpha-helices. *The Journal of Biological Chemistry, 273*, 10602–10608.
Kaneko, T., Freeha, K., Wu, X., Mogi, M., Uji, S., Yokoi, H., et al. (2016). Role of notochord cells and sclerotome-derived cells in vertebral column development in fugu, Takifugu rubripes: Histological and gene expression analyses. *Cell and Tissue Research, 366*(1), 37–49.
Kaneto, S., & Wada, H. (2011). Regeneration of amphioxus oral cirri and its skeletal rods: Implications for the origin of the vertebrate skeleton. *Journal of Experimental Zoology. Part B, Molecular and Developmental Evolution, 316*(6), 409–417.
Kang, Y. J., Stevenson, A. K., Yau, P. M., & Kollmar, R. (2008). Sparc protein is required for normal growth of zebrafish otoliths. *Journal of the Association for Research in Otolaryngology, 9*(4), 436–451.

Kasprick, D. S., Kish, P. E., Junttila, T. L., Ward, L. A., Bohnsack, B. L., & Kahana, A. (2011). Microanatomy of adult zebrafish extraocular muscles. *PLoS One*, *6*(11), e27095.
Kawasaki, K., Suzuki, T., & Weiss, K. M. (2004). Genetic basis for the evolution of vertebrate mineralized tissue. *Proceedings of the National Academy of Sciences of the United States of America*, *101*(31), 11356–11361.
Kim, D. J., Christofidou, E. D., Keene, D. R., Milde, H. M., & Adams, J. C. (2015). Intermolecular interactions of thrombospondins drive their accumulation in extracellular matrix. *Molecular Biology of the Cell*, *26*, 2640–2654. + co-first authors.
Kim, S. Y., Ondhia, N., Vidgen, D., Malaval, L., Ringuette, M., & Kalnins, V. I. (1997). Spatiotemporal distribution of SPARC/osteonectin in developing and mature chicken retina. *Experimental Eye Research*, *65*(5), 681–689.
Klenotic, P. A., Zhang, C., & Lin, Z. (2016). Emerging roles of CCN proteins in vascular development and pathology. *Journal of Cell Communication and Signaling*, *10*(3), 251–257.
Kocot, K. M., Aguilera, F., McDougall, C., Jackson, D. J., & Degnan, B. M. (2016). Sea shell diversity and rapidly evolving secretomes: Insights into the evolution of biomineralization. *Frontiers in Zoology*, *13*, 23.
Koehler, A., Desser, S., Chang, B., MacDonald, J., Tepass, U., & Ringuette, M. (2009). Molecular evolution of SPARC: Absence of the acidicmodule and expression in the endodermof the starlet sea anemone, Nematostella vectensis. *Development Genes and Evolution*, *219*, 509–521.
Krupska, I., Bruford, E. A., & Chaqour, B. (2015). Eyeing the Cyr61/CTGF/NOV (CCN) group of genes in development and diseases: Highlights of their structural likenesses and functional dissimilarities. *Human Genomics*, *9*, 24.
Kuo, T. H., Kowalko, J. E., DiTommaso, T., Nyambi, M., Montoro, D. T., Essner, J. J., et al. (2015). TALEN-mediated gene editing of the thrombospondin-1 locus in axolotl. *Regeneration (Oxford, England)*, *2*(1), 37–43.
Kyriakides, T. R., Zhu, Y. H., Smith, L. T., Bain, S. D., Yang, Z., Lin, M. T., et al. (1998). Mice that lack thrombospondin 2 display connective tissue abnormalities that are associated with disordered collagen fibrillogenesis, an increased vascular density, and a bleeding diathesis. *The Journal of Cell Biology*, *140*(2), 419–430.
Kzhyshkowska, J., Workman, G., Cardó-Vila, M., Arap, W., Pasqualini, R., Gratchev, A., et al. (2006). Novel function of alternatively activated macrophages: Stabilin-1-mediated clearance of SPARC. *Journal of Immunology*, *176*(10), 5825–5832.
Lambi, A. G., Pankratz, T. L., Mundy, C., Gannon, M., Barbe, M. F., Richtsmeier, J. T., et al. (2012). The skeletal site-specific role of connective tissue growth factor in prenatal osteogenesis. *Developmental Dynamics*, *241*, 1944–1959.
Lane, T. F., Iruela-Arispe, M. L., Johnson, R. S., & Sage, E. H. (1994). SPARC is a source of copper-binding peptides that stimulate angiogenesis. *The Journal of Cell Biology*, *125*(4), 929–943.
Lane, T. F., & Sage, E. H. (1994). The biology of SPARC, a protein that modulates cell-matrix interactions. *The FASEB Journal*, *8*(2), 163–173.
Lang, D.M., Monzon-Mayor, M., Del Mar Romero-Aleman, M., Yanes, C., Santos, E., Pesheva, P. (2008). Tenascin-R and axon growth-promoting molecules are up-regulated in the regenerating visual pathway of the lizard (Gallotia galloti). Developmental Neurobiology 68(7), 899–16.
Latinkic, B. V., Mercurio, S., Bennett, B., Hirst, E. M., Xu, Q., Lau, L. F., et al. (2003). Xenopus Cyr61 regulates gastrulation movements and modulates Wnt signalling. *Development*, *130*(11), 2429–2441.
Lau, L. F. (2016). Cell surface receptors for CCN proteins. *Journal of Cell Communication and Signaling*, *10*(2), 121–127.
Lawler, J., Duquette, M., Whittaker, C. A., Adams, J. C., McHenry, K., & DeSimone, D. W. (1993). Identification and characterization of thrombospondin-4, a

new member of the thrombospondin gene family. *The Journal of Cell Biology*, *120*(4), 1059–1067.

Lawler, J., Sunday, M., Thibert, V., Duquette, M., George, E. L., Rayburn, H., et al. (1998). Thrombospondin-1 is required for normal murine pulmonary homeostasis and its absence causes pneumonia. *The Journal of Clinical Investigation*, *101*(5), 982–992.

Lawler, J., Weinstein, R., & Hynes, R. O. (1988). Cell attachment to thrombospondin: The role of ARG-GLY-ASP, calcium, and integrin receptors. *The Journal of Cell Biology*, *107*(6 Pt. 1), 2351–2361.

Leask, A. (2013). CCN2: A novel, specific and valid target for anti-fibrotic drug intervention. *Expert Opinion on Therapeutic Targets*, *17*(9), 1067–1071.

Lehane, D. B., McKie, N., Russell, R. G., & Henderson, I. W. (1999). Cloning of a fragment of the osteonectin gene from goldfish, Carassius auratus: Its expression and potential regulation by estrogen. *General and Comparative Endocrinology*, *114*, 80–87.

Li, Y., & Tower, J. (2009). Adult-specific over-expression of the Drosophila genes magu and hebe increases life span and modulates late-age female fecundity. *Molecular Genetics and Genomics*, *281*(2), 147–162.

Liao, E. Y., Peng, Y. Q., Zhou, H. D., Mackie, E. J., Li, J., Hu, P. A., et al. (2004a). Gene symbol: WISP3. Disease: Spondyloepiphyseal dysplasia tarda with progressive arthropathy. *Human Genetics*, *115*(2), 169.

Liao, E. Y., Peng, Y. Q., Zhou, H. D., Mackie, E. J., Li, J., Hu, P. A., et al. (2004b). Gene symbol: WISP3. Disease: Spondyloepiphyseal dysplasia tarda with progressive arthropathy. *Human Genetics*, *115*(2), 174.

Lillis, A. P., Van Duyn, L. B., Murphy-Ullrich, J. E., & Strickland, D. K. (2008). LDL receptor-related protein 1: Unique tissue-specific functions revealed by selective gene knockout studies. *Physiological Reviews*, *88*(3), 887–918.

Lin, C. Y., Huang, C. C., Wang, W. D., Hsiao, C. D., Cheng, C. F., Wu, Y. T., et al. (2013). Low temperature mitigates cardia bifida in zebrafish embryos. *PLoS One*, *8*(7), e69788.

Looso, M., Michel, C. S., Konzer, A., Bruckskotten, M., Borchardt, T., Krüger, M., et al. (2012). Spiked-in pulsed in vivo labeling identifies a new member of the CCN family in regenerating newt hearts. *Journal of Proteome Research*, *11*(9), 4693–4704.

López-Murcia, F. J., Terni, B., & Llobet, A. (2015). SPARC triggers a cell-autonomous program of synapse elimination. *Proceedings of the National Academy of Sciences of the United States of America*, *112*(43), 13366–13371.

Lowy, C. M., & Oskarsson, T. (2015). Tenascin C in metastasis: A view from the invasive front. *Cell Adhesion & Migration*, *9*(1–2), 112–124.

Mackie, E. J., Tucker, R. P., Halfter, W., Chiquet-Ehrismann, R., & Epperlein, H. H. (1988). The distribution of tenascin coincides with pathways of neural crest cell migration. *Development*, *102*(1), 237–250.

Magerd, S., Asuvapongpatana, S., Vanichviriyakit, R., Chotwiwatthanakun, C., & Weerachatyanukul, W. (2013). Characterization of the thrombospondin (TSP)-II gene in Penaeus monodon and a novel role of TSP-like proteins in an induction of shrimp sperm acrosome reaction. *Molecular Reproduction and Development*, *80*(5), 393–402.

Maloney, J. P., Stearman, R. S., Bull, T. M., Calabrese, D. W., Tripp-Addison, M. L., Wick, M. J., et al. (2012). Loss-of-function thrombospondin-1 mutations in familial pulmonary hypertension. *American Journal of Physiology. Lung Cellular and Molecular Physiology*, *302*(6), L541–54.

Martin, E. E., Huang, W., Anwar, T., Arellano-Garcia, C., Burman, B., Guan, J. L., et al. (2017). MMTV-cre;Ccn6 knockout mice develop tumors recapitulating human metaplastic breast carcinomas. *Oncogene*, *36*(16), 2275–2285.

Martinek, N., Shahab, J., Saathoff, M., & Ringuette, M. (2008). Haemocyte-derived SPARC is required for collagen-IV-dependent stability of basal laminae in Drosophila embryos. *Journal of Cell Science*, *121*(Pt. 10), 1671–1680.

Martinek, N., Zou, R., Berg, M., Sodek, J., & Ringuette, M. (2002). Evolutionary conservation and association of SPARC with the basal lamina in Drosophila. *Development Genes and Evolution*, *212*(3), 124–133.

Mason, R. M. (2013). Fell-Muir lecture: Connective tissue growth factor (CCN2)—A pernicious and pleiotropic player in the development of kidney fibrosis. *International Journal of Experimental Pathology*, *94*(1), 1–16.

Mason, I. J., Murphy, D., Münke, M., Francke, U., Elliott, R. W., & Hogan, B. L. (1986). Developmental and transformation-sensitive expression of the Sparc gene on mouse chromosome 11. *The EMBO Journal*, *5*(8), 1831–1837.

Mateus, R., Pereira, T., Sousa, S., de Lima, J. E., Pascoal, S., Saúde, L., et al. (2012). In vivo cell and tissue dynamics underlying zebrafish fin fold regeneration. *PLoS One*, 7(12), e51766.

Maurer, P., Sasaki, T., Mann, K., Göhring, W., Schwarzbauer, J. E., & Timpl, R. (1997). Structural and functional characterization of the extracellular calcium-binding protein BM-40/secreted protein, acidic, rich in cysteine/osteonectin from the nematode Caenorhabditis elegans. *European Journal of Biochemistry*, *248*(1), 209–216.

McCarthy, J. J., Parker, A., Salem, R., Moliterno, D. J., Wang, Q., Plow, E. F., et al. (2004). Large scale association analysis for identification of genes underlying premature coronary heart disease: Cumulative perspective from analysis of 111 candidate genes. *Journal of Medical Genetics*, *41*(5), 334–341.

McCusker, C., Bryant, S. V., & Gardiner, D. M. (2015). The axolotl limb blastema: Cellular and molecular mechanisms driving blastema formation and limb regeneration in tetrapods. *Regeneration (Oxf)*. *2*(2), 54–71.

McKenzie, P., Chadalavada, S. C., Bohrer, J., & Adams, J. C. (2006). Phylogenomic analysis of vertebrate thrombospondins reveals fish-specific paralogues, ancestral gene relationships and a tetrapod innovation. *BMC Evolutionary Biology*, *6*, 33.

McNeill, B., Vulesevic, B., Ostojic, A., Ruel, M., & Suuronen, E. J. (2015). Collagen matrix-induced expression of integrin αVβ3 in circulating angiogenic cells can be targeted by matricellular protein CCN1 to enhance their function. *The FASEB Journal*, *29*(4), 1198–1207.

Mège, R. M., Nicolet, M., Pinçon-Raymond, M., Murawsky, M., & Rieger, F. (1992). Cytotactin is involved in synaptogenesis during regeneration of the frog neuromuscular system. *Developmental Biology*, *149*(2), 381–394.

Melvin, V. S., Feng, W., Hernandez-Lagunas, L., Artinger, K. B., & Williams, T. (2013). A morpholino-based screen to identify novel genes involved in craniofacial morphogenesis. *Developmental Dynamics*, *242*, 817–831.

Mendoza-Londono, R., Fahiminiya, S., Majewski, J., Care4Rare Canada Consortium, Tétreault, M., Nadaf, J., et al. (2015). Recessive osteogenesis imperfecta caused by missense mutations in SPARC. *American Journal of Human Genetics*, *96*(6), 979–985.

Mercer, S. E., Odelberg, S. J., & Simon, H. G. (2013). A dynamic spatiotemporal extracellular matrix facilitates epicardial-mediated vertebrate heart regeneration. *Developmental Biology*, *382*(2), 457–469.

Merkes, C., Turkalo, T. K., Wilder, N., Park, H., Wenger, L. W., Lewin, S. J., et al. (2015). Ewing sarcoma ewsa protein regulates chondrogenesis of Meckel's cartilage through modulation of Sox9 in zebrafish. *PLoS One*, *10*(1), e0116627.

Midwood, K. S., Chiquet, M., Tucker, R. P., & Orend, G. (2016). Tenascin-C at a glance. *Journal of Cell Science*, *129*(23), 4321–4327.

Minear, M. A., Crosslin, D. R., Sutton, B. S., Connelly, J. J., Nelson, S. C., Gadson-Watson, S., et al. (2011). Polymorphic variants in tenascin-C (TNC) are associated with atherosclerosis and coronary artery disease. *Human Genetics*, *129*(6), 641–654.

Mitashov, V. I., Arsanto, J. P., Markitantova, Y. V., & Thouveny, Y. (1995). Remodeling processes during neural retinal regeneration in adult urodeles: An immunohistochemical survey. *The International Journal of Developmental Biology*, *39*(6), 993–1003.

Miyamoto, H., Endo, H., Hashimoto, N., Limura, K., Isowa, Y., Kinoshita, S., et al. (2013). The diversity of shell matrix proteins: Genome-wide investigation of the pearl oyster, Pinctada fucata. *Zoological Science*, *30*(10), 801–816.

Mo, F. E., Muntean, A. G., Chen, C. C., Stolz, D. B., Watkins, S. C., & Lau, L. F. (2002). CYR61 (CCN1) is essential for placental development and vascular integrity. *Molecular and Cellular Biology*, *22*(24), 8709–8720.

Mokalled, M. H., Patra, C., Dickson, A. L., Endo, T., Stainier, D. Y., & Poss, K. D. (2016). Injury-induced ctgfa directs glial bridging and spinal cord regeneration in zebrafish. *Science*, *354*(6312), 630–634.

Mommaerts, H., Esguerra, C. V., Hartmann, U., Luyten, F. P., & Tylzanowski, P. (2014). Smoc2 modulates embryonic myelopoiesis during zebrafish development. *Developmental Dynamics*, *243*(11), 1375–1390.

Morris, A. H., & Kyriakides, T. R. (2014). Matricellular proteins and biomaterials. *Matrix Biology*, *37*, 183–191.

Morrissey, M. A., Jayadev, R., Miley, G. R., Blebea, C. A., Chi, Q., Ihara, S., et al. (2016). SPARC promotes cell invasion in vivo by decreasing type IV collagen levels in the basement membrane. *PLoS Genetics*, *12*(2), e1005905.

Mosher, D. F., & Adams, J. C. (2012). Adhesion-modulating/ matricellular ECM protein families: A structural, functional and evolutionary appraisal. *Matrix Biology*, *31*, 155–161.

Murphy-Ullrich, J. E., & Höök, M. (1989). Thrombospondin modulates focal adhesions in endothelial cells. *The Journal of Cell Biology*, *109*(3), 1309–1319.

Murphy-Ullrich, J.E., Lane, T.F., Pallero, M,A., Sage, E.H. (1995). SPARC mediates focal adhesion disassembly in endothelial cells through a follistatin-like region and the Ca(2+)-binding EF-hand. Journal of Cellular Biochemistry 57(2), 341–50.

Murphy-Ullrich, J. E., Lightner, V. A., Aukhil, I., Yan, Y. Z., Erickson, H. P., & Höök, M. (1991). Focal adhesion integrity is downregulated by the alternatively spliced domain of human tenascin. *The Journal of Cell Biology*, *115*(4), 1127–1136.

Murphy-Ullrich, J. E., & Sage, E. H. (2014). Revisiting the matricellular concept. *Matrix Biology*, *37*, 1–14.

Myers, R. B., Rwayitare, K., Richey, L., Lem, J., & Castellot, J. J., Jr. (2012). CCN5 expression in mammals. III. Early embryonic mouse development. *Journal of Cell Communication and Signaling*, *6*(4), 217–223.

Naba, A., Clauser, K. R., Hoersch, S., Liu, H., Carr, S. A., & Hynes, R. O. (2012). The matrisome: In silico definition and in vivo characterization by proteomics of normal and tumor extracellular matrices. *Molecular & Cellular Proteomics*, *11*(4). M111.014647.

Naba, A., Clauser, K. R., Lamar, J. M., Carr, S. A., & Hynes, R. O. (2014). Extracellular matrix signatures of human mammary carcinoma identify novel metastasis promoters. *eLife*, *3*, e01308.

Naba, A., Pearce, O. M. T., Del Rosario, A., Ma, D., Ding, H., Rajeeve, V., et al. (2017). Characterization of the extracellular matrix of normal and diseased tissues using proteomics. *Journal of Proteome Research*, *16*(8), 3083–3091.

Nagasawa-Masuda, A., & Terai, K. (2017). Yap/Taz transcriptional activity is essential for vascular regression via Ctgf expression and actin polymerization. *PLoS One*, *12*(4), e0174633.

Nakamura, Y., Weidinger, G., Liang, J. O., Aquilina-Beck, A., Tamai, K., Moon, R. T., et al. (2007). The CCN family member Wisp3, mutant in progressive pseudorheumatoid dysplasia, modulates BMP and Wnt signaling. *The Journal of Clinical Investigation*, *117*(10), 3075–3086.

Neely, G. G., Hess, A., Costigan, M., Keene, A. C., Goulas, S., Langeslag, M., et al. (2010). A genome-wide Drosophila screen for heat nociception identifies α2δ3 as an evolutionarily conserved pain gene. *Cell*, *143*(4), 628–638.

Neely, G. G., Kuba, K., Cammarato, A., Isobe, K., Amann, S., Zhang, L., et al. (2010). A global in vivo Drosophila RNAi screen identifies NOT3 as a conserved regulator of heart function. *Cell*, *141*(1), 142–153.
Nemoto, Y., Chatani, M., Inohaya, K., Hiraki, Y., & Kudo, A. (2008). Expression of marker genes during otolith development in medaka. *Gene Expression Patterns*, *8*(2), 92–95.
Neumuller, R. A., Richter, C., Fischer, A., Novatchkova, M., Neumuller, K. G., & Knoblich, J. A. (2011). Genome-wide analysis of self-renewal in Drosophila neural stem cells by transgenic RNAi. *Cell Stem Cell*, *8*, 580–593.
Norman, M., Vuilleumier, R., Springhorn, A., Gawlik, J., & Pyrowolakis, G. (2016). Pentagone internalises glypicans to fine-tune multiple signalling pathways. *eLife 5*, e13301. https://doi.org/10.7554/eLife.13301.
Norose, K., Clark, J. I., Syed, N. A., Basu, A., Heber-Katz, E., Sage, E. H., et al. (1998). SPARC deficiency leads to early-onset cataractogenesis. *Investigative Ophthalmology & Visual Science*, *39*(13), 2674–2680.
Nudelman, F. (2015). Nacre biomineralisation: A review on the mechanisms of crystal nucleation. *Seminars in Cell & Developmental Biology*, *46*, 2–10.
O'Neill, L. A., Golenbock, D., & Bowie, A. G. (2013). The history of toll-like receptors—Redefining innate immunity. *Nature Reviews Immunology*, *13*, 453–460.
Okada, I., Hamanoue, H., Terada, K., Tohma, T., Megarbane, A., Chouery, E., et al. (2011). SMOC1 is essential for ocular and limb development in humans and mice. *American Journal of Human Genetics*, *88*(1), 30–41.
Okumura, T., Kim, Y. K., Kawazoe, I., Yamano, K., Tsutsui, N., & Aida, K. (2006). Expression of vitellogenin and cortical rod proteins during induced ovarian development by eyestalk ablation in the kuruma prawn, Marsupenaeus japonicus. *Comparative Biochemistry and Physiology. Part A, Molecular & Integrative Physiology*, *143*(2), 246–253.
Onda, H., Goldhamer, D. J., & Tassava, R. A. (1990). An extracellular matrix molecule of newt and axolotl regenerating limb blastemas and embryonic limb buds: Immunological relationship of MT1 antigen with tenascin. *Development*, *108*(4), 657–668.
Onda, H., Poulin, M. L., Tassava, R. A., & Chiu, I. M. (1991). Characterization of a newt tenascin cDNA and localization of tenascin mRNA during newt limb regeneration by in situ hybridization. *Developmental Biology*, *148*(1), 219–232.
Otey, C. A., & Burridge, K. (1990). Patterning of the membrane cytoskeleton by the extracellular matrix. *Seminars in Cell Biology*, *1*(5), 391–399.
Özbek, S., Balasubramainian, P., Chiquet-Ehrismann, R., Tucker, R. P., & Adams, J. C. (2010). The evolution of extracellular matrix. *Molecular Biology of the Cell*, *21*, 4300–4306.
Pacifici, M., Oshima, O., Fisher, L. W., Young, M. F., Shapiro, I. M., & Leboy, P. S. (1990). Changes in osteonectin distribution and levels are associated with mineralization of the chicken tibial growth cartilage. *Calcified Tissue International*, *47*(1), 51–61.
Parichy, D. M. (1996). When neural crest and placodes collide: Interactions between melanophores and the lateral lines that generate stripes in the salamander Ambystoma tigrinum tigrinum (Ambystomatidae). *Developmental Biology*, *175*(2), 283–300.
Parichy, D. M. (2001). Homology and evolutionary novelty in the deployment of extracellular matrix molecules during pigment pattern formation in the salamanders Taricha torosa and T. rivularis (Salamandridae). *The Journal of Experimental Zoology*, *291*(1), 13–24.
Pashay Ahi, E., Walker, B. S., Lassiter, C. S., & Jónsson, Z. O. (2016). Investigation of the effects of estrogen on skeletal gene expression during zebrafish larval head development. *PeerJ*, *4*, e1878.
Payne, S. L., Peacock, H. M., & Vickaryous, M. K. (2017). Blood vessel formation during tail regeneration in the leopard gecko (Eublepharis macularius): The blastema is not avascular. *Journal of Morphology*, *278*(3), 380–389.

Pazin, D. E., & Albrecht, K. H. (2009). Developmental expression of Smoc1 and Smoc2 suggests potential roles in fetal gonad and reproductive tract differentiation. *Developmental Dynamics*, *238*(11), 2877–2890.

Peacock, H. M., Gilbert, E. A., & Vickaryous, M. K. (2015). Scar-free cutaneous wound healing in the leopard gecko, Eublepharis macularius. *Journal of Anatomy*, *227*(5), 596–610.

Pérez-Moreno, J. J., Espina-Zambrano, A. G., García-Calderón, C. B., & Estrada, B. (2017). Kon-tiki enhances PS2 integrin adhesion and localizes its ligand, thrombospondin, in the myotendinous junction. *Journal of Cell Science*, *130*(5), 950–962.

Perrimon, N., Pitsouli, C., & Shilo, B. Z. (2012). Signaling mechanisms controlling cell fate and embryonic patterning. *Cold Spring Harbor Perspectives in Biology*, *4*(8), a005975.

Peterson, A. J., & O'Connor, M. B. (2013). Activin receptor inhibition by Smad2 regulates Drosophila wing disc patterning through BMP-response elements. *Development*, *140*(3), 649–659.

Pocha, S. M., & Montell, D. J. (2014). Cellular and molecular mechanisms of single and collective cell migrations in Drosophila: Themes and variations. *Annual Review of Genetics*, *48*, 295–318.

Pongsomboon, S., Wongpanya, R., Tang, S., Chalorsrikul, A., & Tassanakajon, A. (2008). Abundantly expressed transcripts in the lymphoid organ of the black tiger shrimp, Penaeus monodon, and their implication in immune function. *Fish & Shellfish Immunology*, *25*(5), 485–493.

Portela, M., Casas-Tinto, S., Rhiner, C., López-Gay, J. M., Domínguez, O., Soldini, D., et al. (2010). Drosophila SPARC is a self-protective signal expressed by loser cells during cell competition. *Developmental Cell*, *19*(4), 562–573.

Preechaphol, R., Leelatanawit, R., Sittikankeaw, K., Klinbunga, S., Khamnamtong, B., Puanglarp, N., et al. (2007). Expressed sequence tag analysis for identification and characterization of sex-related genes in the giant tiger shrimp Penaeus monodon. *Journal of Biochemistry and Molecular Biology*, *40*(4), 501–510.

Purcell, L., Gruia-Gray, J., Scanga, S., & Ringuette, M. (1993). Developmental anomalies of Xenopus embryos following microinjection of SPARC antibodies. *The Journal of Experimental Zoology*, *265*(2), 153–164.

Raines, E. W., Lane, T. F., Iruela-Arispe, M. L., Ross, R., & Sage, E. H. (1992). The extracellular glycoprotein SPARC interacts with platelet-derived growth factor (PDGF)-AB and -BB and inhibits the binding of PDGF to its receptors. *Proceedings of the National Academy of Sciences of the United States of America*, *89*(4), 1281–1285.

Rainger, J., van Beusekom, E., Ramsay, J. K., McKie, L., Al-Gazali, L., Pallotta, R., et al. (2011). Loss of the BMP antagonist, SMOC-1, causes Ophthalmo-acromelic (Waardenburg Anophthalmia) syndrome in humans and mice. *PLoS Genetics*, *7*(7), e1002114.

Reinhard, J., Roll, L., & Faissner, A. (2017). Tenascins in retinal and optic nerve neurodegeneration. *Frontiers in Integrative Neuroscience*, *11*, 30.

Renn, J., Schaedel, M., Volff, J. N., Goerlich, R., Schartl, M., & Winkler, C. (2006). Dynamic expression of sparc precedes formation of skeletal elements in the Medaka (Oryzias latipes). *Gene*, *372*, 208–218.

Rentz, T. J., Poobalarahi, F., Bornstein, P., Sage, E. H., & Bradshaw, A. D. (2007). SPARC regulates processing of procollagen I and collagen fibrillogenesis in dermal fibroblasts. *The Journal of Biological Chemistry*, *282*(30), 22062–22071.

Rienks, M., & Papageorgiou, A. P. (2016). Novel regulators of cardiac inflammation: Matricellular proteins expand their repertoire. *Journal of Molecular and Cellular Cardiology*, *91*, 172–178.

Ringuette, M., Damjanovski, S., & Wheeler, D. (1991). Expression of SPARC/osteonectin in tissues of bony and cartilaginous vertebrates. *Biochemistry and Cell Biology*, *69*(4), 245–250.

Riou, J. F., Shi, D. L., Chiquet, M., & Boucaut, J. C. (1988). Expression of tenascin in response to neural induction in amphibian embryos. *Development*, *104*, 511–524.
Riou, J. F., Shi, D. L., Chiquet, M., & Boucaut, J. C. (1990). Exogenous tenascin inhibits mesodermal cell migration during amphibian gastrulation. *Developmental Biology*, *137*(2), 305–317.
Risher, W. C., & Eroglu, C. (2012). Thrombospondins as key regulators of synaptogenesis in the central nervous system. *Matrix Biology*, *31*(3), 170–177.
Röll, S., Seul, J., Paulsson, M., & Hartmann, U. (2006). Testican-1 is dispensable for mouse development. *Matrix Biology*, *25*(6), 373–381.
Rosset, E. M., & Bradshaw, A. D. (2016). SPARC/osteonectin in mineralized tissue. *Matrix Biology*, *52–54*, 78–87.
Rotllant, J., Liu, D., Yan, Y. L., Postlethwait, J. H., Westerfield, M., & Du, S. J. (2008). Sparc (Osteonectin) functions in morphogenesis of the pharyngeal skeleton and inner ear. *Matrix Biology*, *27*(6), 561–572.
Saga, Y., Yagi, T., Ikawa, Y., Sakakura, T., & Aizawa, S. (1992). Mice develop normally without tenascin. *Genes & Development*, *6*, 1821–1831.
Sage, E. H. (1992). Modulation of endothelial cell shape by SPARC does not involve chelation of extracellular Ca2+ and Mg2+. *Biochemistry and Cell Biology*, *70*(1), 56–62.
Sage, E. H., & Bornstein, P. (1991). Extracellular proteins that modulate cell-matrix interactions. SPARC, tenascin, and thrombospondin. *The Journal of Biological Chemistry*, *266*(23), 14831–14834.
Sage, H., Vernon, R. B., Funk, S. E., Everitt, E. A., & Angello, J. (1989). SPARC, a secreted protein associated with cellular proliferation, inhibits cell spreading in vitro and exhibits Ca+2-dependent binding to the extracellular matrix. *The Journal of Cell Biology*, *109*, 341–356.
Sallin, P., de Preux Charles, A. S., Duruz, V., Pfefferli, C., & Jaźwińska, A. (2015). A dual epimorphic and compensatory mode of heart regeneration in zebrafish. *Developmental Biology*, *399*(1), 27–40.
Scherberich, A., Tucker, R. P., Samandari, E., Brown-Luedi, M., Martin, D., & Chiquet-Ehrismann, R. (2004). Murine tenascin-W: A novel mammalian tenascin expressed in kidney and at sites of bone and smooth muscle development. *Journal of Cell Science*, *117*, 571–581.
Schmidt, J. G., Andersen, E. W., Ersbøll, B. K., & Nielsen, M. E. (2016). Muscle wound healing in rainbow trout (Oncorhynchus mykiss). *Fish & Shellfish Immunology*, *48*, 273–284.
Schnorrer, F., Schönbauer, C., Langer, C. C., Dietzl, G., Novatchkova, M., Schernhuber, K., et al. (2010). Systematic genetic analysis of muscle morphogenesis and function in Drosophila. *Nature*, *464*(7286), 287–291.
Schwarzbauer, J. E., & Spencer, C. S. (1993). The Caenorhabditis elegans homologue of the extracellular calcium binding protein SPARC/osteonectin affects nematode body morphology and mobility. *Molecular Biology of the Cell*, *14*(9), 941–952.
Schweitzer, J., Becker, T., Lefebvre, J., Granato, M., Schachner, M., & Becker, C. G. (2005). Tenascin-C is involved in motor axon outgrowth in the trunk of developing zebrafish. *Developmental Dynamics*, *234*(3), 550–566.
Sebé-Pedrós, A., Irimia, M., Del Campo, J., Parra-Acero, H., Russ, C., Nusbaum, C., et al. (2013). Regulated aggregative multicellularity in a close unicellular relative of metazoa. *eLife*, *2*, e01287.
Segade, F., Cota, C., Famiglietti, A., Cha, A., & Davidson, B. (2016). Fibronectin contributes to notochord intercalation in the invertebrate chordate, Ciona intestinalis. *EvoDevo*, 7(1), 21.
Seifert, A. W., Monaghan, J. R., Voss, S. R., & Maden, M. (2012). Skin regeneration in adult axolotls: A blueprint for scar-free healing in vertebrates. *PLoS One*, 7(4), e32875.

Shahab, J., Baratta, C., Scuric, B., Godt, D., Venken, K. J., & Ringuette, M. J. (2015). Loss of SPARC dysregulates basal lamina assembly to disrupt larval fat body homeostasis in Drosophila melanogaster. *Developmental Dynamics*, *244*(4), 540–552.

Singh, S. K., Stogsdill, J. A., Pulimood, N. S., Dingsdale, H., Kim, Y. H., Pilaz, L. J., et al. (2016). Astrocytes assemble thalamocortical synapses by bridging NRX1α and NL1 via Hevin. *Cell*, *164*(1–2), 183–196.

Somasekhar, T., & Nordlander, R. H. (1995). Differential distributions of HNK-1 and tenascin immunoreactivity during innervation of myotomal muscle in Xenopus. *Brain Research. Developmental Brain Research*, *88*(1), 53–67.

Stavri, S., & Zarnescu, O. (2013). The expression of alkaline phosphatase, osteopontin, osteocalcin, and chondroitin sulfate during pectoral fin regeneration in Carassius auratus gibelio: A combined histochemical and immunohistochemical study. *Microscopy and Microanalysis*, *19*(1), 233–242.

Stenina, O. I., Topol, E. J., & Plow, E. F. (2007). Thrombospondins, their polymorphisms, and cardiovascular disease. *Arteriosclerosis, Thrombosis, and Vascular Biology*, *27*(9), 1886–1894.

Subramanian, A., & Schilling, T. F. (2014). Thrombospondin-4 controls matrix assembly during development and repair of myotendinous junctions. *eLife*, *18*, 3.

Subramanian, A., Wayburn, B., Bunch, T., & Volk, T. (2007). Thrombospondin-mediated adhesion is essential for the formation of the myotendinous junction in Drosophila. *Development*, (7), 1269–1278.

Sullivan, M. M., Puolakkainen, P. A., Barker, T. H., Funk, S. E., & Sage, E. H. (2008). Altered tissue repair in hevin-null mice: Inhibition of fibroblast migration by a matricellular SPARC homolog. *Wound Repair and Regeneration*, *16*(2), 310–319.

Sun, Y. D., Zhao, X. F., Kang, C. J., & Wang, J. X. (2006). Molecular cloning and characterization of Fc-TSP from the Chinese shrimp Fennerpenaeus chinensis. *Molecular Immunology*, *43*(8), 1202–1210.

Swarup, S., & Verheyen, E. M. (2012). Wnt/Wingless signaling in Drosophila. *Cold Spring Harbor Perspectives in Biology*, *4*(6), a007930.

Sweetwyne, M. T., & Murphy-Ullrich, J. E. (2012). Thrombospondin1 in tissue repair and fibrosis: TGF-β-dependent and independent mechanisms. *Matrix Biology*, *31*(3), 178–186.

Tan, K., & Lawler, J. (2009). The interaction of thrombospondins with extracellular matrix proteins. *Journal of Cell Communication and Signaling*, *3*(3–4), 177–187.

Tarr, J. T., Visser, T. G., Moon, J. E., Hendesi, H., Barbe, M. F., Bradley, J. P., et al. (2017). The pivotal role of CCN2 in mammalian palatogenesis. *Journal of Cell Communication and Signaling*, *11*(1), 25–37.

Terauchi, M., Yamagishi, T., Hanyuda, T., & Kawai, H. (2017). Genome-wide computational analysis of the secretome of brown algae (Phaeophyceae). *Marine Genomics*, *32*, 49–59.

Terry, N. A., Tulina, N., Matunis, E., & DiNardo, S. (2006). Novel regulators revealed by profiling Drosophila testis stem cells within their niche. *Developmental Biology*, *294*, 246–257.

Theisen, E. R., Pishas, K. I., Saund, R. S., & Lessnick, S. L. (2016). Therapeutic opportunities in Ewing sarcoma: EWS-FLI inhibition via LSD1 targeting. *Oncotarget*, 7(14), 17616–17630.

Thomas, J. T., Canelos, P., Luyten, F. P., & Moos, M., Jr. (2009). Xenopus SMOC-1 inhibits bone morphogenetic protein signaling downstream of receptor binding and is essential for postgastrulation development in Xenopus. *The Journal of Biological Chemistry*, *284*(28), 18994–19005.

Thomas, J. T., Dollins, D., Andrykovich, K. R., Chu, T., Stultz, B. G., Hursh, D. A., et al. (2017). SMOC can act as both an antagonist and an expander of BMP signaling. *eLife*, *6*, pii: e17935.

Thomas, F. H., Wilson, H., Silvestri, A., & Fraser, H. M. (2008). Thrombospondin-1 expression is increased during follicular atresia in the primate ovary. *Endocrinology, 149*(1), 185–192.

Tingaud-Sequeira, A., Chauvigné, F., Lozano, J., Agulleiro, M. J., Asensio, E., & Cerdà, J. (2009). New insights into molecular pathways associated with flatfish ovarian development and atresia revealed by transcriptional analysis. *BMC Genomics, 10*, 434.

Tongiorgi, E. (1999). Tenascin-C expression in the trunk of wild-type, cyclops and floating head zebrafish embryos. *Brain Research Bulletin, 48*(1), 79–88.

Tongiorgi, E., Bernhardt, R. R., Zinn, K., & Schachner, M. (1995). Tenascin-C mRNA is expressed in cranial neural crest cells, in some placodal derivatives, and in discrete domains of the embryonic zebrafish brain. *Journal of Neurobiology, 28*(3), 391–407.

Topol, E. J., McCarthy, J., Gabriel, S., Moliterno, D. J., Rogers, W. J., Newby, L. K., et al. (2001). Single nucleotide polymorphisms in multiple novel thrombospondin genes may be associated with familial premature myocardial infarction. *Circulation, 104*(22), 2641–2644.

Torres-Nuñez, E., Sobrino, C., Neale, P. J., Ceinos, R. M., Du, S., & Rotllant, J. (2012). Molecular response to ultraviolet radiation exposure in fish embryos: Implications for survival and morphological development. *Photochemistry and Photobiology, 88*(3), 701–707.

Torres-Núñez, E., Suarez-Bregua, P., Cal, L., Cal, R., Cerdá-Reverter, J. M., & Rotllant, J. (2015). Molecular cloning and characterization of the matricellular protein Sparc/osteonectin in flatfish, Scophthalmus maximus, and its developmental stage-dependent transcriptional regulation during metamorphosis. *Gene, 568*(2), 129–139.

Tucker, R. P. (2004). The thrombospondin type 1 repeat superfamily. *The International Journal of Biochemistry & Cell Biology, 36*(6), 969–974.

Tucker, R. P., & Adams, J. C. (2014). Adhesion networks of Cnidaria: A post-genomic view. *International Review of Cell and Molecular Biology, 308*, 323–377.

Tucker, R. P., & Chiquet-Ehrismann, R. (2009). Evidence for the evolution of tenascin and fibronectin early in the chordate lineage. *The International Journal of Biochemistry & Cell Biology, 41*(2), 424–434.

Tucker, R. P., Drabikowski, K., Hess, J. F., Ferralli, J., Chiquet-Ehrismann, R., & Adams, J. C. (2006). Phylogenetic analysis of the tenascin gene family: Evidence of origin early in the chordate lineage. *BMC Evolutionary Biology, 6*, 60.

Tucker, R. P., Hess, J. F., Gong, Q., Garvey, K., Shibata, B., & Adams, J. C. (2013). A thrombospondin in the anthozoan *Nematostella vectensis* is associated with the nervous system and upregulated during regeneration. *Biology Open, 2*, 217–226.

Umbhauer, M., Riou, J. F., Smith, J. C., & Boucaut, J. C. (1994). Control of somitic expression of tenascin in Xenopus embryos by myogenic factors and Brachyury. *Developmental Dynamics, 200*(4), 269–277.

Umbhauer, M., Riou, J. F., Spring, J., Smith, J. C., & Boucaut, J. C. (1992). Expression of tenascin mRNA in mesoderm during Xenopus laevis embryogenesis: The potential role of mesoderm patterning in tenascin regionalization. *Development, 116*(1), 147–157.

Urry, L. A., Whittaker, C. A., Duquette, M., Lawler, J., & DeSimone, D. W. (1998). Thrombospondins in early Xenopus embryos: Dynamic patterns of expression suggest diverse roles in nervous system, notochord, and muscle development. *Developmental Dynamics, 211*(4), 390–407.

Vega-Macaya, F., Manieu, C., Valdivia, M., Mlodzik, M., & Olguín, P. (2016). Establishment of the muscle-tendon junction during thorax morphogenesis in Drosophila requires the rho-kinase. *Genetics, 204*(3), 1139–1149.

Vieira, F. A., Pinto, P. I., Guerreiro, P. M., & Power, D. M. (2012). Divergent responsiveness of the dentary and vertebral bone to a selective estrogen-receptor modulator (SERM) in the teleost Sparus auratus. *General and Comparative Endocrinology, 179*(3), 421–427.

Vieira, F. A., Thorne, M. A., Stueber, K., Darias, M., Reinhardt, R., Clark, M. S., et al. (2013). Comparative analysis of a teleost skeleton transcriptome provides insight into its regulation. *General and Comparative Endocrinology*, *191*, 45–58.

Vilmos, P., Gaudenz, K., Hegedus, Z., & Marsh, J. L. (2001). The twisted gastrulation family of proteins, together with the IGFBP and CCN families, comprise the TIC superfamily of cysteine rich secreted factors. *Molecular Pathology*, *54*(5), 317–323.

Vincent, T. L., Woolfson, D. N., & Adams, J. C. (2013). Prediction and analysis of higher-order coiled-coils: Insights from proteins of the extracellular matrix, tenascins and thrombospondins. *The International Journal of Biochemistry & Cell Biology*, *45*, 2392–2401.

Vuilleumier, R., Springhorn, A., Patterson, L., Koidl, S., Hammerschmidt, M., Affolter, M., et al. (2010). Control of Dpp morphogen signalling by a secreted feedback regulator. *Nature Cell Biology*, *12*, 611–617.

Wagener, J., Yang, W., Kazuschke, K., Winterhager, E., & Gellhaus, A. (2013). CCN3 regulates proliferation and migration properties in Jeg3 trophoblast cells via ERK1/2, Akt and Notch signalling. *Molecular Human Reproduction*, *19*(4), 237–249.

Wang, D., Li, S., & Li, F. (2013). Screening of genes regulated by Relish in Chinese shrimp Fenneropenaeus chinensis. *Developmental and Comparative Immunology*, *41*(2), 209–216. https://doi.org/10.1016/j.dci.2013.06.003.

Wang, W. Z., Oeschger, F. M., Montiel, J. F., García-Moreno, F., Hoerder-Suabedissen, A., Krubitzer, L., et al. (2011). Comparative aspects of subplate zone studied with gene expression in sauropsids and mammals. *Cerebral Cortex*, *21*(10), 2187–2203.

Wargelius, A., Fjelldal, P. G., & Hansen, T. (2005). Heat shock during early somitogenesis induces caudal vertebral column defects in Atlantic salmon (Salmo salar). *Development Genes and Evolution*, *215*, 350–357.

Weber, A. L., Khan, G. F., Magwire, M. M., Tabor, C. L., Mackay, T. F., & Anholt, R. R. (2012). Genome-wide association analysis of oxidative stress resistance in Drosophila melanogaster. *PLoS One*, 7(4), e34745.

Weber, P., Montag, D., Schachner, M., & Bernhardt, R. R. (1998). Zebrafish tenascin-W, a new member of the tenascin family. *Journal of Neurobiology*, *35*(1), 1–16.

Weigele, J., Franz-Odendaal, T. A., & Hilbig, R. (2015). Expression of SPARC and the osteopontin-like protein during skeletal development in the cichlid fish Oreochromis mossambicus. *Developmental Dynamics*, *244*(8), 955–972.

Weigele, J., Franz-Odendaal, T. A., & Hilbig, R. (2016). Not all inner ears are the same: Otolith matrix proteins in the inner ear of sub-adult cichlid fish, Oreochromis mossambicus, reveal insights into the biomineralization process. *The Anatomical Record*, *299*(2), 234–245.

Werner, G. D., Gemmell, P., Grosser, S., Hamer, R., & Shimeld, S. M. (2013). Analysis of a deep transcriptome from the mantle tissue of Patella vulgata Linnaeus (Mollusca: Gastropoda: Patellidae) reveals candidate biomineralising genes. *Marine Biotechnology (New York, N.Y.)*, *15*(2), 230–243.

Whited, J. L., Lehoczky, J. A., Austin, C. A., & Tabin, C. J. (2011). Dynamic expression of two thrombospondins during axolotl limb regeneration. *Developmental Dynamics*, *240*(5), 1249–1258.

Williams, F., Tew, H. A., Paul, C. E., & Adams, J. C. (2014). The predicted secretomes of *Monosiga brevicollis* and *Capsaspora owczarzaki*, close unicellular relatives of metazoans, reveal new insights into the evolution of the metazoan extracellular matrix. *Matrix Biology*, *37*, 60–68.

Williamson, D. A., Parrish, E. P., & Edelman, G. M. (1991a). Distribution and expression of two interactive extracellular matrix proteins, cytotactin and cytotactin-binding proteoglycan, during development of Xenopus laevis. I. Embryonic development. *Journal of Morphology*, *209*(2), 189–202.

Williamson, D. A., Parrish, E. P., & Edelman, G. M. (1991b). Distribution and expression of two interactive extracellular matrix proteins, cytotactin and cytotactin-binding proteoglycan, during development of Xenopus laevis. II. Metamorphosis. *Journal of Morphology, 209*(2), 203–213.

Wolfstetter, G., & Holz, A. (2012). The role of LamininB2 (LanB2) during mesoderm differentiation in Drosophila. *Cellular and Molecular Life Sciences, 69*(2), 267–282.

Wu, F. R., Zhou, L. Y., Nagahama, Y., & Wang, D. S. (2009). Duplication and distinct expression patterns of two thrombospondin-1 isoforms in teleost fishes. *Gene Expression Patterns, 9*(6), 436–443.

Xie, Y., Li, F., Wang, B., Li, S., Wang, D., Jiang, H., et al. (2010). Screening of genes related to ovary development in Chinese shrimp Fenneropenaeus chinensis by suppression subtractive hybridization. *Comparative Biochemistry and Physiology. Part D, Genomics & Proteomics, 5*(2), 98–104.

Yamano, K., Qiu, G. F., & Unuma, T. (2004). Molecular cloning and ovarian expression profiles of thrombospondin, a major component of cortical rods in mature oocytes of penaeid shrimp, Marsupenaeus japonicus. *Biology of Reproduction, 70*(6), 1670–1678.

Yan, D., Neumüller, R. A., Buckner, M., Ayers, K., Li, H., Hu, Y., et al. (2014). A regulatory network of Drosophila germline stem cell self-renewal. *Developmental Cell, 28*(4), 459–473.

Ytteborg, E., Baeverfjord, G., Torgersen, J., Hjelde, K., & Takle, H. (2010). Molecular pathology of vertebral deformities in hyperthermic Atlantic salmon (Salmo salar). *BMC Physiology, 10*, 12.

Yu, Y. M., Cristofanilli, M., Valiveti, A., Ma, L., Yoo, M., Morellini, F., et al. (2011). The extracellular matrix glycoprotein tenascin-C promotes locomotor recovery after spinal cord injury in adult zebrafish. *Neuroscience, 183*, 238–250.

Zaidi, A. U., Kafitz, K. W., Greer, C. A., & Zielinski, B. S. (1998). The expression of tenascin-C along the lamprey olfactory pathway during embryonic development and following axotomy-induced replacement of the olfactory receptor neurons. *Brain Research. Developmental Brain Research, 109*(2), 157–168.

Zeng, X., Han, L., Singh, S. R., Liu, H., Neumüller, R. A., Yan, D., et al. (2015). Genome-wide RNAi screen identifies networks involved in intestinal stem cell regulation in Drosophila. *Cell Reports, 10*(7), 1226–1238.

Zhao, Y., Zhao, F., Zong, L., Zhang, P., Guan, L., Zhang, J., et al. (2013). Exome sequencing and linkage analysis identified tenascin-C (TNC) as a novel causative gene in nonsyndromic hearing loss. *PLoS One, 8*(7), e69549.

Zheng, Q., Wang, Y., Vargas, E., & DiNardo, S. (2011). magu is required for germline stem cell self-renewal through BMP signaling in the Drosophila testis. *Developmental Biology, 357*(1), 202–210.

Zhou, F., Zheng, L., Zhang, D., Huang, J., Qiu, L., Yang, Q., et al. (2011). Molecular cloning, characterization and expression analysis of thrombospondin gene from Penaeus monodon. *Marine Genomics, 4*(2), 121–128.

Zweers, M. C., Bristow, J., Steijlen, P. M., Dean, W. B., Hamel, B. C., Otero, M., et al. (2011). Haploinsufficiency of TNXB is associated with hypermobility type of Ehlers-Danlos syndrome. *American Journal of Human Genetics, 73*(1), 214–217.

Zwicker, J. I., Peyvandi, F., Palla, R., Lombardi, R., Canciani, M. T., Cairo, A., et al. (2006). The thrombospondin-1 N700S polymorphism is associated with early myocardial infarction without altering von Willebrand factor multimer size. *Blood, 108*(4), 1280–1283.

CHAPTER THREE

Collagen Fibril Assembly and Function

David F. Holmes[1], Yinhui Lu, Tobias Starborg, Karl E. Kadler[1]
Wellcome Trust Centre for Cell-Matrix Research, Faculty of Biology, Medicine and Health, Manchester Academic Health Science Centre, University of Manchester, Manchester, United Kingdom
[1]Corresponding authors: e-mail address: david.f.holmes@manchester.ac.uk; karl.kadler@manchester.ac.uk

Contents

Abstract

Collagen fibrils are the major mechanical component in the extracellular matrix of a broad range of multicellular animals from echinoderms to vertebrates where they provide a stable framework for tissues. They form the key tension-resisting element of a complex fiber-composite system that has a tissue-specific hierarchical structure linked to mechanical demands. Remarkably, these tissues are self-maintaining and avoid

Current Topics in Developmental Biology, Volume 130
ISSN 0070-2153
https://doi.org/10.1016/bs.ctdb.2018.02.004

fatigue failure over the lifetime of the animal. Collagen fibrils can assemble spontaneously from purified solutions of collagen molecules. In developing tissues, however, in addition to the intrinsic self-assembly properties, there is cellular machinery that regulates fibril nucleation, spatial orientation, and fibril size, according to the tissue and stage of development. The intricate mechanisms underlying the generation of a collagen fibril network of defined architecture and mechanical properties are now becoming apparent. Impairment of this system leads ultimately to mechanical failure or tissue fibrosis.

1. INTRODUCTION

Collagens are a large family of triple helical proteins with functions pertaining to cell–matrix interactions and tissue structure and function. There are 28 distinct collagens in vertebrates (Huxley-Jones, Robertson, & Boot-Handford, 2007; Kadler, Baldock, Bella, & Boot-Handford, 2007; Mienaltowski & Birk, 2014), ~200 in *C. elegans* (Johnstone, 2000), and additional collagens in invertebrates (Exposito, Valcourt, Cluzel, & Lethias, 2010; Thurmond & Trotter, 1994; Trotter & Koob, 1989), bacteria (see Ghosh et al., 2012 and references therein), and viruses (Legendre, Santini, Rico, Abergel, & Claverie, 2011; Rasmussen, Jacobsson, & Bjorck, 2003). There are several excellent reviews on collagens (examples are Bella, 2016; Bella & Hulmes, 2017; Mienaltowski & Birk, 2014) and therefore these topics are only briefly covered here.

1.1 Collagens Are Triple Helical Molecules

The individual polypeptide chains of collagen have a repeating Gly-X-Y motif in which glycine occurs at every third residue position and X and Y are frequently occupied by the imino acids proline and hydroxyproline (see Bella, Eaton, Brodsky, & Berman, 1994; Brodsky & Persikov, 2005; Brodsky & Ramshaw, 1997 and reviewed by Bella, 2016). The polypeptide chains in collagens are termed α-chains; α-chains in different collagen molecules are denoted by a Roman numeral. Therefore, the α-chains in type I collagen are denoted α1(I) and α2(I) and are encoded by the genes COL1A1 and COL1A2. Collagens can be homotrimers and heterotrimers. For example, type I collagen is usually a heterotrimer of two α1(I) chains and a single α2(I) chain, i.e., $[\alpha1(I)]_2,\alpha2(I)$. Type I collagen can also occur as a homotrimer of three individual α1(I) chains, which occurs in some variants of osteogenesis imperfecta (Deak, van der Rest, & Prockop, 1985;

Kadler, Hojima, & Prockop, 1992) and in some carcinomas (Pucci-Minafra et al., 1998). Type II collagen is a homotrimer of three α1(II) chains, i.e., $[\alpha 1(II)]_3$.

1.2 Fibrillar Collagens

Most, and possibly all, collagens participate in higher-order assemblies including networks, filaments, microfibrils, or fibrils (for review, see Mienaltowski & Birk, 2014). Here, we are concerned only with the collagens that assemble into fibrils. These include types I, II, III, V, XI, XXIV, and XXVII. The fibrils are the primary tensile element and form the mechanical basis of bony, cartilaginous, fibrous, and tubular structures. Fibrillar collagens are synthesized as soluble precursor procollagen molecules (Bellamy & Bornstein, 1971) that contain globular "propeptides" at each end of the triple helix. These are proteolytically removed by procollagen N- and C-proteinases to produce collagen (Hojima, McKenzie, van der Rest, & Prockop, 1989; Hojima, van der Rest, & Prockop, 1985; Njieha, Morikawa, Tuderman, & Prockop, 1982; Tuderman & Prockop, 1982) (Fig. 1). The final processed collagens have uninterrupted triple helices of ~300 nm in length flanked by short "telopeptides" that contain critical

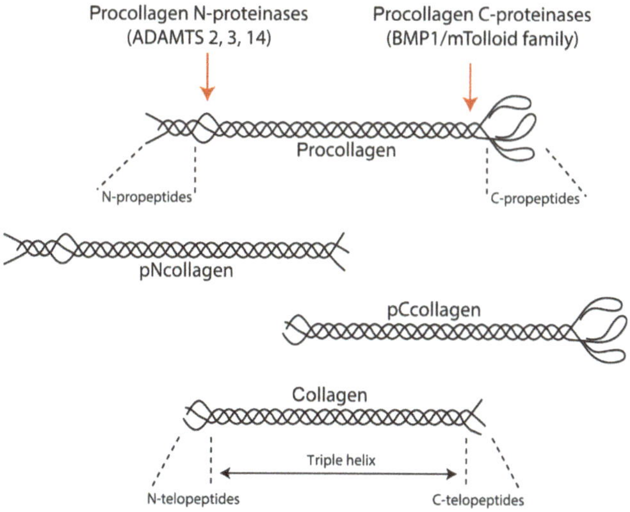

Fig. 1 Schematic showing the conversion of procollagen to collagen by cleavage of the terminal propeptides by the procollagen N- and C-proteinases. The cleavage intermediates are pNcollagen (retaining the N-propeptides) and pCcollagen (retaining the C-propeptides). The final collagen molecule contains telopeptides (none Gly-X-Y peptides) that contain critical binding sites for fibril formation.

binding sites for collagen fibril assembly (Prockop & Fertala, 1998; see below). Types XXIV and XXVII were identified by genome sequencing and were included in the fibril-forming subfamily on the basis of protein domain structure (Koch et al., 2003) and the presence of type XXVII collagen in thin fibrils (Plumb et al., 2007).

1.3 Fibrils Are Molecular Complexes

Collagen fibrils are comprised of more than one collagen type (Hansen & Bruckner, 2003) and, as far as is known, always have additional molecules bound to their surfaces including glycoproteins, proteoglycans, and plasma membrane receptors including integrins, discoidin domain-containing receptors (DDRs), and mannose receptors (Di Lullo, Sweeney, Korkko, Ala-Kokko, & San Antonio, 2002; Jokinen et al., 2004; Orgel, San Antonio, & Antipova, 2011; Sweeney et al., 2008). Although the fibrils are copolymers of collagens, type I collagen and type II collagen do not appear to assemble into the same fibril. The mechanistic basis of this exclusivity is unknown. Thus, fibrils are either "predominately type I collagen" or "predominately type II collagen." Predominately type I collagen fibrils often have minor quantities of type III and V collagens (in addition to type I collagen) and occur in fibrous, vascular, and calcified tissues. Predominately type II collagen fibrils occur in cartilaginous tissues and often contain minor quantities of type IX and XI collagens, in addition to type II collagen.

2. FIBRIL STRUCTURAL HIERARCHY—FROM MOLECULES TO FIBRIL ARRAYS

Collagen displays a structural hierarchy in tissues from molecules to spatially organized arrays of fibrils that can be bundled into larger fascicles that can contain several hundreds to thousands of fibrils. The fibril diameters, the fibril volume fraction (i.e., the fraction of the tissue occupied by fibrils), and the spatial arrangement of fibrils are dependent on the tissue and stage of development (Craig & Parry, 1981a). Single collagen molecules can be readily visualized by rotary shadowing and electron microscopy (Mould, Holmes, Kadler, & Chapman, 1985; Mould & Hulmes, 1987) as shown in Fig. 2A or by atomic force microscopy (AFM) (Yamamoto et al., 2000). Collagen fibrils can be mechanically isolated from connective tissues and visualized unstained by annular dark-field scanning transmission electron microscopy (AD-STEM) or by transmission electron microscopy (TEM) after negative staining (Holmes, Graham, Trotter, & Kadler, 2001)

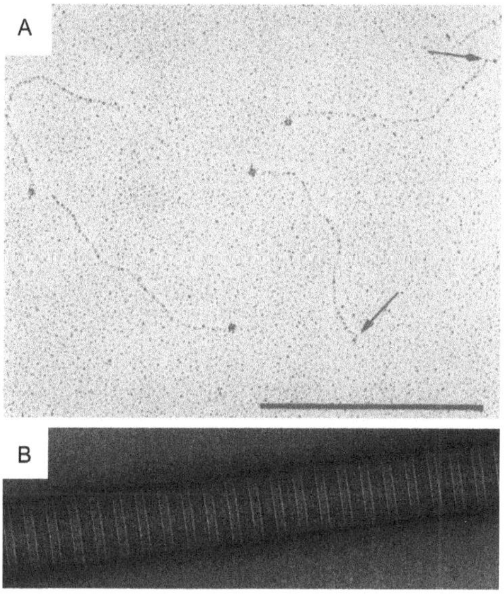

Fig. 2 From molecules to fibrils. (A) TEM image of procollagen molecules after rotary shadowing with platinum. The C-propeptide appears as a globular domain at one end of the molecule and the N-propeptide as a tick-like extension at the other end (marked with *arrows*). Scale bar = 300 nm. (B) TEM image of a collagen fibril mechanically dispersed from mouse Achilles tendon and negatively stained with 2% uranyl acetate. The fibril shows the characteristic 67 nm axial periodicity and "gap–overlap" structure. The fibril has a diameter of ~185 nm equivalent to ~1300 molecules in transverse section.

as shown in Fig. 2B. Mass per unit length measurements can be made from ADF-STEM and this allows the number of collagen molecules in the transverse section of a fibril to be calculated (Holmes, Graham, et al., 2001). The spatial arrangement of collagen fibrils can be observed by TEM of tissue sections as shown in Figs. 3 and 4 for embryonic vertebrate tendon and embryonic vertebrate cornea, respectively. The recent established method of serial block face imaging by scanning electron microscopy (Starborg et al., 2013) allows the 3D tissue architecture to be observed over extended volumes (Fig. 5).

Uniform fibril diameters and a uniform spatial arrangement of collagen fibrils are found in a range of connective tissues. Examples include embryonic tendon, which typically contains fibrils of diameter 30–35 nm (Parry, Barnes, & Craig, 1978). The fibrils are arranged in bundles with a regular (near-hexagonal) array of fibrils (Fig. 3D) aligned along the long axis of the tendon. Although the parallel alignment of fibrils is maintained in adult

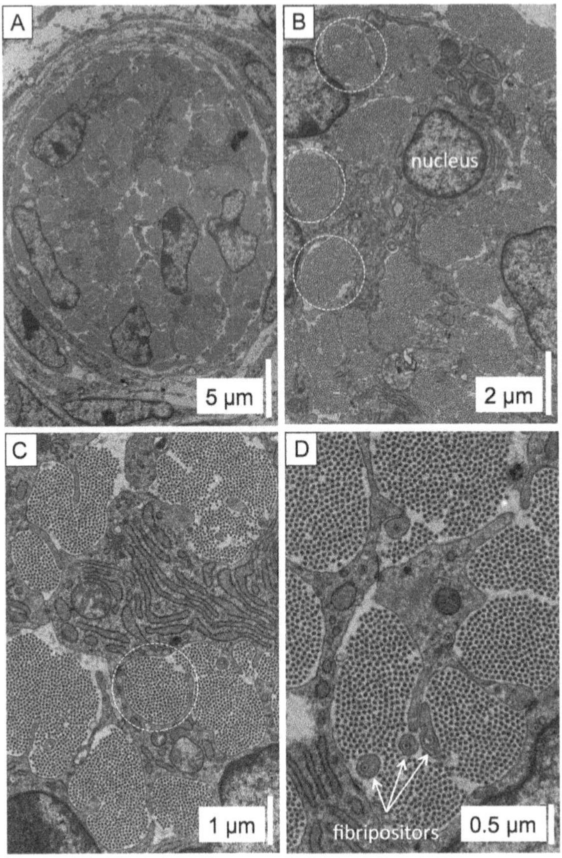

Fig. 3 Transmission electron microscope images of thin (~70 nm) sections of embryonic mouse-tail tendon cut transversely (at 90°) to the tendon long axis. (A) The field of view shows one of the four tendon bundles in the tail. (B) A higher power (smaller field of view) image showing bundles (three bundles highlighted with *dashed circles*) of collagen fibrils residing between cells. Nuclei of individual cells are apparent. (C) A higher power view showing bundles of collagen fibrils, which are approximately circular (*dashed circle*) in cross section and situated up against the plasma membrane of cells. Endoplasmic reticulum and mitochondria are apparent at this magnification. (D) At higher power the bundles are seen to contain collagen fibrils that are closely packed and parallel to the long axis of the tendon. Fibripositors (*arrows*) are seen at this magnification.

tendon, the initial uniformity of diameter and spatial order seen in the embryonic tendon is lost as fibrils grow in diameter and the fibril volume fraction increases in order to provide increased mechanical stiffness and strength in adult tendon. The corneal stratum, in contrast, maintains a stable, uniform fibril diameter distribution and spatial arrangement into maturity,

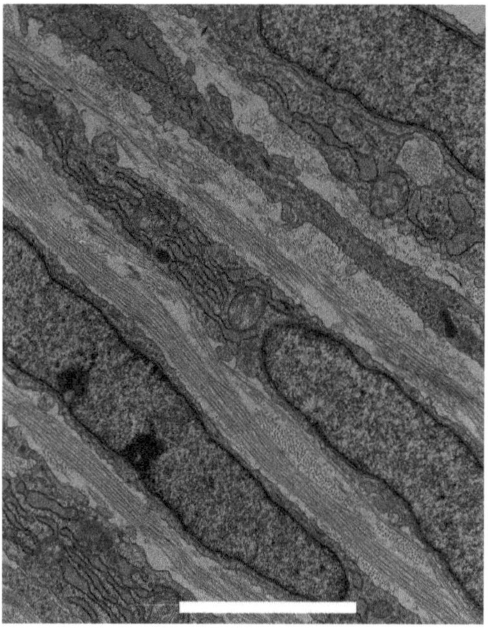

Fig. 4 Transmission electron microscope image of a thin (~70 nm) section of embryonic E15.5 mouse cornea cut transversely (at 90°) to curvilinear axis of the tissue. Collagen fibrils are seen in longitudinal and transverse section. Bar, 3 μm.

to preserve the optical transparency of the tissue. Even in this case the lateral size of the collagen fibrils is not precise at the molecular level, as in the case of muscle thick filament (Miroshnichenko, Balanuk, & Nozdrenko, 2000), but shows an inter- and intrafibrillar variation (Holmes & Kadler, 2005). Linked to transparency and mechanical strength, the fibrils of the cornea show a plywood-like arrangement with alternating lamellae of uniformly spaced fibrils, each lamella rotated by ~90° with respect to the neighboring lamellae as shown in Fig. 4. Noteworthy, the collagen fibrils in tendon and cornea are predominately type I collagen fibrils. Therefore, although the basic biochemical composition is similar, the 3D organization and diameter control mechanisms are different in the two tissues.

A fibril crimp structure has been observed in a number of tissues, with tendon the most studied (Gathercole & Keller, 1991; Raspanti, Congiu, & Guizzardi, 2001). The crimp structure has been observed directly by serial block face-scanning electron microscopy in embryonic mouse-tail tendon and in this case takes the form of a spiral, nearly always left-handed (Kalson et al., 2015). In this study the crimp wavelength was observed to be

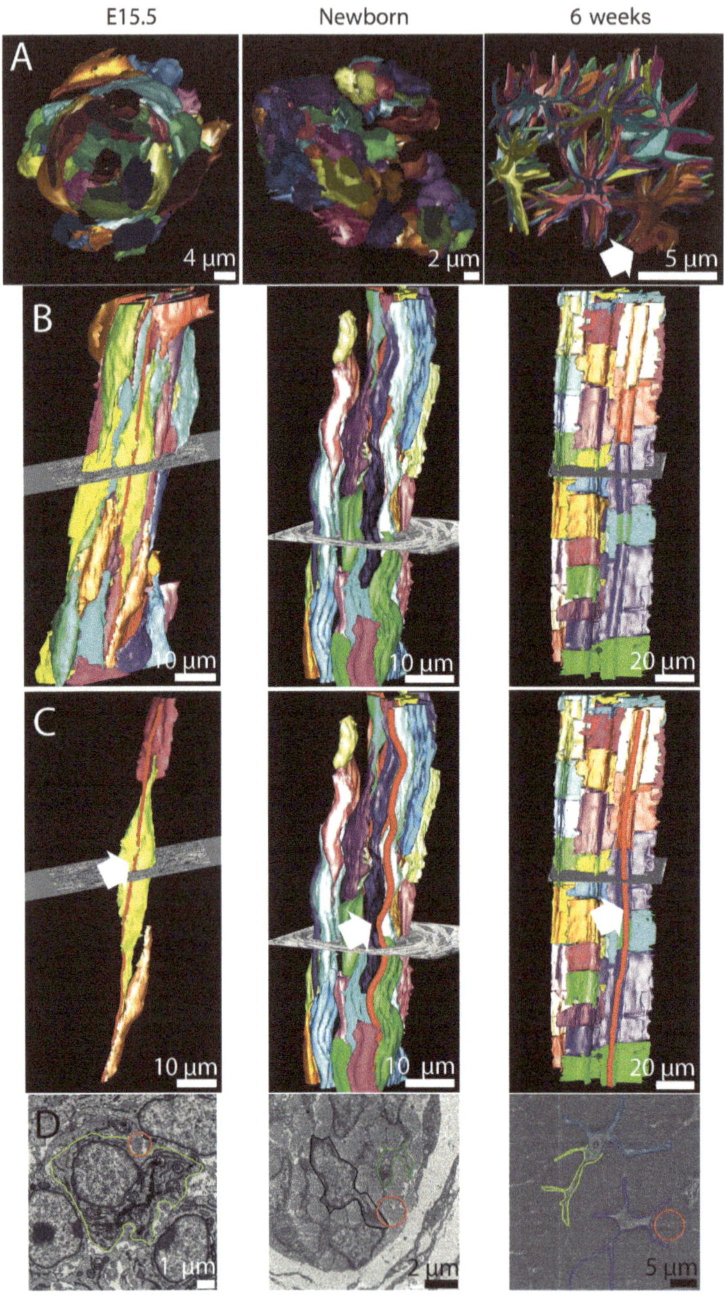

Fig. 5 Three-dimensional reconstruction from serial block face-scanning electron microscopy of mouse-tail tendon from embryonic to postnatal stages of development. Individual cells are depicted in *different colors*. (A) Transverse view of reconstructed tendon fibroblasts. During development the ECM expands (*arrow*) and the cells change shape from cylindrical to stellate in cross section. (B) Longitudinal views showing cells stacked end-on-end. (C) Crimped cells and bundles pronounced at birth. (D) Serial block face images corresponding to the EM images shown in (C). A fibril bundle is *circled red*. Cell membranes are also *outlined*.

~14 μm at E15.5 and ~100 μm at 6 week postnatal. The formation of this structural feature has been linked to cell-derived forces on the extracellular matrix in the embryonic tissue that cause a buckling of collagen fibrils (Herchenhan et al., 2012).

A D-periodic microfibrillar substructure of ~4 nm in diameter is evident within collagen fibrils using TEM (Holmes et al., 2001). Current fibril structure models based on high-angle X-ray diffraction of rat-tail tendon also involve a 5-molecular-stranded microfibril (see fibril structure in Section 6). Other, larger, subfibrils have also been observed in collagen fibrils from tissue by scanning electron microscopy and atomic force microscopy (Raspanti et al., 2001). Treatment of tissue with fibril-destabilizing agents has been observed to liberate subfibrils typically of diameter ~25 nm (Zhao, Weinhold, Lee, & Dahners, 2011).

The underlying mechanisms responsible for the formation of these collagen fibril structures are discussed in the following sections including the fibril assembly pathway and regulation of fibril diameter. In general the assembly process is governed by the intrinsic self-assembly properties of the collagen molecules combined with tissue-specific cell regulation of fibril nucleation, growth, and orientation. The relative contributions of these molecular and cell factors in collagen fibril array formation have been the subject of continued debate. For example, liquid crystalline ordering of procollagen has been proposed as a determinant of three-dimensional extracellular matrix architecture which would involve a minimal cell involvement in establishing the collagen fibril spatial architecture (Martin et al., 2000).

3. COLLAGEN FIBRIL FORMATION AS A SELF-ASSEMBLY PROCESS

It has long been known that collagen fibrils can form by a self-assembly process from a purified solution of collagen molecules in a warm, neutral buffer (Gross & Kirk, 1958; Wood & Keech, 1960). The acid-extracted collagen is fully processed and lacks the N- and C-propeptide extensions of the initial cellular product, i.e., procollagen. Importantly the molecules retain extrahelical telopeptides at the ends of the triple helical molecule when it is extracted with acetic acid solution (Capaldi & Chapman, 1982), as opposed to pepsin extraction which leads to partial loss of telopeptides. The telopeptides (typically 16 and 25 amino acids long at the N- and C-ends of the type I collagen, respectively) are critical in directing the assembly pathway in the assembly of *D*-periodic fibrils in vitro. Partial loss of either telopeptide leads to a less efficient assembly process with different intermediates (Capaldi & Chapman, 1982; Veis, 1981). Extensive loss of the N-telopeptides leads to the formation of fibrils with a *D*-periodic symmetric banding pattern where the molecules are in antiparallel array (Leibovich & Weiss, 1970). In contrast, extensive loss of the C-telopeptides leads to *D*-periodic cigar-shaped assemblies (tactoids) (Leibovich & Weiss, 1970). Despite the telopeptides only representing ~4% of the molecular mass of the collagen molecule, they are critical in promoting a kinetically efficient assembly route leading to *D*-periodic, polarized fibrils.

Interestingly, the fibril assembly pathway in vitro has been shown to depend on the initiating steps used to establish the solution conditions for assembly (Holmes, Capaldi, & Chapman, 1981), as shown in Fig. 6. If the cold, acid solution of collagen is prewarmed and then neutralized (warm start), numerous "early fibrils" which are tapered and show a distinct *D*-periodic (~67 nm) band pattern are observed in the first stages of reconstitution when the turbidity is still near zero (Fig. 6C). In contrast, if the collagen solution in acetic acid is first neutralized and then warmed (neutral start), then an accumulation of filamentous aggregates is seen initially (Fig. 6D) with *D*-periodic fibrils appearing later in the assembly process. Simultaneous neutralization and warming shows similar "early fibril" intermediates, which implies that the filamentous intermediate assemblies are nucleated in the transient cold neutral solution. The observation of the critical influence of the initiating procedure provided an explanation of why different intermediate stages in fibril reconstitution were reported in different laboratories despite using similar collagen solutions (Gelman, Williams, & Piez, 1979; Williams, Gelman, Poppke, & Piez, 1978). Of note, the early fibrils observed using the warm-start procedure showed a well-defined size and shape with an initial diameter limit of ~20 nm attained for the fibrils of length ~90 *D*-periods (~6 μm). Subsequent interfibrillar fusion can then explain the final diameter range (~30–70 nm). This early diameter limitation is supported by fibril seeding experiments where early fibrils are added to a dilute collagen solution, favoring further growth by accretion and minimizing interfibrillar fusion, where the fibrils continued to grow at uniform diameter (Haworth & Chapman, 1977).

3.1 To What Extent Is the Reconstitution of Collagen Fibril From Purified Solutions Representative of Fibril Assembly In Vivo?

There is a correspondence between the early fibrils produced from an acid-soluble collagen solution using the warm-start reconstitution process and

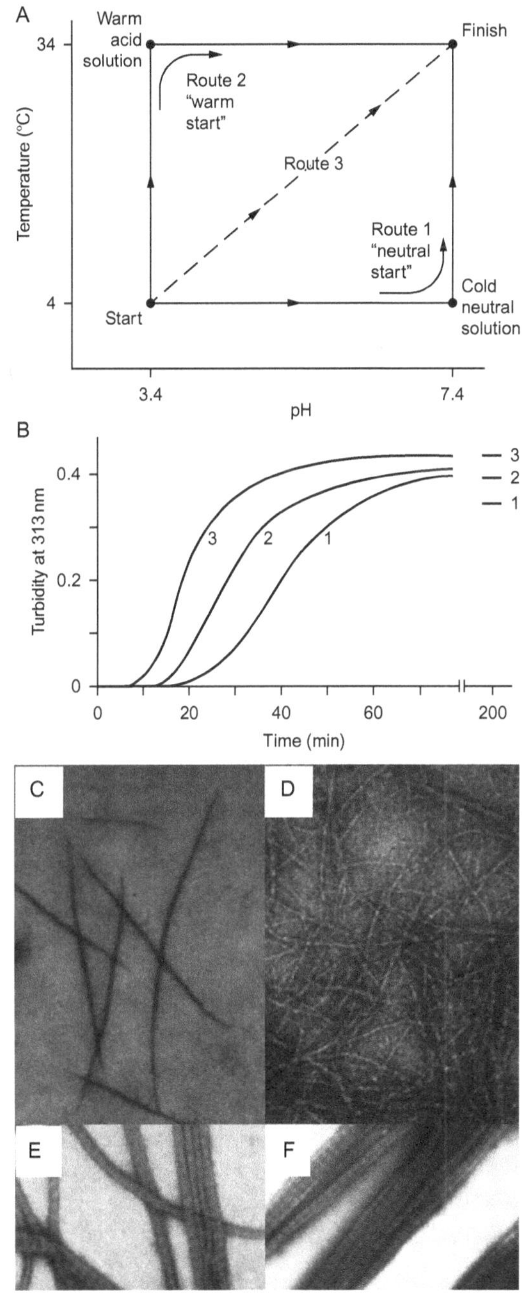

Fig. 6 Alternative assembly routes observed in the reconstitution of type I collagen fibrils from purified solution. (A) Diagram to show three initiation routes to transfer the collagen from a cold acid solution to a warm neutral solution. Routes 1 and 2 involve

(Continued)

those observed in embryonic tissues (Birk, Zycband, Winkelmann, & Trelstad, 1989; see below), but some features of the early fibril seen in tissue are lacking. These include the occurrence of bipolar fibrils (Holmes, Lowe, & Chapman, 1994) that can limit interfibrillar fusion and influence fibril tip shape (Holmes, Graham, & Kadler, 1998). In an attempt to replicate the collagen fibril assembly process that occurs in tissues a cell-free fibril assembly system was developed starting with procollagen with retained N- and C-propeptides (Kadler, Hojima, & Prockop, 1987). Isolation and purification of the procollagen N- and C-proteinases that cleave the propeptides made it possible to study collagen fibril assembly in vitro by cleavage of procollagen, as occurs in vivo (Kadler et al., 1987; Kadler, Hojima, & Prockop, 1988). This system also allowed collagen fibril assembly to be studied in vitro in the absence of lysyl oxidase-derived cross-links (Eyre, Weis, & Wu, 2008). The fibrils exhibited a critical concentration of assembly (Kadler et al., 1987), which is analogous to the self-formation of inorganic crystals. The assembly process was limited by microunfolding of the collagen molecules (Kadler et al., 1988). The fibrils grew from pointed tips in which the N-termini of the collagen molecules pointed toward the tip (therefore the fibrils were N,N-bipolar fibrils; Kadler, Hojima, & Prockop, 1990) (Fig. 7A). Studies of collagen fibrils formed by cleavage of procollagen in vitro showed that the tips are the sites of diameter regulation (Holmes et al., 1998; see below), that fibrils bear close resemblance to fibrils formed in vivo (Holmes, Watson, Chapman, & Kadler, 1996), and that the tips of fibrils are paraboloidal in shape (Holmes, Chapman, Prockop, & Kadler, 1992; see below).

3.2 Unipolar and Bipolar Fibrils

Of further interest, fibrils formed by cleavage of procollagen with the N- and C-proteinases in vitro are N,N-bipolar and thereby contain a central region

Fig. 6—Cont'd a warming step and a neutralization step. Route 3 is a combined single step achieved by adding a warm buffer solution to the initial collagen solution. (B) Turbidimetric curves to show a comparison in the kinetics of fibril formation between the three routes. (C) Typical early fibrils found in the lag period after the route 2 (warm start) initiation procedure. These are compact, clearly D-periodic, and have smoothly tapered tips. (D) Typical loose fibrous assemblies of collagen that accumulate during the lag phase after the route 1 (neutral start) initiation procedure. (E and F) TEM images of the final collagen fibrils using the warm-start and neutral-start initiation procedures, respectively. The D-period, apparent in the negatively stained fibrils, is ~67 nm.

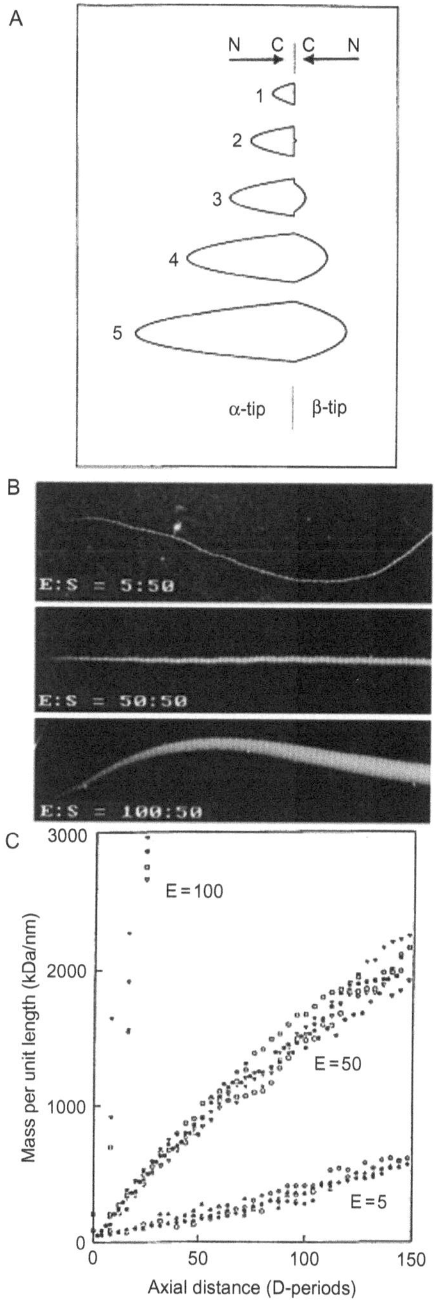

Fig. 7 Growth mode of collagen fibrils starting from purified procollagen in the "cell-free system." Fibril assembly is initiated by mixing pCcollagen, retaining the C-propeptide, with procollagen C-proteinase, which releases the C-propeptide. (A) The resultant fibrils are N,N-bipolar with a slender (α-) and coarse (β-) tip, as shown in the schematic. (B) Dark-field STEM images of unstained α-tips formed at different enzyme:substrate (E:S) ratios. The 67 nm D-periodicity is apparent in the unstained fibrils. (C) The mass slopes of the α-tips are well defined and dependent on the E:S ratio.

of antiparallel molecular packing, outside of which the molecules were in polarized array with the their N-termini directed toward the ends of the molecules. N,N-bipolar collagen fibrils were subsequently found in vertebrate tendon along with unipolar collagen fibrils (i.e., all the collagen molecules were oriented in the same direction). By coincidence N,N-bipolar collagen fibrils were also observed at the same time, as the exclusive form of collagen fibril, in the mutable tissues of echinoderms (Thurmond & Trotter, 1994) (see below).

3.3 Type II Collagen Fibrils Have Also Been Reconstituted In Vitro

Type II collagen alone was found to produce thick D-periodic fibrils, whereas a mixture of type II and XI collagen or type I, IX, and XI gave thin uniform, *D*-periodic fibrils ~20 nm in diameter (Hansen & Bruckner, 2003). Collagen XI has a critical role in the assembly of collagen II-containing fibrils; it nucleates the self-assembly and limits lateral growth of the collagen-II fibrils (Blaschke, Eikenberry, Hulmes, Galla, & Bruckner, 2000); and cho/cho mice that lack the α1(XI) chain have a complete absence of collagen XI from cartilage and developed chondrodysplasia accompanied by the presence of extremely thick collagen fibrils (Seegmiller, Fraser, & Sheldon, 1971). These studies demonstrate that the collagen fibrils of cartilage are composed of collagen types II, IX, and XI and occur as both thick (~50–80 nm in diameter) and thin fibrils (~20 nm in diameter) (Holmes & Kadler, 2006).

4. COLLAGEN FIBRIL GROWTH REGULATION MODELS

A frequently asked question related to collagen fibril assembly is "What limits fibril diameter?" Extensive studies spanning many decades have addressed this question using solutions of purified collagen in vitro, in the absence of cells. The conclusion is that collagen fibril growth most probably occurs in three stages: (i) nucleation, (ii) growth by accretion of monomer (or small oligomers) onto the fibril surface, and (iii) interfibrillar fusion. This appears to be the "intrinsic" self-assembly pathway. However, collagen fibrils in vivo are heterotypic polymers comprised of core elements (e.g., the fibril-forming collagens) and surface-associated molecules including nonfibril-forming collagens, glycoproteins, proteoglycans, and receptors. Therefore, fibril formation in vivo most probably involves extrinsic control of the intrinsic process to account for the high degree of diameter regulation and the tissue-specific and elaborate 3D organization of collagen fibrils in tissues.

4.1 Molecular Accretion

A range of quantitative models has been proposed over several decades as the basis of an intrinsic fibril diameter control mechanism. These all consider the case of fibril growth by molecular accretion and are listed in chronological order in Table 1. The shape and size of the final fibril can be either an equilibrium state of minimum free energy or be determined by kinetic factors. In the latter group, the interface-controlled models involve a rate-limiting step when molecules accrete onto the fibril surface, which depends on structural interactions at the fibril surface. In the case of diffusion-limited growth the rate-limiting step is the diffusion of collagen molecules to the fibril surface.

An early model involved a cumulative molecular strain with increasing fibril diameter (Chapman, 1966). This type of model has also been proposed for diameter regulation in the growth of fibrin fibrils (Weisel & Litvinov, 2013). Another equilibrium model was subsequently proposed on the basis of the nontriple helical telopeptide regions of the molecule generating a positive free-energy contribution on binding to the fibril surface and this predicted a minimum free energy at a specific fibril diameter (Haworth, 1972). Subsequently a third type of diameter regulation model was based on the (transient) retention of the N-propeptide of the collagen molecule during fibril growth. This model predicted preferred fibril diameters in accord with those in unimodal fibril diameter distribution observed in tissue (Parry & Craig, 1979). It is also applicable to the case of corneal collagen fibrils composed of type I and type V collagen molecules (Birk, Fitch, Babiarz, & Linsenmayer, 1988) where the collagen V component has a long-term retention of the N-propeptide domain (Linsenmayer et al., 1993). The other models involve kinetic factors in determining the fibril shape and size. These different types of kinetic models all predict a linear mass profile for the collagen fibril tips as observed in a range of collagen fibrils by STEM mass-mapping studies (Holmes et al., 1992, 1998). The simplest model assumes that the rate of molecular accretion onto the fibril surface is diffusion limited, rather than by the activation energy barrier of molecular binding to the fibril surface, and used a 3D computer growth simulation of accretion of single molecules (Parkinson et al., 1995). The first of the

Table 1 Summary of Collagen Fibril Growth Control Models

Source	Model	Type	No. of Independent Parameters	Experimental Support
Chapman (1966)	Cumulative strain	Equilibrium	N/A	Electron microscopy
Haworth (1972)	Negative surface free energy	Equilibrium	N/A	Electron microscopy
Chapman (1989)	Retained N-propeptide on the surface of type I-containing fibrils	Equilibrium	1	Preferred fibril diameters in unimodal distributions
Silver, Miller, Harrison, and Prockop (1992)	Computer simulation with variable accretion rates	Interface controlled	>10	Linear mass profiles of fibril tips
Parkinson, Kadler, and Brass (1995)	DLA (diffusion-limited aggregation)	Diffusion limited	2	Linear mass profiles of fibril tips
Trotter, Kadler, and Holmes (2000)	SNAP (surface nucleation and propagation)	Three surface nucleation sites on N,N-bipolar fibrils	2	Linear mass profiles of fibril tips Matches growth curve features of N,N-bipolar fibrils of echinoderms

N/A, not applicable.

interface-controlled kinetic models was also a computer-generated 3D simulation of fibril growth involving multiple accretion rate parameters where the accretion rate depended on the local fibril diameter of the growing tip (Silver et al., 1992). Subsequently, a new form of diameter limitation model was proposed for N,N-bipolar fibrils involving three axial sites for surface nucleation (Trotter et al., 2000). This surface nucleation hypothesis was proposed following the experimental determination of fibril growth curves for N,N-bipolar fibrils extending up to 440 μm in length, as described in more detail below.

4.2 Fibril Fusion

Data from serial section reconstruction electron microscopy of tendon have been used to propose that the increase in fibril diameter and length within a relatively mature tendon is a result of linear and lateral fusion of fibril segments. The increase in fibril length was interpreted as being the result of a postdepositional regulated assembly of segments via an end-to-end fusion to form mature fibrils (Birk, Zycband, Woodruff, Winkelmann, & Trelstad, 1997). This model predicts multiple polarity changes along the fibril length based on the exclusive presence of N,N-bipolar fibrils. An important observation was that both unipolar and N,N-bipolar fibrils are present in embryonic tendon and end-to-end fusion of collagen fibrils requires a C-end (Graham, Holmes, Watson, & Kadler, 2000). C-ends only occur in unipolar fibrils; in N,N-bipolar fibrils the C-termini of all the molecules are buried within the body of the fibril. Thus, in a starting population of both unipolar and N,N-bipolar fibrils, and with subsequent removal of C-ends by fusion of unipolar fibrils with other unipolar fibril C-ends or the N-ends of N,N-bipolar fibrils, a stable population of N,N-bipolar fibrils is attained (Fig. 9). These fibrils could potentially continue to grow in the absence of fusion both axially and laterally by molecular accretion and provide a basis for the postnatal growth of collagen fibrils, as reported in a 3D EM study (Kalson et al., 2015; see Section 5).

4.3 Lessons From Echinoderms

In the mutable tissues of the echinoderms, collagen fibrils grow in the absence of interfibrillar fusion. Instead, they grow entirely by molecular accretion onto a single fibril nucleus. The growth of these fibril types has been studied in detail over a large length range 14–444 μm for sea cucumber dermis (Trotter, Chapman, Kadler, & Holmes, 1998) and 37–431 μm for sea urchin ligament (Trotter et al., 2000) and sets of growth curves have been assembled using STEM mass mapping. These fibrils are exclusively N,N-bipolar and have a local region midway along their length with antiparallel packing (polarity transition region) and unidirectional molecules outside this zone such that the N-termini of the molecules point toward the tips (Fig. 9). Importantly, the polarity transition region occurs midway along the fibril length. Similar growth curves could be generated by a simple model involving surface nucleation and propagation ("SNAP" model) (see Section 4.1). The model involves just two adjustable kinetic parameters to fit the experimental growth curves, and predicts the characteristic slope transitions in the mass profiles of the tip regions (Fig. 8). This example of collagen fibril

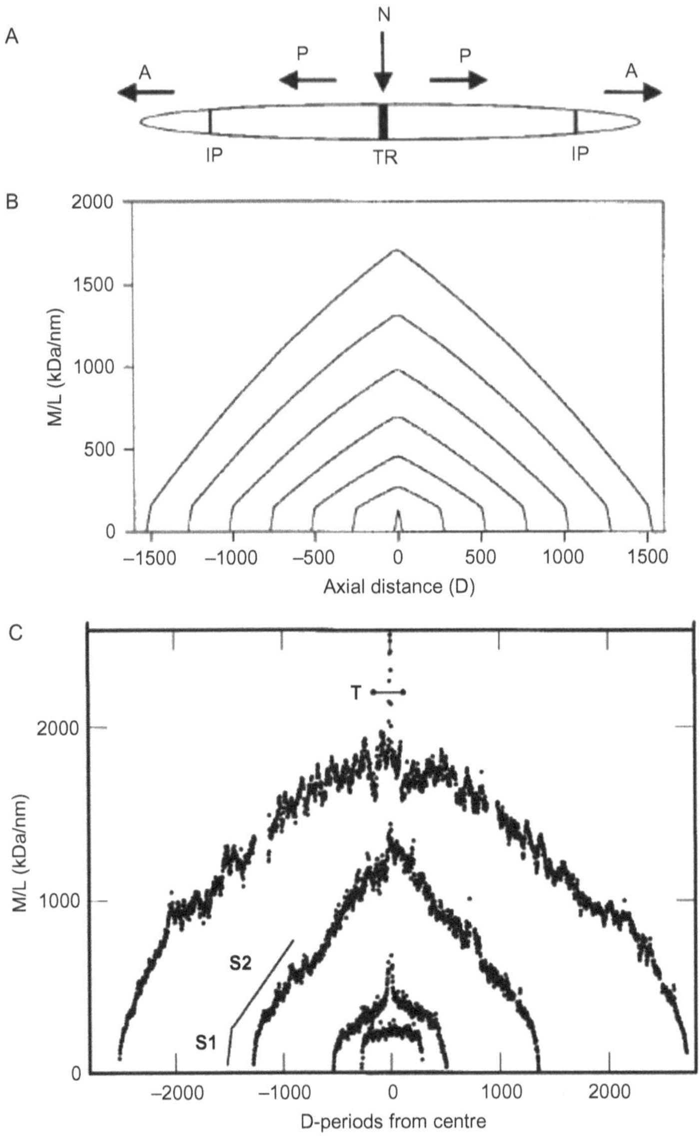

Fig. 8 Surface nucleation and propagation (SNAP) model to simulate the growth characteristics of N,N-bipolar collagen fibrils from the mutable tissues of echinoderms. (A) Schematic showing the growth features of the SNAP model. The model is based on three unique sites along the fibril—the central localized region of antiparallel collagen molecules, the polarity transition region (PTR), and the two ends of the fibrils. Growth waves are initiated by nucleation at the PTR followed by propagation down the length of the fibrils. Only three kinetic parameters are assumed: the rate of nucleation (N) at the PTR, the propagation rate (P) down the fibril, and the tip elongation rate (A). (B) Two independent parameters (N/P and P/A) determine the predicted shape of the growth curves to match the experimental data shown in (C).

growth establishes the concept of surface nucleation as a feasible major determinant of fibril size and shape. It is also applicable to the growth of type I collagen N,N-bipolar fibrils in the cell-free system and to N,N-bipolar fibrils that occur, along with unipolar fibrils, in vertebrate tissues. In the latter case it provides a growth regulation mechanism to explain the continued slow increase in fibril diameter in the postnatal vertebrate tendon (see below).

5. COLLAGEN FIBRIL FORMATION IN VIVO: EXTRINSIC CONTROL OF FIBRIL FORMATION

Although collagen molecules can spontaneously self-assemble into fibrils in vitro, additional factors must exist in vivo to account for the tissue-specific alignment of fibrils. Furthermore, fibrils in vivo exhibit regulation of diameter depending on tissue and stage of development (Craig & Parry, 1981b; Parry et al., 1978; Parry & Craig, 1977). The in vivo regulation of collagen fibril formation has been studied for over a century, and although enormous progress has been made, the mechanisms of collagen fibril assembly and three-dimensional organization in vivo remain elusive.

A primary function of collagen fibrils is to stress-shield cells from mechanical forces that would destroy isolated cells. Cells in vivo can be exposed to different forces ranging from compression to tension and shear. Therefore, the number, length, diameter, and three-dimensional arrangement of collagen fibrils in vivo are tuned to the mechanical requirements of individual tissues. For example, narrow fibrils are arranged in close-packed orthogonal lattices in cornea to generate a tissue that is both mechanically resistant to hydrodynamic pressure and is transparent. Furthermore, fibrils with a broad distribution of diameters arranged in close-packed parallel crimped fibers occur in tendon to transmit forces generated by muscle. At the other extreme, an open network of narrow fibrils occurs in cartilage and vitreous humor to resist swelling pressure. The three-dimensional arrangement of fibrils and fibril number are established in the embryo in preparation for the stresses of postnatal life. All collagen fibrils are relatively short and narrow in the embryo but can, depending on tissue, increase in length and diameter to match the growing skeleton and changing mechanical requirements.

5.1 Site of Fibril Assembly

The earliest reports on the existence of collagen fibrils date back to the early 1900s. Mallory described a "fibrillar substance" produced by connective tissue cells (Mallory, 1903). A breakthrough came in 1940 when Mary Stearns published her observations of fibroblasts "secreting collagen fibers" (Stearns, 1940). She published drawings of fibers growing at the cell surface. Almost 40 years later, Trelstad and Hayashi used TEM to show collagen fibrils in invaginations of the plasma membrane of embryonic fibroblasts (Trelstad & Hayashi, 1979). A decade later these observations were extended using high-voltage TEM of cornea and embryonic chick tendon (Birk & Trelstad, 1984, 1985, 1986; Trelstad & Birk, 1985). With improvements in image analysis methods it became possible to generate three-dimensional reconstructions from TEM images, and later serial block face-scanning electron microscopy (Starborg et al., 2013), to image collagen fibrils contained within plasma membrane structures called "fibripositors" (Canty et al., 2004). Fibripositors are actin-dependent (Canty et al., 2006) invaginations of the plasma membrane that are powered by nonmuscle myosin II to transport newly assembled collagen fibrils (Kalson et al., 2013).

5.2 Regulation of Collagen Fibril Number

Intuitively, tissues containing larger numbers of fibrils with the same diameter distributions and 3D organization might be expected to be stronger and stiffer. A mechanism must exist to regulate the number of fibrils in any given tissue, but no homeostatic mechanism has been studied. Gene knockout studies in mice have shown that tenascin X and type V collagen are critical regulators of fibril number; the type V collagen-deficient mouse lacks collagen fibrils in tendon and the heterozygous mouse contains approximately half the number of fibrils (Wenstrup et al., 2004); furthermore, the tenascin X knockout mouse has approximately 50% of the collagen fibrils compared to wild-type control animals (Mao et al., 2002). From the limited number of studies that have been performed, it appears that fibril numbers in the mouse Achilles tendon increase steadily throughout embryonic development and reach a final limit at birth. Fibrils then grow steadily in length and diameter during the subsequent tissue growth period (Kalson et al., 2015).

5.3 Fibril Length Regulation

Collagen fibrils range in length from a few microns to centimeters (Craig, Birtles, Conway, & Parry, 1989) and therefore have molecular weights in the tera Dalton range (based on calculations described by Chapman, 1989). Direct measurement of fibril length has been largely restricted to "early fibrils" (or "fibril segments") found in embryonic tissues where entire fibrils can be observed in the length range ~2–200 μm (Birk & Zycband, 1994; Graham et al., 2000). Entire fibrils of greater length cannot be generally isolated from vertebrate tissues or viewed in 3D reconstructions. The mutable tissues of echinoderms are an exception and entire collagen fibrils of up to ~3 mm in length can be observed in tissue dispersions (Trotter & Koob, 1989). Collagen fibrils can grow in length by either accretion onto the tips or by end-to-end fusion (as described above). The presence of N,N-bipolar fibrils is a potential mechanism to limit this latter process in vertebrate tissues (Fig. 9). The mutable tissues of echinoderms contain exclusively

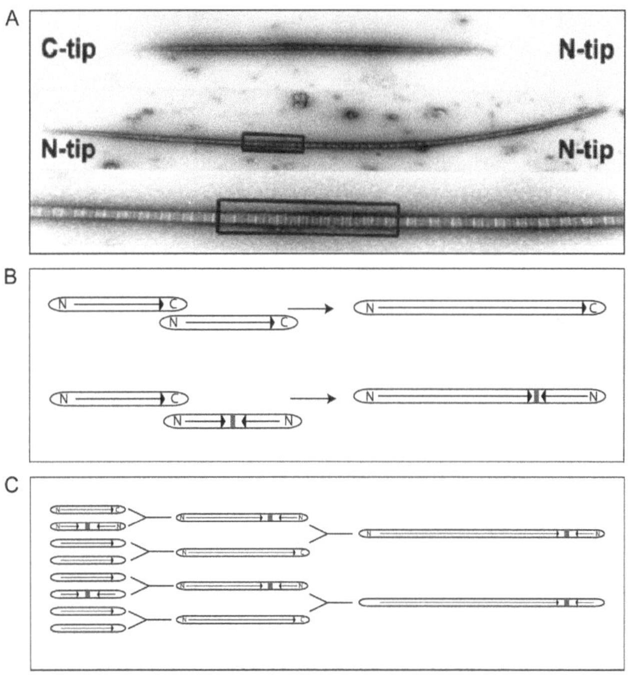

Fig. 9 The possible involvement of N,N-bipolar fibrils in limiting end-to-end fibril fusion. (A) A unipolar and N,N-bipolar fibril obtained from chick embryonic tendon and negatively stained with uranyl acetate. The polarity transition region of antiparallel molecular packing (marked by a *rectangle*) is shown enlarged. (B) End-to-end fusion requires the C-end of a unipolar fibril; N–N tip fusion is not observed. A mixed population of unipolar and N,N-bipolar fibrils would finally become, after linear fusion, a stable population of N, N-bipolar incapable of further fusion.

N,N-bipolar fibrils and these fibrils do not undergo interfibrillar fusion but grow exclusively by an accretion process in a near-symmetric manner (see Section 4.3). An average length of collagen fibrils in mature tissues can be estimated by counting the number of fibril ends either in sections or in fibril suspensions after mechanically dispersion (Craig et al., 1989; Starborg et al., 2013). Using this latter method the average length of collagen fibrils in bovine cornea has been estimated as ~600 μm (Holmes & Kadler, 2005), although the length distribution remains unknown. Recently 3D reconstructions of the murine stapedius tendon have been achieved and in this shortest tendon type (length ~300 μm) the collagen fibrils appear to run the entire length of the tendon (Svensson et al., 2017).

5.4 Fibril Diameter Regulation

Collagen fibril diameters vary depending on species, tissue, stage of development (Craig et al., 1989; Parry et al., 1978), and in response to injury and repair (Pingel et al., 2014), and typically show characteristic unimodal (Canty et al., 2004) or bimodal (Goh et al., 2012) distributions. Typically, unimodal diameter distributions show a mean diameter of a multiple of 8 nm (Parry & Craig, 1979). This has been corrected to a multiple of 11 nm allowing for shrinkage during dehydration for electron microscopy (Chapman, 1989).

Measurement of fibril diameters by TEM is fairly routine and is often used to help characterize the phenotype of gene knockout studies in mice. Thus it has become clear that fibril diameters and the shape of fibril transverse profiles are affected by mutations in genes encoding type I, II, III, and V collagen as well as fibril-associated collagens with interrupted triple helices (FACITs) that bind to the surfaces of collagen fibrils, e.g., type XII and type XIV collagen (Young, Zhang, Koch, & Birk, 2002), proteoglycans that interact with fibrils, e.g., decorin (Danielson et al., 1997), lumican (Chakravarti et al., 1998), fibromodulin (Hedlund, Mengarelli-Widholm, Heinegard, Reinholt, & Svensson, 1994; Svensson et al., 1999), osteoglycin (Tasheva et al., 2002), keratocan (Liu, Birk, Hassell, Kane, & Kao, 2003), and biglycan (Heegaard et al., 2007) (for review, see Kalamajski & Oldberg, 2010), and enzymes required for posttranslational modification of collagen α-chains, e.g., prolyl 4-hydroxylase (Mussini, Hutton, & Udenfriend, 1967), lysyl hydroxylases (Takaluoma et al., 2007), and lysyl oxidases (Maki et al., 2002).

The retention of N-propeptides has also been linked to diameter limitation. Examples are the type I/V collagen fibrils of the cornea where the N-propeptides of the type V collagen are retained (Linsenmayer et al., 1993) and the thin type II/XI fibrils of cartilage where the type XI collagen retains its N-propeptide (Blaschke et al., 2000). The collagen N-propeptide is constrained to remain on the fibril surface during fibril assembly as is demonstrated in the formation of the D-periodic sheet assembles of type I pNcollagen (Hulmes et al., 1989). Thus, collagen fibrillogenesis is a precisely regulated process in which the mechanisms that maintain the appropriate number, size, and organization of collagen fibrils in adult tissues appear to be sensitive to a range of cell-mediated factors.

6. COLLAGEN FIBRIL STRUCTURE

Collagen fibril structure has been addressed primarily by two parallel approaches: electron microscopy and low-angle X-ray diffraction. The axial stain pattern of type I collagen fibrils has been extensively studied and interpreted in terms of the axially projected molecular structure in a periodic array with molecules axially staggered by 67 nm to generate the D-periodicity (Chapman & Hardcastle, 1974; Chapman, Tzaphlidou, Meek, & Kadler, 1990; Meek, Chapman, & Hardcastle, 1979). This approach which explains the 1D fibril structure has been more recently extended to the heterotypic fibrils of cartilage and vitreous that contain collagen types II, IX, and XI (Bos et al., 2001).

Elucidating the 3D structure of collagen fibrils has proved a major challenge for over 5 decades. There is a broad range of collagen fibril forms in tissue with various collagen compositions and diameters are previously indicated. Common structural features exist across the range of fibrils, but there is accumulated evidence of substantial 3D structural variation.

Rat-tail tendon which contains large fibrils with an extensive natural cross-link network produces the most informative low-angle X-ray diffraction patterns (Brodsky et al., 1978). The fibril structure in this tissue has a limited crystalline order and the degree of lateral packing order varies dependent on the axial position in the D-period. Zones of high lateral order are found in the cross-linked regions involving the ends of the molecule. Interpretation of this X-ray diffraction data led to a number of structural models over several decades involving alternative microfibril sizes (with 2, 4, 5, 7, or

8 molecular strands) (Hulmes & Miller, 1979). By the late 1980s a quasi-hexagonal packing scheme with a 5-stranded microfibrillar grouping of molecules was preferred over other models (Hulmes & Miller, 1979). Recently improved diffraction patterns have been generated using synchrotron X-ray sources and higher resolution models have been produced for the lateral structure of the crystalline regions of collagen fibrils from rat-tail tendon (Orgel, Irving, Miller, & Wess, 2006; Orgel et al., 2001; Orgel, Wess, & Miller, 2000). Elucidation of 3D structural aspects from other collagen fibrils from different tissues and different heterotypic composition is largely dependent on electron imaging methods including electron tomography as described below. In addition, both scanning electron microscopy and atomic force microscopy have given valuable information on the surface structure of collagen fibrils from tissue, revealing subfibrils of various diameters and fibril crimp morphology (Franchi et al., 2007, 2008; Raspanti et al., 2006; Raspanti, Reguzzoni, Protasoni, & Basso, 2018; Raspanti, Viola, Sonaggere, Tira, & Tenni, 2007). Further fibril surface structural information and the interfibril networks that occur in tissue have been obtained using the quick-freeze-deep-etch method generating replicas that can be viewed in the TEM (Hirsch, Noske, Prenant, & Renard, 1999; Hirsch, Prenant, & Renard, 2001).

Electron tomography of negatively stained collagen fibrils from mature bovine cornea has provided 3D structural information (Fig. 10) revealing microfibrils ($\sim$4 nm) diameter following a helical path around the long axis of the fibril (Marini et al., 2001). The microfibrils appeared to have a tilt of $\sim 15°$ consistent with a constant tilt rather than a constant pitch model (Raspanti, Ottani, & Ruggeri, 1989). Three axial sites of maximal lateral order were identified, corresponding to the "D-band" of the gap region together with those corresponding to the N- and C-ends of the molecule. Globular macromolecules were also evident on the fibril surface at specific axial sites within the D-period. Combined data from a variety of EM methods (STEM mass measurements, TEM axial stain pattern analysis, and TEM 2D images of negatively stained fibrils) have been analyzed to provide a 3D model for the structure of the thin heterotypic collagen fibrils of embryonic cartilage, containing collagen types I, IX, and XI (Holmes & Kadler, 2006) (Fig. 11). The data were consistent with a central core of $4 \times$ 5-stranded microfibrils (possibly formed of type XI collagen molecules) with a tilt of 2.5–3.5° surrounded by a sheath of $10 \times$ 5-stranded microfibrils (possibly formed of type II collagen molecules) with a tilt of $\sim 2.5°$. New high-resolution cryoelectron microscopy techniques (Frank, 2017; Wan & Briggs, 2016) now offer an opportunity for obtaining improved

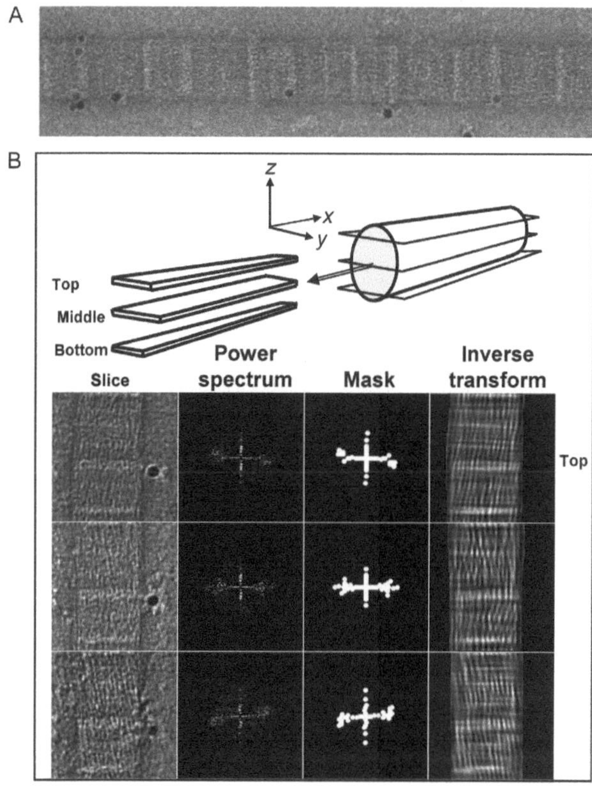

Fig. 10 Electron tomography of negatively stained corneal collagen fibrils. (A) Collagen fibril from bovine cornea after negative staining with uranyl acetate and the addition of colloidal gold particles as alignment markers. Tilt series (±60°) of similar fibril samples were acquired for tomographic 3D reconstruction. (B) Visualization of a ~4 nm microfibrillar structure in the 3D reconstruction. Longitudinal virtual slices (*x–y*) through the 3-D reconstruction sampling the fibril in the top, middle, and bottom zones, as indicated schematically. The raw slice images are shown in the *left-hand column* together with the power spectra (*second column*) and the power spectra masks (*third column*) that include the main peak intensities. The Fourier-filtered images obtained by using these masks are shown in the *right-hand column*. A filamentous substructure is apparent in the original images and this is enhanced by Fourier filtering. The filaments show a predominant tilt of about 115° and 215° in the upper and lower zones of the fibril, respectively. The tilt direction changes rapidly in the central zone. Both tilt components can be seen in the *middle slice*.

3D structural information from a range of collagen fibrils of different types from different tissues at different stages of development. Ultimately this would lead to a clarification of the mechanisms that regulate fibril growth (Figs. 12–15).

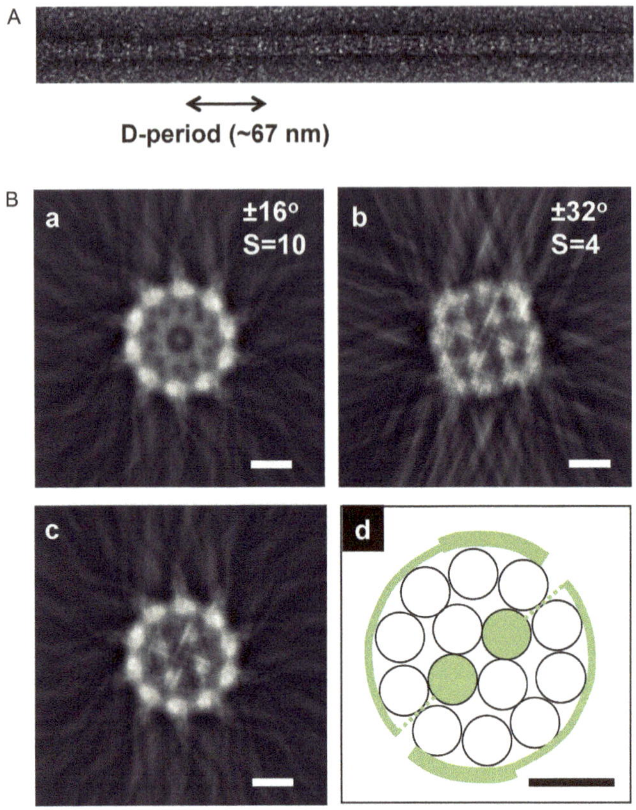

Fig. 11 Structural model for thin cartilage collagen fibrils. (A) Negatively stained thin collagen fibril from embryonic chick sternum. Experimental reconstructions B(a)–(c) of the transverse section of a negatively stained thin cartilage fibril together with proposed model structure are shown in B(d). Images B(a) and (b) were obtained by r-weighted backprojection from angle-limited tilt series using S = 10 and S = 4 rotational symmetry, respectively. The transverse reconstructions were averaged over 8 nm along the fibril axis. The outer core image in B(a) has been combined with the inner core image in B(b) to generate the composite image in B(c). The diameter of the fibril was measured as 15 nm and the center-to-center microfibrillar spacings were measured as 3.8 and 3.9 nm for the outer and inner cores, respectively. B(d) Schematic model shows the transverse structure of the thin cartilage fibril. This transverse section corresponds to the axial location of the nontriple helical component of the type XI N-propeptide. The structural components are 5-stranded microfibrils of collagen type II (*open circles*) and collagen type XI (*green*). Each microfibril of type XI collagen would result in one N-propeptide per D-period. The hypothetical circumferential extent of these domains is shown. The *dotted lines* show the projection of the tilted minor triple helix of the N-propeptides. (Scale bars, 5 nm.)

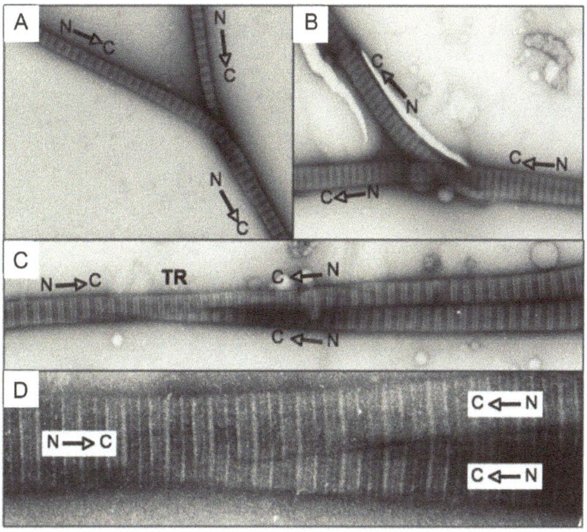

Fig. 12 Examination of fibril polarity across points of anastomosis. Fibrils were extracted by Dounce homogenization in gentamicin, followed by negative staining in 2% uranyl acetate. (A) Branch point where the branch arms are aligned with the carboxyl termini pointing toward the fusion point. (B) Branch point where each branch is aligned with the amino termini pointing toward the fusion point. (C and enlargement D) Anastomosing fibril where the fusion point occurs at a transition region (TR). Therefore, each branch limb and the single shaft are aligned so that the carboxyl termini point toward the point of fusion.

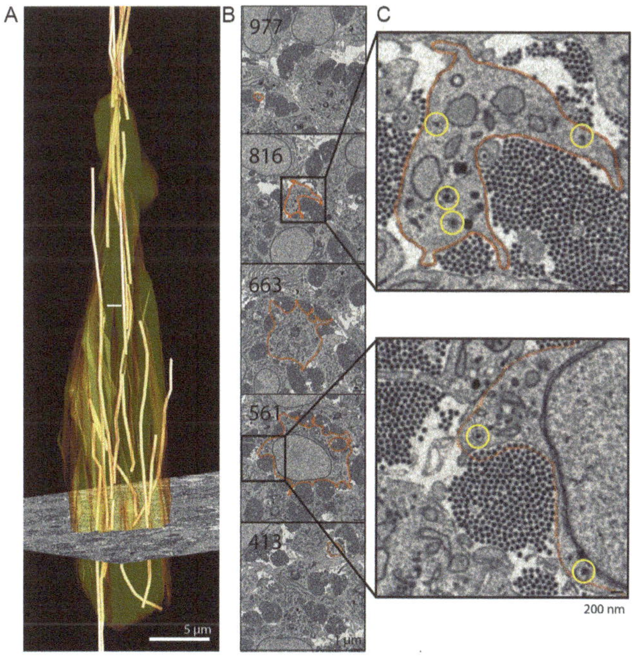

Fig. 13 Serial block face-scanning electron microscopy study of the cell–matrix interface in embryonic chick tendon. (A) A single cell (*gold*) showing collagen fibrils (*yellow*) associated with the plasma membrane. An EM image is superimposed. (B) Block face images corresponding to cuts 413, 561, 663, 816, and 977. *Orange*, plasma membrane. (C) Higher power views of cuts 561 and 816 showing collagen fibrils in fibripositors (*yellow circles*).

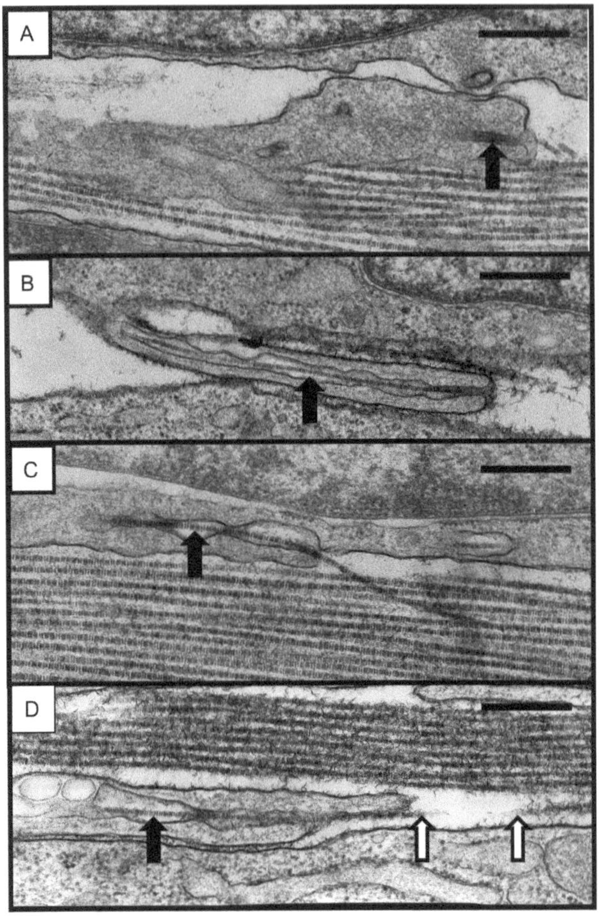

Fig. 14 Transmission electron microscopy of embryonic chick metatarsal tendon. Longitudinal sections show fibrils (*arrow*) contained within fibripositors. Bar, 500 nm.

7. FUTURE DIRECTIONS

Major advances have been made in understanding key aspects of the formation of tissue-specific arrays of collagen fibrils, as described in this review. It is now apparent that a complex and highly regulated cellular machinery is at play in vivo and this operates in combination with intrinsic and specific self-assembly properties of the collagen fibrils. Multiscale 3D imaging studies are now in progress involving high-resolution cryoelectron tomography, serial block face-scanning electron microscopy, confocal light microscopy, and correlative approaches using a combination of these

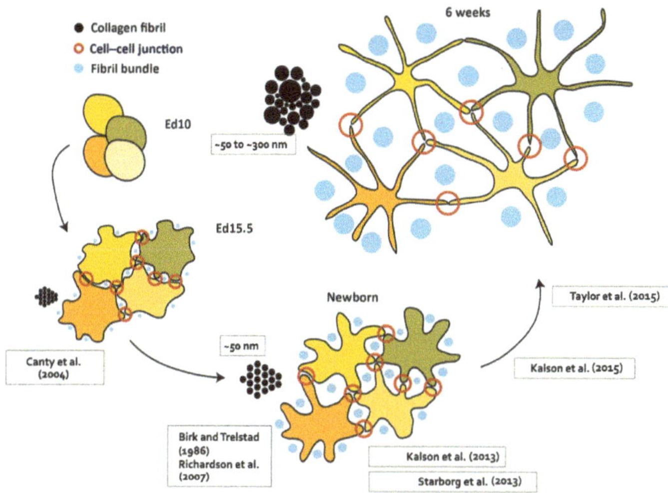

Fig. 15 Model of murine tendon development from embryonic day 10 through to 6 weeks postnatal. Early embryonic tendon tissue is populated by progenitor cells. Collagen fibrils begin to appear in channels formed by cell–cell adhesions by E15.5. Channels become more pronounced as matrix expansion occurs. Inflation of the matrix in the channels and persistence of cell–cell contacts result in greater distance between cell bodies and larger fibril bundles in postnatal tissues.

techniques. Studies so far have been largely limited to static snapshots of the cell–extracellular matrix structures, but the development of super-resolution live cell imaging techniques now offers the opportunity of dynamic studies where the sequence of events involved in fibril nucleation, growth, and fusion can be directly observed. This is expected to reveal not only the molecular processes involved in fibril deposition and growth in tissue but also the cell-based mechanisms of maintenance and repair that must operate to preserve long-term mechanical function in a healthy extracellular matrix.

ACKNOWLEDGMENTS

The authors are grateful to generous research funding from the Wellcome Trust (110126/Z/15/Z, 203128/Z/16/Z) and the European Union Marie Curie RISE program (project number #690850). We also acknowledge the support of the Electron Microscopy Faculty in the Faculty of Biology, Medicine and Health, at the University of Manchester.

REFERENCES

Bella, J. (2016). Collagen structure: New tricks from a very old dog. *The Biochemical Journal*, *473*, 1001–1025.

Bella, J., Eaton, M., Brodsky, B., & Berman, H. M. (1994). Crystal and molecular structure of a collagen-like peptide at 1.9 A resolution. *Science*, *266*, 75–81.

Bella, J., & Hulmes, D. J. (2017). Fibrillar collagens. *Sub-Cellular Biochemistry*, *82*, 457–490.

Bellamy, G., & Bornstein, P. (1971). Evidence for procollagen, a biosynthetic precursors of collagen. *Proceedings of the National Academy of Sciences of the United States of America*, *68*, 1138–1142.

Birk, D. E., Fitch, J. M., Babiarz, J. P., & Linsenmayer, T. F. (1988). Collagen type I and type V are present in the same fibril in the avian corneal stroma. *The Journal of Cell Biology*, *106*, 999–1008.

Birk, D. E., & Trelstad, R. L. (1984). Extracellular compartments in matrix morphogenesis: Collagen fibril, bundle, and lamellar formation by corneal fibroblasts. *The Journal of Cell Biology*, *99*, 2024–2033.

Birk, D. E., & Trelstad, R. L. (1985). Fibroblasts create compartments in the extracellular space where collagen polymerizes into fibrils and fibrils associate into bundles. *Annals of the New York Academy of Sciences*, *460*, 258–266.

Birk, D. E., & Trelstad, R. L. (1986). Extracellular compartments in tendon morphogenesis: Collagen fibril, bundle, and macroaggregate formation. *Journal of Cell Biology*, *103*, 231–240.

Birk, D. E., & Zycband, E. (1994). Assembly of the tendon extracellular matrix during development. *Journal of Anatomy*, *184*(Pt. 3), 457–463.

Birk, D. E., Zycband, E. I., Winkelmann, D. A., & Trelstad, R. L. (1989). Collagen fibrillogenesis in situ: Fibril segments are intermediates in matrix assembly. *Proceedings of the National Academy of Sciences of the United States of America*, *86*, 4549–4553.

Birk, D. E., Zycband, E. I., Woodruff, S., Winkelmann, D. A., & Trelstad, R. L. (1997). Collagen fibrillogenesis in situ: Fibril segments become long fibrils as the developing tendon matures. *Developmental Dynamics*, *208*, 291–298.

Blaschke, U. K., Eikenberry, E. F., Hulmes, D. J., Galla, H. J., & Bruckner, P. (2000). Collagen XI nucleates self-assembly and limits lateral growth of cartilage fibrils. *The Journal of Biological Chemistry*, *275*, 10370–10378.

Bos, K. J., Holmes, D. F., Kadler, K. E., McLeod, D., Morris, N. P., & Bishop, P. N. (2001). Axial structure of the heterotypic collagen fibrils of vitreous humour and cartilage. *Journal of Molecular Biology*, *306*, 1011–1022.

Brodsky, B., Hukins, D. W., Hulmes, D. J., Miller, A., White, S., & Woodhead-Galloway, J. (1978). Low angle X-ray diffraction studies on stained rat tail tendons. *Biochimica et Biophysica Acta*, *535*, 25–32.

Brodsky, B., & Persikov, A. V. (2005). Molecular structure of the collagen triple helix. *Advances in Protein Chemistry*, *70*, 301–339.

Brodsky, B., & Ramshaw, J. A. (1997). The collagen triple-helix structure. *Matrix Biology*, *15*, 545–554.

Canty, E. G., Lu, Y., Meadows, R. S., Shaw, M. K., Holmes, D. F., & Kadler, K. E. (2004). Coalignment of plasma membrane channels and protrusions (fibripositors) specifies the parallelism of tendon. *Journal of Cell Biology*, *165*, 553–563.

Canty, E. G., Starborg, T., Lu, Y., Humphries, S. M., Holmes, D. F., Meadows, R. S., et al. (2006). Actin filaments are required for fibripositor-mediated collagen fibril alignment in tendon. *Journal of Biological Chemistry*, *281*, 38592–38598.

Capaldi, M. J., & Chapman, J. A. (1982). The C-terminal extrahelical peptide of type I collagen and its role in fibrillogenesis in vitro. *Biopolymers*, *21*, 2291–2313.

Chakravarti, S., Magnuson, T., Lass, J. H., Jepsen, K. J., LaMantia, C., & Carroll, H. (1998). Lumican regulates collagen fibril assembly: Skin fragility and corneal opacity in the absence of lumican. *The Journal of Cell Biology*, *141*, 1277–1286.

Chapman, J. A. (1966). In G. E. W. Wolstenholme & M. O'Connor (Eds.), Discussion in *Principles of Biomolecular Organization* (pp. 129–130). London: CIBA Foundation Symposium, Churchill.

Chapman, J. A. (1989). The regulation of size and form in the assembly of collagen fibrils in vivo. *Biopolymers, 28*, 1367–1382.
Chapman, J. A., & Hardcastle, R. A. (1974). The staining pattern of collagen fibrils. II. A comparison with patterns computer-generated from the amino acid sequence. *Connective Tissue Research, 2*, 151–159.
Chapman, J. A., Tzaphlidou, M., Meek, K. M., & Kadler, K. E. (1990). The collagen fibril—A model system for studying the staining and fixation of a protein. *Electron Microscopy Reviews, 3*, 143–182.
Craig, A. S., Birtles, M. J., Conway, J. F., & Parry, D. A. (1989). An estimate of the mean length of collagen fibrils in rat tail-tendon as a function of age. *Connective Tissue Research, 19*, 51–62.
Craig, A. S., & Parry, D. A. (1981a). Growth and development of collagen fibrils in immature tissues from rat and sheep. *Proceedings of the Royal Society of London B: Biological Sciences, 212*, 85–92.
Craig, A. S., & Parry, D. A. (1981b). Collagen fibrils of the vertebrate corneal stroma. *Journal of Ultrastructure Research, 74*, 232–239.
Danielson, K. G., Baribault, H., Holmes, D. F., Graham, H., Kadler, K. E., & Iozzo, R. V. (1997). Targeted disruption of decorin leads to abnormal collagen fibril morphology and skin fragility. *Journal of Cell Biology, 136*, 729–743.
Deak, S. B., van der Rest, M., & Prockop, D. J. (1985). Altered helical structure of a homotrimer of alpha 1(I)chains synthesized by fibroblasts from a variant of osteogenesis imperfecta. *Collagen and Related Research, 5*, 305–313.
Di Lullo, G. A., Sweeney, S. M., Korkko, J., Ala-Kokko, L., & San Antonio, J. D. (2002). Mapping the ligand-binding sites and disease-associated mutations on the most abundant protein in the human, type I collagen. *The Journal of Biological Chemistry, 277*, 4223–4231.
Exposito, J. Y., Valcourt, U., Cluzel, C., & Lethias, C. (2010). The fibrillar collagen family. *International Journal of Molecular Sciences, 11*, 407–426.
Eyre, D. R., Weis, M. A., & Wu, J. J. (2008). Advances in collagen cross-link analysis. *Methods, 45*, 65–74.
Franchi, M., Fini, M., Quaranta, M., De Pasquale, V., Raspanti, M., Giavaresi, G., et al. (2007). Crimp morphology in relaxed and stretched rat Achilles tendon. *Journal of Anatomy, 210*, 1–7.
Franchi, M., Raspanti, M., Dell'Orbo, C., Quaranta, M., De Pasquale, V., Ottani, V., et al. (2008). Different crimp patterns in collagen fibrils relate to the subfibrillar arrangement. *Connective Tissue Research, 49*, 85–91.
Frank, J. (2017). Advances in the field of single-particle cryo-electron microscopy over the last decade. *Nature Protocols, 12*, 209–212.
Gathercole, L. J., & Keller, A. (1991). Crimp morphology in the fibre-forming collagens. *Matrix, 11*, 214–234.
Gelman, R. A., Williams, B. R., & Piez, K. A. (1979). Collagen fibril formation. Evidence for a multistep process. *The Journal of Biological Chemistry, 254*, 180–186.
Ghosh, N., McKillop, T. J., Jowitt, T. A., Howard, M., Davies, H., Holmes, D. F., et al. (2012). Collagen-like proteins in pathogenic E. coli strains. *PLoS One, 7*, e37872.
Goh, K. L., Holmes, D. F., Lu, Y., Purslow, P. P., Kadler, K. E., Bechet, D., et al. (2012). Bimodal collagen fibril diameter distributions direct age-related variations in tendon resilience and resistance to rupture. *Journal of Applied Physiology, 113*, 878–888.
Graham, H. K., Holmes, D. F., Watson, R. B., & Kadler, K. E. (2000). Identification of collagen fibril fusion during vertebrate tendon morphogenesis. The process relies on unipolar fibrils and is regulated by collagen-proteoglycan interaction. *Journal of Molecular Biology, 295*, 891–902.
Gross, J., & Kirk, D. (1958). The heat precipitation of collagen from neutral salt solutions: Some rate-regulating factors. *The Journal of Biological Chemistry, 233*, 355–360.

Hansen, U., & Bruckner, P. (2003). Macromolecular specificity of collagen fibrillogenesis: Fibrils of collagens I and XI contain a heterotypic alloyed core and a collagen I sheath. *The Journal of Biological Chemistry*, *278*, 37352–37359.

Haworth, R. A. (1972). *A study of the growth of collagen fibrils in vitro using electron microscope autoradiography*. PhD thesis (p. 95). Department of Medical Biophysics, University of Manchester.

Haworth, R. A., & Chapman, J. A. (1977). A study of the growth of normal and iodinated collagen fibrils in vitro using electron microscope autoradiography. *Biopolymers*, *16*, 1895–1906.

Hedlund, H., Mengarelli-Widholm, S., Heinegard, D., Reinholt, F. P., & Svensson, O. (1994). Fibromodulin distribution and association with collagen. *Matrix Biology*, *14*, 227–232.

Heegaard, A. M., Corsi, A., Danielsen, C. C., Nielsen, K. L., Jorgensen, H. L., Riminucci, M., et al. (2007). Biglycan deficiency causes spontaneous aortic dissection and rupture in mice. *Circulation*, *115*, 2731–2738.

Helseth, D. L., Jr., & Veis, A. (1981). Collagen self-assembly in vitro. Differentiating specific telopeptide-dependent interactions using selective enzyme modification and the addition of free amino telopeptide. *The Journal of Biological Chemistry*, *256*, 7118–7128.

Herchenhan, A., Kalson, N. S., Holmes, D. F., Hill, P., Kadler, K. E., & Margetts, L. (2012). Tenocyte contraction induces crimp formation in tendon-like tissue. *Biomechanics and Modeling in Mechanobiology*, *11*, 449–459.

Hirsch, M., Noske, W., Prenant, G., & Renard, G. (1999). Fine structure of the developing avian corneal stroma as revealed by quick-freeze, deep-etch electron microscopy. *Experimental Eye Research*, *69*, 267–277.

Hirsch, M., Prenant, G., & Renard, G. (2001). Three-dimensional supramolecular organization of the extracellular matrix in human and rabbit corneal stroma, as revealed by ultrarapid-freezing and deep-etching methods. *Experimental Eye Research*, *72*, 123–135.

Hojima, Y., McKenzie, J. A., van der Rest, M., & Prockop, D. J. (1989). Type I procollagen N-proteinase from chick embryo tendons. Purification of a new 500-kDa form of the enzyme and identification of the catalytically active polypeptides. *The Journal of Biological Chemistry*, *264*, 11336–11345.

Hojima, Y., van der Rest, M., & Prockop, D. J. (1985). Type I procollagen carboxyl-terminal proteinase from chick embryo tendons. Purification and characterization. *The Journal of Biological Chemistry*, *260*, 15996–16003.

Holmes, D. F., Capaldi, M. J., & Chapman, J. A. (1981). Reconstitution of collagen fibrils in vitro; the assembly process depends on the initiating procedure. *International Journal of Biological Macromolecules*, *8*, 161–166.

Holmes, D. F., Chapman, J. A., Prockop, D. J., & Kadler, K. E. (1992). Growing tips of type I collagen fibrils formed in vitro are near-paraboloidal in shape, implying a reciprocal relationship between accretion and diameter. *Proceedings of the National Academy of Sciences of the United States of America*, *89*, 9855–9859.

Holmes, D. F., Gilpin, C. J., Baldock, C., Ziese, U., Koster, A. J., & Kadler, K. E. (2001). Corneal collagen fibril structure in three dimensions: Structural insights into fibril assembly, mechanical properties, and tissue organization. *Proceedings of the National Academy of Sciences of the United States of America*, *98*, 7307–7312.

Holmes, D. F., Graham, H. K., & Kadler, K. E. (1998). Collagen fibrils forming in developing tendon show an early and abrupt limitation in diameter at the growing tips. *Journal of Molecular Biology*, *283*, 1049–1058.

Holmes, D. F., Graham, H. K., Trotter, J. A., & Kadler, K. E. (2001). STEM/TEM studies of collagen fibril assembly. *Micron*, *32*, 273–285.

Holmes, D. F., & Kadler, K. E. (2005). The precision of lateral size control in the assembly of corneal collagen fibrils. *Journal of Molecular Biology*, *345*, 773–784.

Holmes, D. F., & Kadler, K. E. (2006). The 10+4 microfibril structure of thin cartilage fibrils. *Proceedings of the National Academy of Sciences of the United States of America, 103*, 17249–17254.
Holmes, D. F., Lowe, M. P., & Chapman, J. A. (1994). Vertebrate (chick) collagen fibrils formed in vivo can exhibit a reversal in molecular polarity. *Journal of Molecular Biology, 235*, 80–83.
Holmes, D. F., Watson, R. B., Chapman, J. A., & Kadler, K. E. (1996). Enzymic control of collagen fibril shape. *Journal of Molecular Biology, 261*, 93–97.
Hulmes, D. J., Kadler, K. E., Mould, A. P., Hojima, Y., Holmes, D. F., Cummings, C., et al. (1989). Pleomorphism in type I collagen fibrils produced by persistence of the procollagen N-propeptide. *Journal of Molecular Biology, 210*, 337–345.
Hulmes, D. J., & Miller, A. (1979). Quasi-hexagonal molecular packing in collagen fibrils. *Nature, 282*, 878–880.
Huxley-Jones, J., Robertson, D. L., & Boot-Handford, R. P. (2007). On the origins of the extracellular matrix in vertebrates. *Matrix Biology: Journal of the International Society for Matrix Biology, 26*, 2–11.
Johnstone, I. L. (2000). Cuticle collagen genes. Expression in Caenorhabditis elegans. *Trends in Genetics, 16*, 21–27.
Jokinen, J., Dadu, E., Nykvist, P., Kapyla, J., White, D. J., Ivaska, J., et al. (2004). Integrin-mediated cell adhesion to type I collagen fibrils. *The Journal of Biological Chemistry, 279*, 31956–31963.
Kadler, K. E., Baldock, C., Bella, J., & Boot-Handford, R. P. (2007). Collagens at a glance. *Journal of Cell Science, 120*, 1955–1958.
Kadler, K. E., Hojima, Y., & Prockop, D. J. (1987). Assembly of collagen fibrils de novo by cleavage of the type I pC-collagen with procollagen C-proteinase. Assay of critical concentration demonstrates that collagen self-assembly is a classical example of an entropy-driven process. *The Journal of Biological Chemistry, 262*, 15696–15701.
Kadler, K. E., Hojima, Y., & Prockop, D. J. (1988). Assembly of type I collagen fibrils de novo. Between 37 and 41 degrees C the process is limited by micro-unfolding of monomers. *The Journal of Biological Chemistry, 263*, 10517–10523.
Kadler, K. E., Hojima, Y., & Prockop, D. J. (1990). Collagen fibrils in vitro grow from pointed tips in the C- to N-terminal direction. *The Biochemical Journal, 268*, 339–343.
Kalamajski, S., & Oldberg, A. (2010). The role of small leucine-rich proteoglycans in collagen fibrillogenesis. *Matrix Biology, 29*, 248–253.
Kalson, N. S., Lu, Y., Taylor, S. H., Starborg, T., Holmes, D. F., & Kadler, K. E. (2015). A structure-based extracellular matrix expansion mechanism of fibrous tissue growth. *eLife, 4*, e05958.
Kalson, N. S., Starborg, T., Lu, Y., Mironov, A., Humphries, S. M., Holmes, D. F., et al. (2013). Nonmuscle myosin II powered transport of newly formed collagen fibrils at the plasma membrane. *Proceedings of the National Academy of Sciences of the United States of America, 110*, E4743–4752.
Koch, M., Laub, F., Zhou, P., Hahn, R. A., Tanaka, S., Burgeson, R. E., et al. (2003). Collagen XXIV, a vertebrate fibrillar collagen with structural features of invertebrate collagens: Selective expression in developing cornea and bone. *The Journal of Biological Chemistry, 278*, 43236–43244.
Legendre, M., Santini, S., Rico, A., Abergel, C., & Claverie, J. M. (2011). Breaking the 1000-gene barrier for Mimivirus using ultra-deep genome and transcriptome sequencing. *Virology Journal, 8*, 99.
Leibovich, S. J., & Weiss, J. B. (1970). Electron microscope studies of the effects of endo- and exopeptidase digestion on tropocollagen. A novel concept of the role of terminal regions in fibrillogenesis. *Biochimica et Biophysica Acta, 214*, 445–454.

Linsenmayer, T. F., Gibney, E., Igoe, F., Gordon, M. K., Fitch, J. M., Fessler, L. I., et al. (1993). Type V collagen: Molecular structure and fibrillar organization of the chicken alpha 1(V) NH2-terminal domain, a putative regulator of corneal fibrillogenesis. *The Journal of Cell Biology*, *121*, 1181–1189.

Liu, C. Y., Birk, D. E., Hassell, J. R., Kane, B., & Kao, W. W. (2003). Keratocan-deficient mice display alterations in corneal structure. *The Journal of Biological Chemistry*, *278*, 21672–21677.

Maki, J. M., Rasanen, J., Tikkanen, H., Sormunen, R., Makikallio, K., Kivirikko, K. I., et al. (2002). Inactivation of the lysyl oxidase gene Lox leads to aortic aneurysms, cardiovascular dysfunction, and perinatal death in mice. *Circulation*, *106*, 2503–2509.

Mallory, F. B. (1903). A hitherto undescribed fibrillar substance produced by connective-tissue cells. *Journal of Medical Science*, *10*, 334–341.

Mao, J. R., Taylor, G., Dean, W. B., Wagner, D. R., Afzal, V., Lotz, J. C., et al. (2002). Tenascin-X deficiency mimics Ehlers-Danlos syndrome in mice through alteration of collagen deposition. *Nature Genetics*, *30*, 421–425.

Marini, M., Frabetti, F., Canaider, S., Dini, L., Falcieri, E., & Poirier, G. G. (2001). Modulation of caspase-3 activity by zinc ions and by the cell redox state. *Experimental Cell Research*, *266*, 323–332.

Martin, R., Farjanel, J., Eichenberger, D., Colige, A., Kessler, E., Hulmes, D. J., et al. (2000). Liquid crystalline ordering of procollagen as a determinant of three-dimensional extracellular matrix architecture. *Journal of Molecular Biology*, *301*, 11–17.

McBride, D. J., Jr., Kadler, K. E., Hojima, Y., & Prockop, D. J. (1992). Self-assembly into fibrils of a homotrimer of type I collagen. *Matrix*, *12*, 256–263.

Meek, K. M., Chapman, J. A., & Hardcastle, R. A. (1979). The staining pattern of collagen fibrils. Improved correlation with sequence data. *The Journal of Biological Chemistry*, *254*, 10710–10714.

Mienaltowski, M. J., & Birk, D. E. (2014). Structure, physiology, and biochemistry of collagens. *Advances in Experimental Medicine and Biology*, *802*, 5–29.

Miroshnichenko, N. S., Balanuk, I. V., & Nozdrenko, D. N. (2000). Packing of myosin molecules in muscle thick filaments. *Cell Biology International*, *24*, 327–333.

Mould, A. P., Holmes, D. F., Kadler, K. E., & Chapman, J. A. (1985). Mica sandwich technique for preparing macromolecules for rotary shadowing. *Journal of Ultrastructure Research*, *91*, 66–76.

Mould, A. P., & Hulmes, D. J. (1987). Surface-induced aggregation of type I procollagen. *Journal of Molecular Biology*, *195*, 543–553.

Mussini, E., Hutton, J. J., Jr., & Udenfriend, S. (1967). Collagen proline hydroxylase in wound healing, granuloma formation, scurvy, and growth. *Science*, *157*, 927–929.

Njieha, F. K., Morikawa, T., Tuderman, L., & Prockop, D. J. (1982). Partial purification of a procollagen C-proteinase. Inhibition by synthetic peptides and sequential cleavage of type I procollagen. *Biochemistry*, *21*, 757–764.

Orgel, J. P., Irving, T. C., Miller, A., & Wess, T. J. (2006). Microfibrillar structure of type I collagen in situ. *Proceedings of the National Academy of Sciences of the United States of America*, *103*, 9001–9005.

Orgel, J. P., Miller, A., Irving, T. C., Fischetti, R. F., Hammersley, A. P., & Wess, T. J. (2001). The in situ supermolecular structure of type I collagen. *Structure*, *9*, 1061–1069.

Orgel, J. P., San Antonio, J. D., & Antipova, O. (2011). Molecular and structural mapping of collagen fibril interactions. *Connective Tissue Research*, *52*, 2–17.

Orgel, J. P., Wess, T. J., & Miller, A. (2000). The in situ conformation and axial location of the intermolecular cross-linked non-helical telopeptides of type I collagen. *Structure*, *8*, 137–142.

Parkinson, J., Kadler, K. E., & Brass, A. (1995). Simple physical model of collagen fibrillogenesis based on diffusion limited aggregation. *Journal of Molecular Biology*, *247*, 823–831.

Parry, D. A., Barnes, G. R., & Craig, A. S. (1978). A comparison of the size distribution of collagen fibrils in connective tissues as a function of age and a possible relation between fibril size distribution and mechanical properties. *Proceedings of the Royal Society of London B: Biological Sciences*, *203*, 305–321.
Parry, D. A., & Craig, A. S. (1977). Quantitative electron microscope observations of the collagen fibrils in rat-tail tendon. *Biopolymers*, *16*, 1015–1031.
Parry, D. A., & Craig, A. S. (1979). Electron microscope evidence for an 80 A unit in collagen fibrils. *Nature*, *282*, 213–215.
Pingel, J., Lu, Y., Starborg, T., Fredberg, U., Langberg, H., Nedergaard, A., et al. (2014). 3-D ultrastructure and collagen composition of healthy and overloaded human tendon: Evidence of tenocyte and matrix buckling. *Journal of Anatomy*, *224*, 548–555.
Plumb, D. A., Dhir, V., Mironov, A., Ferrara, L., Poulsom, R., Kadler, K. E., et al. (2007). Collagen XXVII is developmentally regulated and forms thin fibrillar structures distinct from those of classical vertebrate fibrillar collagens. *The Journal of Biological Chemistry*, *282*, 12791–12795.
Prockop, D. J., & Fertala, A. (1998). Inhibition of the self-assembly of collagen I into fibrils with synthetic peptides. Demonstration that assembly is driven by specific binding sites on the monomers. *The Journal of Biological Chemistry*, *273*, 15598–15604.
Pucci-Minafra, I., Andriolo, M., Basirico, L., Alessandro, R., Luparello, C., Buccellato, C., et al. (1998). Absence of regular alpha2(I) collagen chains in colon carcinoma biopsy fragments. *Carcinogenesis*, *19*, 575–584.
Rasmussen, M., Jacobsson, M., & Bjorck, L. (2003). Genome-based identification and analysis of collagen-related structural motifs in bacterial and viral proteins. *The Journal of Biological Chemistry*, *278*, 32313–32316.
Raspanti, M., Congiu, T., & Guizzardi, S. (2001). Tapping-mode atomic force microscopy in fluid of hydrated extracellular matrix. *Matrix Biology*, *20*, 601–604.
Raspanti, M., Ottani, V., & Ruggeri, A. (1989). Different architectures of the collagen fibril: Morphological aspects and functional implications. *International Journal of Biological Macromolecules*, *11*, 367–371.
Raspanti, M., Protasoni, M., Manelli, A., Guizzardi, S., Mantovani, V., & Sala, A. (2006). The extracellular matrix of the human aortic wall: Ultrastructural observations by FEG-SEM and by tapping-mode AFM. *Micron*, *37*, 81–86.
Raspanti, M., Reguzzoni, M., Protasoni, M., & Basso, P. (2018). Not only tendons: The other architecture of collagen fibrils. *International Journal of Biological Macromolecules*, *107*(Pt. B), 1668–1674. https://doi.org/10.1016/j.ijbiomac.2017.10.037.
Raspanti, M., Viola, M., Sonaggere, M., Tira, M. E., & Tenni, R. (2007). Collagen fibril structure is affected by collagen concentration and decorin. *Biomacromolecules*, *8*, 2087–2091.
Seegmiller, R., Fraser, F. C., & Sheldon, H. (1971). A new chondrodystrophic mutant in mice. Electron microscopy of normal and abnormal chondrogenesis. *The Journal of Cell Biology*, *48*, 580–593.
Silver, D., Miller, J., Harrison, R., & Prockop, D. J. (1992). Helical model of nucleation and propagation to account for the growth of type I collagen fibrils from symmetrical pointed tips: A special example of self-assembly of rod-like monomers. *Proceedings of the National Academy of Sciences of the United States of America*, *89*, 9860–9864.
Starborg, T., Kalson, N. S., Lu, Y., Mironov, A., Cootes, T. F., Holmes, D. F., et al. (2013). Using transmission electron microscopy and 3View to determine collagen fibril size and three-dimensional organization. *Nature Protocols*, *8*, 1433–1448.
Stearns, M. L. (1940). Studies on the development of connective tissue in transparent chambers in the rabbit's ear. II. *American Journal of Anatomy*, *67*, 55–97.
Svensson, L., Aszodi, A., Reinholt, F. P., Fassler, R., Heinegard, D., & Oldberg, A. (1999). Fibromodulin-null mice have abnormal collagen fibrils, tissue organization, and altered lumican deposition in tendon. *The Journal of Biological Chemistry*, *274*, 9636–9647.

Svensson, R. B., Herchenhan, A., Starborg, T., Larsen, M., Kadler, K. E., Qvortrup, K., et al. (2017). Evidence of structurally continuous collagen fibrils in tendons. *Acta Biomaterialia, 50*, 293–301.
Sweeney, S. M., Orgel, J. P., Fertala, A., McAuliffe, J. D., Turner, K. R., Di Lullo, G. A., et al. (2008). Candidate cell and matrix interaction domains on the collagen fibril, the predominant protein of vertebrates. *The Journal of Biological Chemistry, 283*, 21187–21197.
Takaluoma, K., Hyry, M., Lantto, J., Sormunen, R., Bank, R. A., Kivirikko, K. I., et al. (2007). Tissue-specific changes in the hydroxylysine content and cross-links of collagens and alterations in fibril morphology in lysyl hydroxylase 1 knock-out mice. *The Journal of Biological Chemistry, 282*, 6588–6596.
Tasheva, E. S., Koester, A., Paulsen, A. Q., Garrett, A. S., Boyle, D. L., Davidson, H. J., et al. (2002). Mimecan/osteoglycin-deficient mice have collagen fibril abnormalities. *Molecular Vision, 8*, 407–415.
Thurmond, F. A., & Trotter, J. A. (1994). Native collagen fibrils from echinoderms are molecularly bipolar. *Journal of Molecular Biology, 235*, 73–79.
Trelstad, R. L., & Birk, D. E. (1985). The fibroblast in morphogenesis and fibrosis: Cell topography and surface-related functions. *Ciba Foundation Symposium, 114*, 4–19.
Trelstad, R. L., & Hayashi, K. (1979). Tendon collagen fibrillogenesis: Intracellular subassemblies and cell surface changes associated with fibril growth. *Developmental Biology, 71*, 228–242.
Trotter, J. A., Chapman, J. A., Kadler, K. E., & Holmes, D. F. (1998). Growth of sea cucumber collagen fibrils occurs at the tips and centers in a coordinated manner. *Journal of Molecular Biology, 284*, 1417–1424.
Trotter, J. A., Kadler, K. E., & Holmes, D. F. (2000). Echinoderm collagen fibrils grow by surface-nucleation-and-propagation from both centers and ends. *Journal of Molecular Biology, 300*, 531–540.
Trotter, J. A., & Koob, T. J. (1989). Collagen and proteoglycan in a sea urchin ligament with mutable mechanical properties. *Cell and Tissue Research, 258*, 527–539.
Tuderman, L., & Prockop, D. J. (1982). Procollagen N-proteinase. Properties of the enzyme purified from chick embryo tendons. *European Journal of Biochemistry, 125*, 545–549.
Wan, W., & Briggs, J. A. (2016). Cryo-electron tomography and subtomogram averaging. *Methods in Enzymology, 579*, 329–367.
Weisel, J. W., & Litvinov, R. I. (2013). Adaptation of fibrous biopolymers to recurring increasing strains. *Proceedings of the National Academy of Sciences of the United States of America, 110*, 12164–12165.
Wenstrup, R. J., Florer, J. B., Brunskill, E. W., Bell, S. M., Chervoneva, I., & Birk, D. E. (2004). Type V collagen controls the initiation of collagen fibril assembly. *The Journal of Biological Chemistry, 279*, 53331–53337.
Williams, B. R., Gelman, R. A., Poppke, D. C., & Piez, K. A. (1978). Collagen fibril formation. Optimal in vitro conditions and preliminary kinetic results. *The Journal of Biological Chemistry, 253*, 6578–6585.
Wood, G. C., & Keech, M. K. (1960). The formation of fibrils from collagen solutions. 1. The effect of experimental conditions: Kinetic and electron-microscope studies. *The Biochemical Journal, 75*, 588–598.
Yamamoto, S., Nakamura, F., Hitomi, J., Shigeno, M., Sawaguchi, S., Abe, H., et al. (2000). Atomic force microscopy of intact and digested collagen molecules. *Journal of Electron Microscopy, 49*, 423–427.
Young, B. B., Zhang, G., Koch, M., & Birk, D. E. (2002). The roles of types XII and XIV collagen in fibrillogenesis and matrix assembly in the developing cornea. *Journal of Cellular Biochemistry, 87*, 208–220.
Zhao, T., Weinhold, P. S., Lee, N. Y., & Dahners, L. E. (2011). Some observations on the subfibrillar structure of collagen fibrils as noted during treatment with NKISK and cathepsin G with mechanical agitation. *Journal of Electron Microscopy, 60*, 177–182.

CHAPTER FOUR

Basement Membranes in Development and Disease

Rei Sekiguchi[1], Kenneth M. Yamada[1]
Cell Biology Section, National Institute of Dental and Craniofacial Research, National Institutes of Health, Bethesda, MD, United States
[1]Corresponding authors: e-mail address: rei.sekiguchi@nih.gov; kyamada@dir.nidcr.nih.gov

Contents

Current Topics in Developmental Biology, Volume 130
ISSN 0070-2153
https://doi.org/10.1016/bs.ctdb.2018.02.005

Abstract

The basement membrane is a thin but dense, sheet-like specialized type of extracellular matrix that has remarkably diverse functions tailored to individual tissues and organs. Tightly controlled spatial and temporal changes in its composition and structure contribute to the diversity of basement membrane functions. These different basement membranes undergo dynamic transformations throughout animal life, most notably during development. Numerous developmental mechanisms are regulated or mediated by basement membranes, often by a combination of molecular and mechanical processes. A particularly important process involves cell transmigration through a basement membrane because of its link to cell invasion in disease. While developmental and disease processes share some similarities, what clearly distinguishes the two is dysregulation of cells and extracellular matrices in disease. With its relevance to many developmental and disease processes, the basement membrane is a vitally important area of research that may provide novel insights into biological mechanisms and development of innovative therapeutic approaches. Here we present a review of developmental and disease dynamics of basement membranes in *Caenorhabditis elegans*, *Drosophila*, and vertebrates.

1. INTRODUCTION

The extracellular matrix (ECM) is a noncellular component of multicellular organisms that plays essential roles in animal development and throughout life (Naba et al., 2016). Constantly undergoing remodeling, the ECM provides highly dynamic microenvironments. Two types of ECM—interstitial connective tissue and basement membrane—have both overlapping and distinct biological functions through their bidirectional interactions with cells and tissues. Both interstitial connective tissue and basement membrane ECMs can modulate cell proliferation, differentiation, angiogenesis, branching morphogenesis, tissue repair, and homeostasis (Bonnans, Chou, & Werb, 2014; Rozario & DeSimone, 2010). Through binding and sequestering of soluble growth factors in the presence of appropriate cell-mediated forces or proteolytic degradation, both types of ECM can also enable spatial–temporal regulation of receptor–ligand interactions. ECMs provide adhesive scaffolds and sometimes even concentration gradients for migratory cells. Furthermore, ECMs can generate and transduce mechanical signals. Through their interactions with cell surface receptors, ECMs modulate a remarkably wide range of intracellular signaling processes (DeSimone & Mecham, 2013; Hynes & Yamada, 2012).

On the other hand, interstitial connective tissue and basement membrane ECMs have many unique characteristics. Interstitial connective tissue

ECMs can range from a gel-like scaffold comprised of collagen I, fibronectin, and/or cartilage proteoglycans to tough, dense sheets or tendons (DeSimone & Mecham, 2013; Jayadev & Sherwood, 2017). Other components can include a wide variety of glycoproteins, proteoglycans, other types of collagen except type IV, and integrins (transmembrane heterodimeric receptors). The interstitial connective tissue ECM provides structure to spaces between cells and modulates intercellular and intertissue interactions via integrins and other cell surface ECM receptors.

In contrast, the basement membrane is an ultrathin, dense, and sheet-like ECM that is associated with virtually all organized cells (Yurchenco & Patton, 2009). The basement membrane underlies epithelial and endothelial cells and surrounds muscle, fat, and Schwann cells. Its primary structural elements consist of two polymeric networks comprised of laminin and type IV collagen, which are interconnected with nidogen, perlecan, and other molecules (Yurchenco, 2011).

The basement membrane is essential for animal development. It provides tissue integrity, elasticity, and biochemical and mechanical signaling, while facilitating intracellular and intercellular interactions. Mutations in basement membrane components lead to a variety of detrimental conditions affecting multiple organs and structures across embryonic and postnatal stages. During development, the basement membrane displays the opposing but complementary traits of rigidity and plasticity—it defines tissue boundaries and protects tissues from mechanical damage while exhibiting a high degree of plasticity that permits growth, morphogenesis, and cellular and tissue interactions (McClatchey et al., 2016). The basement membrane achieves these seemingly contradictory tasks through its physical pliability and versatility in tissue- and time-specific structures and functions. The focus of this review will be on basement membranes in development: their functions, spatial and temporal variations in biochemical and biophysical properties, mechanisms by which they coordinate tissue morphogenesis, and the relevance of such developmental mechanisms to pathological conditions, including cancer invasion and congenital or acquired disorders.

2. FUNCTIONS

The basement membrane provides tissues with a wide array of functions that include tissue separation, barrier, provision of an adhesive substrate and signaling platform for migration, polarization, differentiation, tissue shaping, and growth.

2.1 Tissue Separation and Barrier

While the basement membrane comprises a mere fraction of total ECM mass, the stable interlocking lattice of cross-linked type IV collagen, and laminin forms a major structural barrier to transmigration of most cells (except leukocytes) in normal physiology and pathological conditions (Kalluri, 2003).

Basement membrane permeability varies greatly throughout the body. In the developing kidney, two basement membranes—one associated with the vascular endothelium and another with the podocyte epithelium—merge to form the glomerular basement membrane (Fig. 1F) (Deen, 2004). The glomerular basement membrane serves as the kidney's filtration barrier with selective molecular permeability. With its controlled porosity based on size and charge, the glomerular basement membrane is impermeable to plasma proteins while permitting low-molecular weight substances to pass readily. Lung alveolar development also involves a merger of alveolar and capillary basement membranes, which establishes a gas exchange surface (Fig. 1C) (Rodeck & Whittle, 2009).

In brain development, the blood–brain barrier is formed at the interface between the vascular system and the brain by merging capillary endothelial and parenchymal basement membranes (Fig. 1A). The blood–brain barrier has particularly restrictive permeability and strictly controls the exchange and transport of molecules between the two tissues (Engelhardt, Vajkoczy, & Weller, 2017). Integrity of the embryonic and neonatal blood–brain barrier is largely regulated by laminin $\alpha4$ in the endothelial basement membrane, and genetic deletion of laminin $\alpha4$ leads to hemorrhage (Thyboll et al., 2002). Cellular sources of laminin for regulating the blood–brain barrier vary across time. Barrier integrity is mainly regulated by pericyte-derived laminin during development, but by astrocytic laminin in adulthood (Yao, Chen, Norris, & Strickland, 2014). Mice defective in pericytic laminin occasionally manifest hydrocephalus and blood–brain barrier breakdown accompanied by reductions in pericytes, AQP4, and tight junction proteins (Gautam, Zhang, & Yao, 2016). In adulthood, pathological conditions such as stroke and Alzheimer's disease can cause disruption of blood–brain barrier integrity. Basement membrane proteins are reduced in stroke whereas in Alzheimer's disease, basement membrane thickening is accompanied by increased collagen IV deposition and amyloid-β accumulation (Thomsen, Routhe, & Moos, 2017).

The basement membrane separates tissues through adhesions mediated by hemidesmosomes at the epithelial–mesenchymal interface. Laminin, a

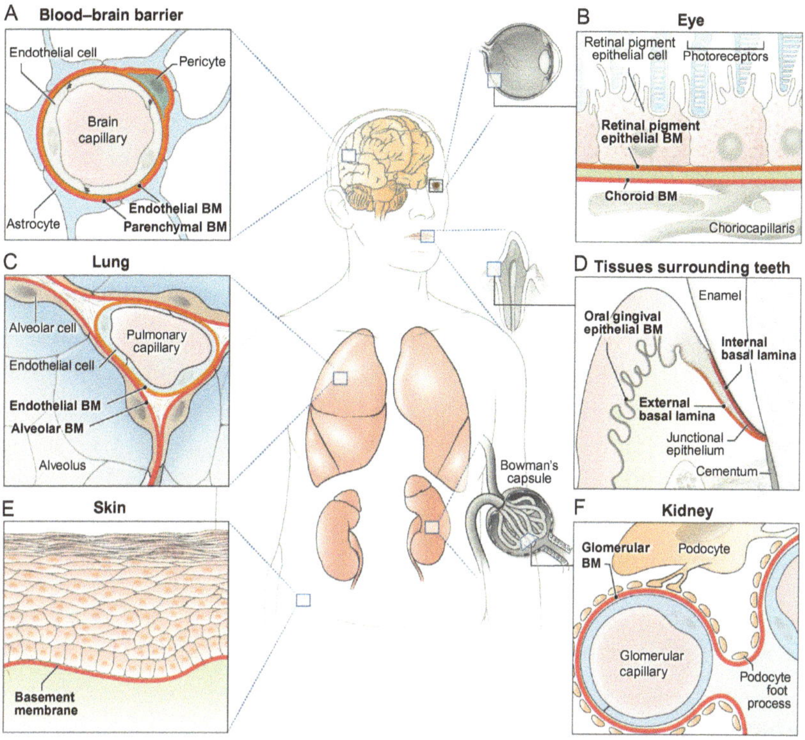

Fig. 1 Schematics of basement membranes (BMs) in various tissues. (A) Blood–brain barrier: endothelial and parenchymal BMs surround the brain capillary. (B) Eye: retinal pigment epithelium BM, choroid BM, and an intermediate collagenous zone comprise Bruch's membrane. (C) Lung: endothelial and alveolar BMs merge to provide a site for gas exchange. (D) Tissues surrounding teeth: junctional epithelium adheres to the enamel via the internal basal lamina and to the mesenchyme via the external basal lamina. The oral gingival epithelium BM is adjacent to the oral epithelium. (E) Skin: cutaneous BM underlies basal epithelial cells. (F) Kidney: during development, endothelial and podocyte-derived epithelial BMs merge to form the glomerular BM. Figures are not drawn to scale.

key component of hemidesmosomes, contributes to cell–matrix adhesion by interacting with cell surface integrins and anchoring fibrils predominantly comprised of collagen VII (Hohenester & Yurchenco, 2013). The formation of a stable adhesion complex between the two tissue layers protects tissues from destabilizing shear forces. The epidermal basement membrane and associated molecules collectively build a protective skin barrier (Fig. 1E). For example, collagen VII anchoring fibrils connect the epidermal basement membrane to the dermis (Rousselle et al., 1997). Collagen VII originates within the basement membrane where it binds to laminin, projecting into the papillary dermis where anchoring plaques interweave with collagen I and III fibrils (Breitkreutz, Koxholt, Thiemann, & Nischt, 2013). Collagen VII mutations result in recessive dystrophic epidermolysis bullosa characterized by blistering and infection due to a perturbed basement membrane barrier and impaired wound healing (Guerra, Odorisio, Zambruno, & Castiglia, 2017). Sadly, the majority of those affected by the severe generalized subtype also develop lethal cutaneous squamous cell carcinoma (Fine et al., 2014).

Basement membranes serve as an important anatomical landmark for the clinical diagnostic classification of tumors. Conditions associated with more favorable prognoses, i.e., dysplasia and carcinoma in situ, represent confinement of the primary neoplasm by an intact basement membrane. Once neoplastic cells breach the basement membrane, however, they are considered to be invasive and have the capacity to metastasize (Moasser, 2013). Therefore, basement membranes have important barrier functions in both development and disease.

2.2 Cell Adhesion and Migration

Basement membranes provide an adhesive substrate for cells, and they are linked functionally to the actin cytoskeleton via integrins or other ECM receptors to mediate cell attachment and migration, as well as modulating intracellular signaling pathways. Adhesion to basement membranes via cell surface receptors allows cells to mechanosense local stiffness, stiffness gradients, and other physical cues, which ultimately affect cellular behavior (Hynes, 1992). During *Drosophila* development, the receptor–ligand interplay between integrins and laminins in the basement membrane regulates follicle cell migration. Whereas integrin levels in follicle cells remain relatively stable, laminin expression in the basement membrane increases over time. The onset and speed of follicle cell migration are determined by this balance between integrin and laminin levels. Laminin-mutant eggs display altered cell migration and disrupted tissue shaping of developing follicles (Diaz de la Loza et al., 2017). Similarly, peripheral nerve establishment in mice requires laminin α5-dependent migration of neural crest cells, which differentiate into the peripheral nervous system and glial cells as they complete migration (Coles, Gammill, Miner, & Bronner-Fraser, 2006).

An in vitro model of cell migration demonstrates the profound effects of ECM dimensionality (2D vs 3D) on cellular behavior (Hakkinen, Harunaga, Doyle, & Yamada, 2011). Human foreskin fibroblasts in a 3D basement membrane extract (Matrigel) lose directionality and fail to migrate, yet the same substrate in a 2D configuration analogous to a basement membrane sheet allows highly efficient migration. These changes in cell migration patterns can be attributed to differences in ECM dimensionality. Even though useful for cell culture of epithelial cells, immersion of cells in a 3D basement membrane extract does not accurately simulate the basement membrane in vivo. A better representation of an in vivo environment is the 2D basement membrane extract model, which allows efficient cell migration. Indeed, cells in vivo migrate efficiently adjacent to the basement membrane. In the mouse submandibular gland, outer bud epithelial cells adjacent to the basement membrane show the highest rates of motility (Daley et al., 2017; Hsu et al., 2013). Motility of the outer bud cells is myosin II- and integrin α6β1 dependent, which suggests cell–ECM interaction. These findings highlight the importance of interactions between basement membranes, cell surface receptors, and cellular processes in regulating cell migratory behavior.

2.3 Polarity

During early morphogenesis, the basement membrane coordinates epithelial tissue organization by modulating apical polarity. Epithelial polarity is established through sorting of plasma membrane proteins to apical and basolateral surfaces, organizing polarity proteins and lipids at the plasma membrane, and utilizing adhesion molecules as positional cues in interactions with other epithelial cells and the adjacent basement membrane (Tanos & Rodriguez-Boulan, 2008). Loss of epithelial polarity is observed in pathological conditions such as cancer, and the extent of loss often correlates with tumor aggressiveness. Evidence exists for a potential role of myoepithelial cell-derived laminin α chains in regulation of neoplastic mammary gland epithelial polarity (Slade, Coope, Gomm, & Coombes, 1999).

Laminin is a key molecule for establishing apical polarity during tooth development (Fukumoto et al., 2006). Laminin α5 is a subunit of the major laminins in the tooth germ basement membrane, LM-511 and LM-521. Laminin α5-null mice display altered localization patterns of integrin α6β4 (the laminin α5 receptor) and hypoplastic tooth germs with reduced proliferation of the dental epithelium and loss of basal cell polarity. Importantly, the enamel knot (the signaling center for tooth morphogenesis) is defective in these mutant mice, with reduced sonic hedgehog (SHH) and fibroblast growth factor 4 (FGF4) (Fukumoto et al., 2006). Disruption in polarity formation is also evident in mammary epithelial cells when basement membrane stability is perturbed by loss of collagen IV or cell–basement membrane adhesions (Plachot et al., 2009).

Polarity formation is mediated through bidirectional interactions between the basement membrane and epithelial cells. Basal epithelial cells provide positional cues for establishing spatially restricted organization of the basement membrane (Gervais et al., 2016). The basal epithelial cells synthesize basement membrane proteins and organize their basal deposition, which requires expression of the polarity protein, PAR-1b. In the embryonic mouse submandibular salivary gland, establishment of basement membrane organization relies on basal expression of PAR-1b in the epithelium (Daley et al., 2012). Through reciprocal interactions with the epithelium, the basement membrane contributes to the establishment of apical polarity, cellular organization, and coherent tissue architecture.

2.4 Tissue Shaping

The basement membrane mediates tissue remodeling through molecular and mechanical processes. Unlike localized biochemical intercellular signaling, mechanical stress can exert global tissue-shaping forces due to the transmission of tension through interconnected cells (Legoff, Rouault, & Lecuit, 2013). Involvement of the basement membrane in tissue shaping is the best exemplified in *Drosophila* wing and egg chamber development. Collagen IV knockdown attenuates Dpp signaling (the *Drosophila* homolog of bone morphogenetic protein—BMP) and reduces *Drosophila* wing size (Ma, Cao, Dai, & Pastor-Pareja, 2017). During egg chamber development, the initially spherical egg chamber elongates in response to constrictive forces applied by the surrounding basement membrane. This constricting collagen IV, along with laminin and perlecan, is organized in a robust circumferential fibril-like orientation (Haigo & Bilder, 2011; Isabella & Horne-Badovinac, 2016). Secreted in site-specific fashion by Rab10 into the pericellular space between egg follicle cells, this fibril-like "molecular corset" of basement membrane proteins has been thought to arise from the collective rotation of follicle cells in a direction that coincides with the orientation of deposited collagen IV protein (Haigo & Bilder, 2011; Isabella & Horne-Badovinac, 2016). However, more recent findings suggest the involvement of an additional tissue-shaping mechanism independent of collective tissue rotation. Mutation of the atypical cadherin *Fat2* results in egg chambers that fail to rotate; nonetheless, mutant eggs still display aligned collagen IV and proper egg chamber elongation (Aurich & Dahmann, 2016).

Though these findings demonstrate the role of basement membranes in tissue shaping, some important questions remain to be answered: What is the biological significance of collective epithelial migration? What establishes and regulates the fibril-like circumferential collagen IV alignment? How does fibril-like collagen IV biophysically compare to collagen I fibrils? How do these findings in *Drosophila* translate to mammalian systems?

2.5 Signaling

Although many developmental mechanisms are known to be regulated by molecular signals, a combination of mechanical and regulatory gene processes can often provide a more complete mechanistic blueprint. For instance, pattern formation of avian skin follicles is initiated by mechanical compression of the epidermis by the contracting dermis, followed by β-catenin translocation to the nucleus in the compressed epithelial cells. New follicles emerge as the primordium basement membrane buckles increasingly (Shyer et al., 2017). Downstream genes associated with follicle formation are activated upon creation of cell aggregates by such dermal contraction. Since tissue morphogenesis is influenced by the level of tension in cells and the basement membrane (Moore et al., 2005), basement membrane buckling itself may also be involved in this mechanical crosstalk that regulates pattern formation of skin follicles.

During development and throughout animal life, basement membranes can function as signaling platforms by sequestering growth factors and other ligands. Perlecan, agrin, and collagen XVIII bind to many growth factors via heparan sulfate glycosaminoglycan chains. By signaling through cell surface receptors, growth factors such as FGFs, BMPs, vascular endothelial growth factor (VEGF), and transforming growth factor-β (TGF-β) regulate stem cell maintenance, cell migration, proliferation, and survival (Jayadev & Sherwood, 2017). Furthermore, stem cell maintenance can be achieved directly by laminin; a combination of laminin and E-cadherin is capable of maintaining clonal survival and self-renewal of human embryonic stem cells (Rodin et al., 2014).

In addition to growth factors, a large number of ECM-modifying enzymes that mediate proteolysis, sulfation, and glycosylation act at basement membranes to modulate signaling (Yoshizaki & Yamada, 2013). Protease-mediated degradation of basement membranes can activate signaling pathways, not only by releasing bound growth factors, but also by releasing or exposing cryptic fragments of basement membrane proteins with signaling functions. Mouse submandibular salivary gland branching morphogenesis is enhanced by membrane-type 2 matrix metalloprotease (MT2-MMP) which releases bioactive collagen IV noncollagenous 1 (NC1) domains (Rebustini et al., 2009). MMP-9 also cleaves collagen IV and releases a collagen IV fragment that regulates angiogenesis (Hamano et al., 2003). MMP-2 cleaves LM-111 (laminin-1), exposing a laminin fragment that mediates the epithelial-to-mesenchymal transition (EMT) in embryonic stem cells (Horejs et al., 2014).

Moreover, intact laminin and collagen IV can directly activate signaling pathways. Collagen IV modulates avian lung development through regulation of lung epithelial differentiation, myofibroblast proliferation, differentiation, and migration (Loscertales et al., 2016). By interacting with dystroglycan and integrin family receptors, laminin regulates cell survival, migration, differentiation, and proliferation (Bonnans et al., 2014). Laminin–integrin interactions create dynamic cell–ECM links that involve signaling pathway induction and intracellular cytoskeleton organization (Berrier & Yamada, 2007; DeSimone & Mecham, 2013; Hynes, 1999; Kim, Turnbull, & Guimond, 2011; Yurchenco, 2011). The signaling activated by laminin–cell membrane receptor binding can control gene transcription and chromatin remodeling of gene promoters (Domogatskaya, Rodin, & Tryggvason, 2012). In mammary epithelia, functional differentiation leading to milk production is determined by growth hormones and interaction with neighboring cells as well as the basement membrane. Induction of β-casein expression in single cells depends on the types of substrate in which they are embedded (Streuli, Bailey, & Bissell, 1991). Single cells in a collagen I substrate are unable to express β-casein unless they are permitted cell–cell contact. In contrast, single cells in a laminin-rich basement membrane substrate are capable of expressing β-casein. The binding of LM-111 to dystroglycan leads to growth hormone-induced activation of STAT5 and increases in β-casein and insulin growth factor-1 (IGF-1) expression (Leonoudakis et al., 2010). These findings show that the basement membrane–cell surface receptor interaction modulates growth hormone induction and functional differentiation of mammary epithelia.

Basement membrane receptors are spatially and temporally regulated throughout development. For example, isoforms of syndecan, a nonintegrin ECM receptor, switch their binding specificity to growth factors. Through its heparan- and chondroitin-sulfate glycosaminoglycan side chains, syndecan binds to a variety of growth factors that include FGF, VEGF, epidermal growth factor, hepatocyte growth factor, and platelet-derived growth factor (PDGF) (Carey, 1997). In the embryonic mouse neuroepithelium, syndecan-1 switches its binding ligand from FGF2 to FGF1 in a stage-dependent manner (Nurcombe, Ford, Wildschut, & Bartlett, 1993).

Basement membrane-mediated signaling can modulate tissue morphogenesis by sensitization of cells to growth factors. During *Drosophila* renal tubule development, targeted deposition of the basement membrane that ensheaths renal tubules is achieved by hemocytes (macrophages) (Bunt et al., 2010). Deposited collagen IV enhances sensitivity of the tubule cells to localized Dpp guidance signals, thereby guiding directional morphogenesis of renal tubules. Collagen IV mediates activation of this Dpp signaling pathway in a subset of tubule cells, which enables stereotypic positioning and outgrowth of *Drosophila* renal tubules (Bunt et al., 2010). More recently, migrating macrophages have been identified as major contributors to de novo basement membrane formation in the *Drosophila* embryo (Matsubayashi et al., 2017). The basement membrane of the *Drosophila* embryo exhibits a hierarchical pattern of deposition of laminin, followed by collagen IV, and finally perlecan. Absence of laminin severely impairs subsequent collagen IV and perlecan incorporation (Matsubayashi et al., 2017), which suggests that environmental cues from initial laminin deposition are likely necessary for matrix formation.

The basement membrane also contributes to neuronal extensions in *Drosophila* by enhancing the response of extending axons to environmental signals that guide their growth (Isabella & Horne-Badovinac, 2015). For example, perturbation of laminin results in axon guidance defects (Garcia-Alonso, Fetter, & Goodman, 1996). Syndecan, as a coreceptor for Slit (Johnson et al., 2004) also provides guidance cues for developing axons. Slit is an extracellular protein that serves as a repulsive signal for axons that express the Robo receptor (Brose et al., 1999). Syndecan binding to Slit stabilizes the Slit–Robo interaction (Hussain et al., 2006). Taken together, basement membranes have diverse signaling functions through interactions with cell receptors, release of cryptic fragments upon proteolytic breakdown, and mechanical processes combined with molecular signaling.

3. COMPOSITION

Present in nearly all human tissues, basement membranes are spatially and temporally customized, with associated proteins that display time- and tissue-specific expression patterns. Databases of ECM molecules are provided by the Matrixome Project (Manabe et al., 2008), the Matrisome

Project (Hynes & Naba, 2012), and the Human Protein Atlas (Uhlen et al., 2015). The Matrixome Project offers a high-resolution atlas that focuses on spatial and temporal localization of mouse basement membrane proteins. Based on the immunohistochemical analyses of proteins following large-scale in silico and in vitro screening, this project has compiled a comprehensive atlas of more than 40 basement membrane proteins at http://www.matrixome.com/bm (Manabe et al., 2008). The Matrisome Project lists databases that contain the ECM compositions of 14 types of tissue and tumor at http://matrisomeproject.mit.edu (Hynes & Naba, 2012; Naba et al., 2016). The Human Protein Atlas provides a map of the human tissue proteome based on transcriptomics combined with immunohistochemistry (Uhlen et al., 2015).

Basement membranes emerged in the evolution of animal multicellularity. Basement membrane proteins are highly conserved among mammals (Ozbek, Balasubramanian, Chiquet-Ehrismann, Tucker, & Adams, 2010). Particularly ancient molecules are the core basement membrane proteins collagen IV, laminin, nidogen, and perlecan (Yoshizaki & Yamada, 2013). These ECM proteins are commonly comprised of repeated domains and evolved by exon shuffling (Engel, 1996; Patthy, 1999). The small number of core basement membrane molecules is disproportionate to the diversity of their functions. This complexity of functional diversity stems from the addition of other proteins, posttranslational modifications, protein–protein interactions, RNA splicing variants, alternative promoters, and spatiotemporal expression of, or functional changes in, basement membrane components and membrane receptors (Jayadev & Sherwood, 2017; Yoshizaki & Yamada, 2013). The combination of these mechanisms generates biochemically and biophysically unique basement membrane structures, which are associated with specialized biological functions tailored to individual tissues and organs.

3.1 Collagen IV

Collagen IV builds a network of triple-helical molecules, which most commonly consists of one α2 and two α1 subunits (Yurchenco, 2011). This network binds integrins, provides surrounding tissues with tensile strength, and tethers laminins, proteoglycans, and growth factors (Fidler et al., 2017). Collagen IV and laminin generate two independent but interconnected polymeric networks that serve as the foundation for basement membranes (Fig. 2C). The collagen IV network is assembled in response to an

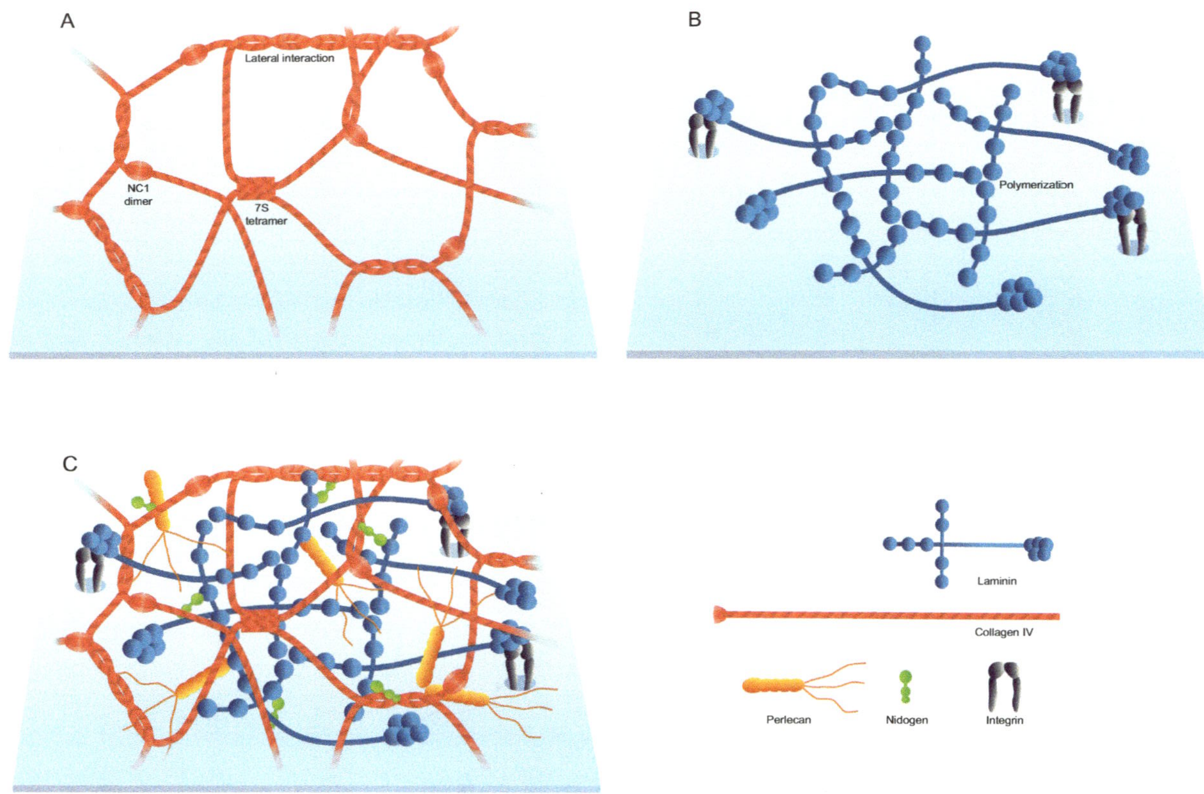

Fig. 2 Basic structure of the BM. (A) Collagen IV network, (B) laminin network, and (C) schematic representation of the BM with its major components.

extracellular Cl^- ion signal, which triggers a conformational switch in NC1 domains (Cummings et al., 2016). The network assembly process of collagen IV trimers involves three modes of interactions: NC1 dimerization, N-terminal (7S) tetramerization, and lateral associations (Fig. 2A) (Timpl, Wiedemann, van Delden, Furthmayr, & Kuhn, 1981; Yurchenco & Ruben, 1987). Found primarily in basement membranes, collagen IV has unique characteristics that distinguish it from other types of collagen. Owing to the absence of a glycine at every third residue, collagen IV is unable to form a tight collagen helical structure (Abreu-Velez & Howard, 2012). Furthermore, its NC1 domain remains attached at the C-terminus. This restricts NC1 domain interactions, permitting only head-to-head fiber connections with the flat NC1 surfaces facing each other. As a result, collagen IV networks are structurally flatter and more pliable (Abreu-Velez & Howard, 2012). Mutations in collagen IV α3, α4, and α5 give rise to the kidney disease Alport's syndrome and, in some cases, hemorrhagic stroke (Hudson, Tryggvason, Sundaramoorthy, & Neilson, 2003).

3.2 Laminin

Laminins are a family of large heterotrimeric multidomain proteins that consist of three chains, α, β, and γ (Aumailley et al., 2005; Miner & Yurchenco, 2004). At least 16 different isoforms are present in vertebrates (Yurchenco, 2011). The first ECM proteins identified in the mammalian preimplantation embryo are the laminin subunits α1, α5, β1, and γ1 (Miner, Li, Mudd, Go, & Sutherland, 2004). While laminin β1 and γ1 subunits are nearly ubiquitous and shared by many laminin isoforms, laminin α subunits have more specific spatiotemporal distribution patterns (Pickering, Cunliffe, Van Eeden, & Borycki, 2017). Laminins generate honeycomb-like networks connected in a spot-welding-like manner by adhesive aggregates containing perlecan (Behrens et al., 2012).

Laminins assume a cross or Y shape. The short arms are comprised of α, β, or γ chains, with LN domains located at the end of each arm. LN domains provide a site for laminin polymerization (Fig. 2B). LG domains at the end of the long arm (coiled–coil) bind to a cell surface through sulfated glycolipids, integrins, and α-dystroglycan (Yurchenco, 2011). The LG domain–cell surface receptor binding promotes laminin polymerization and therefore enhances network assembly. The LG1–3 cluster provides the principal binding sites for a variety of integrins such as α6β1, α6β4, α7β1, and α3β1 (Nishiuchi et al., 2006). Laminin–integrin interactions create dynamic cell–ECM links that involve signaling pathway induction and intracellular cytoskeleton organization (Berrier & Yamada, 2007; DeSimone & Mecham, 2013; Hynes, 1999; Kim et al., 2011; Yurchenco, 2011).

Laminins have various cell type-specific expression patterns, and laminin mutations can cause embryonic lethality or severe diseases affecting multiple organs (Domogatskaya et al., 2012). LM-211 (laminin-2) is prominent in pancreas, muscle, and the nervous system, whereas LM-311 (laminin-6) and -321 (laminin-7) are found in lung and skin (Mecham, 2011). LM-332 (laminin-5) is the most abundant laminin expressed by keratinocytes (Botta et al., 2012). LM-332 binds tightly to the noncollagenous domain of collagen VII of anchoring fibrils (Rousselle et al., 1997) and the laminin-specific integrin $\alpha6\beta4$ (Nishiuchi et al., 2006). Integrin $\alpha6\beta4$ binds to intracellular plectin, another important component of hemidesmosomes (Litjens, de Pereda, & Sonnenberg, 2006). Disruption of this cell–matrix adhesion complex comprised of laminin, integrin, and anchoring fibrils causes epidermolysis bullosa (Pulkkinen & Uitto, 1999). LM-221 (laminin-4), -421 (laminin-9), and -521 (laminin-11) are found in the neuromuscular junction, where lack of laminin $\alpha4$ leads to failure of innervation due to mismatched nerve–muscle connections of the active zones at the axon terminus and the junctional folds of the muscle endplate (Patton et al., 2001). LM-511 is widely expressed in skin, intestine, lung, kidney, and salivary gland ((laminin-10) Mecham, 2011). Laminin $\alpha5$ regulates transformation of the neural tube into the emerging central nervous system during early embryonic mouse neurulation (Miner, Cunningham, & Sanes, 1998).

3.3 Nidogen

Nidogens are monomeric glycoproteins that contribute to a linkage of polymeric networks with stable affinity to the laminin and collagen IV (Fig. 2C) (Fox et al., 1991; Mayer et al., 1993; Yurchenco, 2011). Expressed mainly by mesenchymal cells, nidogen 1 and nidogen 2 are ubiquitous basement membrane proteins (Nischt et al., 2007). Both nidogen 1 and nidogen 2 promote basement membrane stability by contributing, at least in part, to the connection of laminin and collagen IV networks (Has & Nyström, 2015). Nidogens consist of three globular domains termed G1–G3 (4–5 nm in diameter). The domains are interconnected by a flexible segment between G1 and G2 and a stiff rod between G2 and G3. G2 (collagen and proteoglycan-binding domain) and G3 (laminin-binding domain) are separated by 15 nm, which permits various orientations and interactions with other basement membrane proteins (Fox et al., 1991).

Nidogen-deficient mice generally manifest mild phenotypes in skin without effects on formation of the basement membrane. In contrast, a loss of endothelial basement membrane integrity results from substantial reductions in collagen IV, laminin-411 (laminin-8), and perlecan (Mokkapati et al., 2008). Mice lacking both nidogen-1 and -2 can have cardiac and pulmonary abnormalities associated with basement membrane assembly defects (Bader et al., 2005) and in some cases, altered FGF distribution, syndactyly (digit fusion), and failure to form the limb bud ectodermal basement membrane (Bose et al., 2006). Strikingly, a severe phenotype results from specific ablation of the nidogen-binding site in laminin γ1 (Willem et al., 2002). These mice exhibit renal agenesis, impaired lung development, and neonatal lethality, suggesting the importance of laminin–nidogen interaction in development and the potential role of the nidogen-binding site of laminin γ1 in other developmental mechanisms. Laminin γ1–nidogen interaction is also known to regulate pial development (Halfter, Dong, Yip, Willem, & Mayer, 2002). In addition to the nidogens, other binding molecules such as perlecan likely have compensatory and redundant functions that are independent of each other, since the basement membrane has many essential life functions.

3.4 Heparan Sulfate Proteoglycans

Heparan sulfate proteoglycans are glycoproteins with one or more heparan sulfate chains, which are a type of glycosaminoglycan (GAG) chain (Domogatskaya et al., 2012). The heparan sulfate proteoglycans present in the basement membrane are perlecan, agrin, and collagen XVIII. These glycated proteins define basement membrane structure in collaboration with other matrix proteins. Owing to the hygroscopic properties of glycosaminoglycans, perlecan and agrin increase the volume of matrix (Domogatskaya et al., 2012). Collagen XVIII is a hybrid collagen-proteoglycan expressed at the initiation of lung and kidney organogenesis (Patel, Pineda, & Hoffman, 2017). Inhibition of collagen XVIII disrupts lung and kidney branching morphogenesis (Karihaloo et al., 2001; Lin et al., 2001).

Heparan sulfate binds to many growth factors including FGF with high affinity. Association and dissociation of heparan sulfate–ligand interactions can create a gradient of FGF ligand through mass action involving adjacent binding sites. In addition, heparan sulfate can approximate two adjacent proteins, facilitating growth factor–receptor tyrosine kinase interaction, such as for FGF and FGFR (Sarrazin, Lamanna, & Esko, 2011). FGF10–FGFR2b signaling is essential for early embryonic development of multiple organs, including the lung, trachea (Min et al., 1998), mammary gland (Mailleux et al., 2002), lacrimal gland (Entesarian et al., 2005), and salivary gland (De Moerlooze et al., 2000; Ohuchi et al., 2000). Heparan sulfate alters the level of FGF10–FGFR2b affinity: when present in a ternary complex with heparan sulfate, the affinity between the receptor and the ligand increases (Kan, Wu, Wang, & McKeehan, 1999). Proteolysis by heparanase (an endoglucuronidase that cleaves heparan sulfate) releases the growth factor, which in turn promotes neural differentiation of murine embryonic stem cells (Xiong et al., 2017) and enhances branching morphogenesis of the mouse salivary gland (Patel et al., 2007). Notably, heparanase promotes not only development but also tumor progression in the mammary gland (Boyango et al., 2018). This illustrates an example of the same mechanism mediating both development and disease.

3.4.1 Perlecan

Perlecan is a secreted heparan sulfate proteoglycan that is a potential linker of laminin and collagen IV networks (Fig. 2C). Perlecan is a large (> 200 nm), evolutionarily old (>550 million years) ECM protein (Farach-Carson, Warren, Harrington, & Carson, 2014). Both epidermal keratinocytes and dermal fibroblasts express perlecan (Sher et al., 2006). Perlecan binds multiple growth factors and basement membrane components including laminin and collagen IV (Battaglia, Mayer, Aumailley, & Timpl, 1992). By sequestering growth factors such as FGF7, perlecan promotes keratinocyte survival and differentiation (Sher et al., 2006). Furthermore, perlecan interacts with integrin $\alpha 2\beta 1$, α-dystroglycan, and fibrillin-1; fibrillin-1 generates microfibrils that anchor the epidermal basement membrane to the papillary matrix (Herzog et al., 2004; Woodall et al., 2008). Perlecan and agrin establish collateral associations between basement membrane proteins and cell surface receptors by linking nidogen and laminin to integrins, dystroglycan, and sulfated glycolipids. The chain of bound proteins involving dystroglycan is termed a dystrophin–glycoprotein complex that extends from basement membrane proteins to F-actin through dystroglycan and dystrophin/utrophin (Yurchenco, 2011).

Perlecan is indispensable for formation of the epidermal basement membrane (Costell et al., 1999). Perlecan is especially important in corneal epithelial development (Inomata et al., 2012) and maintenance of basement membrane integrity in regions with high mechanical stress, including the contracting myocardium and expanding brain vesicles (Costell et al., 1999).

Moreover, perlecan has a distinct role in tissue development independent of other core basement membrane proteins. Though the basement membrane is absent in cartilage, perlecan is expressed in cartilage and helps to mediate cartilage development (Yoshizaki & Yamada, 2013).

3.5 FRAS/FREM

Located in the sublamina densa, FRAS1, FREM1, and FREM2 form a ternary complex known as the Fraser complex (Petrou, Makrygiannis, & Chalepakis, 2008). This protein complex controls stabilization of cell–matrix adhesions during embryonic development. Deficiency in FRAS1, FREM1, or FREM2 is the etiological origin of Fraser syndrome, which manifests as hidden eyes (cryptophthalmos), syndactyly, embryonic skin blistering, and renal agenesis or dysplasia with water transport deficits. Loss of *Fras1* in mice leads to arrest of ureteric bud growth due to severe deficiency of glial cell line-derived neutrophic factor (GDNF)—the inducer of the budding of ureteric buds from the Wolffian duct (Pitera, Scambler, & Woolf, 2008).

The importance of this protein complex in embryonic development is linked to the spatially and temporally restricted expression of the basement membrane protein nephronectin, which is a ligand for integrin α8β1 (Sato et al., 2009). The Fraser complex anchors nephronectin to the ureteric bud basement membrane and therefore stabilizes its binding to integrin α8β1 (Kiyozumi et al., 2012). Nephronectin-null mice frequently display renal agenesis or hypoplasia (Linton, Martin, & Reichardt, 2007) whereas α8-integrin mutations in humans can cause renal agenesis associated with compromised epithelial–mesenchymal interaction (Humbert et al., 2014). These studies reveal bound basement membrane and transmembrane receptor proteins that act in concert to direct spatiotemporal orchestration of mammalian organogenesis.

4. SPATIAL AND TEMPORAL VARIATIONS IN BASEMENT MEMBRANE THICKNESS AND COMPOSITION

Conventional transmission electron microscopy measurements made with desiccated tissues estimate basement membrane thickness to be less than 100 nm (Bluemink, Van Maurik, & Lawson, 1976; Cutler, 1977; Lehtonen, 1975). These measurements are most likely gross underestimations. More realistic measurements by atomic force microscopy on hydrated chick, mouse, and human basement membranes identify their thickness to be at least twice as thick (Jayadev & Sherwood, 2017). The basement membrane requires sufficient thickness to allow network formation by laminin molecules (80 nm in length) and collagen IV molecules (400 nm in length). Variations in basement membrane thickness most likely result from differences in tissue hydration levels and inherent tissue and species differences (Candiello, Cole, & Halfter, 2010; Halfter et al., 2015).

Basement membrane density and thickness vary across time and space. Fibril-like aggregation of collagen IV in the embryonic mouse salivary gland becomes increasingly denser and thicker over time (see Harunaga, Doyle, & Yamada, 2014). Even though electron microscopy measurements on desiccated tissues likely underestimate the thickness of basement membranes, findings from classical studies consistently demonstrate region-specific variations. In developing lung (Bluemink et al., 1976), kidney (Lehtonen, 1975), mammary gland (Paine & Lewis, 2017), and submandibular salivary gland (Bernfield & Banerjee, 1982; Cutler, 1977), the basement membrane adjacent to regions of active growth and tissue expansion is thinner and discontinuous compared to the relatively static regions away from the bud tip.

Interestingly, the thinner basement membrane at the actively growing regions of the salivary gland exhibits microperforations through which epithelial bleb-like protrusions commonly extend (Harunaga et al., 2014). This observation is consistent with classical electron microscopy studies of the embryonic lung (Bluemink et al., 1976), kidney (Lehtonen, 1975), tooth (Burgess & Katchburian, 1982; Slavkin & Bringas, 1976), and salivary gland (Bernfield & Banerjee, 1982). These studies capture direct epithelial–mesenchymal contact through perforations, with cytoplasmic protrusions extending from epithelial or mesenchymal cells. As discussed below, the spatial specificity of this observation of cell interactions associated with active growth supports the idea that direct cellular contact may be a mode of epithelial–mesenchymal interaction.

This regional difference in basement membrane thickness coincides with the level of ECM turnover in the embryonic mouse submandibular gland. The basement membrane at active branching regions is subject to extensive remodeling. Glycosaminoglycans are degraded much more rapidly at the distal end bud region than in the interlobular clefts (Bernfield & Banerjee, 1982). Local alterations in the mechanical compliance of the basement membrane may also account for the regional differences in its morphological appearance. Tissue morphogenesis and patterning are controlled by both molecular signals and alterations in the mechanical compliance of the basement membrane as they respond to cellular forces (Mammoto & Ingber, 2010). Changes in cytoskeletal tension mediated by Rho-signaling through Rho-associated kinase have a profound impact on embryonic lung development. Inhibiting cellular tension results in disruption of the basement membrane, blood vessels, and bud formation, whereas increasing cell tension enhances lung branching and capillary development (Moore et al., 2005). Thus, increased mechanical compliance and distensibility associated with enhanced degradation and the outward expansive force from the growing organ may underlie thinning of the basement membrane at regions of active growth.

4.1 Spatial Variations in Mammary Gland Basement Membrane Composition and Density

Basement membrane composition and density vary within the developing mammary gland. During puberty, the terminal end bud emerges at the tip of the mammary duct (Sternlicht, 2006). The terminal end bud is the regulatory control point for proliferation, branching, angiogenesis, and pubertal branching morphogenesis (Huebner, Lechler, & Ewald, 2014; Paine & Lewis, 2017). The tip of the terminal end bud represents the active site of branching morphogenesis, whereas the neck of the terminal end bud is the site of ductal morphogenesis and tissue stabilization (Silberstein & Daniel, 1982). Cells in the terminal end bud show spatially limited and reversible reductions in intercellular adhesion and polarity (Ewald et al., 2012), which correlate with a high degree of morphological plasticity in the region. Basement membrane accumulation toward the neck of terminal end buds contributes to sculpting of the elongating ducts by constriction (Jayadev & Sherwood, 2017). The basement membrane at the bulbous tip of the terminal end bud is very thin (104 nm), while it is thicker in the neck region of the bud (1.4 μm) (Paine & Lewis, 2017). The thin basement membrane at the tip of the terminal end buds consists primarily of laminin, collagen IV, and hyaluronic acid. In contrast, the basement membrane at the neck of the terminal end bud is thicker, forming a more defined meshwork of LM-111 and LM-332, collagen IV, and heparan sulfate proteoglycans (Fata, Werb, & Bissell, 2004; Keely, Wu, & Santoro, 1995). The variations in basement membrane composition and thickness of the terminal end bud likely facilitate epithelial–mesenchymal interaction and morphogenesis.

4.2 Variations in Laminin Expression in Development and Sjogren's Syndrome

Present in skin, kidney, lung, thymus, brain, gastrointestinal tract, and lung, LM-332 (laminin-5) displays uneven spatial and temporal distribution throughout epithelial basement membranes. LM-332 is enriched within hemidesmosomes, but its expression decreases during hair morphogenesis (Nanba, Hieda, & Nakanishi, 2000). Time-dependent expression of different types of laminin is also observed in the subendothelial basement membrane. The subendothelial basement membrane provides a substrate for β-cells in pancreatic islets, podocytes in renal glomeruli, and alveolar epithelial cells in the lung (Hallmann et al., 2005). Two main laminin isoforms, LM-411 (laminin-8) and LM-511 (laminin-10), are commonly present in the subendothelial basement membrane. When the basement membrane undergoes active remodeling during embryonic angiogenesis, LM-411 is the predominant laminin expressed in the vascular endothelial basement membrane. LM-511 emerges postnatally when the subendothelial basement membrane is stabilized (Hallmann et al., 2005). These time- and space-specific expression patterns of laminin suggest distinct functions for each isoform during development.

Laminin expression in the embryonic mouse submandibular salivary gland varies temporally and spatially. The salivary gland at embryonic day 13, which marks the onset of active branching, shows expression of laminin α1 and α5 surrounding the distal end bud and α1, α3, and α5 chains along the duct (Kadoya & Yamashina, 2005). This laminin distribution changes later in embryonic development, when α2 and α4 emerge adjacent to the end bud, while α3 increases around myoepithelial cells that surround acini, the saliva-secreting units.

The basement membrane of labial salivary glands in individuals with Sjogren's syndrome shows structural disorganization and altered expression of collagen IV and laminin (Molina et al., 2006). The epithelial acinar basement membrane of Sjogren's syndrome patients can display reduced laminins (LM-111 and LM-211), accompanied by decreased integrin (α1β1 and α2β1) receptors that bind to laminin (Laine et al., 2008). Basement membrane disorganization may be associated with loss of structural integrity and atrophy of salivary acini, ductal cell hyperplasia, and impaired secretory function. Lack of basement membrane integrity makes the salivary gland more susceptible to lymphocytic invasion in Sjogren's syndrome patients (Hayashi, 2011; Molina et al., 2006). Thus, basement membrane structure and composition are greatly altered in disease, suggesting that the basement membrane plays an important role beyond development.

4.3 Variations in Glomerular and Retinal Capillary Basement Membrane Thickness

Two basement membranes (retinal pigment epithelial and choroid) in Bruch's membrane maintain integrity of the retinal pigment epithelium, which separates the retina from the choriocapillaris (Fig. 1B) (Strauss, 2005). Retinal pigment epithelium supports photoreceptor health, and disruption of its integrity is often a disease manifestation. Age-dependent thickening of the basement membrane is associated with pathology in the retina, but not in the kidney. In the kidney, thickening of the glomerular basement membrane during aging does not generally affect hydraulic permeability of the glomerular capillary wall (Neumann, Kellner, Kuhn, Stolte, & Schurek, 2004). In contrast, age-related thickening of retinal capillary basement membranes is likely linked to senile retinopathies (Nagata, Katz, & Robison, 1986). Retinal vascular basement membrane thickening is associated with human microaneurysms during aging—a hallmark of retinal vascular disease. Small microaneurysms show basement membrane thickening and upregulation of collagen IV, laminin, fibronectin, nidogen, and perlecan. On the other hand, large microaneurysms display loss of basement membrane integrity and increase in MMP-9 and plasminogen activator inhibitor-1 (Lopez-Luppo et al., 2017).

Other conditions linked to ocular basement membrane thickening are diabetes mellitus and drusen accumulation. Diabetic retinopathy is associated with basement membrane thickening, stiffening, and compositional changes (To et al., 2013). With aging, progressive accumulation of drusen comprised of cellular debris can overwhelm removal and result in thickened basement membrane (Salvi, Akhtar, & Currie, 2006). Basal laminar drusen primarily accumulates in the macular area. With progressive accumulation, drusen is associated with the development of age-related macular degeneration.

Ocular basement membranes have sidedness. The inner epithelial face of the human adult ocular basement membrane is twofold stiffer than the outer stromal side (Halfter et al., 2013). Laminin is enriched on the stiffer inner epithelial-facing side, whereas collagen IV is enriched on the stromal side, with a distinct distribution of the C- and N-terminal domains in the basement membrane. This unique sidedness of basement membrane structure may lead to spatial and temporal variations in levels of susceptibility to basement membrane thickening and degradation. Biological implications of basement membrane sidedness in development and aging remain to be identified.

4.4 Temporal Variations in Glomerular Basement Membrane Composition

Along with their adjacent organs, basement membranes undergo dramatic transformation throughout development, with altered composition and structure. For instance, laminin and collagen isoforms switch during progression from embryonic development to adult stages in the human glomerular basement membrane. The basement membrane during kidney development is formed by the stage-specific assembly of a common laminin γ1 and different α and β subunits (Skorecki et al., 2015). Components of the fetal glomerular basement membrane include collagen α1 and α2 (IV) and laminin α1 and β1, which are gradually replaced by collagen α3–α5(IV) and laminin α5 and β2 subunits (Abrahamson, St John, Stroganova, Zelenchuk, & Steenhard, 2013; Harvey et al., 1998; Miner et al., 1997). This developmental switch in collagen and laminin isoforms is crucial for kidney development and maturation. Laminin α5-mutant mice show glomerular basement membrane breakdown and renal agenesis or defective glomerulogenesis because of a failed transition from laminin α1 to laminin α5 (Miner & Li, 2000). A laminin γ1 loss-of-function mutation in the murine ureteric bud causes renal agenesis as a result of failure of basement membrane formation and compromised integrin-based and growth factor-based (FGF2, WNT11, and GDNF/RET) signaling pathways (Yang et al., 2011). Laminin β2 mutation is the etiological cause of Pierson syndrome, which is characterized by neurological, ocular, and renal deficits (Miner, Go, Cunningham, Patton, & Jarad, 2006).

4.5 Spatial and Temporal Variations in Dental Basement Membrane Composition

During tooth morphogenesis, the dental basement membrane undergoes dynamic structural and chemical transformation involving degradation and calcification. Epithelial and mesenchymal cells in the tooth bud differentiate into their respective secretory cells (ameloblasts and odontoblasts) along the basement membrane (He et al., 2010). Many basement membrane proteins show site-specific expression in the embryonic day 16.5 mouse tooth bud. Laminin subunits α5 and β1, collagens IV and XVIII, nidogen-1, perlecan, agrin, and other proteins are widely expressed throughout the basement membranes of the tooth bud and oral epithelium (Fig. 3A) (Manabe et al., 2008). Laminin γ1 and nidogen-2 are expressed specifically around the entire tooth bud (Fig. 3B). FRAS1 and FREM1 are localized to the basement

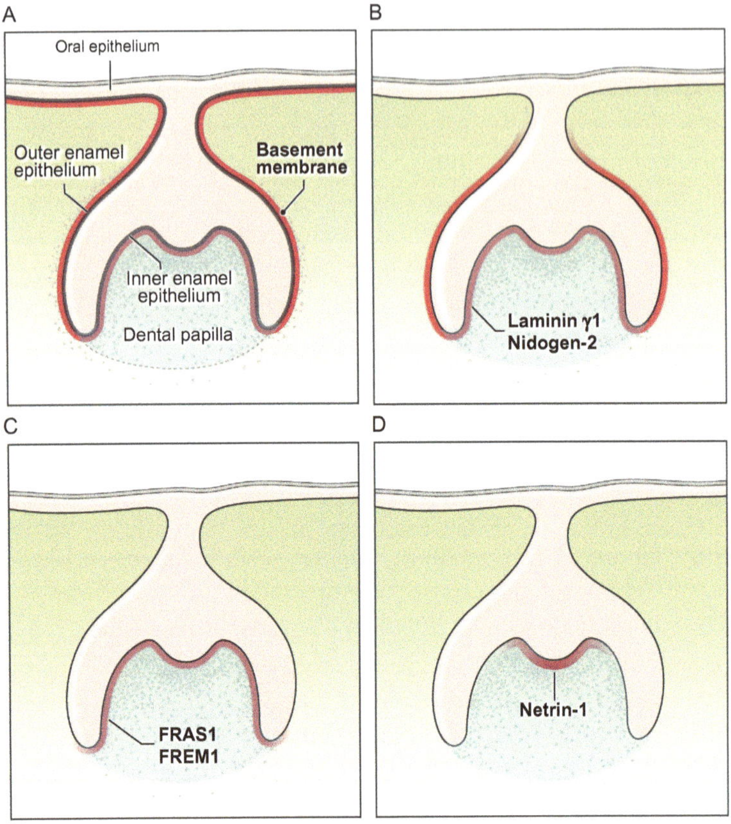

Fig. 3 The BM of embryonic mouse tooth bud. (A) The basement membrane underlies the oral epithelium and surrounds inner and outer enamel epithelia of the enamel organ. The dental papilla lies beneath the inner enamel epithelium and gives rise to the dentin and pulp. (B) Laminin $\gamma 1$ and nidogen-2 are expressed around the tooth bud. (C) FRAS1 and FREM1 show a spatially specific expression pattern along the inner enamel epithelium. (D) A gradient expression pattern of netrin-1 is found subjacent to the primary enamel knot.

membrane subjacent to the inner enamel epithelium (Fig. 3C). Other proteins such as netrin-1 show a gradient expression pattern along the inner enamel epithelium; the expression level of netrin-1 is the highest adjacent to the primary enamel knot, from which it gradually declines laterally (Fig. 3D) (Manabe et al., 2008).

A similar gradient pattern is found in odontoblast differentiation. Odontoblast differentiation follows a spatial gradient that begins adjacent to the enamel knot and proceeds in a direction away from the enamel knot (Bloch-Zupan, Sedano, & Scully, 2012). Gradients of SHH and BMP are thought to regulate cell proliferation and differentiation (Li et al., 2015; Seppala, Fraser, Birjandi, Xavier, & Cobourne, 2017). By binding to heparan sulfate proteoglycans, netrin-1 may modulate morphogen gradients (Manabe et al., 2008).

As ameloblasts and odontoblasts secrete enamel and dentin matrices, respectively, the basement membrane is degraded and replaced with interdigitations of calcified enamel and dentin (Sahlberg, Reponen, Tryggvason, & Thesleff, 1992; Thesleff et al., 1981). This interface forms the dentin–enamel junction, which joins two unique matrices that have different biomechanical properties (Chun, Choi, & Lee, 2014). Composed of mineralized connective tissue, dentin is similar to bone in its composition and nanostructure. On the other hand, enamel primarily consists of hydroxyapatite and represents the hardest tissue in the body (Gibson et al., 2001). Accordingly, the dentin–enamel junction serves as a crack arrest barrier for the brittle enamel, preventing the two distinct matrices from dislodging (Imbeni, Kruzic, Marshall, Marshall, & Ritchie, 2005). However, how the dentin–enamel junction, which is devoid of a basement membrane, establishes the effective adhesion needed to withstand enormous occlusal loads is not completely understood. The extent of basement membrane degradation in teeth—whether it is complete or partial—has been questioned because presence of collagens IV and VII at the dentin–enamel junction has been reported in adult teeth (McGuire, Gorski, Dusevich, Wang, & Walker, 2014; McGuire, Walker, Dusevich, Wang, & Gorski, 2014). The dentin–enamel junction becomes a vulnerable site for enamel delamination in irradiated teeth of head and neck cancer patients, where radiation likely affects biochemical and structural integrity at the dentin–enamel junction. Further research is required to identify detailed mechanisms maintaining the integrity of the dentin–enamel junction in health and postirradiation.

4.6 Differences in Basement Membranes in Tissues Surrounding Teeth

A tooth is surrounded by supporting dento-gingival tissues that consist of junctional epithelium, basement membranes, and gingival fibers (Newman, Takei, Klokkevold, & Carranza, 2011). Oral gingival and junctional epithelia surrounding the tooth are associated with three distinct basal laminae (Figs. 1D and 4A). The oral gingival epithelium basement membrane consists of typical basement membrane components, including collagen IV, nidogen, LM-332 (laminin-5), LM-311 (laminin-6), LM-321 (laminin-7), LM-511 (laminin-10), and LM-521 (laminin-11) (Oksonen, Sorokin,

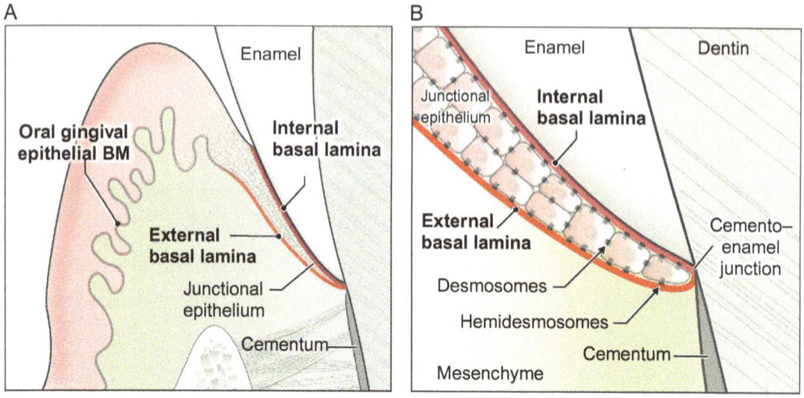

Fig. 4 Basic structure of dental supporting basal laminae. (A) Schematic illustration of basal laminae around a tooth. (B) Desmosomes mediate intercellular adhesions. Hemidesmosomes mediate attachment of internal and external basal laminae to enamel and mesenchyme, respectively.

Virtanen, & Hormia, 2001). Located adjacent to the tooth surface, the junctional epithelium is sandwiched between two unique basement membranes: the internal and external basal laminae. The internal basal lamina mediates attachment to the tooth surface, whereas the external basal lamina attaches to the gingival connective tissue (Fig. 4B) (Newman et al., 2011). Compared to the oral gingival epithelium, the junctional epithelium contains relatively few desmosomes and less E-cadherin, with wide intercellular spaces permissive for leukocyte migration (Hatakeyama et al., 2006). The internal and external basal laminae undergo continuous rapid remodeling because the adjacent junctional epithelium displays high turnover as a means of defense against bacterial infection (Dabija-Wolter, Bakken, Cimpan, Johannessen, & Costea, 2013). Forming an epithelial attachment to the tooth surface through basal laminae, the junctional epithelium serves as the first line of defense against bacterial plaque; its disruption and detachment below the cemento–enamel junction signify initiation of periodontitis (Kinane, 2001; Marks, McKee, Zalzal, & Nanci, 1994). The composition of the external basal lamina—collagen IV, LM-332, LM-311, and LM511, and nidogen-1—is similar to that of the oral gingival epithelium (Oksonen et al., 2001), whereas the internal basal lamina is primarily comprised of LM-332 and collagen VIII (Larjava, Koivisto, Hakkinen, & Heino, 2011). These differences in composition suggest that the two types of basal lamina may derive in part from different cell populations and may have evolved along different pathways to function in their unique local microenvironments.

5. BASEMENT MEMBRANE MICROPERFORATIONS

Although composed of a structurally dense sheet of interwoven networks, basement membranes in some tissues have microperforations. A combination of covalent and noncovalent forces coalesce the interconnected laminin and collagen IV meshwork into a molecular sieve with pores (Kalluri, 2003). Pores or microperforations in the basement membrane exist in a wide range of shapes and sizes (5 nm–8 μm in diameter) (Hironaka, Makino, Yamasaki, & Ota, 1993; Takeuchi & Gonda, 2004). The size and form appear to be spatially and temporally determined. Proper sizes of basement membrane pores or microperforations are crucial for regulation of permeability and intercellular interactions—and in certain tissues, permissibility for cell migration.

Alterations in basement membrane pore size can be a pathological manifestation. For instance, disruption/enlargement of the extremely tiny (~5–10 nm in diameter) (Hironaka et al., 1993) pores of the glomerular basement membrane is the underlying cause of proteinuria in the kidney disease Alport's syndrome, which results from collagen IV mutations (Barker et al., 1990; Longo et al., 2002).

Basement membrane microperforations are normal physiological findings in bronchial airway (0.75–3.85 μm in diameter) (Howat, Holmes, Holgate, & Lackie, 2001) and in the intestinal villi and lymph nodules of ileal Peyer's patches (~1–8 μm in diameter) (Takeuchi & Gonda, 2004). Basement membrane microperforations in these tissues facilitate lymphocyte migration, especially in the small intestine, where the abundance of microperforations corresponds to the degree of lymphocytic infiltration.

The transient presence of basement membrane microperforations in the embryonic mouse submandibular salivary gland correlates with the time and location of maximal epithelial expansion (Harunaga et al., 2014). Microperforations are most prominent at the beginning of the active branching phase, when cell proliferation and repetitive clefting and branching substantially increase the organ surface area (Fig. 5). The tip of the end bud has both the largest size and density of microperforations. The largest microperforations (4 μm in diameter) are still smaller than the size of a cell. With 90% of the microperforations being even smaller (<2.5 μm in diameter), they are not permissive to cell transmigration. Both the size and density of microperforations dramatically decrease further away from the tip. Microperforations eventually disappear toward the equator of the bud, which is a

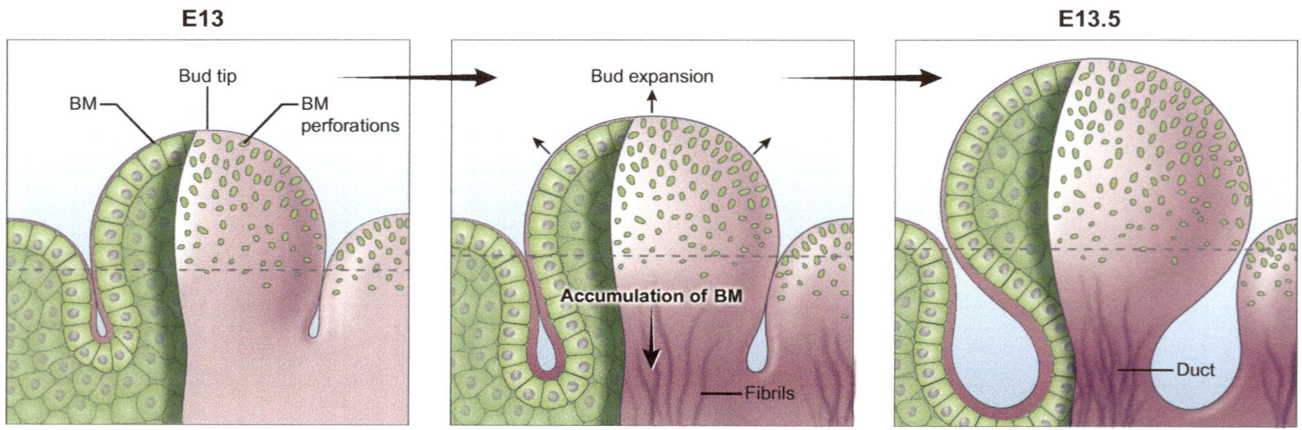

Fig. 5 BM microperforations in embryonic mouse submandibular salivary gland. Microperforations are prominent at the tip of epithelial buds. These perforations gradually disappear toward the equator (*dotted line*). As the growing buds expand rapidly, the number of perforations peaks at embryonic day (E) 13.5. Basement membrane translocation away from the bud tip is accompanied by accumulation of fibril-like basement membrane proteins at the forming secondary duct.

site of basement membrane accumulation (Fig. 5). Protease and myosin II-dependent global translocation of the basement membrane occurs rearward toward the duct, in a direction opposite to epithelial outgrowth. The long axis of the oblong microperforations is oriented parallel to the presumptive pulling force that drives global basement membrane translocation. Fibril-like basement membrane aggregates have varying thickness and direction, but their general orientation also runs parallel to the long axis of the end bud. Thus, the general orientation of fibril-like basement membranes also coincides with the axis of the basement membrane translocation (Harunaga et al., 2014), which suggests a possible link between pore shape, orientation of fibril-like deposition, and global basement membrane translocation.

Consistent with findings of classical electron microscopy studies on epithelial–mesenchymal inductive interactions (Bluemink et al., 1976; Cutler, 1977; Lehtonen, 1975), cellular protrusions from basal epithelial cells are observed exclusively at the tip of the end bud in the embryonic mouse lung and submandibular gland. The protrusions appear to expand the basement membrane microperforations as they protrude as far as 5 μm beyond the basement membrane (Harunaga et al., 2014). Individual microperforations undergo local dilation and contraction, a process dependent on actomyosin-mediated rapid physical distortion of subjacent cells and protease activity. Unlike *Caenorhabditis elegans* anchor (uterine) cell invasion, which involves physical widening of a gap by the cell and laminin accumulation at the edge of the gap (Hagedorn et al., 2013; Matus et al., 2014), the mouse submandibular salivary gland does not exhibit basement membrane displacement or accumulation around the periphery of the microperforation. Microperforations in the salivary gland basement membrane are not enlarged to a size sufficient to permit cell crossing. Rather, the salivary gland basement membrane as a whole becomes distensible in response to cellular forces; the amount of distensibility and the size of perforations are just sufficient for extension of cellular processes (Harunaga et al., 2014). Similarly, epithelial–mesenchymal cell contact is found during tooth morphogenesis. Differentiating ameloblasts lining the inner enamel epithelium extend cellular processes through the basement membrane, which penetrate into developing uncalcified dentin (Burgess & Katchburian, 1982). Cytoplasmic processes from odontoblasts in the tooth bud mesenchyme subsequently penetrate the basement membrane to contact preameloblasts (Slavkin & Bringas, 1976). A series of reciprocal inductions between the two apposed tissues is essential for sequential differentiation of the epithelial and mesenchymal cells into their respective matrix-secreting cells (Simmer et al., 2010). In addition to growth factors secreted by the oral epithelium (Tucker & Sharpe, 2004), this cellular contact may be another mechanism of cross-tissue interactions. The implications of direct cell–cell contact, the detailed mechanisms of microperforation formation in mammalian organs, the function(s) of microperforations, and the source of global basement membrane translocation are important research questions to be explored in future studies.

6. BASEMENT MEMBRANE TRANSMIGRATION AND INVASION

6.1 Basement Membrane Transmigration

Basement membranes in many tissues are a permissive substrate for certain migrating cells. In tissues where leukocyte trafficking is prevalent, such as in lymph nodes, the basement membrane has preexisting gaps at specific locations that facilitate routine cell transmigration (Nourshargh & Alon, 2014). Neutrophils can migrate between the circulation and the subendothelial space in response to injury. Upon sterile thermal hepatic injury, neutrophils migrate from the vasculature into the subendothelial space and then later reverse migrate back into the circulation (Wang et al., 2017). However, the basement membranes of many other tissues contain pores that are substantially smaller than the diameters of cells in physiological conditions. Consequently, enlargement of basement membrane pores is required for cells to traverse the barrier, by means of expansion, disassembling, stretching, proteolytic breakdown, or finding regions with reduced density and cross-linking (Rowe & Weiss, 2008; Wang et al., 2006). Nonetheless, many cells frequently transmigrate through basement membranes during embryonic development, such as during gastrulation (Nakaya, Sukowati, Wu, & Sheng, 2008), and disease, as in inflammatory disease (Sorokin, 2010) and metastatic cancer (Steeg, 2003).

Transmigration through mammalian basement membranes is often not solely a simple mechanical process. Rather, it may evoke molecular signaling that alters gene transcription in transmigrating cells. During leukocyte trafficking, leukocytes traverse beyond the endothelium and through the vascular basement membrane (Nourshargh & Alon, 2014). Their vascular wall breaching involves extensive alterations of both the migrating leukocytes and the matrix components through reciprocal signaling. Breaching of the vascular wall and basement membrane prime the transmigrating leukocytes for clearing pathogens and neoplastic cells with enhanced survival and increased effector functions (Nourshargh, Hordijk, & Sixt, 2010). Though the mechanistic role of the basement membrane in the priming process has yet to be elucidated, cell transmigration through the basement membrane is most likely a combined molecular and mechanical process that is tightly regulated during development and leukocyte transmigration, which becomes dysregulated in disease.

6.2 Basement Membrane Invasion During Tumor Progression

Normal developmental processes are often hijacked during tumor progression. Reduced epithelial cell polarity, EMT, and attenuation of intercellular and cell–matrix adhesion during tumor progression partially recapitulate developmental processes (Ewald et al., 2012; Tanos & Rodriguez-Boulan, 2008). While tumor progression and normal morphogenetic processes may share some regulatory mechanisms, what clearly differentiates tumor progression from developmental processes is transcriptional dysregulation. Because the basement membrane serves as a barrier against malignant cell dissemination, basement membrane breaching and dysregulation are hallmarks of many malignant tumors (Nguyen-Ngoc et al., 2012). Tumor progression involves alterations of ECM along with transcriptional and posttranscriptional acquisition of invasive cellular machinery by tumor cells (Ozanne, Spence, McGarry, & Hennigan, 2006).

Invadopodia are membrane-associated F-actin-based structures that can help mediate cell invasion through surface expression of proteolytic enzymes that locally degrade ECM (Gimona, Buccione, Courtneidge, & Linder, 2008). Cellular manifestation of invadopodia can be a part of normal morphogenetic processes such as axon guidance (Santiago-Medina, Gregus, Nichol, O'Toole, & Gomez, 2015) and can also be induced in nonmalignant cells by a dense fibrillar collagen matrix (Artym et al., 2015). However, acquisition of invadopodia by tumor cells reflects a transcriptional and signaling switch. Malignant cells often insert invadopodia into the basement membrane for local degradation as they establish adhesion to the underlying stroma (Friedl & Wolf, 2003). Transcriptional and signaling switches confer on tumor cells the ability to extend invadopodia, perforate, and invade through the basement membrane. The invasive cells can destroy the basement membrane as they invade with local protease activity, consistent with a prominent characteristic of tumor histology: a reduction or complete absence of the basement membrane (Tosios, Kapranos, & Papanicolaou,

1998). The presence of such basement membrane-invasive cells portends a poorer prognosis (Hanahan & Weinberg, 2000).

This invasive behavior of neoplastic cells may initially be attenuated or blocked by basement membranes. The barrier function of basement membranes derives from its tightly cross-linked matrix and niche formation that prevents carcinoma cell dissemination. Both normal and malignant mammary epithelia initially form cellular protrusions and disseminate in a collagen I substrate (Nguyen-Ngoc et al., 2012). However, a striking difference arises in subsequent behavior of these cells. Malignant mammary epithelium continues to disseminate without organizing structured basement membranes. On the other hand, normal mammary epithelium ceases to disseminate as it progressively organizes basement membranes. Deletion of P-cadherin, an adhesion gene that is downregulated in breast carcinoma cells, induces epithelial cell dissemination into collagen I but not in Matrigel. These findings demonstrate that the malignant epithelium preferentially disseminates in collagen I and suggest that breaks in basement membranes may induce invasion (Nguyen-Ngoc et al., 2012).

Normal nonmalignant epithelial cells depend on prosurvival signaling from laminin that defines their niche (Chiarugi & Giannoni, 2008). Leaving their ECM niche can result in apoptosis due to lack of survival signaling. Thus, remaining in contact with their natural niche is crucial for basal epithelial cell survival. However, tumor cells that undergo malignant transformation can survive outside of their niche by actively altering their transcription pattern and basement membrane composition. Malignantly transformed cells can overexpress the basement membrane protein they depend on or alter their intrinsic integrin signaling to adjust to a foreign ECM (Chiarugi & Giannoni, 2008). For example, a mechanism commonly employed by carcinoma cells is overexpression of LM-332—a mechanism observed in esophageal, colon, and squamous cell carcinomas (Marinkovich, 2007). Survival signals from LM-332 can enable carcinoma cells to leave their natural niche and metastasize. Furthermore, angiogenesis is stimulated by laminin and perlecan, providing essential nutrients and oxygen for survival of disseminated cancer cells.

Invasion and proliferation can be mutually incompatible cell states as demonstrated in mice (Gil-Henn et al., 2013) and *C. elegans* (Sherwood & Sternberg, 2003). The *C. elegans* model of basement membrane invasion utilizes a hermaphrodite developmental process in which anchor cells transmigrate through juxtaposed uterine and ventral epidermal basement membranes to initiate uterine–vulval connection (Sherwood & Sternberg, 2003).

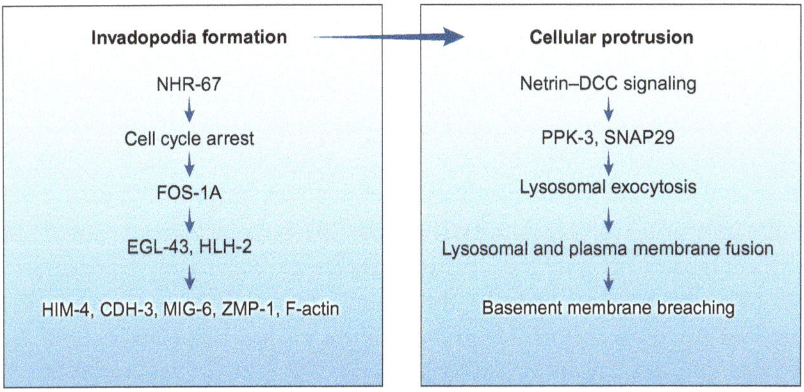

Fig. 6 Factors involved in *C. elegans* BM transmigration. The first stage is initiated by cell cycle arrest and activation of a transcriptional cascade that leads to invadopodia formation. The second stage involves netrin–DCC signaling that directs growth of a single large cellular protrusion, allowing the anchor cell to breach the juxtaposed uterine and vulval basement membranes.

Transcription factor *nhr-67/tlx*-directed G1 cell cycle arrest is required for invadopodia formation, proinvasive gene transcription, and basement membrane invasion (Fig. 6) (Matus et al., 2015). G1 cell cycle arrest promotes the expression of FOS-1A (the vertebrate FOS protooncogene) in *C. elegans*. FOS-1A in turn triggers a transcriptional cascade through transcription factors HLH-2 (E proteins) (Schindler & Sherwood, 2011) and EGL-43 (EVII) (Rimann & Hajnal, 2007). FOS-1A regulates proinvasive genes, including CDH-3 (protocadherin), HIM-4 (hemicentin), ZMP-1 (MMP) (Sherwood, Butler, Kramer, & Sternberg, 2005), MIG-6 (papilin), and F-actin (Schindler & Sherwood, 2011). This requirement for cell cycle arrest is partially echoed in a developmental process underlying mammalian tooth morphogenesis. Throughout incisor placode and bud morphogenesis, a subset of nonproliferative epithelial cells remains in the G1 phase (Ahtiainen, Uski, Thesleff, & Mikkola, 2016). These cells localize to the emerging enamel knot that exhibits restricted expression of SHH, BMP, FGF, and cyclin-dependent kinase inhibitor P21. The neighboring epithelial cells surrounding this nonproliferating emerging enamel knot contribute to the growth of the tooth bud (Ahtiainen et al., 2016).

C. elegans anchor cell transmigration takes place in two stages: first the generation of initially tiny basement membrane perforations by small invadopodia, and second subsequent local gap enlargement by proteolysis and physical force from a large cellular protrusion (Fig. 6). Deciphering the mechanism of the second stage that clears a path for tissue entry is important because of its similarity to tumor cell transmigration and invasion during cancer metastasis. The first stage involving invadopodia formation is dependent on integrin, which promotes netrin receptor (DCC) localization to the anchor cell membrane in contact with the basement membrane (Hagedorn et al., 2009, 2013). Netrin–DCC signaling directs the second stage via creation of a single large cellular protrusion through the exocyst and SNARE-mediated fusion of lysosomal membranes to fuel local plasma membrane expansion (Naegeli et al., 2017). This outward expansion of the invasive protrusion is anchored and promoted by F-actin and dystroglycan, which form a diffusion barrier at the neck of the protrusion. This transient large anchor cell protrusion and MMP protease activity expand the hole by a combination of local basement membrane displacement and degradation, enabling the cell to transmigrate (Naegeli et al., 2017).

Cell invasion in development and tumors can correlate with EMT (Micalizzi, Farabaugh, & Ford, 2010; Thiery, Acloque, Huang, & Nieto, 2009). EMT is promoted by the transcription factors SNAIL, SLUG, TWIST, and ZEB1/2 (Hanahan & Weinberg, 2011). EMT in development is well exemplified by neural crest cell migration. First induced in the ectodermal germ layer during gastrulation, neural crest cells undergo EMT involving SNAI2, ZEB2, and TWIST after neural tube closure (Simões-Costa & Bronner, 2015). Upon acquisition of migratory behavior and transformation into multipotent progenitors, neural crest cells leave the central nervous system and migrate through the basement membrane along highly stereotypical paths. Along with EMT, degradation of the basement membrane by proteases is crucial for neural crest cell migration (Theveneau & Mayor, 2011). However, transcription factors alone do not specify invasive cellular behavior, since their effects on protease activity and invadopodia formation are indirect (Eckert et al., 2011; Jorda et al., 2005). In addition, EMT and invadopodia are not essential for all instances of basement membrane breaching. During development and metastatic carcinoma, cells can collectively migrate or invade through the basement membrane without undergoing complete EMT (Nieto, Huang, Jackson, & Thiery, 2016). Thus, additional mechanisms may as well regulate carcinoma cell invasion.

During tumor progression, dysregulation of proteases such as MMPs facilitate basement membrane degradation and invasion (Wiseman & Werb, 2002). However, clinical trials of MMP inhibitors have been largely unsuccessful. Some of the potential reasons relate to issues with drug specificity, side effects that limited doses that could be tolerated, study design of clinical trials with respect to the stage of cancer, as well as study design of many initial in vitro preclinical studies. The initial in vitro studies using cancer cell lines that overexpress certain MMP family members may not accurately reflect in vivo dynamics, where the primary source of proteases in carcinoma is apparently nonneoplastic stromal cells (Madsen & Bugge, 2015; Nielsen et al., 2001). The cellular source of proteases indeed matters because the activity of released enzymes significantly differs between cell types (Kessenbrock, Plaks, & Werb, 2010).

The basement membrane initially physically separates neoplastic epithelial cells from nonmalignant stromal cells. The likelihood that the stromal cells are the major source of proteases in carcinomas raises a fundamental question: How do neoplastic epithelial cells initiate basement membrane breaching? One possibility is activation of stromal cells by inflammatory signals such as interleukins-6 and -8 from neoplastic cells. Another potential mechanism is localized loss of the adhesion receptor dystroglycan. Loss of dystroglycan in basal epiblast cells during chick gastrulation allows epiblast cells undergoing EMT to cross through the basement membrane into the developing embryo (Nakaya, Sukowati, Alev, Nakazawa, & Sheng, 2011). Additionally, mechanical forces may mediate basement membrane breaching. During postimplantation embryonic mouse development, localized constriction at the distal tip of the developing embryo by the surrounding maternal uterus appears to rupture the basement membrane. Basement membrane rupture in turn permits epiblast cell transmigration and initiation of establishment of the anterior–posterior axis (Hiramatsu et al., 2013). Similarly, tumor cell invasion may involve a combination of molecular signaling, degradation of basement membrane components, global mechanical hydrostatic force from the growing tumor, and local breaching force exerted by an invasive cellular protrusion. The mechanism of this initial basement membrane breaching remains a crucial gap in our understanding of tumor progression.

7. PERSPECTIVES AND FUTURE DIRECTIONS

Advances in high-throughput omics and bioinformatics techniques have enabled the identification of novel ECM genes and proteins at unprecedented speed. However, the relevance of many of these newly discovered genes to development and disease remains to be unraveled. Accordingly, time course proteomic and functional analyses of basement membrane proteins from embryonic development through adulthood and disease will help identify their roles in both development and disease.

In carcinoma progression, the basement membrane serves as an important barrier that spatially confines malignant epithelial cells prior to invasion. Identification of mechanisms underlying basement membrane invasion will provide valuable insight into cancer progression. One of the important topics that require further investigation is the source of specific ECM-degrading proteases and their contributions in initiation of basement membrane breaching. Also not clear is whether initial protease-induced breaks in the basement membrane are enlarged by tumor cell protrusions. For example, do tumor cells use cellular extensions that expand rapidly due to focal exocytosis of lysosomes to enlarge initially tiny basement membrane holes analogous to the mechanism used by *C. elegans* anchor cells? Another challenge will be the full elucidation of the biophysical, transcriptional, and post-translational changes affecting the basement membrane at different stages of cancer.

Identification of complex developmental and disease mechanisms will require a better understanding of not only molecular signaling but also biomechanical processes. Further investigations are needed to characterize the biophysical properties of the different basement membranes in relation to development and disease. If technically feasible, measurement of local stiffness vs tension within the basement membranes of embryonic tissues will provide valuable insight, which might be achieved by atomic force microscopy (Casuso, Rico, & Scheuring, 2011) or fluorescence resonance energy transfer measurements of tension sensors (Grashoff et al., 2010). The levels of stiffness or tension may well vary across developmental stages, throughout the body, and in different regions of the basement membrane within the same organ. Variations in tension, stiffness, and patterns of microperforation may alter the levels of cell adhesion, the extent of epithelial–mesenchymal interactions, patterns of cell migration, and/or the mode of tissue remodeling and morphogenesis. Live imaging of tissue explants will be valuable for structural and functional characterization of basement membrane constituents in developing, mature, and pathological or genetically perturbed tissues. Temporally and spatially controlled gene perturbation will be crucial because many developmental and disease mechanisms are precisely controlled at the single cell level.

The ultimate goal of future research will be to develop effective therapeutic approaches, e.g., to help prevent basement membrane breaching in cancer. Additional future therapeutic approaches could benefit individuals afflicted by conditions in which the basement membrane plays a pivotal role in pathogenesis and disease progression. For example, gene therapy for laminin α2-deficient congenital muscular dystrophy (Qiao et al., 2005) and genetically modified epidermal stem cell therapy for epidermolysis bullosa (Hirsch et al., 2017; Mavilio et al., 2006) represent potentially promising future directions for the treatment of congenital disorders. Another recent advance in regenerative medicine research is laminin-guided human pluripotent stem cell differentiation into endothelial progenitor cells, e.g., using a fragment of LM-411 (Ohta et al., 2016). LM-411 found predominantly in the vascular endothelial basement membrane seems to instruct cell fate determination and differentiation into endothelial cells. Because some laminin isoforms are cell type specific, this method may provide a means of guiding differentiation toward specific cell types when combined with chemical differentiation protocols. With its relevance to many developmental and disease processes, the basement membrane provides an exciting and vitally important area for further research.

ACKNOWLEDGMENTS

We thank Shaohe Wang for assistance with the artwork. We would also like to thank Thomas Bugge for advice on carcinoma cell invasion. Work in the authors' laboratory is supported by the Intramural Research Program of the NIH, NIDCR.

REFERENCES

Abrahamson, D. R., St John, P. L., Stroganova, L., Zelenchuk, A., & Steenhard, B. M. (2013). Laminin and type IV collagen isoform substitutions occur in temporally and spatially distinct patterns in developing kidney glomerular basement membranes. *The Journal of Histochemistry and Cytochemistry*, *61*, 706–718.

Abreu-Velez, A. M., & Howard, M. S. (2012). Collagen IV in normal skin and in pathological processes. *North American Journal of Medical Sciences*, *4*, 1–8.

Ahtiainen, L., Uski, I., Thesleff, I., & Mikkola, M. L. (2016). Early epithelial signaling center governs tooth budding morphogenesis. *The Journal of Cell Biology*, *214*, 753–767.

Artym, V. V., Swatkoski, S., Matsumoto, K., Campbell, C. B., Petrie, R. J., Dimitriadis, E. K., et al. (2015). Dense fibrillar collagen is a potent inducer of invadopodia via a specific signaling network. *The Journal of Cell Biology*, *208*, 331–350.

Aumailley, M., Bruckner-Tuderman, L., Carter, W. G., Deutzmann, R., Edgar, D., Ekblom, P., et al. (2005). A simplified laminin nomenclature. *Matrix Biology*, *24*, 326–332.

Aurich, F., & Dahmann, C. (2016). A mutation in fat2 uncouples tissue elongation from global tissue rotation. *Cell Reports*, *14*, 2503–2510.

Bader, B. L., Smyth, N., Nedbal, S., Miosge, N., Baranowsky, A., Mokkapati, S., et al. (2005). Compound genetic ablation of nidogen 1 and 2 causes basement membrane defects and perinatal lethality in mice. *Molecular and Cellular Biology*, *25*, 6846–6856.

Barker, D. F., Hostikka, S. L., Zhou, J., Chow, L. T., Oliphant, A. R., Gerken, S. C., et al. (1990). Identification of mutations in the COL4A5 collagen gene in Alport syndrome. *Science*, *248*, 1224–1227.

Battaglia, C., Mayer, U., Aumailley, M., & Timpl, R. (1992). Basement-membrane heparan sulfate proteoglycan binds to laminin by its heparan sulfate chains and to nidogen by sites in the protein core. *European Journal of Biochemistry, 208*, 359–366.

Behrens, D. T., Villone, D., Koch, M., Brunner, G., Sorokin, L., Robenek, H., et al. (2012). The epidermal basement membrane is a composite of separate laminin- or collagen IV-containing networks connected by aggregated perlecan, but not by nidogens. *The Journal of Biological Chemistry, 287*, 18700–18709.

Bernfield, M., & Banerjee, S. D. (1982). The turnover of basal lamina glycosaminoglycan correlates with epithelial morphogenesis. *Developmental Biology, 90*, 291–305.

Berrier, A. L., & Yamada, K. M. (2007). Cell–matrix adhesion. *Journal of Cellular Physiology, 213*, 565–573.

Bloch-Zupan, A., Sedano, H. O., & Scully, C. (2012). *Dento/Oro/craniofacial anomalies and genetics*. Elsevier.

Bluemink, J. G., Van Maurik, P., & Lawson, K. A. (1976). Intimate cell contacts at the epithelial/mesenchymal interface in embryonic mouse lung. *Journal of Ultrastructure Research, 55*, 257–270.

Bonnans, C., Chou, J., & Werb, Z. (2014). Remodelling the extracellular matrix in development and disease. *Nature Reviews Molecular Cell Biology, 15*, 786–801.

Bose, K., Nischt, R., Page, A., Bader, B. L., Paulsson, M., & Smyth, N. (2006). Loss of nidogen-1 and -2 results in syndactyly and changes in limb development. *The Journal of Biological Chemistry, 281*, 39620–39629.

Botta, A., Delteil, F., Mettouchi, A., Vieira, A., Estrach, S., Négroni, L., et al. (2012). Confluence switch signaling regulates ECM composition and the plasmin proteolytic cascade in keratinocytes. *Journal of Cell Science, 125*, 4241–4252.

Boyango, I., Barash, U., Fux, L., Naroditsky, I., Ilan, N., & Vlodavsky, I. (2018). Targeting heparanase to the mammary epithelium enhances mammary gland development and promotes tumor growth and metastasis. *Matrix Biology, 65*, 91–103.

Breitkreutz, D., Koxholt, I., Thiemann, K., & Nischt, R. (2013). Skin basement membrane: The foundation of epidermal integrity—BM functions and diverse roles of bridging molecules nidogen and perlecan. *BioMed Research International, 2013*, 179784.

Brose, K., Bland, K. S., Wang, K. H., Arnott, D., Henzel, W., Goodman, C. S., et al. (1999). Slit proteins bind Robo receptors and have an evolutionarily conserved role in repulsive axon guidance. *Cell, 96*, 795–806.

Bunt, S., Hooley, C., Hu, N., Scahill, C., Weavers, H., & Skaer, H. (2010). Hemocyte-secreted type IV collagen enhances BMP signaling to guide renal tubule morphogenesis in Drosophila. *Developmental Cell, 19*, 296–306.

Burgess, A. M., & Katchburian, E. (1982). Morphological types of epithelial–mesenchymal cell contacts in odontogenesis. *Journal of Anatomy, 135*, 577–584.

Candiello, J., Cole, G. J., & Halfter, W. (2010). Age-dependent changes in the structure, composition and biophysical properties of a human basement membrane. *Matrix Biology, 29*, 402–410.

Carey, D. J. (1997). Syndecans: Multifunctional cell-surface co-receptors. *The Biochemical Journal, 327*(Pt 1), 1–16.

Casuso, I., Rico, F., & Scheuring, S. (2011). Biological AFM: Where we come from—Where we are—Where we may go. *Journal of Molecular Recognition, 24*, 406–413.

Chiarugi, P., & Giannoni, E. (2008). Anoikis: A necessary death program for anchorage-dependent cells. *Biochemical Pharmacology, 76*, 1352–1364.

Chun, K. J., Choi, H. H., & Lee, J. Y. (2014). Comparison of mechanical property and role between enamel and dentin in the human teeth. *Journal of Dental Biomechanics, 5*, 1758736014520809.

Coles, E. G., Gammill, L. S., Miner, J. H., & Bronner-Fraser, M. (2006). Abnormalities in neural crest cell migration in laminin α5 mutant mice. *Developmental Biology, 289*, 218–228.

Costell, M., Gustafsson, E., Aszodi, A., Morgelin, M., Bloch, W., Hunziker, E., et al. (1999). Perlecan maintains the integrity of cartilage and some basement membranes. *The Journal of Cell Biology, 147*, 1109–1122.

Cummings, C. F., Pedchenko, V., Brown, K. L., Colon, S., Rafi, M., Jones-Paris, C., et al. (2016). Extracellular chloride signals collagen IV network assembly during basement membrane formation. *The Journal of Cell Biology, 213*, 479–494.

Cutler, L. S. (1977). Intercellular contacts at the epithelial–mesenchymal interface of the developing rat submandibular gland in vitro. *Journal of Embryology and Experimental Morphology, 39*, 71–77.

Dabija-Wolter, G., Bakken, V., Cimpan, M. R., Johannessen, A. C., & Costea, D. E. (2013). In vitro reconstruction of human junctional and sulcular epithelium. *Journal of Oral Pathology & Medicine, 42*, 396–404.

Daley, W. P., Gervais, E. M., Centanni, S. W., Gulfo, K. M., Nelson, D. A., & Larsen, M. (2012). ROCK1-directed basement membrane positioning coordinates epithelial tissue polarity. *Development, 139*, 411–422.

Daley, W. P., Matsumoto, K., Doyle, A. D., Wang, S., DuChez, B. J., Holmbeck, K., et al. (2017). Btbd7 is essential for region-specific epithelial cell dynamics and branching morphogenesis in vivo. *Development, 144*, 2200–2211.

De Moerlooze, L., Spencer-Dene, B., Revest, J. M., Hajihosseini, M., Rosewell, I., & Dickson, C. (2000). An important role for the IIIb isoform of fibroblast growth factor receptor 2 (FGFR2) in mesenchymal–epithelial signalling during mouse organogenesis. *Development, 127*, 483–492.

Deen, W. M. (2004). What determines glomerular capillary permeability? *The Journal of Clinical Investigation, 114*, 1412–1414.

DeSimone, D. W., & Mecham, R. (2013). *Extracellular matrix in development.* Berlin/ Heidelberg: Springer.

Diaz de la Loza, M. C., Diaz-Torres, A., Zurita, F., Rosales-Nieves, A. E., Moeendarbary, E., Franze, K., et al. (2017). Laminin levels regulate tissue migration and anterior–posterior polarity during egg morphogenesis in Drosophila. *Cell Reports, 20*, 211–223.

Domogatskaya, A., Rodin, S., & Tryggvason, K. (2012). Functional diversity of laminins. *Annual Review of Cell and Developmental Biology, 28*, 523–553.

Eckert, M. A., Lwin, T. M., Chang, A. T., Kim, J., Danis, E., Ohno-Machado, L., et al. (2011). Twist1-induced invadopodia formation promotes tumor metastasis. *Cancer Cell, 19*, 372–386.

Engel, J. (1996). Domain organizations of modular extracellular matrix proteins and their evolution. *Matrix Biology, 15*, 295–299.

Engelhardt, B., Vajkoczy, P., & Weller, R. O. (2017). The movers and shapers in immune privilege of the CNS. *Nature Immunology, 18*, 123–131.

Entesarian, M., Matsson, H., Klar, J., Bergendal, B., Olson, L., Arakaki, R., et al. (2005). Mutations in the gene encoding fibroblast growth factor 10 are associated with aplasia of lacrimal and salivary glands. *Nature Genetics, 37*, 125–127.

Ewald, A. J., Huebner, R. J., Palsdottir, H., Lee, J. K., Perez, M. J., Jorgens, D. M., et al. (2012). Mammary collective cell migration involves transient loss of epithelial features and individual cell migration within the epithelium. *Journal of Cell Science, 125*, 2638–2654.

Farach-Carson, M. C., Warren, C. R., Harrington, D. A., & Carson, D. D. (2014). Border patrol: Insights into the unique role of perlecan/heparan sulfate proteoglycan 2 at cell and tissue borders. *Matrix Biology, 34*, 64–79.

Fata, J. E., Werb, Z., & Bissell, M. J. (2004). Regulation of mammary gland branching morphogenesis by the extracellular matrix and its remodeling enzymes. *Breast Cancer Research, 6*, 1–11.

Fidler, A. L., Darris, C. E., Chetyrkin, S. V., Pedchenko, V. K., Boudko, S. P., Brown, K. L., et al. (2017). Collagen IV and basement membrane at the evolutionary dawn of metazoan tissues. *eLife, 6*.

Fine, J. D., Bruckner-Tuderman, L., Eady, R. A., Bauer, E. A., Bauer, J. W., Has, C., et al. (2014). Inherited epidermolysis bullosa: Updated recommendations on diagnosis and classification. *Journal of the American Academy of Dermatology, 70,* 1103–1126.

Fox, J. W., Mayer, U., Nischt, R., Aumailley, M., Reinhardt, D., Wiedemann, H., et al. (1991). Recombinant nidogen consists of three globular domains and mediates binding of laminin to collagen type IV. *The EMBO Journal, 10,* 3137–3146.

Friedl, P., & Wolf, K. (2003). Tumour-cell invasion and migration: Diversity and escape mechanisms. *Nature Reviews Cancer, 3,* 362–374.

Fukumoto, S., Miner, J. H., Ida, H., Fukumoto, E., Yuasa, K., Miyazaki, H., et al. (2006). Laminin alpha5 is required for dental epithelium growth and polarity and the development of tooth bud and shape. *The Journal of Biological Chemistry, 281,* 5008–5016.

Garcia-Alonso, L., Fetter, R. D., & Goodman, C. S. (1996). Genetic analysis of laminin A in Drosophila: Extracellular matrix containing laminin a is required for ocellar axon pathfinding. *Development, 122,* 2611–2621.

Gautam, J., Zhang, X., & Yao, Y. (2016). The role of pericytic laminin in blood brain barrier integrity maintenance. *Scientific Reports, 6,* 36450.

Gervais, E. M., Sequeira, S. J., Wang, W., Abraham, S., Kim, J. H., Leonard, D., et al. (2016). Par-1b is required for morphogenesis and differentiation of myoepithelial cells during salivary gland development. *Organogenesis, 12,* 194–216.

Gibson, C. W., Yuan, Z. A., Hall, B., Longenecker, G., Chen, E., Thyagarajan, T., et al. (2001). Amelogenin-deficient mice display an amelogenesis imperfecta phenotype. *The Journal of Biological Chemistry, 276,* 31871–31875.

Gil-Henn, H., Patsialou, A., Wang, Y., Warren, M. S., Condeelis, J. S., & Koleske, A. J. (2013). Arg/Abl2 promotes invasion and attenuates proliferation of breast cancer in vivo. *Oncogene, 32,* 2622–2630.

Gimona, M., Buccione, R., Courtneidge, S. A., & Linder, S. (2008). Assembly and biological role of podosomes and invadopodia. *Current Opinion in Cell Biology, 20,* 235–241.

Grashoff, C., Hoffman, B. D., Brenner, M. D., Zhou, R., Parsons, M., Yang, M. T., et al. (2010). Measuring mechanical tension across vinculin reveals regulation of focal adhesion dynamics. *Nature, 466,* 263–266.

Guerra, L., Odorisio, T., Zambruno, G., & Castiglia, D. (2017). Stromal microenvironment in type VII collagen-deficient skin: The ground for squamous cell carcinoma development. *Matrix Biology, 63,* 1–10.

Hagedorn, E. J., Yashiro, H., Ziel, J. W., Ihara, S., Wang, Z., & Sherwood, D. R. (2009). Integrin acts upstream of netrin signaling to regulate formation of the anchor cell's invasive membrane in C. elegans. *Developmental Cell, 17,* 187–198.

Hagedorn, E. J., Ziel, J. W., Morrissey, M. A., Linden, L. M., Wang, Z., Chi, Q., et al. (2013). The netrin receptor DCC focuses invadopodia-driven basement membrane transmigration in vivo. *The Journal of Cell Biology, 201,* 903–913.

Haigo, S. L., & Bilder, D. (2011). Global tissue revolutions in a morphogenetic movement controlling elongation. *Science, 331,* 1071–1074.

Hakkinen, K. M., Harunaga, J. S., Doyle, A. D., & Yamada, K. M. (2011). Direct comparisons of the morphology, migration, cell adhesions, and actin cytoskeleton of fibroblasts in four different three-dimensional extracellular matrices. *Tissue Engineering Part A, 17,* 713–724.

Halfter, W., Dong, S., Yip, Y. P., Willem, M., & Mayer, U. (2002). A critical function of the pial basement membrane in cortical histogenesis. *The Journal of Neuroscience, 22,* 6029–6040.

Halfter, W., Monnier, C., Muller, D., Oertle, P., Uechi, G., Balasubramani, M., et al. (2013). The bi-functional organization of human basement membranes. *PLoS One, 8,* e67660.

Halfter, W., Oertle, P., Monnier, C. A., Camenzind, L., Reyes-Lua, M., Hu, H., et al. (2015). New concepts in basement membrane biology. *The FEBS Journal, 282,* 4466–4479.
Hallmann, R., Horn, N., Selg, M., Wendler, O., Pausch, F., & Sorokin, L. M. (2005). Expression and function of laminins in the embryonic and mature vasculature. *Physiological Reviews, 85,* 979–1000.
Hamano, Y., Zeisberg, M., Sugimoto, H., Lively, J. C., Maeshima, Y., Yang, C., et al. (2003). Physiological levels of tumstatin, a fragment of collagen IV α3 chain, are generated by MMP-9 proteolysis and suppress angiogenesis via αVβ3 integrin. *Cancer Cell, 3,* 589–601.
Hanahan, D., & Weinberg, R. A. (2000). The hallmarks of cancer. *Cell, 100,* 57–70.
Hanahan, D., & Weinberg, R. A. (2011). Hallmarks of cancer: The next generation. *Cell, 144,* 646–674.
Harunaga, J. S., Doyle, A. D., & Yamada, K. M. (2014). Local and global dynamics of the basement membrane during branching morphogenesis require protease activity and actomyosin contractility. *Developmental Biology, 394,* 197–205.
Harvey, S. J., Zheng, K., Sado, Y., Naito, I., Ninomiya, Y., Jacobs, R. M., et al. (1998). Role of distinct type IV collagen networks in glomerular development and function. *Kidney International, 54,* 1857–1866.
Has, C., & Nyström, A. (2015). Chapter Four—Epidermal basement membrane in health and disease. In J. H. Miner (Ed.), *Current topics in membranes* (pp. 117–170). Academic Press.
Hatakeyama, S., Yaegashi, T., Oikawa, Y., Fujiwara, H., Mikami, T., Takeda, Y., et al. (2006). Expression pattern of adhesion molecules in junctional epithelium differs from that in other gingival epithelia. *Journal of Periodontal Research, 41,* 322–328.
Hayashi, T. (2011). Dysfunction of lacrimal and salivary glands in Sjögren's syndrome: Nonimmunologic injury in Preinflammatory phase and mouse model. *Journal of Biomedicine & Biotechnology, 2011,* 407031.
He, P., Zhang, Y., Kim, S. O., Radlanski, R. J., Butcher, K., Schneider, R. A., et al. (2010). Ameloblast differentiation in the human developing tooth: Effects of extracellular matrices. *Matrix Biology, 29,* 411–419.
Herzog, C., Has, C., Franzke, C. W., Echtermeyer, F. G., Schlotzer-Schrehardt, U., Kroger, S., et al. (2004). Dystroglycan in skin and cutaneous cells: Beta-subunit is shed from the cell surface. *The Journal of Investigative Dermatology, 122,* 1372–1380.
Hiramatsu, R., Matsuoka, T., Kimura-Yoshida, C., Han, S. W., Mochida, K., Adachi, T., et al. (2013). External mechanical cues trigger the establishment of the anterior–posterior axis in early mouse embryos. *Developmental Cell, 27,* 131–144.
Hironaka, K., Makino, H., Yamasaki, Y., & Ota, Z. (1993). Renal basement membranes by ultrahigh resolution scanning electron microscopy. *Kidney International, 43,* 334–345.
Hirsch, T., Rothoeft, T., Teig, N., Bauer, J. W., Pellegrini, G., De Rosa, L., et al. (2017). Regeneration of the entire human epidermis using transgenic stem cells. *Nature, 551,* 327–332.
Hohenester, E., & Yurchenco, P. D. (2013). Laminins in basement membrane assembly. *Cell Adhesion & Migration, 7,* 56–63.
Horejs, C. M., Serio, A., Purvis, A., Gormley, A. J., Bertazzo, S., Poliniewicz, A., et al. (2014). Biologically-active laminin-111 fragment that modulates the epithelial-to-mesenchymal transition in embryonic stem cells. *Proceedings of the National Academy of Sciences of the United States of America, 111,* 5908–5913.
Howat, W. J., Holmes, J. A., Holgate, S. T., & Lackie, P. M. (2001). Basement membrane pores in human bronchial epithelium: A conduit for infiltrating cells? *The American Journal of Pathology, 158,* 673–680.

Hsu, J. C., Koo, H., Harunaga, J. S., Matsumoto, K., Doyle, A. D., & Yamada, K. M. (2013). Region-specific epithelial cell dynamics during branching morphogenesis. *Developmental Dynamics*, *242*, 1066–1077.

Hudson, B. G., Tryggvason, K., Sundaramoorthy, M., & Neilson, E. G. (2003). Alport's syndrome, Goodpasture's syndrome, and type IV collagen. *The New England Journal of Medicine*, *348*, 2543–2556.

Huebner, R. J., Lechler, T., & Ewald, A. J. (2014). Developmental stratification of the mammary epithelium occurs through symmetry-breaking vertical divisions of apically positioned luminal cells. *Development*, *141*, 1085–1094.

Humbert, C., Silbermann, F., Morar, B., Parisot, M., Zarhrate, M., Masson, C., et al. (2014). Integrin alpha 8 recessive mutations are responsible for bilateral renal agenesis in humans. *American Journal of Human Genetics*, *94*, 288–294.

Hussain, S. A., Piper, M., Fukuhara, N., Strochlic, L., Cho, G., Howitt, J. A., et al. (2006). A molecular mechanism for the heparan sulfate dependence of slit-robo signaling. *The Journal of Biological Chemistry*, *281*, 39693–39698.

Hynes, R. O. (1992). Integrins: Versatility, modulation, and signaling in cell adhesion. *Cell*, *69*, 11–25.

Hynes, R. O. (1999). Cell adhesion: Old and new questions. *Trends in Cell Biology*, *9*, M33–37.

Hynes, R. O., & Naba, A. (2012). Overview of the matrisome—An inventory of extracellular matrix constituents and functions. *Cold Spring Harbor Perspectives in Biology*, *4*, a004903.

Hynes, R. O., & Yamada, K. M. (2012). *Extracellular matrix biology*. Cold Spring Harbor Laboratory Press.

Imbeni, V., Kruzic, J. J., Marshall, G. W., Marshall, S. J., & Ritchie, R. O. (2005). The dentin–enamel junction and the fracture of human teeth. *Nature Materials*, *4*, 229–232.

Inomata, T., Ebihara, N., Funaki, T., Matsuda, A., Watanabe, Y., Ning, L., et al. (2012). Perlecan-deficient mutation impairs corneal epithelial structure. *Investigative Ophthalmology & Visual Science*, *53*, 1277–1284.

Isabella, A. J., & Horne-Badovinac, S. (2015). Building from the ground up: Basement membranes in Drosophila development. *Current Topics in Membranes*, *76*, 305–336.

Isabella, A. J., & Horne-Badovinac, S. (2016). Rab10-mediated secretion synergizes with tissue movement to build a polarized basement membrane architecture for organ morphogenesis. *Developmental Cell*, *38*, 47–60.

Jayadev, R., & Sherwood, D. R. (2017). Basement membranes. *Current Biology*, *27*, R207–r211.

Johnson, K. G., Ghose, A., Epstein, E., Lincecum, J., O'Connor, M. B., & Van Vactor, D. (2004). Axonal heparan sulfate proteoglycans regulate the distribution and efficiency of the repellent slit during midline axon guidance. *Current Biology*, *14*, 499–504.

Jorda, M., Olmeda, D., Vinyals, A., Valero, E., Cubillo, E., Llorens, A., et al. (2005). Upregulation of MMP-9 in MDCK epithelial cell line in response to expression of the snail transcription factor. *Journal of Cell Science*, *118*, 3371–3385.

Kadoya, Y., & Yamashina, S. (2005). Salivary gland morphogenesis and basement membranes. *Anatomical Science International*, *80*, 71–79.

Kalluri, R. (2003). Basement membranes: Structure, assembly and role in tumour angiogenesis. *Nature Reviews Cancer*, *3*, 422–433.

Kan, M., Wu, X., Wang, F., & McKeehan, W. L. (1999). Specificity for fibroblast growth factors determined by heparan sulfate in a binary complex with the receptor kinase. *The Journal of Biological Chemistry*, *274*, 15947–15952.

Karihaloo, A., Karumanchi, S. A., Barasch, J., Jha, V., Nickel, C. H., Yang, J., et al. (2001). Endostatin regulates branching morphogenesis of renal epithelial cells and ureteric bud.

Proceedings of the National Academy of Sciences of the United States of America, 98, 12509–12514.
Keely, P. J., Wu, J. E., & Santoro, S. A. (1995). The spatial and temporal expression of the alpha 2 beta 1 integrin and its ligands, collagen I, collagen IV, and laminin, suggest important roles in mouse mammary morphogenesis. *Differentiation, 59*, 1–13.
Kessenbrock, K., Plaks, V., & Werb, Z. (2010). Matrix metalloproteinases: Regulators of the tumor microenvironment. *Cell, 141*, 52–67.
Kim, S. H., Turnbull, J., & Guimond, S. (2011). Extracellular matrix and cell signalling: The dynamic cooperation of integrin, proteoglycan and growth factor receptor. *The Journal of Endocrinology, 209*, 139–151.
Kinane, D. F. (2001). Causation and pathogenesis of periodontal disease. *Periodontology 2000, 2000*(25), 8–20.
Kiyozumi, D., Takeichi, M., Nakano, I., Sato, Y., Fukuda, T., & Sekiguchi, K. (2012). Basement membrane assembly of the integrin alpha8beta1 ligand nephronectin requires Fraser syndrome-associated proteins. *The Journal of Cell Biology, 197*, 677–689.
Laine, M., Virtanen, I., Porola, P., Rotar, Z., Rozman, B., Poduval, P., et al. (2008). Acinar epithelial cell laminin-receptors in labial salivary glands in Sjogren's syndrome. *Clinical and Experimental Rheumatology, 26*, 807–813.
Larjava, H., Koivisto, L., Hakkinen, L., & Heino, J. (2011). Epithelial integrins with special reference to oral epithelia. *Journal of Dental Research, 90*, 1367–1376.
Legoff, L., Rouault, H., & Lecuit, T. (2013). A global pattern of mechanical stress polarizes cell divisions and cell shape in the growing Drosophila wing disc. *Development, 140*, 4051–4059.
Lehtonen, E. (1975). Epithelio–mesenchymal interface during mouse kidney tubule induction in vivo. *Journal of Embryology and Experimental Morphology, 34*, 695–705.
Leonoudakis, D., Singh, M., Mohajer, R., Mohajer, P., Fata, J. E., Campbell, K. P., et al. (2010). Dystroglycan controls signaling of multiple hormones through modulation of STAT5 activity. *Journal of Cell Science, 123*, 3683–3692.
Li, J., Feng, J., Liu, Y., Ho, T. V., Grimes, W., Ho, H. A., et al. (2015). BMP–SHH signaling network controls epithelial stem cell fate via regulation of its niche in the developing tooth. *Developmental Cell, 33*, 125–135.
Lin, Y., Zhang, S., Rehn, M., Itaranta, P., Tuukkanen, J., Heljasvaara, R., et al. (2001). Induced repatterning of type XVIII collagen expression in ureter bud from kidney to lung type: Association with sonic hedgehog and ectopic surfactant protein C. *Development, 128*, 1573–1585.
Linton, J. M., Martin, G. R., & Reichardt, L. F. (2007). The ECM protein nephronectin promotes kidney development via integrin alpha8beta1-mediated stimulation of Gdnf expression. *Development, 134*, 2501–2509.
Litjens, S. H. M., de Pereda, J. M., & Sonnenberg, A. (2006). Current insights into the formation and breakdown of hemidesmosomes. *Trends in Cell Biology, 16*, 376–383.
Longo, I., Porcedda, P., Mari, F., Giachino, D., Meloni, I., Deplano, C., et al. (2002). COL4A3/COL4A4 mutations: From familial hematuria to autosomal-dominant or recessive Alport syndrome. *Kidney International, 61*, 1947–1956.
Lopez-Luppo, M., Nacher, V., Ramos, D., Catita, J., Navarro, M., Carretero, A., et al. (2017). Blood vessel basement membrane alterations in human retinal microaneurysms during aging. *Investigative Ophthalmology & Visual Science, 58*, 1116–1131.
Loscertales, M., Nicolaou, F., Jeanne, M., Longoni, M., Gould, D. B., Sun, Y., et al. (2016). Type IV collagen drives alveolar epithelial-endothelial association and the morphogenetic movements of septation. *BMC Biology, 14*, 59.
Ma, M., Cao, X., Dai, J., & Pastor-Pareja, J. C. (2017). Basement membrane manipulation in Drosophila wing discs affects Dpp retention but not growth mechanoregulation. *Dev Cell, 42*, 97–106.e104.

Madsen, D. H., & Bugge, T. H. (2015). The source of matrix-degrading enzymes in human cancer: Problems of research reproducibility and possible solutions. *The Journal of Cell Biology, 209*, 195–198.

Mailleux, A. A., Spencer-Dene, B., Dillon, C., Ndiaye, D., Savona-Baron, C., Itoh, N., et al. (2002). Role of FGF10/FGFR2b signaling during mammary gland development in the mouse embryo. *Development, 129*, 53–60.

Mammoto, T., & Ingber, D. E. (2010). Mechanical control of tissue and organ development. *Development (Cambridge, England), 137*, 1407–1420.

Manabe, R., Tsutsui, K., Yamada, T., Kimura, M., Nakano, I., Shimono, C., et al. (2008). Transcriptome-based systematic identification of extracellular matrix proteins. *Proceedings of the National Academy of Sciences of the United States of America, 105*, 12849–12854.

Marinkovich, M. P. (2007). Tumour microenvironment: Laminin 332 in squamous-cell carcinoma. *Nature Reviews Cancer, 7*, 370–380.

Marks, S. C., Jr., McKee, M. D., Zalzal, S., & Nanci, A. (1994). The epithelial attachment and the dental junctional epithelium: Ultrastructural features in porcine molars. *The Anatomical Record, 238*, 1–14.

Matsubayashi, Y., Louani, A., Dragu, A., Sánchez-Sánchez, B. J., Serna-Morales, E., Yolland, L., et al. (2017). A moving source of matrix components is essential for de novo basement membrane formation. *Current Biology, 27*, 3526–3534.

Matus, D. Q., Chang, E., Makohon-Moore, S. C., Hagedorn, M. A., Chi, Q., & Sherwood, D. R. (2014). Cell division and targeted cell cycle arrest opens and stabilizes basement membrane gaps. *Nature Communications, 5*, 4184.

Matus, D. Q., Lohmer, L. L., Kelley, L. C., Schindler, A. J., Kohrman, A. Q., Barkoulas, M., et al. (2015). Invasive cell fate requires G1 cell-cycle arrest and histone deacetylase-mediated changes in gene expression. *Developmental Cell, 35*, 162–174.

Mavilio, F., Pellegrini, G., Ferrari, S., Di Nunzio, F., Di Iorio, E., Recchia, A., et al. (2006). Correction of junctional epidermolysis bullosa by transplantation of genetically modified epidermal stem cells. *Nature Medicine, 12*, 1397–1402.

Mayer, U., Nischt, R., Poschl, E., Mann, K., Fukuda, K., Gerl, M., et al. (1993). A single EGF-like motif of laminin is responsible for high affinity nidogen binding. *The EMBO Journal, 12*, 1879–1885.

McClatchey, S. T., Wang, Z., Linden, L. M., Hastie, E. L., Wang, L., Shen, W., et al. (2016). Boundary cells restrict dystroglycan trafficking to control basement membrane sliding during tissue remodeling. *eLife, 5*.

McGuire, J. D., Gorski, J. P., Dusevich, V., Wang, Y., & Walker, M. P. (2014). Type IV collagen is a novel DEJ biomarker that is reduced by radiotherapy. *Journal of Dental Research, 93*, 1028–1034.

McGuire, J. D., Walker, M. P., Dusevich, V., Wang, Y., & Gorski, J. P. (2014). Enamel organic matrix: Potential structural role in enamel and relationship to residual basement membrane constituents at the dentin enamel junction. *Connective Tissue Research, 55*, 33–37.

Mecham, R. (2011). *The extracellular matrix: An overview*. Springer Berlin Heidelberg.

Micalizzi, D. S., Farabaugh, S. M., & Ford, H. L. (2010). Epithelial–mesenchymal transition in cancer: Parallels between normal development and tumor progression. *Journal of Mammary Gland Biology and Neoplasia, 15*, 117–134.

Min, H., Danilenko, D. M., Scully, S. A., Bolon, B., Ring, B. D., Tarpley, J. E., et al. (1998). Fgf-10 is required for both limb and lung development and exhibits striking functional similarity to Drosophila branchless. *Genes & Development, 12*, 3156–3161.

Miner, J. H., Cunningham, J., & Sanes, J. R. (1998). Roles for laminin in embryogenesis: Exencephaly, syndactyly, and placentopathy in mice lacking the laminin α5 chain. *The Journal of Cell Biology, 143*, 1713–1723.

Miner, J. H., Go, G., Cunningham, J., Patton, B. L., & Jarad, G. (2006). Transgenic isolation of skeletal muscle and kidney defects in laminin beta2 mutant mice: Implications for Pierson syndrome. *Development, 133*, 967–975.
Miner, J. H., & Li, C. (2000). Defective glomerulogenesis in the absence of laminin alpha5 demonstrates a developmental role for the kidney glomerular basement membrane. *Developmental Biology, 217*, 278–289.
Miner, J. H., Li, C., Mudd, J. L., Go, G., & Sutherland, A. E. (2004). Compositional and structural requirements for laminin and basement membranes during mouse embryo implantation and gastrulation. *Development, 131*, 2247–2256.
Miner, J. H., Patton, B. L., Lentz, S. I., Gilbert, D. J., Snider, W. D., Jenkins, N. A., et al. (1997). The laminin α chains: Expression, developmental transitions, and chromosomal locations of α1–5, identification of heterotrimeric laminins 8–11, and cloning of a novel α3 isoform. *The Journal of Cell Biology, 137*, 685–701.
Miner, J. H., & Yurchenco, P. D. (2004). Laminin functions in tissue morphogenesis. *Annual Review of Cell and Developmental Biology, 20*, 255–284.
Moasser, M. M. (2013). Neoplasia. In G. D. Hammer & S. J. McPhee (Eds.), *Pathophysiology of disease: An introduction to clinical medicine.* New York, NY: McGraw-Hill Education.
Mokkapati, S., Baranowsky, A., Mirancea, N., Smyth, N., Breitkreutz, D., & Nischt, R. (2008). Basement membranes in skin are differently affected by lack of nidogen 1 and 2. *The Journal of Investigative Dermatology, 128*, 2259–2267.
Molina, C., Alliende, C., Aguilera, S., Kwon, Y. J., Leyton, L., Martinez, B., et al. (2006). Basal lamina disorganisation of the acini and ducts of labial salivary glands from patients with Sjogren's syndrome: Association with mononuclear cell infiltration. *Annals of the Rheumatic Diseases, 65*, 178–183.
Moore, K. A., Polte, T., Huang, S., Shi, B., Alsberg, E., Sunday, M. E., et al. (2005). Control of basement membrane remodeling and epithelial branching morphogenesis in embryonic lung by rho and cytoskeletal tension. *Developmental Dynamics, 232*, 268–281.
Naba, A., Clauser, K. R., Ding, H., Whittaker, C. A., Carr, S. A., & Hynes, R. O. (2016). The extracellular matrix: Tools and insights for the "omics" era. *Matrix Biology, 49*, 10–24.
Naegeli, K., Hastie, E., Garde, A., Wang, Z., Keeley, D., Gordon, K. L., et al. (2017). Cell invasion in vivo via rapid exocytosis of a transient lysosome-derived membrane domain. *Developmental Cell, 43*, 403–417.
Nagata, M., Katz, M. L., & Robison, W. G., Jr. (1986). Age-related thickening of retinal capillary basement membranes. *Investigative Ophthalmology & Visual Science, 27*, 437–440.
Nakaya, Y., Sukowati, E. W., Alev, C., Nakazawa, F., & Sheng, G. (2011). Involvement of dystroglycan in epithelial–mesenchymal transition during chick gastrulation. *Cells, Tissues, Organs, 193*, 64–73.
Nakaya, Y., Sukowati, E. W., Wu, Y., & Sheng, G. (2008). RhoA and microtubule dynamics control cell-basement membrane interaction in EMT during gastrulation. *Nature Cell Biology, 10*, 765–775.
Nanba, D., Hieda, Y., & Nakanishi, Y. (2000). Remodeling of desmosomal and hemidesmosomal adhesion systems during early morphogenesis of mouse pelage hair follicles. *Journal of Investigative Dermatology, 114*, 171–177.
Neumann, K. H., Kellner, C., Kuhn, K., Stolte, H., & Schurek, H. J. (2004). Age-dependent thickening of glomerular basement membrane has no major effect on glomerular hydraulic conductivity. *Nephrology, Dialysis, Transplantation, 19*, 805–811.
Newman, M. G., Takei, H., Klokkevold, P. R., & Carranza, F. A. (2011). *Carranza's clinical periodontology.* Elsevier Health Sciences.
Nguyen-Ngoc, K. V., Cheung, K. J., Brenot, A., Shamir, E. R., Gray, R. S., Hines, W. C., et al. (2012). ECM microenvironment regulates collective migration

and local dissemination in normal and malignant mammary epithelium. *Proceedings of the National Academy of Sciences of the United States of America*, *109*, E2595–2604.

Nielsen, B. S., Sehested, M., Duun, S., Rank, F., Timshel, S., Rygaard, J., et al. (2001). Urokinase plasminogen activator is localized in stromal cells in ductal breast cancer. *Laboratory Investigation*, *81*, 1485–1501.

Nieto, M. A., Huang, R. Y., Jackson, R. A., & Thiery, J. P. (2016). EMT: 2016. *Cell*, *166*, 21–45.

Nischt, R., Schmidt, C., Mirancea, N., Baranowsky, A., Mokkapati, S., Smyth, N., et al. (2007). Lack of nidogen-1 and -2 prevents basement membrane assembly in skin-organotypic coculture. *The Journal of Investigative Dermatology*, *127*, 545–554.

Nishiuchi, R., Takagi, J., Hayashi, M., Ido, H., Yagi, Y., Sanzen, N., et al. (2006). Ligand-binding specificities of laminin-binding integrins: A comprehensive survey of laminin–integrin interactions using recombinant alpha3beta1, alpha6beta1, alpha7beta1 and alpha6beta4 integrins. *Matrix Biology*, *25*, 189–197.

Nourshargh, S., & Alon, R. (2014). Leukocyte migration into inflamed tissues. *Immunity*, *41*, 694–707.

Nourshargh, S., Hordijk, P. L., & Sixt, M. (2010). Breaching multiple barriers: Leukocyte motility through venular walls and the interstitium. *Nature Reviews Molecular Cell Biology*, *11*, 366–378.

Nurcombe, V., Ford, M. D., Wildschut, J. A., & Bartlett, P. F. (1993). Developmental regulation of neural response to FGF-1 and FGF-2 by heparan sulfate proteoglycan. *Science*, *260*, 103–106.

Ohta, R., Niwa, A., Taniguchi, Y., Suzuki, N. M., Toga, J., Yagi, E., et al. (2016). Laminin-guided highly efficient endothelial commitment from human pluripotent stem cells. *Scientific Reports*, *6*, 35680.

Ohuchi, H., Hori, Y., Yamasaki, M., Harada, H., Sekine, K., Kato, S., et al. (2000). FGF10 acts as a major ligand for FGF receptor 2 IIIb in mouse multi-organ development. *Biochemical and Biophysical Research Communications*, *277*, 643–649.

Oksonen, J., Sorokin, L. M., Virtanen, & Hormia, M. (2001). The junctional epithelium around murine teeth differs from gingival epithelium in its basement membrane composition. *Journal of Dental Research*, *80*, 2093–2097.

Ozanne, B. W., Spence, H. J., McGarry, L. C., & Hennigan, R. F. (2006). Invasion is a genetic program regulated by transcription factors. *Current Opinion in Genetics & Development*, *16*, 65–70.

Ozbek, S., Balasubramanian, P. G., Chiquet-Ehrismann, R., Tucker, R. P., & Adams, J. C. (2010). The evolution of extracellular matrix. *Molecular Biology of the Cell*, *21*, 4300–4305.

Paine, I. S., & Lewis, M. T. (2017). The terminal end bud: The little engine that could. *Journal of Mammary Gland Biology and Neoplasia*, *22*, 93–108.

Patel, V. N., Knox, S. M., Likar, K. M., Lathrop, C. A., Hossain, R., Eftekhari, S., et al. (2007). Heparanase cleavage of perlecan heparan sulfate modulates FGF10 activity during ex vivo submandibular gland branching morphogenesis. *Development*, *134*, 4177–4186.

Patel, V. N., Pineda, D. L., & Hoffman, M. P. (2017). The function of heparan sulfate during branching morphogenesis. *Matrix Biology*, *57–58*, 311–323.

Patthy, L. (1999). Genome evolution and the evolution of exon-shuffling—A review. *Gene*, *238*, 103–114.

Patton, B. L., Cunningham, J. M., Thyboll, J., Kortesmaa, J., Westerblad, H., Edstrom, L., et al. (2001). Properly formed but improperly localized synaptic specializations in the absence of laminin [alpha]4. *Nature Neuroscience*, *4*, 597–604.

Petrou, P., Makrygiannis, A. K., & Chalepakis, G. (2008). The Fras1/Frem family of extracellular matrix proteins: Structure, function, and association with Fraser syndrome and the mouse bleb phenotype. *Connective Tissue Research*, *49*, 277–282.

Pickering, J., Cunliffe, V. T., Van Eeden, F., & Borycki, A. G. (2017). Hedgehog signalling acts upstream of laminin alpha1 transcription in the zebrafish paraxial mesoderm. *Matrix Biology*, *62*, 58–74.
Pitera, J. E., Scambler, P. J., & Woolf, A. S. (2008). Fras1, a basement membrane-associated protein mutated in Fraser syndrome, mediates both the initiation of the mammalian kidney and the integrity of renal glomeruli. *Human Molecular Genetics*, *17*, 3953–3964.
Plachot, C., Chaboub, L. S., Adissu, H. A., Wang, L., Urazaev, A., Sturgis, J., et al. (2009). Factors necessary to produce basoapical polarity in human glandular epithelium formed in conventional and high-throughput three-dimensional culture: Example of the breast epithelium. *BMC Biology*, *7*, 77.
Pulkkinen, L., & Uitto, J. (1999). Mutation analysis and molecular genetics of epidermolysis bullosa. *Matrix Biology*, *18*, 29–42.
Qiao, C., Li, J., Zhu, T., Draviam, R., Watkins, S., Ye, X., et al. (2005). Amelioration of laminin-alpha2-deficient congenital muscular dystrophy by somatic gene transfer of miniagrin. *Proceedings of the National Academy of Sciences of the United States of America*, *102*, 11999–12004.
Rebustini, I. T., Myers, C., Lassiter, K. S., Surmak, A., Szabova, L., Holmbeck, K., et al. (2009). MT2-MMP-dependent release of collagen IV NC1 domains regulates submandibular gland branching morphogenesis. *Developmental Cell*, *17*, 482–493.
Rimann, I., & Hajnal, A. (2007). Regulation of anchor cell invasion and uterine cell fates by the egl-43 Evi-1 proto-oncogene in Caenorhabditis elegans. *Developmental Biology*, *308*, 187–195.
Rodeck, C. H., & Whittle, M. J. (2009). *Fetal medicine: Basic science and clinical practice*. Churchill Livingstone.
Rodin, S., Antonsson, L., Niaudet, C., Simonson, O. E., Salmela, E., Hansson, E. M., et al. (2014). Clonal culturing of human embryonic stem cells on laminin-521/E-cadherin matrix in defined and xeno-free environment. *Nature Communications*, *5*, 3195.
Rousselle, P., Keene, D. R., Ruggiero, F., Champliaud, M.-F., Rest, M. V. D., & Burgeson, R. E. (1997). Laminin 5 binds the NC-1 domain of type VII collagen. *The Journal of Cell Biology*, *138*, 719–728.
Rowe, R. G., & Weiss, S. J. (2008). Breaching the basement membrane: Who, when and how? *Trends in Cell Biology*, *18*, 560–574.
Rozario, T., & DeSimone, D. W. (2010). The extracellular matrix in development and morphogenesis: A dynamic view. *Developmental Biology*, *341*, 126–140.
Sahlberg, C., Reponen, P., Tryggvason, K., & Thesleff, I. (1992). Association between the expression of murine 72 kDa type IV collagenase by odontoblasts and basement membrane degradation during mouse tooth development. *Archives of Oral Biology*, *37*, 1021–1030.
Salvi, S. M., Akhtar, S., & Currie, Z. (2006). Ageing changes in the eye. *Postgraduate Medical Journal*, *82*, 581–587.
Santiago-Medina, M., Gregus, K. A., Nichol, R. H., O'Toole, S. M., & Gomez, T. M. (2015). Regulation of ECM degradation and axon guidance by growth cone invadosomes. *Development*, *142*, 486–496.
Sarrazin, S., Lamanna, W. C., & Esko, J. D. (2011). Heparan sulfate proteoglycans. *Cold Spring Harbor Perspectives in Biology*, *3*, a004952.
Sato, Y., Uemura, T., Morimitsu, K., Sato-Nishiuchi, R., Manabe, R., Takagi, J., et al. (2009). Molecular basis of the recognition of nephronectin by integrin alpha8beta1. *The Journal of Biological Chemistry*, *284*, 14524–14536.
Schindler, A. J., & Sherwood, D. R. (2011). The transcription factor HLH-2/E/daughterless regulates anchor cell invasion across basement membrane in C. elegans. *Developmental Biology*, *357*, 380–391.
Seppala, M., Fraser, G., Birjandi, A., Xavier, G., & Cobourne, M. (2017). Sonic hedgehog signaling and development of the dentition. *Journal of Developmental Biology*, *5*, 6.

Sher, I., Zisman-Rozen, S., Eliahu, L., Whitelock, J. M., Maas-Szabowski, N., Yamada, Y., et al. (2006). Targeting perlecan in human keratinocytes reveals novel roles for perlecan in epidermal formation. *The Journal of Biological Chemistry*, *281*, 5178–5187.

Sherwood, D. R., Butler, J. A., Kramer, J. M., & Sternberg, P. W. (2005). FOS-1 promotes basement-membrane removal during anchor-cell invasion in C. elegans. *Cell*, *121*, 951–962.

Sherwood, D. R., & Sternberg, P. W. (2003). Anchor cell invasion into the vulval epithelium in C. elegans. *Developmental Cell*, *5*, 21–31.

Shyer, A. E., Rodrigues, A. R., Schroeder, G. G., Kassianidou, E., Kumar, S., & Harland, R. M. (2017). Emergent cellular self-organization and mechanosensation initiate follicle pattern in the avian skin. *Science*, *357*, 811–815.

Silberstein, G. B., & Daniel, C. W. (1982). Glycosaminoglycans in the basal lamina and extracellular matrix of the developing mouse mammary duct. *Developmental Biology*, *90*, 215–222.

Simmer, J. P., Papagerakis, P., Smith, C. E., Fisher, D. C., Rountrey, A. N., Zheng, L., et al. (2010). Regulation of dental enamel shape and hardness. *Journal of Dental Research*, *89*, 1024–1038.

Simões-Costa, M., & Bronner, M. E. (2015). Establishing neural crest identity: A gene regulatory recipe. *Development (Cambridge, England)*, *142*, 242–257.

Skorecki, K., Chertow, G. M., Marsden, P. A., Taal, M. W., Yu, A. S. L., & Luyckx, V. (2015). *Brenner and Rector's the kidney e-book*. Elsevier Health Sciences.

Slade, M. J., Coope, R. C., Gomm, J. J., & Coombes, R. C. (1999). The human mammary gland basement membrane is integral to the polarity of luminal epithelial cells. *Experimental Cell Research*, *247*, 267–278.

Slavkin, H. C., & Bringas, P., Jr. (1976). Epithelial–mesenchyme interactions during odontogenesis. IV. Morphological evidence for direct heterotypic cell–cell contacts. *Developmental Biology*, *50*, 428–442.

Sorokin, L. (2010). The impact of the extracellular matrix on inflammation. *Nature Reviews Immunology*, *10*, 712–723.

Steeg, P. S. (2003). Metastasis suppressors alter the signal transduction of cancer cells. *Nature Reviews Cancer*, *3*, 55–63.

Sternlicht, M. D. (2006). Key stages in mammary gland development: The cues that regulate ductal branching morphogenesis. *Breast Cancer Research*, *8*, 201.

Strauss, O. (2005). The retinal pigment epithelium in visual function. *Physiological Reviews*, *85*, 845–881.

Streuli, C. H., Bailey, N., & Bissell, M. J. (1991). Control of mammary epithelial differentiation: Basement membrane induces tissue-specific gene expression in the absence of cell–cell interaction and morphological polarity. *The Journal of Cell Biology*, *115*, 1383–1395.

Takeuchi, T., & Gonda, T. (2004). Distribution of the pores of epithelial basement membrane in the rat small intestine. *The Journal of Veterinary Medical Science*, *66*, 695–700.

Tanos, B., & Rodriguez-Boulan, E. (2008). The epithelial polarity program: Machineries involved and their hijacking by cancer. *Oncogene*, *27*, 6939–6957.

Thesleff, I., Barrach, H. J., Foidart, J. M., Vaheri, A., Pratt, R. M., & Martin, G. R. (1981). Changes in the distribution of type IV collagen, laminin, proteoglycan, and fibronectin during mouse tooth development. *Developmental Biology*, *81*, 182–192.

Theveneau, E., & Mayor, R. (2011). Collective cell migration of the cephalic neural crest: The art of integrating information. *Genesis*, *49*, 164–176.

Thiery, J. P., Acloque, H., Huang, R. Y., & Nieto, M. A. (2009). Epithelial–mesenchymal transitions in development and disease. *Cell*, *139*, 871–890.

Thomsen, M. S., Routhe, L. J., & Moos, T. (2017). The vascular basement membrane in the healthy and pathological brain. *Journal of Cerebral Blood Flow and Metabolism*, *37*, 3300–3317.

Thyboll, J., Kortesmaa, J., Cao, R., Soininen, R., Wang, L., Iivanainen, A., et al. (2002). Deletion of the laminin alpha4 chain leads to impaired microvessel maturation. *Molecular and Cellular Biology*, *22*, 1194–1202.

Timpl, R., Wiedemann, H., van Delden, V., Furthmayr, H., & Kuhn, K. (1981). A network model for the organization of type IV collagen molecules in basement membranes. *European Journal of Biochemistry*, *120*, 203–211.

To, M., Goz, A., Camenzind, L., Oertle, P., Candiello, J., Sullivan, M., et al. (2013). Diabetes-induced morphological, biomechanical, and compositional changes in ocular basement membranes. *Experimental Eye Research*, *116*, 298–307.

Tosios, K. I., Kapranos, N., & Papanicolaou, S. I. (1998). Loss of basement membrane components laminin and type IV collagen parallels the progression of oral epithelial neoplasia. *Histopathology*, *33*, 261–268.

Tucker, A., & Sharpe, P. (2004). The cutting-edge of mammalian development; how the embryo makes teeth. *Nature Reviews Genetics*, *5*, 499–508.

Uhlen, M., Fagerberg, L., Hallstrom, B. M., Lindskog, C., Oksvold, P., Mardinoglu, A., et al. (2015). Proteomics. Tissue-based map of the human proteome. *Science*, *347*, 1260419.

Wang, J., Hossain, M., Thanabalasuriar, A., Gunzer, M., Meininger, C., & Kubes, P. (2017). Visualizing the function and fate of neutrophils in sterile injury and repair. *Science*, *358*, 111–116.

Wang, S., Voisin, M. B., Larbi, K. Y., Dangerfield, J., Scheiermann, C., Tran, M., et al. (2006). Venular basement membranes contain specific matrix protein low expression regions that act as exit points for emigrating neutrophils. *The Journal of Experimental Medicine*, *203*, 1519–1532.

Willem, M., Miosge, N., Halfter, W., Smyth, N., Jannetti, I., Burghart, E., et al. (2002). Specific ablation of the nidogen-binding site in the laminin gamma1 chain interferes with kidney and lung development. *Development*, *129*, 2711–2722.

Wiseman, B. S., & Werb, Z. (2002). Stromal effects on mammary gland development and breast cancer. *Science*, *296*, 1046–1049.

Woodall, B. P., Nystrom, A., Iozzo, R. A., Eble, J. A., Niland, S., Krieg, T., et al. (2008). Integrin alpha2beta1 is the required receptor for endorepellin angiostatic activity. *The Journal of Biological Chemistry*, *283*, 2335–2343.

Xiong, A., Kundu, S., Forsberg, M., Xiong, Y., Bergstrom, T., Paavilainen, T., et al. (2017). Heparanase confers a growth advantage to differentiating murine embryonic stem cells, and enhances oligodendrocyte formation. *Matrix Biology*, *62*, 92–104.

Yang, D. H., McKee, K. K., Chen, Z. L., Mernaugh, G., Strickland, S., Zent, R., et al. (2011). Renal collecting system growth and function depend upon embryonic gamma1 laminin expression. *Development*, *138*, 4535–4544.

Yao, Y., Chen, Z. L., Norris, E. H., & Strickland, S. (2014). Astrocytic laminin regulates pericyte differentiation and maintains blood brain barrier integrity. *Nature Communications*, *5*, 3413.

Yoshizaki, K., & Yamada, Y. (2013). Gene evolution and functions of extracellular matrix proteins in teeth. *Orthodontic Waves (English ed)*, *72*, 1–10.

Yurchenco, P. D. (2011). Basement membranes: Cell scaffoldings and signaling platforms. *Cold Spring Harbor Perspectives in Biology*, *3*, 17–43.

Yurchenco, P. D., & Patton, B. L. (2009). Developmental and pathogenic mechanisms of basement membrane assembly. *Current Pharmaceutical Design*, *15*, 1277–1294.

Yurchenco, P. D., & Ruben, G. C. (1987). Basement membrane structure in situ: Evidence for lateral associations in the type IV collagen network. *The Journal of Cell Biology*, *105*, 2559–2568.

CHAPTER FIVE

Extracellular Determinants of Arterial Morphogenesis, Growth, and Homeostasis

Robert P. Mecham*, Francesco Ramirez[†,1]
*Department of Cell Biology and Physiology, Washington University School of Medicine, St. Louis, MO, United States
[†]Department of Pharmacological Sciences, Icahn School of Medicine at Mount Sinai, New York, NY, United States
[1]Corresponding author. e-mail address: francesco.ramirez@mssm.edu

Contents

Abstract

The extracellular matrix (ECM) is a highly heterogeneous mixture of macromolecules capable of self-assembling into tissue-specific suprastructures that constitute the architectural elements supporting organ function. Contrary to the traditional view of being a static scaffold that supports tissue integrity along with cell adhesion and migration, the ECM is an inherently dynamic system that specifies cellular function and defines the limits and patterns of tissue organization. Throughout evolution, the composition and organization of the ECM have changed to accommodate basic and new tissue functions, both in terms of providing structural support and integrating multivalent signals to cells. In this review, we will highlight some of these bidirectional cell–matrix interactions that guide the development of a mechanically compliant vascular system. Specifically, we will focus on studies that have investigated how ECM composition and physical properties influence cell fate decisions associated with vascular tissue development and homeostasis and implicitly, vascular disease.

Current Topics in Developmental Biology, Volume 130
ISSN 0070-2153
https://doi.org/10.1016/bs.ctdb.2018.02.001

1. INTRODUCTION

The origins of metazoan life largely coincide with the appearance of a primitive extracellular matrix (ECM) that has evolved over time to accommodate basic and new tissue functions of increasingly complex organisms (Hynes, 2012). Aside from serving as the structural scaffold for virtually every tissue, the ECM influences cell migration, differentiation, and proliferation by interacting with membrane-associated receptors (e.g., integrins) or by modulating the bioavailability of matrix-bound biochemical signals (e.g., TGFβ). Also of importance is the role that the ECM plays in transmitting mechanical stimuli that instruct cell behavior, in creating physical barriers that restrict or guide cell movement, and in endowing tissues with physical properties that specify cell fate (e.g., topography, stiffness, and porosity). In turn, cells promote matrix modeling and remodeling in response to a plethora of regulatory signals that emanate from the surrounding microenvironment. This bidirectional interaction between cells and ECM (a.k.a., "dynamic reciprocity") represents a crucial determinant of tissue development, growth, and homeostasis (Bornstein & Sage, 2002). Importantly, disruption of these genetically programmed cell–matrix interactions is the common hallmark of several heritable or acquired human diseases of connective tissue.

The ECM is a heterogeneous mixture of macromolecules, including collagens, elastin, noncollagenous glycoproteins, and proteoglycans (Fig. 1). Important for matrix assembly and remodeling are ECM-modifying enzymes, including lysyl oxidases (LOXs) and matrix metalloproteinases (MMPs). Any single ECM component seldom functions in isolation but, instead, is designed to interact with other matrix molecules to form a complex biomaterial that adopts properties of its component parts. ECM proteins are traditionally segregated into those that modulate cell function (e.g., fibronectin and matricellular proteins) and those that constitute the architectural scaffold of tissues (e.g., collagen and elastic fibers). Albeit useful, this functional classification is limiting because it does not consider that components of the architectural matrix can also influence key cellular activities and conversely, that ECM molecules regulating cell function can also modify the assembly and/or organization of architectural matrix components. Additionally, the three-dimensional organization of the ECM can elicit distinct cellular responses by determining how structural and biochemical signals are presented to the cognate cell surface receptors.

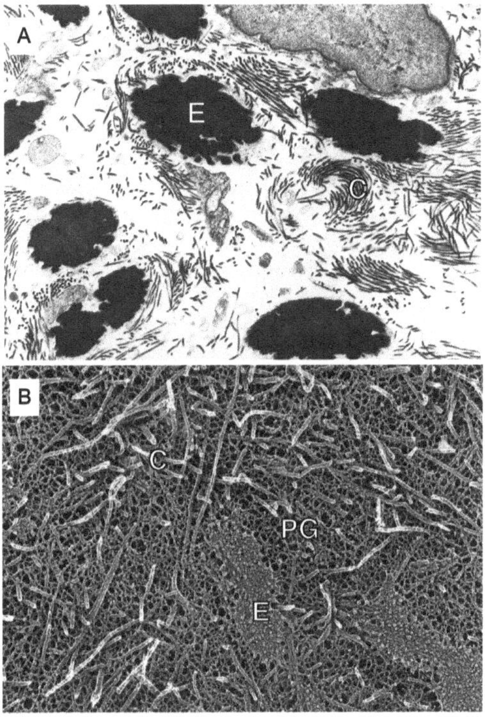

Fig. 1 ECM complexity as visualized by electron microscopy. (A) Transmission electron micrograph of ECM showing elastic fibers (E) and collagen bundles (C) in developing elastic ligament tissue. (B) ECM viewed by quick freeze, deep etch imaging showing elastin and collagen fibers together with a background of proteoglycans which are not visible with transmission electron microscopy.

In this review, we use vascular development to Illustrate how the ECM participates in controlling tissue differentiation and function. The circulatory system has evolved over time from simple diffusion through cells to a complex network of blood vessels that reach all parts of the human body. Evolutionary adaptation of the circulatory design to maintain optimum tissue perfusion was critically important in enabling increased body size, high metabolic rates, and specialized tissue function. As the circulatory design changed through evolution, so did the ECM to meet the physiological requirements of the circulatory circuit. The discussion that follows will focus on two major aspects of arterial development and homeostasis: (1) how ECM biomechanics influence cell fate decisions and (2) how ECM components instruct structural and biochemical regulators of cell differentiation.

2. EVOLUTION OF THE CARDIOVASCULAR SYSTEM

2.1 Invertebrate Vessels

The circulatory systems of all animals are either open or closed. In an open circulatory system, blood (or hemolymph) distributed by a contractile vessel empties into the body cavity where it bathes the cells. Invertebrate vessels consist of a mesoderm-derived epithelium with the basal surface of the cell facing toward the lumen (Davison, Wright, & DeMont, 1995; Shadwick, 1999). These vascular epithelial cells secrete a relatively thick basement membrane (also called basal lamina) that lines the vessel cavity (Fig. 2). Although there are occasional amoebocytic cells on the luminal side of the basement membrane, the luminal surface of invertebrate vessels is acellular.

The basement membrane serves multiple functions, including providing a selective filtration barrier for proteins and small molecules into the stromal space, and affording an adhesion surface for epithelial and endothelial cells. In early development, basement membranes play an important role in providing a boundary between cell populations during processes like gastrulation, the

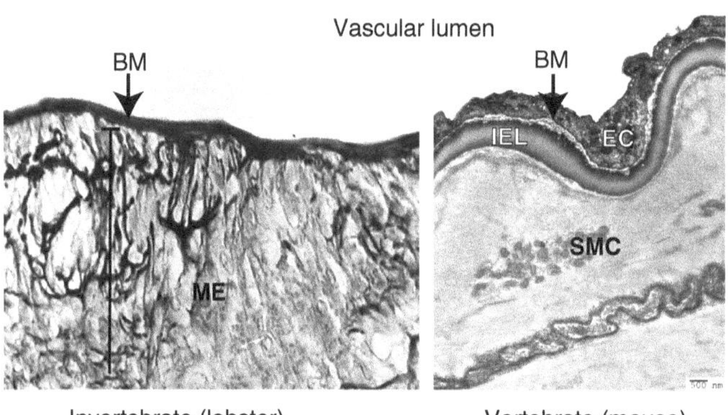

Fig. 2 Vessel wall structure in invertebrates and vertebrates. The arterial lumen in invertebrates is lined by a thick, naked basement membrane (BM, *arrow*) that is produced by the underlying myoepithelial cells (ME) that populate the vessel wall. Lining the vessel lumen in vertebrate vessels is an endothelial monolayer attached to a thin basement membrane (BM, *arrow*) that lies next to the internal elastic lamina (IEL). The arterial wall of vertebrate vessels consists of smooth muscle cells in the medial layer and myofibroblasts in the outer adventitia. *The electron micrograph of the lobster vessel is modified from Davison, I., Wright, G., & DeMont, M. 1995. The structure and physical properties of invertebrate and primitive vertebrate arteries.* The Journal of Experimental Biology. 198, *2185–9216 and used with permission.*

process that forms and positions the three primary germ layers: ectoderm, mesoderm, and endoderm. All basement membranes are composed of a core (architectural) scaffold of interacting proteins that include a crosslinked network of collagen type IV, laminin, nidogen (a.k.a., entactin), and perlecan or aggrecan (Carlin, Jaffe, Bender, & Chung, 1981; Chung & Durkin, 1990; Mayer & Timpl, 1994). By binding laminin near its center and concomitantly interacting with type IV collagen, nidogen indirectly connects these two basement membrane molecules. Perlecan and aggrecan are large heparan sulfate proteoglycans that bind laminin and type IV collagen through their GAG chains and that are involved in charge-dependent molecular sieving, one of the critical functions of basement membrane (Miner, 2011). Other components that are sometimes found associated with basement membranes but may not be intrinsic components of the architectural network include fibronectin, collagen types VI, VII, XV, XIII, and XVIII, fibulin, fibrillins, osteonectin (a.k.a., BM-40 or SPARC), and chondroitin sulfate proteoglycans.

Cells interact with the basement membrane through laminins (Ekblom & Timpl, 1996; Yurchenco, Amenta, & Patton, 2004) (Fig. 3). Laminin is present at the two-cell stage in the mouse embryo, making it one of the first ECM proteins detected during embryogenesis. Sixteen different laminin trimers have been identified in vertebrates, each containing a specific combination of α, β, and γ chains (in mammals, there are five α, four β, and three γ chains). Laminin types are referred to by their α, β, and γ chain composition. For example, laminin 511 contains $\alpha5$, $\beta1$, $\gamma1$ chains (Aumailley et al., 2005). The major laminin in blood vessels is laminin 411. The integrins are the major receptor class linking cells with the basement membrane. The integrin $\beta1$ family includes most of the laminin-binding integrins ($\alpha1\beta1$, $\alpha2\beta1$, $\alpha3\beta1$, $\alpha6\beta1$, $\alpha7\beta1$, $\alpha9\beta1$). Other integrins that bind laminin include $\alpha v\beta3$ and $\alpha6\beta4$. Integrins mostly recognize laminins through the C-terminal globular LG domains of the laminin α chain. In skeletal muscle, dystroglycan, a component of the dystrophin–glycoprotein complex, is the major laminin-binding receptor. Basement membranes can also have an indirect effect on cell behavior by binding and sequestering growth and differentiation factors, such as FGF, PDGF, and TGFβ.

In mature invertebrate vessels, the naked basement membrane that lines the vascular surface serves a filtration function and orients epithelia cell polarity. In vertebrates, there is evidence that the basement membrane is critical for initial vascular tube formation and subsequent tube stabilization (Bowers, Norden, & Davis, 2016), which may also be the case

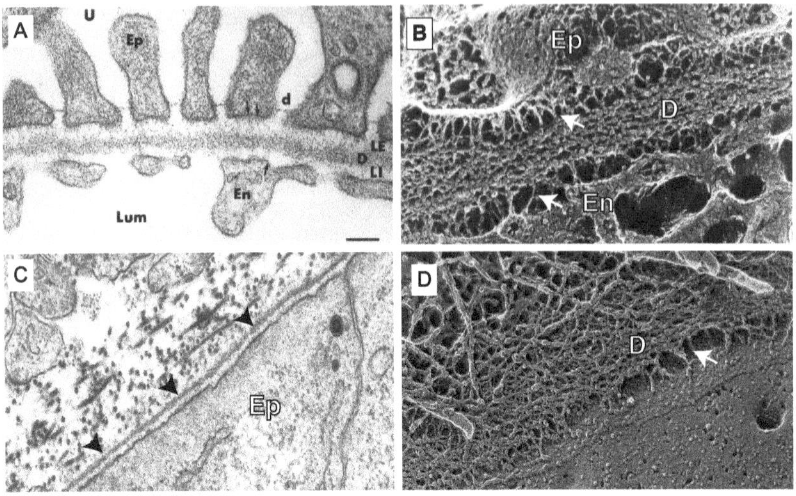

Fig. 3 Basement membrane structure. (A) Transmission electron micrograph showing the mouse kidney glomerular capillary wall. The basement membrane is seen as a dense band, referred to as the lamina densa (D) that separates the capillary endothelial cells (En) from the epithelial podocytes (Ep). On either side of the lamina densa are areas referred to as the lamina lucida externa (LE) and the lamina lucida interna (LI). These spaces are an artifact of tissue processing since deep-etch electron microscopy (B) shows that the LI and LE contain the laminin molecules (*white arrows*) that connect the lamina densa, consisting of mainly type IV collagen, to the cells. Basement membranes can also serve to link cells with the underlying stromal matrix, as shown in (C) and (D). *Panel (A) from Chan, F. L., & Inoue, S. 1994. Lamina lucida of basement membrane: An artefact.* Microscopy Research and Technique. 28, *48–59, used with permission. Images in panels (B) and (D) are courtesy of Dr. John Heuser, Washington University School of Medicine.*

in invertebrates. Vessels in the open circulatory system act as relatively stiff tubes and are not exposed to high intraluminal pressures. To meet these simple pressure-bearing requirements, the ECM in the vessel wall consists mostly of fibrillar collagens that provide the structural support required for maintaining tissue integrity. In higher invertebrates (e.g., lobster) and primitive vertebrates (e.g., lamprey) that are transitioning to a more advanced circulatory system and higher blood pressures, layers of fibrillin-rich microfibrils appear in the wall to provide limited elastic recoil (Davison et al., 1995; Gibbons & Shadwick, 1989; McConnell, Wright, & DeMont, 1996).

2.2 Vertebrate Vessels

The appearance of vertebrate organisms coincided with a dramatic change in cardiovascular design that included a multichambered heart within a closed circulatory system. This unique circulatory network facilitated tissue perfusion over longer distances with greater efficiency than possible with open circulation. Key to this circulatory efficiency was a retooling of the blood vessels driven by the appearance of smooth muscle cells (SMCs), endothelial cells (ECs), and the ECM protein elastin—all of which are unique to vertebrates. In contrast to the naked basement membrane-lined epithelial tubes of invertebrates, the arterial wall of vertebrates typically consists of an endothelial monolayer (the tunica intima) attached to the basement membrane surrounded by multiple concentric rings of SMCs (tunica media) and an outer cellular layer largely consisting of (myo)fibroblasts (tunica adventitia) (Fig. 4). ECs

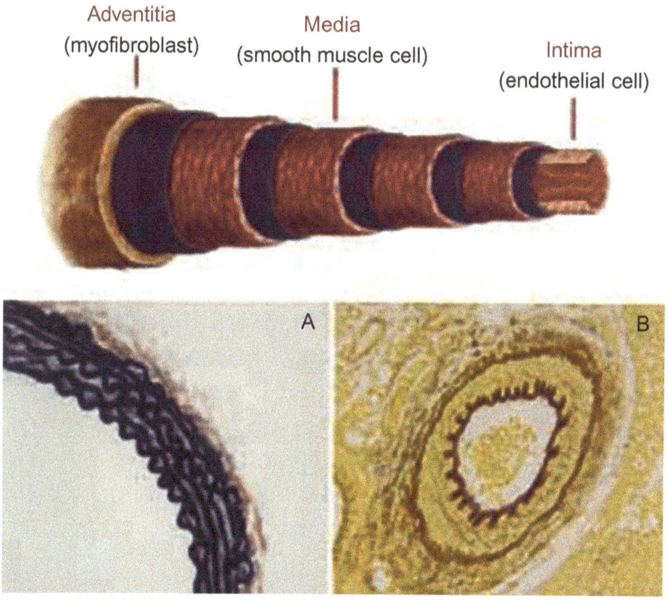

Fig. 4 Layered structure of vertebrate vessel wall and differences between elastic and muscular arteries. The wall of elastic arteries in vertebrates consists of alternating layers (lamellae) of smooth muscle cells between sheets of elastin. The number of layers is determined during development with the highest number of layers found in those vessels subjected to high pulse pressure and hemodynamic forces closest to the heart. Muscular and resistance arteries, in contrast, have an internal and external elastic lamina but little elastin is present in the medial layer. The external elastic lamina is absent or discontinuous in the smallest arteries. Panels (A) and (B) show elastin in the elastic mouse aorta (*stained black* with Verhoeff's van Gieson stain) and in the muscular renal artery (*stained purple* with Hart's stain). *The vessel schematic was modified from D'Armiento, J. (2003). Decreased elastin in vessel walls puts the pressure on.* The Journal of Clinical Investigation, 112, *1308–1310 and used with permission. Histological sections are courtesy of Dr. Carmen Halabi, Washington University School of Medicine.*

not only line the lumen of all vertebrate vessels (arteries and veins), but they also play a major role in vascular patterning through a process that involves movement and self-assembly of mesoderm-derived EC precursors into endothelial tubes (vasculogenesis). Results from several lines of research indicate that vasculogenesis is modulated not only by cell–cell interactions but also by cell–matrix interactions and by growth factors and morphogens sequestered within the ECM (George, Baldwin, & Hynes, 1997; George, Georges-Labouesse, Patel-King, Rayburn, & Hynes, 1993).

Formation of the interconnected tubular network of ECs occurs prior to heart development and onset of circulation (Drake, Hungerford, & Little, 1998; Sato et al., 2010). With the commencement of blood flow, ECs recruit SMC progenitors and induce their differentiation through various signaling pathways. Experimental evidence demonstrating the importance of hemodynamics in vessel wall development has come from numerous studies of vascular remodeling in response to altered pressure and flow. ECs recruit SMCs from surrounding mesenchyme or, in the ascending aorta, pulmonary artery and aortic arch vessels, from neural crest cells (NCCs) and cells derived from the secondary heart field (Sawada, Rateri, Moorleghen, Majesky, & Daugherty, 2017). This recruitment is largely driven by a PDGF-B gradient secreted by the ECs (Hellstrom, Kalen, Lindahl, Abramsson, & Betsholtz, 1999). SMCs are generally defined by the expression of smooth muscle-specific cytoskeletal proteins, such as smooth muscle-α-actin (*Acta2*), smooth muscle myosin heavy chain (*Myh11*), SM22 (*Tagln*), and SM-calponin (*Cnn1*), that change over time and with differentiation state (Owens, 1995). Lineage and fate-mapping studies have identified multiple independent origins for vascular SMC progenitors (Majesky, Dong, Regan, & Hoglund, 2011). Because of their diverse embryological origins, SMCs potentially constitute many distinct lineages that display remarkably diverse phenotypes particularly regarding the types of ECM that they produce. The initial steps in the differentiation and stabilization of what will be SMCs in the vascular wall begin with a direct interaction between ECs and SMC progenitors recruited to the endothelial tube. Evidence suggests that SMC differentiation involves Notch signaling, with Jagged serving as the activating ligand (Baeten & Lilly, 2017). Both Notch and Jagged are transmembrane proteins whose interaction is promoted by cell–cell contact. Jagged1 expression on the surface of ECs apparently engages a Notch receptor on adjacent SMC progenitors, thus resulting in integrin activation and adhesion of the progenitor cell to the subendothelial basement membrane (Scheppke et al., 2012).

The Notch/Jagged signal is then propagated to the next layer of artery wall so that as one muscle layer forms, differentiation of the next layer begins.

Like the inside out differentiation pattern associated with Notch signaling, lineage-tracing experiments in the developing pulmonary artery of mice identified a radial signaling gradient originating from the ECs that resulted in SMC recruitment and differentiation. The effector molecules in this case, however, were suggested to be PDGF-B and cognate receptors (PDGFRs α and β), which display a temporal and spatial expression pattern consistent with a role in SMC differentiation (Greif et al., 2012). As described earlier for Notch signaling, each layer of cells recruited to the wall initiates programmed SMC gene expression and as one muscle layer forms, differentiation of the next layer begins. The adventitial layer forms when the differentiation program arrests early in the formation of the outermost layer. How the Notch and PDGF signaling pathways interact with each other to influence SMC differentiation is not clear. However, Notch is ideally suited to direct SMC maturation and differentiation via cell–cell contact given its ability to induce PDGFRβ and smooth muscle α-actin. The number of smooth muscle layers that will be present in the adult vessel is established relatively late in development (~E14 in the mouse) (Wagenseil et al., 2010, 2005). The fact that vessels nearest to the heart experiencing the greatest wall tension have the highest number of SMC layers indicates that mechanical forces are important determinants of cellular recruitment. Once the appropriate number of lamellae is reached, further increases do not occur. What does change, however, is the phenotype of the vessel wall cells as typified by expression of ECM-related genes.

Fibril-forming collagens (types I, III, and V) and elastic fibers are major architectural components of the vertebrate vessel wall that contribute to the structural integrity and mechanical properties of the mature tissue (Wagenseil & Mecham, 2009). Elastic fibers consist of the protein elastin complexed with 10–15 nm diameter microfibrils (Cleary, Fanning, & Prosser, 1981) (Fig. 5). Microfibrils are unique ECM elements containing the fibrillins as major structural components and other extracellular proteins that modify microfibril function. In addition to facilitating elastin polymerization, microfibrils serve as molecular bridges connecting elastic fibers to adjacent SMCs and the basement membrane underlying ECs (Davis, 1993). Similar collagen-specific interlinks involve collagen VI microfilaments, collagen VIII meshwork and fibril-associated proteoglycan/collagen XIV. Collectively, this interconnected network of extracellular macroaggregates enables transmission of mechanical stimuli from the vessel matrix

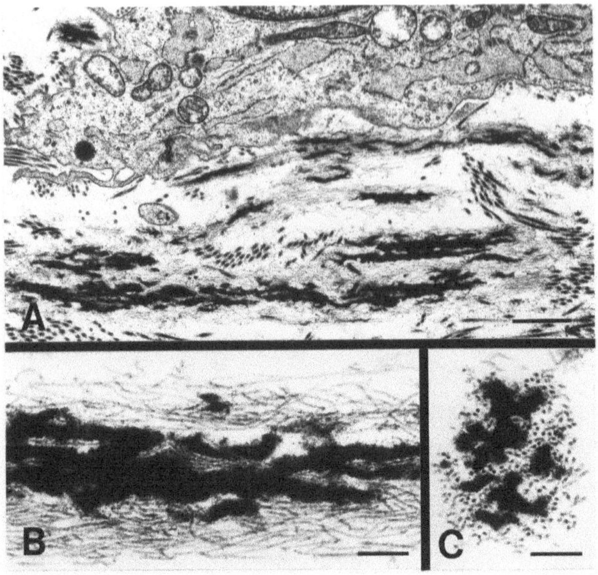

Fig. 5 Elastic fibers consist of elastin and microfibrils. (A) Transmission electron micrograph of developing elastic fibers adjacent to a fibroblast in fetal bovine ligamentum nuchae. Bar, 1 μm. (B) At higher magnification, the elastic fibers are seen to consist of darkly staining amorphous elastin deposited within a bundle of 10 nm microfibrils. Bar, 0.25 μm. (C) In cross section, the microfibrils have a tubular appearance, Bar, 0.25 μm. *From Mecham, R. P., & Davis, E. C. (1994). Elastic fiber structure and assembly. In P. D. Yurchenko, D. E. Birk and R. P. Mecham, eds.* Extracellular matrix assembly and structure, *(pp. 281–314). Academic Press, San Diego.*

to resident cells where they are converted into biochemical signals that support tissue differentiation and homeostasis. As discussed later, fibrillin-rich microfibrils are also involved in modulating endogenous (local) TGFβ bioavailability.

3. ECM COMPOSITION AND MECHANICAL SIGNALING DURING ARTERIAL DEVELOPMENT

3.1 ECM Composition and Elastic Vessel Architecture

SMCs have two major functions depending on where they are located within the vascular tree. In muscular and resistance arteries, the cell's contractile "muscle" properties are important for regulating vascular resistance and, hence, blood flow and pressure, which is a required function in a closed circulatory system. In elastic (a.k.a., conducting) vessels, SMC's secrete an ECM that provides structural integrity and appropriate mechanical function to the vessel wall (Fig. 4). Even though these two SMC populations express similar cytoskeletal marker proteins, their ECM phenotypes are very different.

In the vertebrate closed circulatory system, large arteries became an important component of proper cardiac function by serving as elastic reservoirs, enabling the arterial tree to undergo large volume changes with little change in pressure. Without elastic vessels, the tremendous surge of pressure as blood is ejected from the heart would inhibit its ability to empty, and the pressure in the vessels would fall so rapidly that the heart could not refill, with disastrous consequences for survival. Large elastic arteries store a portion of the stroke volume with each systole and discharge that volume with diastole. This phenomenon, known as the windkessel effect, helps to decrease the load on the heart and to minimize the systolic flow and maximize diastolic flow in the arterioles (Dobrin, 1978). Addition of elastin to the vascular ECM made possible the acquisition of these unique elastic properties. Indeed, the four-chambered vertebrate heart and closed circulation would not have evolved without the evolutionarily appearance of elastin and this major change in vessel function.

The basic morphological plan of large arteries present in all major vertebrate classes consists of repeating concentric layers of SMCs separated by elastic fibers that form interconnected fenestrated sheets, or lamellae. These elastic lamellae are designed to function as an elastic reservoir and to distribute tension uniformly throughout the vessel wall. Between the lamellar layers are bundles of collagen that enable the vessel to become progressively less distensible as the collagen fibers are recruited to support passive wall tension and restrict aortic distention (Wolinsky & Glagov, 1964). Collagen types I and III are both found throughout the vessel wall, with type III being most abundant in the SMC-rich medial layer and type I the most abundant collagen in the fibroblast-rich adventitial layer. The aortic arch develops concurrently with the heart, with significant changes in patterning occurring between mid to late gestation (E11–E14 in the mouse). The complete separation of the extracardiac aorta and pulmonary artery occurs around E13.5 and by E14 the definitive vascular pattern is established. SMC recruitment in the aorta (described earlier) is generally completed at this stage. During recruitment, the ECM around migrating SMC precursors is rich in hyaluronic acid, fibronectin, proteoglycans, and matrix-bound growth factors—a combination that favors cell movement and proliferation (McLean, Mecham, Kelleher, Mariani, & Mecham, 2005). As SMCs organize around endothelial tubes, they downregulate fibronectin production and upregulate hyaluronidase, proteoglycanases, and MMPs that remodel the provisional matrix, thereby allowing the cells to condense into relatively tightly packed concentric layers that define the typical lamellar structure.

Loss-of-function genetic experiments in mice have implicated the matricellular proteins CCN1 and CCN2 in organizing the subendothelial basement membrane and in modulating the interactions between ECs and pericytes/SMCs during vascular development (Schwarzbauer and DeSimone, 2011). As the name itself implies, matricellular proteins are nonstructural components of the ECM that regulate cell–matrix interactions and cell function (Astrof, Crowley, & Hynes, 2007). *Ccn1*$^{-/-}$ mice display severely abnormal and substantially dilated large vessels with fewer SMCs and a disorganized intimal basement membrane. Comparable defects in *Ccn2*$^{-/-}$ mice are instead restricted to microvessels. Analyses of CCN2-deficient mice have demonstrated that this matricellular protein promotes EC recruitment of pericytes by stimulating PDGF-B expression in the former cells and by enhancing PDGF-B-mediated Akt signaling in the latter ones. By regulating fibronectin expression, CCN2 also indirectly regulates vascular basement membrane assembly, a structural requirement for EC-induced pericyte recruitment and retention.

Fibronectin is an important player in many stages of vascular development. This high-molecular-weight glycoprotein functions as a dimer that is expressed by a wide variety of cells in forming and mature tissues (Mosher, 1989; Schwarzbauer & DeSimone, 2011). The primary gene transcript is alternatively spliced to generate multiple mRNAs, each encoding a distinct fibronectin subunit. Sites of alternative splicing lie among the type III repeats (EIIIB/ED-A, EIIIA/ED-B) and in the variable (V) region/IIICS (Schwarzbauer, Tamkun, Lemischka, & Hynes, 1983). EIIIA and EIIIB are single type III modules coded for by single exons that are either included or skipped during splicing. Neither EIIIA nor EIIIB are included in plasma fibronectin, whereas cellular fibronectin monomers can have either, one, or both extra domains. Fibronectin is enriched around dorsal aortae and branchial arch arteries within the anterior trunk of mouse embryos at E10.5 (Astrof & Hynes, 2009). The absence of fibronectin leads to embryonic lethality associated with vascular defects, suggesting a critical role in vasculogenesis (George et al., 1993). Interestingly, while deletion of both EIIIA- and EIIIB-encoding exons resulted in multiple angiogenic and heart defects, there was also a decrease in α-smooth muscle actin-positive cells associated with the dorsal aorta, suggesting that EIIIA and/or EIIIB splice variants are important in the formation of the aortic SMC layer

(Astrof et al., 2007). The functional domains of fibronectin have been defined and are scattered throughout the molecule (Xu & Mosher, 2011). The biological properties of fibronectin necessitate that the protein be assembled into fibrils, which in turn requires the presence of assembly competent cells (Leiss, Beckmann, Giros, Costell, & Fassler, 2008). Fibronectin mediates a variety of cell adhesive and migratory activities by binding to receptors on the cell surface as well as to most ECM proteins. Two major sites of fibronectin that mediate cell adhesion are the cell-binding domain and the alternatively spliced V region. As with laminin, the major receptors for fibronectin are integrins, including α3β1, α5β1, α8β1, αvβ1, αIIbβ3, αvβ3, αvβ5, and αvβ6. These integrins interact with the Arg-Gly-Asp (RGD) sequence in the fibronectin cell-binding domain; by contrast, integrins α4β1 and α4β7 recognize the Leu-Asp-Val (LDV) sequence in the V region (Humphries, Byron, & Humphries, 2006).

3.2 ECM Composition and Elastic Vessel Biomechanics

While clearly important in facilitating SMC recruitment and differentiation and in directing provisional matrix assembly, fibronectin cannot provide the mechanical stability required of a mature blood vessel. Once blood pressure rises and tensional forces within the vessel wall begin to increase, the types of ECM proteins made by SMCs change to better accommodate wall stress. These changes are driven, in large part, by alterations in hemodynamics associated with maturation of the heart. Blood flow is detected in the embryonic mouse heart as early as E8.5 and in the peripheral circulation by E10.5. Blood flow and blood pressure increase rapidly through the later embryonic and early postnatal stages in mice. For example, mean left ventricular outflow velocity doubles from E10.5 to E14.5 and peak systolic left ventricular pressure increases from 2 to 11 mm Hg over the same period. However, the largest increase in cardiac output and mean arterial pressure occurs during the first 30 days after birth, with cardiac output increasing from 5 to 15 mL/min and blood pressure increasing 30–70 mm Hg (Ishiwata, Nakazawa, Pu, Tevosian, & Izumo, 2003; Keller, MacLennan, Tinney, & Yoshigi, 1996).

The change in vessel structure from a primarily cellular artery in early development to a mechanically appropriate elastic artery at maturity requires the stepwise and coordinated expression of numerous matrix components (Le et al., 2015). The provisional matrix that is present in the vessel wall at E14 in the mouse, when blood pressure begins to rise, lacks the strength or elasticity to deal with increasing wall stress and pulsatile flow. In response to changing mechanical signals, SMCs undergo an important phenotypic switch characterized by expression of the major structural matrix proteins (collagens, fibrillins, elastin, etc.) required for proper vessel wall function. Concurrent with the increase in blood pressure and changes in cardiac hemodynamics, expression of these ECM coding genes shows a rapid and steady rise through the first 7–14 days after birth followed by a decrease once blood pressure stabilizes at about P30 to the low levels that persist in the adult (Fig. 6). Most of the structural matrix proteins follow this pattern, which has been dubbed the "matrix phenotype" (McLean et al., 2005). The result is a give-and-take process between ECM assembly and vessel hemodynamics such that incremental increases in blood pressure and wall strength progress concomitantly and one cannot occur without the other. Once the correct pressure (or vessel wall material property) is obtained, elastin, fibrillins, collagens, and other structural matrix proteins are downregulated to maintain wall mechanics for optimized physiological function.

Building or adjusting the vessel wall structure to withstand applied forces requires SMCs to recognize and translate a variety of mechanical signals that originate in the microenvironment. Because the formation of a functional

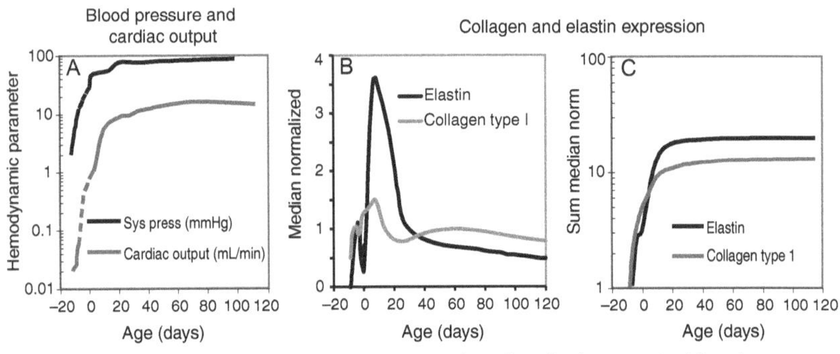

Fig. 6 Elastin and collagen expression are correlated with changes in blood pressure in late embryonic and early postnatal development in mice. (A) Systolic pressure and cardiac output vs age. Both hemodynamic parameters increase steadily until 2–3 weeks after birth. (B) Elastin and type I collagen mRNA levels increase along with blood pressure and are markedly downregulated with blood pressure stabilizes postnatally. (C) The sum of the median normalized elastin and collagen expression vs age. The sum was calculated by totaling all median normalized gene expression from the start of development to the current age. This graph also illustrates that little new elastin or collagen protein accumulates after expression of the two genes are downregulated. *From Wagenseil, J., & Mecham, R. (2009). Vascular extracellular matrix and arterial mechanics.* Physiological Reviews. 89, *957–989, used with permission.*

ECM must occur in an organized sequence, the matrix phenotype is changing throughout the entire period of vessel wall development (Le et al., 2015; Wagenseil & Mecham, 2009). An interesting question is whether there are target material properties that direct the type and amount of ECM produced. The answer appears to be yes. The incremental elastic modulus of the aorta for all species having vessels with distensible elasticity lies in the range of 0.3–1.0 MPa (Gibbons & Shadwick, 1989). To achieve this functional modulus, cells must monitor the changing tensional forces within the wall and adjust the mix of ECM components to produce mechanical properties appropriate for different hemodynamic variables. A universal vessel wall modulus that applies across species and in vessels with different ECM components indicates strong evolutionary pressure to ensure that all elastic arteries have similar mechanical properties at each organism's mean physiological blood pressure.

4. ECM AND ARTERIAL TISSUE ORGANIZATION AND FUNCTION

4.1 Medial Elastogenesis and SMC Differentiation

Structural integrity of and proper cell–matrix interactions within the tunica media are largely responsible for the mechanical compliance of large vessels. Indeed, hierarchical protein–protein interactions that drive fetal assembly and postnatal maturation of elastic fibers are inextricably interconnected with the organization and function of the aortic wall. Biogenesis of elastic fibers (elastogenesis) can be broadly segregated into two distinct assembly processes. One process involves the formation of insoluble elastin from LOX-mediated cross-linking of soluble tropoelastin molecules (Kozel, mecham, & Rosenbloom, 2011). This process is driven by the propensity of tropoelastin to self-assemble (coacervate) under physiological conditions, which serves to align key cross-linking domains. The other assembly process involves the polymerization of fibrillin molecules into 10–15 nm diameter microfibrils (Charbonneau et al., 2003; Ramirez, Sakai, Rifkin, & Dietz, 2007; Sengle & Sakai, 2015). Fibrillin polymers are thought to provide the scaffold that guides tropoelastin alignment and cross-linking in concert with other elastic fiber molecules, such as the fibulins. Despite being part of the same extracellular macroaggregate, loss of function genetic studies in mice have revealed that individual elastic fiber components contribute differently to medial elastogenesis and SMC differentiation.

Elastin, the most abundant component of elastic fibers, regulates key SMC–matrix interactions during arterial development. Loss of elastin gene (*Eln*) expression in mice results in perinatal death due to obstructive arterial disease associated with excessive subendothelial proliferation and accumulation of disorganized SMCs. Arterial occlusion is apparently independent of hemodynamic stress since organ cultures of elastin-deficient arteries undergo occlusion as well. The cell-regulatory function of elastin is further supported by the finding that elastin haploinsufficiency in mice leads to a two- to three-fold increase in the genetically predetermined number of elastic lamellae. This mouse phenotype closely resembles that of individuals afflicted with supravalvular arterial stenosis, a heritable disease similarly associated with elastin haploinsufficiency (Li et al., 1998). Thus, it has been argued that disruption of the biomechanical property of the vessel wall due to decreased elastin content may induce a compensatory increase in the number of elastic lamellae ultimately leading to occlusion of the vessel's lumen. In this view, elastin provides a local autocrine signal that modulates SMC proliferation, migration, and differentiation conceivably by interacting with a cell-surface receptor. While the identity of the elastin receptor has remained elusive, in vitro evidence suggests intracellular involvement of a pertussis toxin-sensitive, G protein-coupled pathway that stimulates $G\alpha_i$ signaling, inhibits adenylate cyclase activity, reduces cyclic adenosine monophosphate levels, and promotes Rho-induced actin polymerization (Karnik et al., 2003). Other cell-surface elastin binding proteins have also been identified as possible signaling receptors (Blood, Sasse, Brodt, & Zetter, 1988; Mecham et al., 1989; Privitera, Prody, Callahan, & Hinek, 1998).

Fibrillins-1 and -2, the major structural constituents of microfibrils in mice (humans have a third fibrillin molecule of undetermined function), exhibit partially overlapping expression profiles during fetal and postnatal development of elastic vessels (Davis, 1993). Fibrillin-2 expression is at its highest during embryogenesis and decreases rapidly to background levels soon after birth. Fibrillin-1 is expressed in both developing and maturing vessel and at consistently higher levels than fibrillin-2. As point of reference, the bulk of elastin is produced postnatally coincident with the surge in cardiac output and mean arterial pressure. Together, these findings and genetic evidence discussed below indicate that (a) fibrillin-2 polymers are involved in initiating elastogenesis in the tunica media; (b) formation of fibrillin-rich microfibrils is required for tropoelastin deposition and cross-linking; and (c) fibrillin-1 imparts force-bearing structural support during postnatal vessel growth, in addition to partially overlapping with fibrillin-2 regulation of early vascular tissue elastogenesis.

Newborn mice with inactive fibrillin-1 genes (*Fbn1*$^{-/-}$ mice) display a disorganized and immature medial layer that cannot sustain cyclic mechanical strain and consequently, they die soon after birth from ruptured aortic aneurysm. However, normal elastin cross-linking in the aorta of these mutant mice implies that fibrillin-2 microfibrils are sufficient to support elastogenesis during fetal life. Embryonic lethality of all *Fbn1*$^{-/-}$*;Fbn2*$^{-/-}$ mice and half of *Fbn1*$^{+/-}$*;Fbn2*$^{-/-}$ mice is associated with severely impaired elastic fiber formation. This finding underscores the specialized function of fibrillin-2 in the initial assembly of the medial matrix (which in part overlaps with fibrillin-1) and with the absolute requirement of continued fibrillin-1 deposition for vessel maturation and function during neonatal life (Torsney, Hu, & Xu, 2005). The presence of fetal fibrillin-2 polymers within the inner core of postnatal fibrillin-1 microfibrils supports these diversified functions (Hu & Xu, 2011). Furthermore, distinct vascular malformations in zebrafish embryos with morpholino-induced silencing of either fibrillin-1 or fibrillin-2 have reiterated the evolutionarily conserved structural/functional properties of these extracellular molecules (Hu et al., 2004; Passman et al., 2008). Fibrillin molecules are also implicated in modulating local TGFβ signals that promote SMC differentiation. Evidence of this "instructive" role includes (a) dysregulated morphogenesis of NCC derivatives in double null mice causing cardiovascular abnormalities (e.g., persistent truncus arteriosus with interrupted aortic arch) that phenocopy those of TGFβ loss-of-function mutants and (b) continued postnatal expression of embryonic genes in fibrillin-1-deficient SMCs, a differentiation defect consistent with abated TGFβ signaling due to impaired growth factor sequestration into the matrix (Torsney et al., 2005).

TGFβ signaling molecules play pivotal roles in SMC differentiation during vascular development as well as phenotypic switching in disease states. Fibrillin-rich microfibrils interact with latent TGFβ-binding proteins (LTBPs), elastic fiber-associated proteins, and key regulators of TGFβ bioavailability (Majesky, Dong, Hoglund, Daum, & Mahoney, 2012). LTBPs 1, 3, and 4 bind the dimeric propeptide of TGFβ noncovalently associated with the bioactive ligand (a.k.a., small latent complex) so as to facilitate folding and secretion of the resulting tripartite complex (a.k.a., large latent complex), which is then tethered to the ECM in part via LTBP association with fibrillin-rich assemblies. Recent findings indicate that LTBPs 3 and 4 association with the matrix of the postnatal aorta depends on fibrillin-1 microfibrils, whereas LTBP-1 association depends on a fibronectin network (Chen et al., 2013). TGFβ sequestration into the ECM is believed to promote spatial distribution and proper concentration of bioactive ligands for either immediate presentation to cells or subsequent release during tissue remodeling/repair. An example of the former mechanism comes from pharmacological studies of mice hypomorphic for fibrillin-1, which develop a severe form of aortic aneurysm like patients afflicted with Marfan syndrome (Psaltis & Simari, 2015). By implying that fibrillin-1 deficiency is associated with loss of TGFβ signaling, these findings demonstrate the importance of fibrillin-1 microfibrils in maintaining baseline TGFβ activity required for proper growth of the postnatal vessel (Psaltis et al., 2014). While yet to be reported for the vascular system, genetic studies have revealed that fibrillins can also regulate local BMP signaling in the developing limbs and differentiating skeletal muscle and bone cells (Chan & Inoue, 1994; D'Armiento, 2003; Mecham & Davis, 1994). BMPs are targeted to the ECM via noncovalent association of the dimeric propeptides with the N-terminal regions of fibrillins, an interaction that causes a conformational change conferring latency to the growth factor (Wohl et al., 2016). Competitive displacement of the BMP prodomains by type II receptors eventually activates growth factor signaling.

The fibulins constitute a seven-member family of ECM proteins that shares structural features with both fibrillins and LTBPs (Papke and Yanagisawa, 2014). Fibulins 4 and 5 are both potent inducer of vascular elastogenesis with the latter gene (*Fbln5*) expressed at higher levels than *Fbln4*. Remarkably, the phenotype of *Fbln4*$^{-/-}$ mice is significantly more severe than that of *Fbln5*$^{-/-}$ mice. Whereas loss of fibulin-4 during embryogenesis results in perinatal death from aortic aneurysm rupture, fibulin-5 deficiency is compatible with postnatal life even in the presence of deteriorating elastic tissues. Proposed functions of fibulin-4 include tethering LOX to tropoelastin, stimulating the interaction between enzyme and substrate, and promoting SMC differentiation. Fibulin-5, on the other hand, is believed to potentiate tropoelastin coacervation, tether the LOX-like 1 (LOXL1) enzyme to the coacervates and inhibit SMC proliferation. Both fibulins bind to fibrillin-1 and fibulin-5 can also form a ternary complex with fibrillin-1 and LTBP-4. In line with the distinct molecular interactions of the two fibrillins, the vascular phenotypes of LOX- and LOXL1-deficient mice resemble those of *Fbln4*$^{-/-}$ and *Fbln5*$^{-/-}$ mice, respectively (Papke and Yanagisawa, 2014).

4.2 ECM and the Vascular Stem-Cell Niche

The ECM is a key component of the microenvironment that influences stem cell fate decisions. This stem-cell "niche" represents a highly specialized microenvironment for each cell type that is in part defined by the ECM contributed by stem cells themselves as well as associated stromal cells. Cells within the niche interact with ECM components through cell-surface receptors (mostly integrins) and with growth and differentiation factors associated with ECM. The biomechanical properties of the ECM also influence the functional diversity of the stem-cell niche (Gattazzo, Urciuolo, & Bonaldo, 2014). Matrix stiffness, porosity, or topography influence cell behavior independent of chemical signals through a process broadly defined as mechanotransduction. By sensing mechanical or physical aspects of their environment, the cytoskeleton becomes an important signal relay system where cell contractility and tension, mediated by matrix receptors on the plasma membranes, provide signals that influence cell fate (Choi, Holle, & Engler, 2013; Watt & Huck, 2013). The details of what makes the stem-cell niche unique from other areas in tissue are still unclear, but stem cells themselves are active players in contributing niche components that maintain a balance between quiescence, self-renewal, and differentiation. ECM components identified in functional stem-cell niches include basement membrane proteins (particularly laminin), collagen VI, fibronectin, fibulin-1, fibrillins, SPARC, heparan and chondroitin proteoglycans, and members of the tenascin family. Since these molecules are also found outside of the stem-cell niche, the question becomes how the information they provide differs in the two matrix compartments? This is an important area of research in matrix biology and a key question for tissue engineering and regenerative medicine.

Like other tissues, the vasculature contains stem cells that reside in a specialized, ECM-rich niche (Majesky et al., 2011). These progenitor cells can differentiate into both SMCs and ECs and contribute to normal cellular renewal (Majesky et al., 2017; Xiao, Wang, Luo, & Xu, 2010). In addition, increasing evidence shows that vascular stem/progenitor cells play a role in numerous cardiovascular diseases where vessel remodeling comes into play, including atherosclerosis and restenosis (Hu & Xu, 2011; Torsney et al., 2005). The vascular progenitor cell niche has been mapped to the adventitia, specifically to the media–adventitia border in a region corresponding to a site of Hedgehog signaling (Hu et al., 2004; Majesky et al., 2012; Passman et al., 2008). Progenitor cells in this compartment express Sca1 and CD34 progenitor markers and have the potential to self-renew or to differentiate in vitro into SMCs, ECs, osteoblasts, chondrocytes, or adipocytes (Chen et al., 2013; Passman et al., 2008). Of interest is the finding of Sca1-positive myeloid progenitors with hematopoietic potential residing in this compartment (Psaltis et al., 2014; Psaltis & Simari, 2015).

Using bone marrow transplant and adoptive transfer approaches, it has been shown that the Sca1-positive cells do not originate from bone marrow, neural crest, or circulating cells. Instead, Majesky et al. (2017) have shown that differentiated SMCs migrate from the arterial media into the adventitia and revert to multipotent vascular progenitor cells through a physiological reprogramming-like process dependent on induction of transcription factor Kruppel-like factor 4 (Klf4). These findings suggest that the local niche of the inner adventitia, defined to a large extent by the ECM composition, directs or stabilizes a vascular progenitor cell pool in the vessel wall that can react to environment clues to become numerous differentiated phenotypes. Which ECM components contribute to this vascular niche, and how they act to maintain cellular progenitor-like properties, remains unknown.

5. CONCLUDING REMARKS

The studies we have reviewed demonstrate that dynamic interactions among ECM proteins, soluble biochemical signals, and resident cells are central to vascular development and homeostasis. As ECM composition regulates cell fate and tissue biomechanics, vascular cells constantly reorganize the ECM to satisfy changing requirements for regulatory signals and time-dependent physical properties. Mutations disrupting this highly integrated cross talk between cellular and structural components of the vessel wall often lead to life-threatening complications; additionally, medial degeneration is the histopathological hallmark of acquired forms of arterial disease. It is therefore safe to predict that much research effort will be devoted in the future to develop new experimental tools that will more effectively than before decipher the complex dynamics of cell–matrix interactions in the forming and mature circulatory system. These future studies are also expected to identify new therapeutic targets for the treatment of human diseases associated with vascular dysfunction.

REFERENCES

Astrof, S., Crowley, D., & Hynes, R. (2007). Multiple cardiovascular defects caused by the absence of alternatively spliced segments of fibronectin. *Developmental Biology*, *311*, 11–24.

Astrof, S., & Hynes, R. (2009). Fibronectins in vascular morphogenesis. *Angiogenesis*, *12*, 165–175.

Aumailley, M., Bruckner-Tuderman, L., Carter, W., Deutzmann, R., Edgar, D., Ekblom, P., et al. (2005). A simplified laminin nomenclature. *Matrix Biology*, *24*, 326–332.

Baeten, J. T., & Lilly, B. (2017). Notch signaling in vascular smooth muscle cells. *Advances in Pharmacology*, *78*, 351–382.
Blood, C., Sasse, J., Brodt, P., & Zetter, B. (1988). Identification of a tumor cell receptor for VGVAPG, an elastin-derived chemotactic peptide. *The Journal of Cell Biology*, *107*, 1987–1993.
Bornstein, P., & Sage, E. H. (2002). Matricellular proteins: Extracellular modulators of cell function. *Current Opinion in Cell Biology*, *14*, 608–616.
Bowers, S. L., Norden, P. R., & Davis, G. E. (2016). Molecular signaling pathways controlling vascular tube morphogenesis and pericyte-induced tube maturation in 3D extracellular matrices. *Advances in Pharmacology*, 77, 241–280.
Carlin, B., Jaffe, R., Bender, B., & Chung, A. E. (1981). Entactin, a novel basal lamina-associated sulfated glycoprotein. *The Journal of Biological Chemistry*, *256*, 5209–5214.
Chan, F. L., & Inoue, S. (1994). Lamina lucida of basement membrane: An artefact. *Microscopy Research and Technique*, *28*, 48–59.
Charbonneau, N. L., Dzamba, B. J., Ono, R. N., Keene, D. R., Corson, G. M., Reinhardt, D. P., et al. (2003). Fibrillins can co-assemble in fibrils, but fibrillin fibril composition displays cell-specific differences. *The Journal of Biological Chemistry*, *278*, 2740–2749.
Chen, Y., Wong, M. M., Campagnolo, P., Simpson, R., Winkler, B., Margariti, A., et al. (2013). Adventitial stem cells in vein grafts display multilineage potential that contributes to neointimal formation. *Arteriosclerosis, Thrombosis, and Vascular Biology*, *33*, 1844–1851.
Choi, Y. S., Holle, A. W., & Engler, A. J. (2013). Engineered ECM microenvironments and their regulation of stem cells. In D. DeSimone & R. Mecham (Eds.), *Extracellular matrix in developoment* (pp. 133–160). Heidelberg: Springer-Verlag.
Chung, A. E., & Durkin, M. E. (1990). Entactin: Structure and function. *American Journal of Respiratory Cell and Molecular Biology*, *3*, 275–282.
Cleary, E., Fanning, J., & Prosser, I. (1981). Possible roles of microfibrils in elastogenesis. *Connective Tissue Research*, *8*, 161–166.
D'Armiento, J. (2003). Decreased elastin in vessel walls puts the pressure on. *The Journal of Clinical Investigation*, *112*, 1308–1310.
Davis, E. C. (1993). Smooth muscle cell to elastic lamina connections in the developing mouse arota: Role in aortic medial organization. *Laboratory Investigation*, *68*, 89–99.
Davison, I., Wright, G., & DeMont, M. (1995). The structure and physical properties of invertebrate and primitive vertebrate arteries. *The Journal of Experimental Biology*, *198*, 2185–9216.
Dobrin, P. (1978). Mechanical properties of arteries. *Physiological Reviews*, *58*, 397–460.
Drake, C. J., Hungerford, J. E., & Little, C. D. (1998). Morphogenesis of the first blood vessels. *Annals of the New York Academy of Sciences*, *857*, 155–179.
Ekblom, P., & Timpl, R. (1996). *The Laminins*. New York: Harwold Academic Publishers.
Gattazzo, F., Urciuolo, A., & Bonaldo, P. (2014). Extracellular matrix: A dynamic microenvironment for stem cell niche. *Biochimica et Biophysica Acta*, *1840*, 2506–2519.
George, E. L., Baldwin, H. S., & Hynes, R. O. (1997). Fibronectins are essential for heart and blood vessel morphogenesis but are dispensable for initial specification of precursor cells. *Blood*, *90*, 3073–3081.
George, E. L., Georges-Labouesse, E. N., Patel-King, R. S., Rayburn, H., & Hynes, R. O. (1993). Defects in mesoderm, neural tube and vascular development in mouse embryos lacking fibronectin. *Development*, *119*, 1079–1091.
Gibbons, C., & Shadwick, R. (1989). Functional similarities in the mechanical design of the aorta in lower vertebrates and mammals. *Experientia*, *45*, 1083 1088.
Greif, D., Kumar, M., Lighthouse, J., Hum, J., An, A., Ding, L., et al. (2012). Radial construction of an arterial wall. *Developmental Cell*, *23*, 482–493.

Hellstrom, M., Kalen, M., Lindahl, P., Abramsson, A., & Betsholtz, C. (1999). Role of PDGF-B and PDGFR-beta in recruitment of vascular smooth muscle cells and pericytes during embryonic blood vessel formation in the mouse. *Development*, *126*, 3047–3055.

Hu, Y., & Xu, Q. (2011). Adventitial biology: Differentiation and function. *Arteriosclerosis, Thrombosis, and Vascular Biology*, *31*, 1523–1529.

Hu, Y., Zhang, Z., Torsney, E., Afzal, A., Davison, F., Metzler, B., et al. (2004). Abundant progenitor cells in the adventitia contribute to atherosclerosis of vein grafts in ApoE-deficient mice. *The Journal of Clinical Investigation*, *113*, 1258–1265.

Humphries, J. D., Byron, A., & Humphries, M. J. (2006). Integrin ligands at a glance. *Journal of Cell Science*, *119*, 3901–3903.

Hynes, R. (2012). The evolution of metazoan extracellular matrix. *The Journal of Cell Biology*, *196*, 671–679.

Ishiwata, T., Nakazawa, M., Pu, W. T., Tevosian, S. G., & Izumo, S. (2003). Developmental changes in ventricular diastolic function correlate with changes in ventricular myoarchitecture in normal mouse embryos. *Circulation Research*, *93*, 857–865.

Karnik, S. K., Brooke, B. S., Bayes-Genis, A., Sorensen, L., Wythe, J. D., Schwartz, R. S., et al. (2003). A critical role for elastin signaling in vascular morphogenesis and disease. *Development*, *130*, 411–423.

Keller, B. B., MacLennan, M. J., Tinney, J. P., & Yoshigi, M. (1996). In vivo assessment of embryonic cardiovascular dimensions and function in day-10.5 to -14.5 mouse embryos. *Circulation Research*, *79*, 247–255.

Kozel, B. A., mecham, R. P., & Rosenbloom, J. (2011). Elastin. In R. P. Mecham (Ed.), *Biology of extracellular matrix* (pp. 267–301). Berlin-Heidelberg: Springer-Verlag.

Le, V. P., Cheng, J. K., Kim, J., Staiculescu, M. C., Ficker, S. W., Sheth, S. C., et al. (2015). Mechanical factors direct mouse aortic remodelling during early maturation. *J. R. Soc. Interface*, *12*, 20141350.

Leiss, M., Beckmann, K., Giros, A., Costell, M., & Fassler, R. (2008). The role of integrin binding sites in fibronectin matrix assembly in vivo. *Current Opinion in Cell Biology*, *20*, 502–507.

Li, D. Y., Faury, G., Taylor, D. G., Davis, E. C., Boyle, W. A., Mecham, R. P., et al. (1998). Novel arterial pathology in mice and humans hemizygous for elastin. *The Journal of Clinical Investigation*, *102*, 1783–1787.

Majesky, M., Dong, X., Hoglund, V., Daum, G., & Mahoney, W. J. (2012). The adventitia: A progenitor cell niche for the vessel wall. *Cells, Tissues, Organs*, *195*, 73–81.

Majesky, M. W., Dong, X. R., Regan, J. N., & Hoglund, V. J. (2011). Vascular smooth muscle progenitor cells: Building and repairing blood vessels. *Circulation Research*, *108*, 365–377.

Majesky, M. W., Horita, H., Ostriker, A., Lu, S., Regan, J. N., Bagchi, A., et al. (2017). Differentiated smooth muscle cells generate a subpopulation of resident vascular progenitor cells in the adventitia regulated by Klf4. *Circulation Research*, *120*, 296–311.

Mayer, U., & Timpl, R. (1994). Nidogen: A versatile binding protein of basement membranes. In P. D. Yurchenco, D. E. Birk, & R. P. Mecham (Eds.), *Extracellular matrix assembly* (pp. 318–416). San Diego: Academic Press.

McConnell, C., Wright, G., & DeMont, M. (1996). The modulus of elasticity of lobster aorta microfibrils. *Experientia*, *52*, 918–921.

McLean, S. E., Mecham, B. H., Kelleher, C. M., Mariani, T. J., & Mecham, R. P. (2005). Extracellular matrix gene expression in developing mouse aorta. In J. H. Miner (Ed.), *Extracellular matrices and development* (pp. 82–128). New York: Elsevier.

Mecham, R. P., & Davis, E. C. (1994). Elastic fiber structure and assembly. In P. D. Yurchenko, D. E. Birk, & R. P. Mecham (Eds.), *Extracellular matrix assembly and structure* (pp. 281–314). San Diego: Academic Press.

Mecham, R., Hinek, A., Entwistle, R., Wrenn, D., Griffin, G., & Senior, R. (1989). Elastin binds to a multifunctional 67-kilodalton peripheral membrane protein. *Biochemistry, 28*, 3716–3722.

Miner, J. H. (2011). Basement Membranes. In R. P. Mecham (Ed.), *The extracellular matrix: An overview* (pp. 117–145). Berlin, Heidelberg: Springer, Berlin Heidelberg.

Mosher, D. F. (1989). Fibronectin. In R. Mecham (Ed.), *Biology of extracellular matrix* (pp. 474). San Diego: Academic Press.

Owens, G. K. (1995). Regulation of differentiation of vascular smooth muscle cells. *Physiological Reviews, 75*, 487–517.

Papke, C. L., & Yanagisawa, H. (2014). Fibulin-4 and fibulin-5 in elastogenesis and beyond: Insights from mouse and human studies. *Matrix Biology, 37*, 142–149.

Passman, J., Dong, X., Wu, S., Maguire, C., Hogan, K., Bautch, V., et al. (2008). A sonic hedgehog signaling domain in the arterial adventitia supports resident Sca1+ smooth muscle progenitor cells. *Proceedings of the National Academy of Sciences of the United States of America, 105*, 9349–9354.

Privitera, S., Prody, C. A., Callahan, J. W., & Hinek, A. (1998). The 67-kDa enzymatically inactive alternatively spliced variant of β-galactosidase is identical to the elastin/laminin binding protein. *The Journal of Biological Chemistry, 273*, 6319–6326.

Psaltis, P. J., Puranik, A. S., Spoon, D. B., Chue, C. D., Hoffman, S. J., Witt, T. A., et al. (2014). Long noncoding RNA MALAT1 regulates endothelial cell function and vessel growth. *Circulation Research, 115*, 364–375.

Psaltis, P. J., & Simari, R. D. (2015). Vascular wall progenitor cells in health and disease. *Circulation Research, 116*, 1392–1412.

Ramirez, F., Sakai, L. Y., Rifkin, D. B., & Dietz, H. C. (2007). Extracellular microfibrils in development and disease. *Cellular and Molecular Life Sciences, 64*, 2437–2446.

Sato, Y., Poynter, G., Huss, D., Filla, M., Czirok, A., Rongish, B., et al. (2010). Dynamic analysis of vascular morphogenesis using transgenic quail embryos. *PLoS One, 5*, e12674.

Sawada, H., Rateri, D. L., Moorleghen, J. J., Majesky, M. W., & Daugherty, A. (2017). Smooth muscle cells derived from second heart field and cardiac neural crest reside in spatially distinct domains in the Media of the Ascending Aorta. *Arteriosclerosis, Thrombosis, and Vascular Biology, 37*, 1722–1726.

Scheppke, L., Murphy, E. A., Zarpellon, A., Hofmann, J. J., Merkulova, A., Shields, D. J., et al. (2012). Notch promotes vascular maturation by inducing integrin-mediated smooth muscle cell adhesion to the endothelial basement membrane. *Blood, 119*, 2149–2158.

Schwarzbauer, J. E., & DeSimone, D. W. (2011). Fibronectins, their fibrillogenesis, and in vivo functions. *Cold Spring Harbor Perspectives in Biology, 3*, a005041.

Schwarzbauer, J. E., Tamkun, J. W., Lemischka, I. R., & Hynes, R. O. (1983). Three different fibronectin mRNAs arise by alternative splicing within the coding region. *Cell, 35*, 421–431.

Sengle, G., & Sakai, L. (2015). The fibrillin microfibril scaffold: A niche for growth factors and mechanosensation. *Matrix Biology, 47*, 3–12.

Shadwick, R. (1999). Mechanical design in arteries. *The Journal of Experimental Biology, 202*, 3305–3313.

Torsney, E., Hu, Y., & Xu, Q. (2005). Adventitial progenitor cells contribute to arteriosclerosis. *Trends in Cardiovascular Medicine, 15*, 64–68.

Wagenseil, J., Ciliberto, C., Knutsen, R., Levy, M., Kovacs, A., & Mecham, R. (2010). The importance of elastin to aortic development in mice. *American Journal of Physiology. Heart and Circulatory Physiology, 299*, H257–64.

Wagenseil, J., & Mecham, R. (2009). Vascular extracellular matrix and arterial mechanics. *Physiological Reviews, 89*, 957–989.

Wagenseil, J. E., Nerurkar, N. L., Knutsen, R. H., Okamoto, R. J., Li, D. Y., & Mecham, R. P. (2005). Effects of elastin Haploinsufficiency on the mechanical behavior of mouse arteries. *American Journal of Physiology. Heart and Circulatory Physiology*, *289*, H1209–1217.

Watt, F. M., & Huck, W. T. (2013). Role of the extracellular matrix in regulating stem cell fate. *Nature Reviews. Molecular Cell Biology*, *14*, 467–473.

Wohl, A. P., Troilo, H., Collins, R. F., Baldock, C., & Sengle, G. (2016). Extracellular regulation of bone morphogenetic protein activity by the microfibril component fibrillin-1. *Journal of Biological Chemistry*, *291*, 12732–12746.

Wolinsky, H., & Glagov, S. (1964). Structural basis for the static mechanical properties of the aortic media. *Circulation Research*, *14*, 400–413.

Xiao, Q., Wang, G., Luo, Z., & Xu, Q. (2010). The mechanism of stem cell differentiation into smooth muscle cells. *Thrombosis and Haemostasis*, *104*, 440–448.

Xu, J., & Mosher, D. F. (2011). Fibronectin and other adhesive glycoproteins. In R. P. Mecham (Ed.), *The extracellular matrix: An overview* (pp. 41–75). Berlin, Heidelberg: Springer, Berlin Heidelberg.

Yurchenco, P. D., Amenta, P. S., & Patton, B. L. (2004). Basement membrane assembly, stability and activities observed through a developmental lens. *Matrix Biology*, *22*, 521–538.

CHAPTER SIX

Structure, Function, and Development of the Tectorial Membrane: An Extracellular Matrix Essential for Hearing

Richard J. Goodyear, Guy P. Richardson[1]
Sussex Neuroscience, School of Life Sciences, University of Sussex, Brighton, United Kingdom
[1]Corresponding author: e-mail address: g.p.richardson@sussex.ac.uk

Contents

Abstract

The tectorial membrane is an extracellular matrix that lies over the apical surface of the auditory epithelia in the inner ears of reptiles, birds, and mammals. Recent studies have shown it is composed of a small set of proteins, some of which are only produced at high levels in the ear and many of which are the products of genes that, when mutated, cause nonsyndromic forms of human hereditary deafness. Quite how the proteins of the tectorial membrane are assembled within the lumen of the inner ear to form a structure

Current Topics in Developmental Biology, Volume 130
ISSN 0070-2153
https://doi.org/10.1016/bs.ctdb.2018.02.006

that is precisely regulated in its size and physical properties along the length of a tonotopically organized hearing organ is a question that remains to be fully answered. In this brief review we will summarize what is known thus far about the structure, protein composition, and function of the tectorial membrane in birds and mammals, describe how the tectorial membrane develops, and discuss major events that have occurred during the evolution of this extracellular matrix.

1. INTRODUCTION

Hearing is a sense that allows animals to detect prey, avoid predators, and communicate with one another. In vertebrates, sounds are detected and encoded as neural signals within the inner ear, a complex labyrinthine structure that develops from the otic placode, a thickening of the head ectoderm that lies adjacent to the rhombencephalon in the early embryo. There are up to eight discrete sensory patches within the inner ear which detect linear acceleration, angular acceleration, or sound. While differing in function these sensory patches all conform to the same basic plan; in most cases they are comprised of a polarized epithelium that contains two cell types, sensory hair cells and nonsensory supporting cells, with a prominent extracellular matrix lying atop of the epithelial apical surface. The hair cells have an apically located, mechanosensory hair bundle, and the overlying matrix generally serves as a structure that delivers the stimulus to the hair bundle. When deflected by gravity, body movements, or motion of the epithelia, the hair bundle, through interactions with the extracellular matrix, generates a receptor potential that, in turn, leads to activation of the afferent nerve fibers contacting the basolateral pole of the hair cell. Sensory information is thereby transmitted to the CNS.

There is considerable diversity in the structure and composition of the extracellular matrices that lie over the different organs in the inner ear (Goodyear & Richardson, 2002). In the vestibular system, a cupula lies over the crista in the ampullae of the semicircular canals, and a mineralized otoconial membrane sits over the maculae of the utricle and the saccule. In the inner ears of birds, reptiles, and mammals, a tectorial membrane (TM) lies over the auditory epithelium: the basilar papilla (in birds and reptiles) and the organ of Corti (in mammals). In this review we will focus on the TMs of birds and mammals, highlighting the major differences in their structure, composition, function, and development. Finally we discuss how and why these changes may have arisen during evolution.

2. THE COCHLEA AND THE MAMMALIAN TM

The cochlea is the mammalian organ of hearing. It is a coiled, fluid-filled tube that is comprised of three chambers, scalae vestibuli, media, and tympani that run along its length. Frequency analysis takes place on the basilar membrane (BM), an extracellular matrix that separates scala media from scala tympani and vibrates in response to sound-induced motions of the cochlear fluids. Movements of the BM are, in turn, detected by the hair cells in the organ of Corti, a neuroepithelium that lies along its medial edge (see Fig. 1A). The BM is graded in stiffness and mass along its length, such that the basal end responds best to high-frequency sounds while the apical end responds best to sounds of low frequency. A sound of a given frequency will set up a traveling wave on the BM, an envelope of vibration that reaches a peak at a position along the cochlea that depends on the frequency. The organ of Corti contains a single row of purely sensory inner hair cells (IHCs) that receive 95% of the cochlear afferent innervation, and three rows of sensorimotor outer hair cells (OHCs) that are innervated by efferent fibers from the brain stem. The OHCs are electromotile and can both detect the sound stimulus and amplify the motion of the BM at low sound pressure levels (Ashmore, 2008; Dallos, 2008). The response of the BM is therefore highly dependent on active feedback from the OHCs; if these cells are lost then cochlear sensitivity decreases by ~1000 × fold and the organ is no longer sharply tuned (Dallos & Harris, 1978).

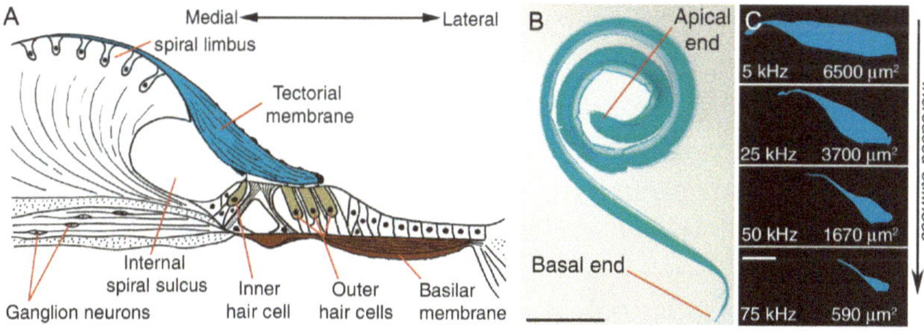

Fig. 1 (A) Diagram illustrating the structure of the mammalian organ of Corti. The tectorial membrane (TM) is colored *blue*, the basilar membrane (BM) is colored *reddish brown*. (B) A tectorial membrane from the left cochlea of a 4-week-old mouse that was obtained by dissection and briefly stained with the cationic dye *Alcian blue*. Scale bar = 500 μm. The tectorial membrane is ~5 mm long. (C) Cross-sectional profiles of the mouse tectorial membrane at four different points along its length. Cross-sectional area at each point is indicated. Scale bar = 50 μm.

The TM is a ribbon-like strip of extracellular matrix (Fig. 1B) that spirals along the length of the organ of Corti. It is attached along its medial edge to the surface of the spiral limbus, stretches across the internal spiral sulcus, and lies over the sensory epithelium, attaching along its lower lateral surface to the tips of the hair bundles of the OHCs. In mature animals, the hair bundles of the IHCs are not attached directly to the TM but lie in the narrow, fluid-filled subtectorial space. The properties of the TM, much like those of the BM, vary monotonically along the length of the cochlea (Fig. 1C). In the mouse, the cross-sectional area of the TM increases by more than 10-fold from the basal, high-frequency end to the apical, low-frequency encoding region of the cochlea, while its stiffness, as measured by a probe applied perpendicular to the upper surface, increases by a similar amount in the opposite direction, i.e., from the apex to the base of the cochlea (Teudt & Richter, 2014).

3. PROTEINS OF THE MAMMALIAN TM

Analysis of the mouse TM by gel electrophoresis reveals the presence of a number of proteins, many of which are sensitive to digestion with bacterial collagenase (see Fig. 2A). These collagenase-sensitive polypeptides are now known to be derived from type II, type IX, and type XI collagen (Richardson, Russell, Duance, & Bailey, 1987; Thalmann et al., 1987), collagens that are typically found in cartilage. Type V collagen may also be present (Richardson et al., 1987). When TMs are digested with bacterial collagenase and run on SDS-polyacrylamide gels under reducing conditions, three broad bands of protein are revealed by Coomassie Brilliant Blue staining. These are largely derived from beta-tectorin (TECTB) and the three disulfide cross-linked subunits (H, M, and L) of alpha-tectorin (TECTA) (Fig. 2A). Under nonreducing conditions two major bands are visible with apparent molecular masses of ~250 kDa (TECTA) and 45 kDa (TECTB). Molecular cloning studies identified TECTA as a large glycoprotein (predicted mass 239 kDa with 33 potential sites for N-glycosylation) comprising three major regions, an N-terminally located entactin G1-like (NIDO) domain, a central zonadhesin (ZA)-like domain containing vWF type D repeats, cysteine-rich C8 regions and trypsin inhibitor-like (TIL) domains, and a C-terminal zona pellucida (ZP) domain (see Fig. 2B; Legan, Rau, Keen, & Richardson, 1997). TECTB (predicted mass 36 kDa with four potential N-glycosylation sites) comprises a single ZP domain (Legan & Richardson, 1997). TECTA and TECTB (Fig. 2B)

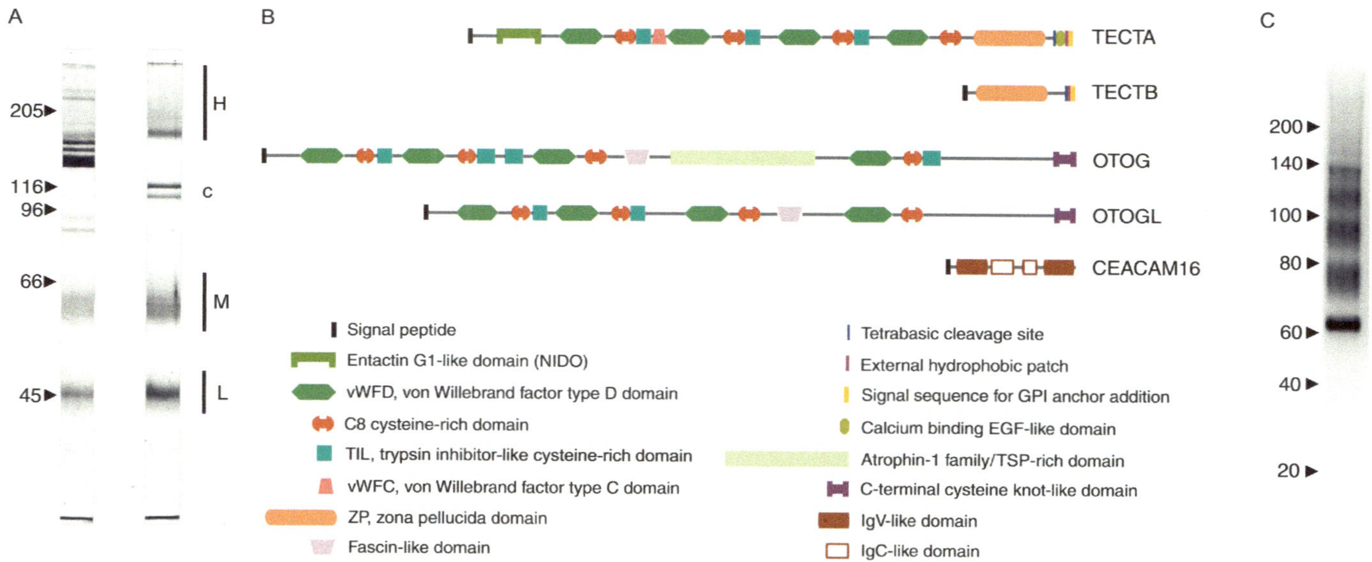

Fig. 2 (A) Coomassie Brilliant Blue-stained polyacrylamide gels of proteins present in the mouse tectorial membrane before (*left*) and after (*right*) digestion with bacterial collagenase (the collagenase is indicated by "c"). 7.5% acrylamide gel with samples run under reducing conditions. Positions of molecular mass markers are indicated on the *left*. *Vertical bars* on the *right* indicate the three, polydisperse protein bands (H, M, and L) remaining following collagenase digestion. H, M, and L are derived from subunits of TECTA; Band L also contains TECTB which comigrates with the L subunit of TECTA. (B) Predicted domain structures of TECTA, TECTB, OTOG, OTOGL, and CEACAM16. (C) Western blot of tectorial membrane proteins stained with antibodies to CEACAM16. 8.25%:12.0% step gradient acrylamide gel run under reducing conditions. CEACAM16 resolves as >6 distinct bands with apparent masses ranging between ~60 and 200 kDa.

are both predicted to be produced as membrane-bound, GPI-anchored precursors that are thought to be released at or while in transit to the plasma membrane by a furin-like endoproteinase acting at a tetrabasic cleavage site located upstream of the signal sequence for GPI-anchor addition. Mice with mutations in the gene encoding TMPRSS1, a transmembrane serine protease (hepsin) that is known to be involved in the release, activation, and subsequent polymerization of the ZP domain protein uromodulin (Brunati et al., 2015), have significant defects in TM structure (Guipponi et al., 2007) that are similar to those seen in mice with missense mutations in the ZP domain of TECTA (Legan et al., 2014, 2005).

The other noncollagenous proteins of the TM (see Fig. 2B) include otogelin (OTOG), otogelin-like (OTOGL), and carcinoma and embryonic antigen-related cell adhesion molecule 16 (CEACAM16) (Cohen-Salmon, El-Amraoui, Leibovici, & Petit, 1997; Yariz et al., 2012; Zheng et al., 2011). These three proteins are not readily visible in Coomassie Brilliant Blue-stained SDS gels of TM proteins. For OTOG and OTOGL this may be due to their large mass (OTOG 313 kDa, OTOGL 258 kDa) and possibly their relatively low abundance, while for CEACAM16 this may be due to the presence of many "isoforms" that are spread across a broad range of molecular masses and are only visible in western blots (see Fig. 2C). OTOG and OTOGL (Fig. 2B) share features in common with TECTA (e.g., vWF type D repeats, C8, and TIL domains) and comprise additional domains not present in TECTA (e.g., a C-terminal cysteine knot). Unlike TECTA and TECTB, there is no evidence that OTOG, OTOGL, and CEACAM16 are produced as membrane-bound precursors. The structure of CEACAM16 (Fig. 2B), with Ig variable-like domains at the N- and the C-terminus separated by two Ig constant-like domains, and the lack of either a C-terminal transmembrane domain or a GPI-anchor, makes it a unique member of the CEACAM protein family (Zebhauser et al., 2005). Furthermore, it does not share structural similarity with any of the other four noncollagenous components of the TM; TECTA, TECTB, OTOG, and OTOGL.

Attachment of the TM to the organ of Corti involves at least two proteins. Adhesion of the TM to the apical surface of the interdental cells in the spiral limbus is mediated by the GPI-anchored glycoprotein otoancorin (OTOA) (Lukashkin et al., 2012; Zwaenepoel et al., 2002), while its attachment to the tips of the tallest stereocilia in the hair bundles of the OHCs requires stereocilin (STRC) (Verpy et al., 2001). STRC and OTOA are distantly related to one another and share homology to mesothelin (Jovine, Park, & Wassarman, 2002; Sathyanarayana, Hahn, Patankar, Pastan, & Lee, 2009), a tumor differentiation antigen that is normally expressed by mesothelial cells and may be involved in cell adhesion (Chang & Pastan, 1996; Hassan, Bera, & Pastan, 2004). Attachment of the TM to the OHCs may also involve Np55, a splice variant of neuroplastin that is only expressed in OHCs (Zeng et al., 2016). The ligands for OTOA and STRC in the TM are unknown, although there is evidence that TECTA is a potential binding partner for OTOA. First, the TM remains attached to the spiral limbus in $Tectb^{-/-}$, $Otog^{-/-}$, and $Ceacam16^{\beta gal/\beta gal}$ mice. Second, the vestigial, purely collagen-based TM seen in $Tecta^{\Delta ENT/\Delta ENT}$ mice lacking functional TECTA, is completely detached from the surface of the organ of Corti and is found adhering, instead, to Reissner's membrane, the epithelium that separates scala vestibuli from scala media.

4. STRUCTURE/ULTRASTRUCTURE OF THE MAMMALIAN TM

The core of the mammalian TM (see Fig. 3A) contains bundles of 20 nm diameter collagen fibrils that are imbedded in a collagenase-resistant, trypsin-sensitive matrix that is composed of sheets formed by two types of fine-diameter (~7 nm) filament, a light- and a dark-staining type that are aligned alternately in parallel and are linked to one another by staggered cross bridges (Hasko & Richardson, 1988). The structure of this striated-sheet matrix is dependent on calcium; in TMs that are fixed following calcium chelation with EGTA, a procedure known to cause swelling of the TM in vitro (Kronester-Frei, 1979; Shah, Freeman, & Weiss, 1995), the filaments are dispersed and striated-sheet matrix is no longer observed (Hasko & Richardson, 1988). If the calcium is replaced following its chelation, the membrane shrinks and the distinctive striated sheets are again visible following fixation and examination in the electron microscope. Although at least two morphological studies have provided evidence that the collagen fibrils interact physically with the noncollagenous filaments of the TM (Andrade, Salles, Grati, Manor, & Kachar, 2016; Tsuprun & Santi, 1997), the protein(s) involved in these interactions remain to be identified.

The striated-sheet matrix merges with a number of distinctive features around the perimeter of the TM (see Fig. 3B–E). These include the covernet

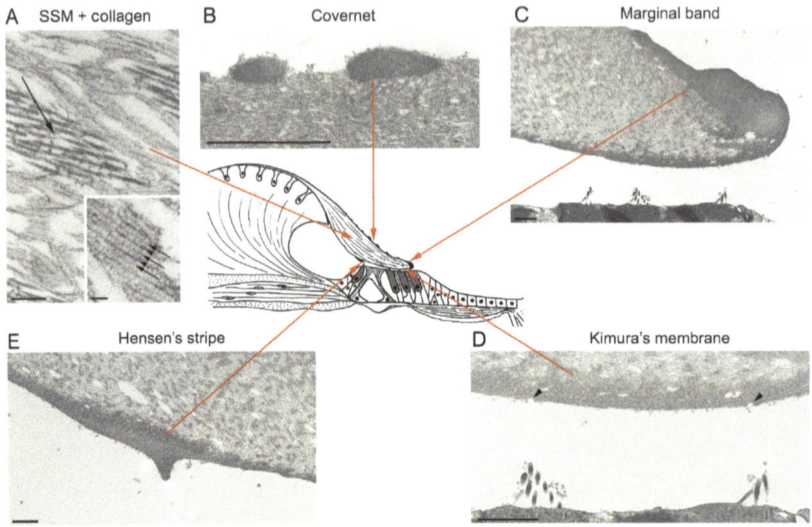

Fig. 3 Core (A) and peripheral (B–E) features of the tectorial membrane in the mouse cochlea. (A) The central core contains radially oriented collagen fibrils (*large arrow*) that are imbedded in striated-sheet matrix (SSM, see *inset*) that is composed of light (*small arrows*) and dark (*small arrowheads*) filaments linked by staggered cross bridges. (B) Covernet fibrils, (C) marginal band, (D) Kimura's membrane, and (E) Hensen's stripe. *Arrowheads* in (D) indicate the attachment sites (imprints) of the tips of the tallest outer hair cell stereocilia. Scale bars = 200 nm in (A), 50 nm in *inset* to (A), 2 μm in (B–E).

fibrils, an anastomosing network of large caliber fibrils that run predominantly along the length of the "upper" surface (Fig. 3B); the marginal band that runs along the lateral edge (shown in cross section in Fig. 3C); Kimura's membrane, a thickening of the "lower" surface into which the tallest stereocilia of the OHCs are imbedded (Fig. 3D); and Hensen's stripe, a V-shaped ridge that runs along much of the length of the TM and lies in close proximity to the hair bundles of the IHCs (Fig. 3E). For the most part these peripheral structures are electron dense, and it is therefore difficult to determine if they are composed of tightly packed filaments and/or granular material. They all react with antibodies directed against TECTA, TECTB, and CEACAM16, but the covernet fibrils, marginal band, and Kimura's membrane are all present in both the *Ceacam16*$^{\beta gal/\beta gal}$ and the *Tectb*$^{-/-}$-null mutant mice (Cheatham et al., 2014; Russell et al., 2007). Hensen's stripe is, however, absent in *Ceacam16*$^{\beta gal/\beta gal}$ and *Tectb*$^{-/-}$ mice (Cheatham et al., 2014; Russell et al., 2007). On the other hand, missense mutations in *Tecta* that cause deafness in humans disrupt the structure of the covernet fibrils, the marginal band, and Hensen's stripe (Legan et al., 2014, 2005; Xia et al., 2010). Thus far, it is not clear what accounts for the structural differences observed between the central and various peripheral noncollagenous features of the TM. There may be variations in the ratios of the different proteins in each region, or local differences in how the constituent proteins are glycosylated or posttranslationally modified.

Evidence from mice with null or functional null mutations in genes encoding TM proteins indicates TECTA, TECTB, and CEACAM16 are all required for formation of the striated-sheet matrix (Cheatham et al., 2014; Legan et al., 2000; Russell et al., 2007). Striated-sheet matrix is of normal appearance in the *Otog*$^{-/-}$ mouse (Simmler et al., 2000), and the ultrastructure of the TM in an *Otogl*-null mutant mouse has yet to be described. TECTA and TECTB both contain a ZP domain (Fig. 2B), a protein polymerization module that is found in many proteins (Jovine, Darie, Litscher, & Wassarman, 2005; Jovine, Qi, Williams, Litscher, & Wassarman, 2002) and may therefore form the filaments of the striated-sheet matrix with CEACAM16 providing cross-linking via its two Ig variable-like domains (Fig. 4A). Whether TECTA and TECTB are each independently able to form homomeric filaments, or if TECTA and TECTB associate to form a single type of heteromeric filament remains to be resolved. However, the lack of a structured interlinker region between the N- and the C-regions of the TECTB ZP domain, and the absence of noncollagenous filaments in the vestigial TM of the *Tecta*$^{\Delta ENT/\Delta ENT}$ mouse suggest TECTB is unlikely to be able to form filaments in the absence of TECTA (Bokhove et al., 2016).

In an alternative molecular model for the striated-sheet matrix of the TM (Kammerer et al., 2012), it has been suggested that CEACAM16 homodimers, stabilized by intermolecular disulfide bridges between the B and the N2 domains, interact homotypically to form the dark zone (i.e., the dark filaments) of the striated-sheet matrix, and that TECTA and TECTB interact heterophylically via their ZP domains to form the light zone (i.e., the light filaments and the staggered cross bridges), with the CEACAM16 homopolymer interacting via its paired N1 domains with the NIDO domain of TECTA (Fig. 4B). While an interesting model there is, as yet, little or no direct evidence for such a structure. Furthermore, the model does not take into account in vitro evidence showing CEACAM16 interacts with both TECTA and TECTB (Cheatham et al., 2014).

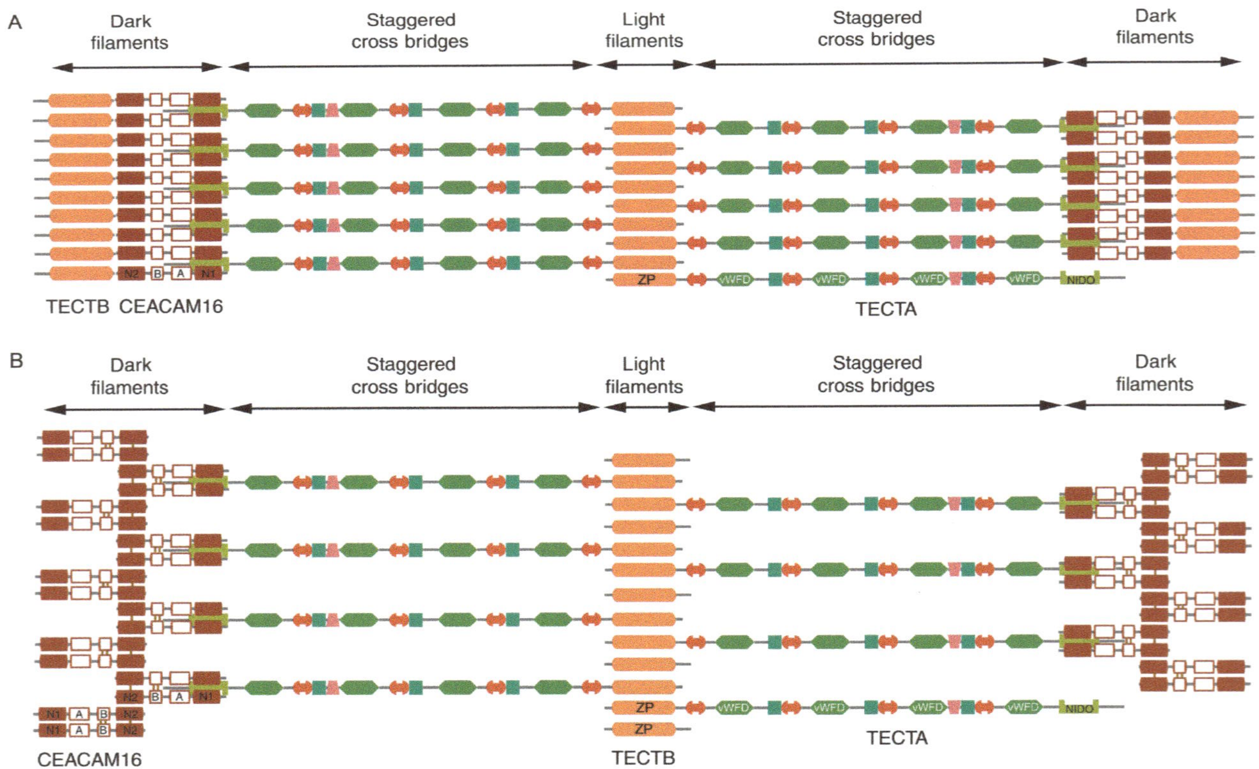

Fig. 4 Hypothetical molecular models for the structure of the striated-sheet matrix. (A) Model in which TECTA and TECTB each form homomeric filaments that are cross linked by CEACAM16. (B) Model proposed by Kammerer et al. (2012) in which TECTA and TECTB interact via their ZP domains to form a heteropolymeric filament.

5. FUNCTION OF THE MAMMALIAN TM

The TM has, over the years, been ascribed a number of roles in hearing. Davis suggested it acted as a rigid plate that slides back-and-forth across the surface of the organ of Corti as the BM vibrates in the transverse direction during acoustic stimulation (Davis, 1965). Later studies suggested it acted as a second resonator, a structure that could undergo compression and expansion in the radial plane and thereby amplify the input to the hair bundles (Neely & Kim, 1986; Zwislocki & Kletsky, 1979), or as an inertial mass, influencing the resonant properties of the hair bundles (Mammano & Nobili, 1993).

Data from ex vivo preparations of the guinea pig cochlea and mutant mice have provided experimental evidence that the TM acts as an inertial mass that can influence the timing and therefore the gain of the amplification provided by the OHCs (Gummer, Hemmert, & Zenner, 1996; Legan et al., 2000). Similar experimental approaches have also shown that the TM can modulate the fluid flow in the subtectorial space, and that it is required to optimize the input to the hair bundles of the IHCs at their best (or characteristic) frequency (Legan et al., 2005; Nowotny & Gummer, 2011).

The radial arrangement of the collagen fibrils generates mechanical anisotropy in the TM, with the stiffness of the TM being greater in the radial as opposed to the longitudinal direction (Abnet & Freeman, 2000; Freeman, Abnet, Hemmert, Tsai, & Weiss, 2003; Gavara & Chadwick, 2009; Richter, Emadi, Getnick, Quesnel, & Dallos, 2007). This property is thought to be important for ensuring that the TM faithfully transmits vibrations between hair bundles in neighboring rows of OHCS, a characteristic that would enhance cochlear amplification by a hair-bundle-based mechanism (Gavara & Chadwick, 2009). In support of this hypothesis, studies in mice lacking the *Col11a2* gene have shown that the density of the radial collagen fibrils in the TM is reduced, that the ratio of the radial to longitudinal shear impedance is decreased by ~50%, and that there is a 30–50 dB increase in the thresholds for both the distortion product otoacoustic emissions and the auditory brain stem responses (Masaki et al., 2009).

In vitro preparations have shown that the TM can propagate radial vibrations as traveling waves along its length, with wavelengths and decay constants that vary according to location and depend upon the porosity and viscoelastic properties of the matrix (Ghaffari, Aranyosi, & Freeman, 2007; Sellon, Ghaffari, Farrahi, Richardson, & Freeman, 2014). In mice lacking TECTB, in which the striated-sheet matrix of the TM fails to form, the sharpness of cochlear tuning determined in vivo is increased without a significant decrease in sensitivity (Russell et al., 2007), and the extent and wavelength of the TM traveling wave measured in vitro are reduced (Ghaffari, Aranyosi, Richardson, & Freeman, 2010). The TM is therefore thought to provide longitudinal coupling within the organ of Corti.

Finally it has been shown that the TM is electrokinetic and is able to move in response to electric potentials of a magnitude similar to those produced by hair cells during mechanotransduction (Ghaffari, Page, Farrahi, Sellon, & Freeman, 2013). This property may also provide a means whereby the TM can contribute to cochlear amplification. Overall these studies have led to the suggestion that the TM may be a multitasking matrix, a structure that serves more than one single function and is indispensable for normal hearing (Lukashkin, Richardson, & Russell, 2010).

6. AVIAN BASILAR PAPILLA AND TM

The hearing organs (basilar papillae, see Fig. 5A) of birds and reptiles are generally shorter than those in mammals, encode a narrower range of frequencies, and have multiple hair cells arrayed across their width. In birds two types of hair cell can be distinguished, tall hair cells lying on the neural (medial) side of the papilla that receive predominantly afferent innervation and short hair cells that are situated on the abneural (lateral) side and are innervated (sometimes exclusively) by efferent fibers. While most of the short hair cells lie over the BM, the tall hair cells sit above the superior cartilaginous plate. The TM of the avian basilar papilla has a wedge-shaped cross-sectional profile. It is attached along its shortest edge to the F-actin-rich homogene cells in the medial wall of the cochlear duct, and along its much broader lower surface to both the hair cells and the supporting cells of the basilar papilla. The tip of each hair bundle is attached to the main body of the TM, and the TM is attached to the surfaces of the supporting cells by the fibrous veils (Goodyear, Holley, & Richardson, 1994), a loosely packed collection of fine filaments (Fig. 5B and C). As such, each hair bundle sits within a dome-shaped indentation within the lower surface of the TM (Fig. 5C).

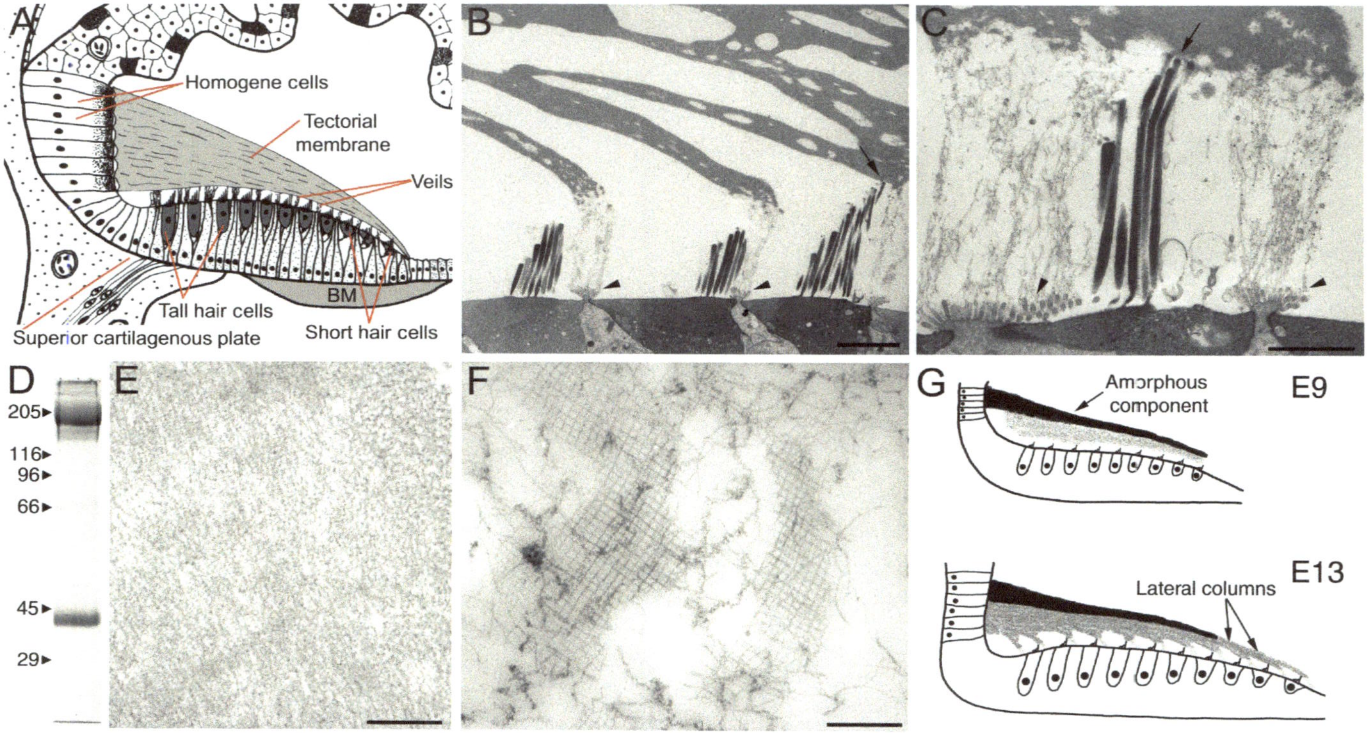

Fig. 5 (A) Diagram illustrating the structure of the avian basilar papilla. BM = basilar membrane. (B and C) Transmission electron micrographs illustrating the structure of the TM, and its attachment to both the hair cell bundles (*arrows*) and the microvilli (*arrowheads*) on the apical surfaces of the supporting cells. Detail of the fibrous veils is shown in (C). (D) Coomassie Brilliant Blue-stained 8.25% polyacrylamide gel of chicken tectorial membrane proteins run under nonreducing conditions. (E and F) Transmission electron micrographs of chicken tectorial membrane fixed before (E) and after (F) incubation in saline containing 1 mM EGTA and 1 mM EDTA for 1 h. (G) Diagrams illustrating formation of the lateral columns of the avian TM. The amorphous component of the TM does not grow in width, but the underlying epithelium does, leading to growth of the lateral columns. Scale bars = 5 μm (B), 1 μm (C), 200 nm (E and F). *Part G is adapted from Shiel, M. J., & Cotanche, D. A. (1990). SEM analysis of the developing tectorial membrane in the chick cochlea.* Hearing Research, *47, 147–157.*

7. PROTEINS OF THE AVIAN TM

Collagens are not present in the avian TM, and the protein composition of this matrix is remarkably simple (Killick, Malenczak, & Richardson, 1992). Two major protein bands with masses of 190 and 43 kDa are visible when chicken TM proteins are analyzed by SDS-PAGE under nonreducing conditions which are, respectively, avian orthologs of TECTA and TECTB (Fig. 5D). While OTOG and OTOGL have not been formally identified as components of the avian TM, both are present in proteomic data sets derived from purified chick utricular hair bundles (Wilmarth et al., 2015) and are presumably derived from the otoconial membrane that overlies the utricular macula. As OTOG and OTOGL are present in both the otoconial and the TMs of the mouse ear, they are also likely to be present in the TM of the bird.

8. STRUCTURE AND ATTACHMENT OF THE AVIAN TM

When the surface of the avian TM is viewed in the scanning electron microscope, two zones can be distinguished, a thin amorphous zone lying adjacent to the homogene cells and a much wider columnar zone covering the lateral area of the papilla (Shiel & Cotanche, 1990). In the columnar zone, large-caliber, coaligned columns of matrix are observed running in the medial-to-lateral direction across the papilla. In thin sections of chemically fixed tissues, the main body of the avian TM appears as a vacuolated, electron-dense fibrogranular matrix within which it is difficult, as for the peripheral features of the mouse TM, to resolve any substructure (Fig. 5E). Likewise, the TM of the chick swells in response to calcium chelation (Freeman, Cotanche, Ehsani, & Weiss, 1994) and in samples that have been fixed for electron microscopy following exposure to divalent cation chelators, fine-diameter (5 nm) filaments can be observed that are often organized into a distinctive square net (Fig. 5F).

9. FUNCTION OF THE AVIAN TM

While the tall and short hair cells in the avian basilar papilla may be considered equivalent, respectively, to the sensory IHCs and the sensorimotor OHCs of the mammalian cochlea, it is unclear whether the short hair cells feed energy back into BM motion. The mechanical response of the basilar papilla in the chick is broadly tuned (Xia, Liu, Raphael, Applegate, & Oghalai, 2016), and frequency tuning in the basilar papilla of birds, as in turtles, may be determined to a large extent by an electrical resonance in the basolateral membranes of the hair cells that amplifies the receptor potentials generated by hair-bundle motion (Fuchs, Nagai, & Evans, 1988). Nonetheless there is evidence that the short hair cells may employ a prestin-based mechanism to produce fast lateral displacement of the TM that could potentially drive the hair bundles of the tall hair cells (Beurg, Tan, & Fettiplace, 2013).

10. CHICK TM DEVELOPMENT

Early studies of TM development in the avian BP using light and transmission electron microscopy indicated the TM begins to form at E7 (i.e., just after the hair cells start to differentiate) and suggested that the majority of the TM was produced by the homogene cells (Cohen-Salmon et al., 1997). A more detailed analysis using SEM and video-enhanced DIC imaging also indicated that the TM was first visible at E7, but concluded that the laterally oriented columns, which can be first distinguished at E9, are produced by the supporting cells within the BP and that the amorphous zone, which is also clearly defined by E9, was produced by the homogene cells (Shiel & Cotanche, 1990). Later in situ hybridization studies, however, have indicated that mRNAs for TECTA and TECTB (Coutinho, Goodyear, Legan, & Richardson, 1999; Goodyear, Killick, Legan, & Richardson, 1996) are not expressed by the homogene cells at any point during development of the avian hearing organ (Coutinho et al., 1999). Possibly these cells express other components of the avian TM (e.g., OTOG and/or OTOGL), or they may simply provide a structure to which the TM can attach, presumably via OTOA which is known to be expressed in the chick inner ear (Wilmarth et al., 2015).

As with the collagen fibrils in the mammalian TM (see below) one can ask what determines the orientation of the lateral columns in the columnar zone of the bird TM. Shiel and Cotanche (Shiel & Cotanche, 1990) speculated that these lateral columns are attached to the medial amorphous zone and that as this zone does not increase much in width over time while the epithelium does, tissue growth is therefore driving the elongation of the columns (see Fig. 5G). While this is certainly plausible, the lateral columns are, on their first appearance at E9, already oriented in a lateral fashion, suggesting other factors are at play.

11. MAMMALIAN TM DEVELOPMENT

The mammalian organ of Corti develops within the dorsal wall of the elongating cochlear duct, a structure composed of polarized, pseudostratified epithelial cells that first emerges from the ventromedial pole of the otocyst at ~E12 in mice and attains its characteristic coiled configuration, albeit not full length, by E15 (Morsli, Choo, Ryan, Johnson, & Wu, 1998). Two regions can be distinguished within the dorsal wall of the cochlear duct from which the organ of Corti develops, the greater and the lesser epithelial ridges (GER and LER, respectively, see Fig. 6 E14.5), regions which differ initially in

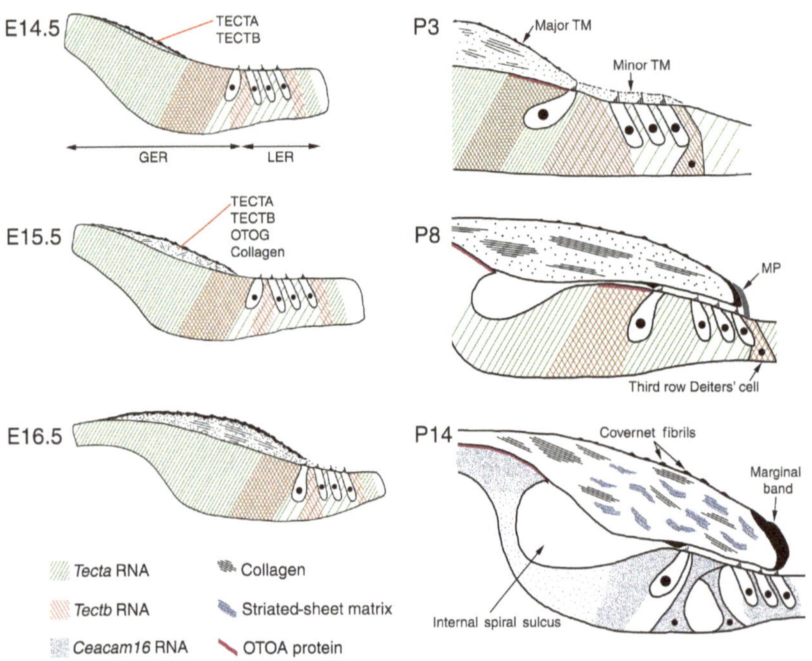

Fig. 6 Diagram summarizing the key stages in the development of the mouse tectorial membrane, expression patterns of *Tecta*, *Tectb*, and *Ceacam16* (hatching density corresponds to relative expression levels), and the distribution of OTOA. (E14.5) Primordial, tectorin-based TM on the surface of the GER. (E15.5) Appearance of OTOG and randomly oriented collagen fibrils. (E16.5) Collagen fibrils increase in density, become coaligned and radially oriented. (P3) The minor TM is forming over the OHCs in the LER. Transient OTOA expression begins on the surface of cells that lie immediately medial to IHCs and express high levels of *Tectb*. (P8) Marginal pillars (MP) are present, the minor TM is detaching from the apical surfaces of the supporting cells, and hair bundles are consolidating stereocilin (STRC)-mediated attachments with Kimura's membrane. (P14) Appearance of striated-sheet matrix within the central core of the TM, Ceacam16 is expressed in supporting cells throughout the organ of Corti, marginal pillars are no longer present.

their thickness. The IHCs with their associated supporting cells and those lying more medial (i.e., the cells of the spiral sulcus and spiral limbus) originate from the GER, while the outer hairs, Dieters' cells, Hensen's cells, and the cells of Claudius and Boettcher originate from the LER.

Many of the key events occurring during TM development were initially documented using a combination of scanning and transmission electron microscopy (Lenoir, Puel, & Pujol, 1987; Lim, 1987; Lim & Anniko, 1985; Rueda, Cantos, & Lim, 1996). These studies described the TM as developing in two parts with the major TM, that contributing to the bulk of the structure, first forming over the GER at E14.5, and the minor TM, that contributing to much of the matrix that lies over the OHCs and their supporting cells, being produced later on (from ~P0 onward) by the LER (see Fig. 6). At E16.5, the TM overlying the GER was described as being composed of amorphous and fibrillar substances, with the latter being identified as the type A protofibrils that were originally described by Kronester-Frei in 1976 and are now known to be collagen fibrils.

12. COLLAGEN FIBRILS: ORIENTATION AND ALIGNMENT

More recent studies have provided further details on the sequence of events occurring during the early stages of TM development (Fig. 6) and have addressed how the collagen fibrils of the TM become coaligned and oriented. In situ hybridization has shown that mRNAs encoding TECTA and TECTB are first expressed at E12.5, as soon as the cochlea emerges, and that expression is initially restricted to the GER (Rau, Legan, & Richardson, 1999). While OTOG protein can be detected at this stage, it has a far more widespread distribution and is found in the cytoplasm of cells that form both the ventral and dorsal walls of the duct (El-Amraoui, Cohen-Salmon, Petit, & Simmler, 2001). The TM that is present on the surface of the GER at E14.5 contains TECTA and TECTB, is thin (~5 μm thick), and is comprised of a lower layer of loosely packed filaments situated immediately adjacent to the epithelium and an upper layer of densely packed filaments that are organized as cords running along the length of the cochlea (Goodyear, Lu, Deans, & Richardson, 2017). OTOG and collagen fibrils are not present in the nascent TM at this stage. Both appear within the TM at E15.5, but the collagen fibrils that form first are tangled and disorganized (Fig. 6 E15.5). Remarkably, within a period of just 24 h, the collagen fibrils become coaligned with one another and radially oriented across the width of the TM by E16.5, replete with the apical offset resembling that seen in mature mice (Fig. 7A).

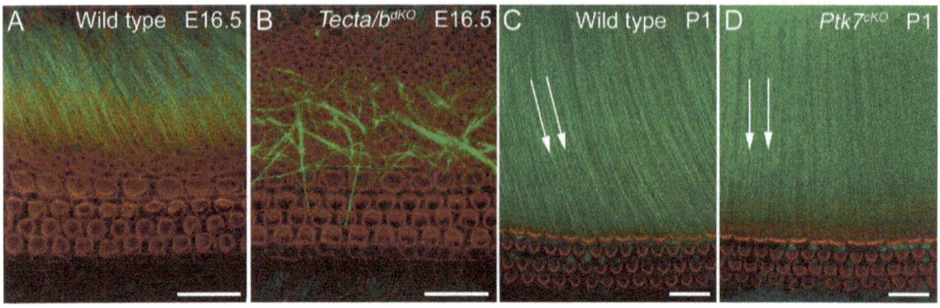

Fig. 7 Collagen fibrils in the tectorial membrane (*green*) with F-actin counterstaining in the underlying cochlear epithelium (*red*). (A) E16.5 wild type, (B) E16.5 *Tecta/b*^{dKO}, (C) P1 wild type, (D) P1 *Ptk7*^{cKO}. A tectorin-based matrix and planar cell polarity signaling are required for normal alignment and orientation of collagen fibrils (*arrows*). Scale bars = 20 μm.

In vitro studies with acid-soluble type I collagen have shown fluid flow (Lee, Lin, Moon, & Lee, 2006), imposed strain (Vader, Kabla, Weitz, & Mahadevan, 2009), and molecular crowding/confinement (Saeidi et al., 2012) can all cause the alignment of collagen fibrils as they polymerize. As the alignment and orientation of collagen fibrils in the developing TM occur in the absence of epithelial cilia (which may or may not be motile), and as it occurs over a 24-h period when the width of the epithelium does not increase to a greater extent than its length, it has been argued that fluid flow and the forces imposed by tissue growth are unlikely to be responsible for determining collagen fibril orientation in the TM (Goodyear et al., 2017). The process occurs, however, while collagen fibril density is increasing rapidly within the narrow space lying between the surface of the GER and the electron-dense cords forming the upper layer of the developing TM. It also fails in *Tecta/b*^{dKO} mice that are doubly homozygous for null mutations in *Tecta* and *Tectb* and lack any confining tectorin-based matrix, with the collagen fibrils forming large spiky structures that are randomly oriented (Fig. 7A and B; Goodyear et al., 2017). On the basis of these observations it has therefore been suggested that molecular crowding plays an important role in determining collagen patterning in the TM (Goodyear et al., 2017).

Confinement does not explain, though, why the fibrils are all oriented across the width, rather than the length, of the TM, or why the fibrils have a distinctive offset toward the apical end of the cochlea. While collagen fibril orientation may not depend on stretch produced by growth of the underlying epithelium, the planar cell polarity protein PRICKLE2 is asymmetrically distributed at the cell–cell junctions of the cells in the GER that lie immediately beneath the middle of the TM (Goodyear et al., 2017) in a fashion that correlates strikingly with the apical offset of the collagen fibrils (Copley, Duncan, Liu, Cheng, & Deans, 2013; Lim, 1972). Furthermore, in conditional knockout mice lacking the planar cell polarity (PCP) genes *Vangl2* and *Ptk7*, in which PRICKLE2 is no longer asymmetrically distributed in the cells of the GER, the apical offset is no longer observed and the collagen fibrils have a strictly radial orientation (Fig. 7C and D; Goodyear et al., 2017). The epithelium can, therefore, via the PCP pathway fine tune the patterning of collagen in the TM. The details of how information is transferred from the PCP pathway in the epithelium to the overlying matrix is, however, a puzzle that remains to be resolved.

13. HENSEN'S STRIPE AND THE MINOR TM

The early studies of David Lim and colleagues also documented the formation of the minor TM, Hensen's stripe, and the transient presence of matrix structures known as marginal pillars during the early postnatal stages of development, P0–P14 (see Fig. 6 P8). In the cat, Hensen's stripe develops from a thin strip of amorphous matrix that lies between the major TM and the apical surfaces of the cells that lie just medial to the IHCs in the GER (Lim, 1987). As described earlier, Hensen's stripe fails to form in a number of mice with null or missense mutation in genes encoding proteins associated with the TM, including the *Tectb*$^{-/-}$ mouse (Russell et al., 2007). The exact stage at which a definitive Hensen's stripe is present in the mouse remains to be determined, but judging from published electron micrographs it appears between P6 and P12 (Rueda et al., 1996). Although OTOA is restricted to the apical surfaces of the interdental cells in mature animals and is known to be required for the attachment of the TM to the spiral limbus (Lukashkin et al., 2012), it also expressed during a brief window of developmental time (from P3 to P10) on the apical membranes of the cells in the GER that lie just medial to the IHCs (Fig. 8 P3, P8; Zwaenepoel et al., 2002), a location immediately adjacent to the amorphous matrix from which Hensen's stripe is thought to form (Lim, 1987). Furthermore, Hensen's stripe is absent in the *Otoa*$^{EGFP/EGFP}$ mice (Lukashkin et al., 2012), suggesting attachment of the TM to the developing inner border cells is required for sculpting this feature of the TM while the more medial central region of the GER recedes to form the inner sulcus.

The minor TM also fails to develop properly in the *Otoa*$^{EGFP/EGFP}$ mice; collagen fibrils, striated-sheet matrix, and covernet fibrils are absent from this region, Kimura's membrane is fenestrated, and the marginal band has an

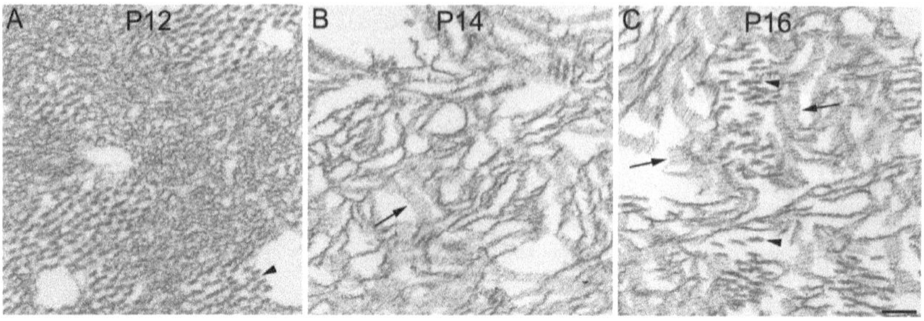

Fig. 8 Transmission electron micrographs illustrating the structure of the matrix in the core of the mouse TM at P12, P14, and P16. Striated-sheet matrix appears between P12 and P14, 2 days after the onset of *Ceacam16* expression at P12. *Arrows* indicate striated-sheet matrix, *arrowheads* indicate collagen fibrils. Scale bar = 200 nm.

abnormal shape (Lukashkin et al., 2012). As OTOA is not expressed in the supporting cell types that form from the LER (i.e., the pillar cells, Dieters' cells, and Hensen's cells), it is unclear, as yet, why the minor TM fails to form correctly in the $Otoa^{EGFP/EGFP}$ mice. *Tectb* is also expressed at high levels in the region of the GER adjacent to the IHCs at P8 (Rau et al., 1999), and the attachment of the developing TM to these cells mediated by OTOA may be required to ensure sufficient protein can be incorporated and allow the minor TM and Hensen's stripe to develop normally.

14. MARGINAL PILLARS

As the TM overlying the inner and IHCs and OHCs develops, it becomes detached from the cells that produce it not only from the border cells in the lateral region of the GER but also from the surfaces of the Dieters' cells and the pillar cells. While it is not known how the minor TM transiently adheres to the supporting cells in the immediate vicinity of the OHCs, large hook-like matrix structures known as marginal pillars emerge (rather like toothpaste from a tube) from the apical surface of the outermost (third) row of Dieters' cells during the development of this region of the TM (Fig. 6 P8). These remarkable structures, which are observed in mice by P6 and then disappear around the onset of hearing at P14, are thought to hold the minor TM in place while it is detaching from the surfaces of both the pillar cells and the first and second rows of Dieters' cells, and as connections between the TM and the hair bundles of the OHCs are being consolidated (Lenoir et al., 1987; Lim, 1987; Rueda et al., 1996). Surprisingly, the molecular composition of the marginal pillars remains to be identified.

15. CROWNS, IMPRINTS, STRC, AND HAIR-BUNDLE ATTACHMENT

Attachment of the OHC hair bundle to the TM is mediated by TM attachment crowns, arrays of electron-dense particles that are located only around the distal tips of the tallest stereocilia (Goodyear, Marcotti, Kros, & Richardson, 2005; Tsuprun & Santi, 2002) and engage with the TM at the sites of the hair-bundle imprints (Fig. 3D), the V-shaped arrays of dimple-like indentations that are present in Kimura's membrane (Kimura, 1966). STRC is known to be required for the attachment of the hair bundles of the OHCs to the TM, and both the attachment crowns and the imprints are absent in mature *Strc*$^{-/-}$ mice (Verpy et al., 2011). Although immature attachment crowns are found at the tips of all three rows of stereocilia during early postnatal development, STRC, when it first appears in the stereocilia at P5, localizes uniquely to the distal tips of the tallest stereocilia, and not to the distal tips of those in the other rows (Verpy et al., 2011). STRC may therefore stabilize and organize the TM attachment crowns at the very tips of the tallest stereocilia, allowing the hair bundles to attach to the minor TM while the filamentous meshwork linking this region to the surfaces of the pillar cells and Dieters' cells is beginning to recede.

16. STRIATED-SHEET MATRIX

Surprisingly, and despite its prominence in the mature TM, the striated-sheet matrix is one of the last structures to appear, materializing at ~P14 in the mouse (Fig. 8) from a denser fibrogranular matrix that pervades throughout the core of the TM prior to this stage (Cheatham et al., 2014). The appearance of striated-sheet matrix lags the onset of *Ceacam16* expression in the nonsensory cells of the organ of Corti and the appearance of CEACAM16 protein in the TM by about 2 days, and it occurs around the onset of hearing (Cheatham et al., 2014). As yet it is not known how CEACAM16 causes the observed structural transformation, or whether additional processing of the polymerized tectorin-based matrix is required to initiate such a change. There is evidence that TECTA is cleaved to produce the three disulfide cross-linked subunits described earlier at a stage between P3 and P10 (G. Richardson, unpublished observations). The exact time at which this occurs and the identity of the protease involved are not yet known, but it is possible that this additional cleavage step is required for the subsequent formation of the characteristic striated-sheet matrix of the TM.

17. EVOLUTION OF THE TM

Clearly, and as discussed earlier, the presence of collagen is one feature that distinguishes the mammalian TM from that of birds and probably also from that of reptiles. The appearance of radially oriented collagen fibrils parallels an increase in length of the TM and accompanies the greater range of frequencies that can be encoded by the cochlea; the collagen fibrils may also help provide the mechanical properties required to amplify BM motion and drive the IHCs efficiently. Type II collagen is produced by many epithelial structures during early development and is secreted basolaterally, localizing to the epithelial–mesenchymal interface where it acts as a morphogenetic signal (Wood, Ashhurst, Corbett, & Thorogood, 1991). Although normally secreted basally, a random change in collagen targeting within the developing GER may have resulted in it being secreted both apically and basally and this could have provided an advantage for hearing that was selected for during evolution. As yet it is not known how the collagens of the TM are directed to the apical surface of the cochlear epithelium during development, and this would be an interesting point to pursue.

A second distinguishing feature of the mammalian TM is CEACAM16. Although data are limited, the presence of this mammalian-specific member of the CEA subfamily of the immunoglobulin superfamily in the mouse TM correlates with existence of striated-sheet matrix, a structure that confers frequency-dependent stiffness properties (Jones, Elliott, Russell, & Lukashkin, 2015). This structure has not been observed in the TM of the chicken and is absent from the TM of the *Ceacam16*$^{\beta gal/\beta gal}$-null mutant mouse (Cheatham et al., 2014). Clearly the ultrastructure of the TM needs to be examined carefully in more species, both in other birds and mammals as well as in reptiles and monotremes. The latter would be particularly interesting to study as *Ceacam16* is present in the genome of the platypus, and yet the hearing organs in this egg-laying mammal are short, have a fairly narrow frequency range, and resemble partially those of birds and reptiles with multiple rows of both IHCs and OHCs (Ladhams & Pickles, 1996; Manley, 2012). Although comparative ultrastructural studies are limited thus far to mice and chickens, there is a good correlation in the mouse between the onset of Ceacam16 expression in the developing cochlear epithelium, the time at which CEACAM16 is detected in the TM, and the appearance of striated-sheet matrix within the central core of the TM.

While CEACAM16 and striated-sheet matrix may be unique to the mouse TM, the advantage this provides for hearing in mammals remains unclear. A study of a *Ceacam16*$^{-/-}$ mouse that was produced on the albino Balb/c background has shown hearing thresholds are elevated for sounds below 10 kHz and above 20 kHz in young mice, and then get progressively worse (increase) with age (Kammerer et al., 2012), a finding consistent with the observation that missense mutations in CEACAM16 cause late onset, progressive forms of deafness in humans (Hofrichter et al., 2015; Wang et al., 2015; Zheng et al., 2011). Possibly, with the detachment of the TM from the surfaces of the supporting cells that has occurred in mammals, the production of a soluble secreted protein such a CEACAM16 was required to stabilize and maintain the matrix of the TM for long periods. Although data from *Ceacam16*$^{\beta gal/\beta gal}$ mice on the pigmented C57Bl/6J background mouse indicate auditory thresholds are nearly normal in young mice despite the lack of striated-sheet matrix, the incidence of spontaneous otoacoustic emissions is vastly increased (Cheatham et al., 2014). CEACAM16 may therefore be required to both produce a matrix with a structure that can effectively dampen spontaneous emissions and reduce unwanted noise in the system. It may also stabilize the tectorin-based striated-sheet matrix of the TM as *Ceacam16* expression can be readily detected at >1 year of age in the mouse, while the expression levels of *Tecta* and *Tectb* decline to extremely low levels prior to the onset of hearing at P14 (Kammerer et al., 2012; Rau et al., 1999). Whether CEACAM16 is required for both the formation of the striated-sheet matrix and its continual maintenance once formed is, however, still an open question and one that would be interesting to address in the near future.

ACKNOWLEDGMENTS

The authors would like to thank Richard Osgood for his critical and helpful comments on the manuscript.

REFERENCES

Abnet, C. C., & Freeman, D. M. (2000). Deformations of the isolated mouse tectorial membrane produced by oscillatory forces. *Hearing Research*, *144*, 29–46.

Andrade, L. R., Salles, F. T., Grati, M., Manor, U., & Kachar, B. (2016). Tectorins crosslink type II collagen fibrils and connect the tectorial membrane to the spiral limbus. *Journal of Structural Biology*, *194*, 139–146.

Ashmore, J. (2008). Cochlear outer hair cell motility. *Physiological Reviews*, *88*, 173–210.

Beurg, M., Tan, X., & Fettiplace, R. (2013). A prestin motor in chicken auditory hair cells: Active force generation in a nonmammalian species. *Neuron*, *79*, 69–81.

Bokhove, M., Nishimura, K., Brunati, M., Han, L., de Sanctis, D., Rampoldi, L., et al. (2016). A structured interdomain linker directs self-polymerization of human uromodulin. *Proceedings of the National Academy of Sciences of the United States of America*, *113*, 1552–1557.

Brunati, M., Perucca, S., Han, L., Cattaneo, A., Consolato, F., Andolfo, A., et al. (2015). The serine protease hepsin mediates urinary secretion and polymerisation of zona pellucida domain protein uromodulin. *eLife*, *4*, e08887.

Chang, K., & Pastan, I. (1996). Molecular cloning of mesothelin, a differentiation antigen present on mesothelium, mesotheliomas, and ovarian cancers. *Proceedings of the National Academy of Sciences of the United States of America*, *93*, 136–140.

Cheatham, M. A., Goodyear, R. J., Homma, K., Legan, P. K., Korchagina, J., Naskar, S., et al. (2014). Loss of the tectorial membrane protein CEACAM16 enhances spontaneous, stimulus-frequency, and transiently evoked otoacoustic emissions. *The Journal of Neuroscience: The Official Journal of the Society for Neuroscience*, *34*, 10325–10338.

Cohen-Salmon, M., El-Amraoui, A., Leibovici, M., & Petit, C. (1997). Otogelin: A glycoprotein specific to the acellular membranes of the inner ear. *Proceedings of the National Academy of Sciences of the United States of America*, *94*, 14450–14455.

Copley, C. O., Duncan, J. S., Liu, C., Cheng, H., & Deans, M. R. (2013). Postnatal refinement of auditory hair cell planar polarity deficits occurs in the absence of Vangl2. *The Journal of Neuroscience: The Official Journal of the Society for Neuroscience*, *33*, 14001–14016.

Coutinho, P., Goodyear, R., Legan, P. K., & Richardson, G. P. (1999). Chick alpha-tectorin: Molecular cloning and expression during embryogenesis. *Hearing Research*, *130*, 62–74.

Dallos, P. (2008). Cochlear amplification, outer hair cells and prestin. *Current Opinion in Neurobiology*, *18*, 370–376.

Dallos, P., & Harris, D. (1978). Properties of auditory nerve responses in absence of outer hair cells. *Journal of Neurophysiology*, *41*, 365–383.

Davis, H. (1965). A model for transducer action in the cochlea. *Cold Spring Harbor Symposia on Quantitative Biology*, *30*, 181–190.

El-Amraoui, A., Cohen-Salmon, M., Petit, C., & Simmler, M. C. (2001). Spatiotemporal expression of otogelin in the developing and adult mouse inner ear. *Hearing Research*, *158*, 151–159.

Freeman, D. M., Abnet, C. C., Hemmert, W., Tsai, B. S., & Weiss, T. F. (2003). Dynamic material properties of the tectorial membrane: A summary. *Hearing Research*, *180*, 1–10.

Freeman, D. M., Cotanche, D. A., Ehsani, F., & Weiss, T. F. (1994). The osmotic response of the isolated tectorial membrane of the chick to isosmotic solutions: Effect of Na+, K+, and Ca2+ concentration. *Hearing Research*, *79*, 197–215.

Fuchs, P. A., Nagai, T., & Evans, M. G. (1988). Electrical tuning in hair cells isolated from the chick cochlea. *The Journal of Neuroscience: The Official Journal of the Society for Neuroscience*, *8*, 2460–2467.

Gavara, N., & Chadwick, R. S. (2009). Collagen-based mechanical anisotropy of the tectorial membrane: Implications for inter-row coupling of outer hair cell bundles. *PLoS One*, *4*, e4877.

Ghaffari, R., Aranyosi, A. J., & Freeman, D. M. (2007). Longitudinally propagating traveling waves of the mammalian tectorial membrane. *Proceedings of the National Academy of Sciences of the United States of America*, *104*, 16510–16515.

Ghaffari, R., Aranyosi, A. J., Richardson, G. P., & Freeman, D. M. (2010). Tectorial membrane travelling waves underlie abnormal hearing in Tectb mutant mice. *Nature Communications*, *1*, 96.

Ghaffari, R., Page, S. L., Farrahi, S., Sellon, J. B., & Freeman, D. M. (2013). Electrokinetic properties of the mammalian tectorial membrane. *Proceedings of the National Academy of Sciences of the United States of America*, *110*, 4279–4284.

Goodyear, R., Holley, M., & Richardson, G. (1994). Visualisation of domains in the avian tectorial and otolithic membranes with monoclonal antibodies. *Hearing Research*, *80*, 93–104.

Goodyear, R., Killick, R., Legan, P. K., & Richardson, G. P. (1996). Distribution of beta-tectorin mRNA in the early posthatch and developing avian inner ear. *Hearing Research*, *96*, 167–178.

Goodyear, R. J., Lu, X., Deans, M. R., & Richardson, G. P. (2017). A tectorin-based matrix and planar-cell-polarity genes are required for normal collagen-fibril orientation in the developing tectorial membrane. *Development*, *144*(21), 3978–3989.

Goodyear, R. J., Marcotti, W., Kros, C. J., & Richardson, G. P. (2005). Development and properties of stereociliary link types in hair cells of the mouse cochlea. *The Journal of Comparative Neurology*, *485*, 75–85.

Goodyear, R. J., & Richardson, G. P. (2002). Extracellular matrices associated with the apical surfaces of sensory epithelia in the inner ear: Molecular and structural diversity. *Journal of Neurobiology*, *53*, 212–227.

Guipponi, M., Tan, J., Cannon, P. Z., Donley, L., Crewther, P., Clarke, M., et al. (2007). Mice deficient for the type II transmembrane serine protease, TMPRSS1/hepsin, exhibit profound hearing loss. *The American Journal of Pathology*, *171*, 608–616.

Gummer, A. W., Hemmert, W., & Zenner, H. P. (1996). Resonant tectorial membrane motion in the inner ear: Its crucial role in frequency tuning. *Proceedings of the National Academy of Sciences of the United States of America*, *93*, 8727–8732.

Hasko, J. A., & Richardson, G. P. (1988). The ultrastructural organization and properties of the mouse tectorial membrane matrix. *Hearing Research*, *35*, 21–38.

Hassan, R., Bera, T., & Pastan, I. (2004). Mesothelin: A new target for immunotherapy. *Clinical Cancer Research: An Official Journal of the American Association for Cancer Research*, *10*, 3937–3942.

Hofrichter, M. A., Nanda, I., Graf, J., Schroder, J., Shehata-Dieler, W., Vona, B., et al. (2015). A novel de novo mutation in CEACAM16 associated with postlingual hearing impairment. *Molecular Syndromology*, *6*, 156–163.

Jones, G. P., Elliott, S. J., Russell, I. J., & Lukashkin, A. N. (2015). Modified protein expression in the tectorial membrane of the cochlea reveals roles for the striated sheet matrix. *Biophysical Journal*, *108*, 203–210.

Jovine, L., Darie, C. C., Litscher, E. S., & Wassarman, P. M. (2005). Zona pellucida domain proteins. *Annual Review of Biochemistry*, *74*, 83–114.

Jovine, L., Park, J., & Wassarman, P. M. (2002). Sequence similarity between stereocilin and otoancorin points to a unified mechanism for mechanotransduction in the mammalian inner ear. *BMC Cell Biology*, *3*, 28.

Jovine, L., Qi, H., Williams, Z., Litscher, E., & Wassarman, P. M. (2002). The ZP domain is a conserved module for polymerization of extracellular proteins. *Nature Cell Biology*, *4*, 457–461.

Kammerer, R., Ruttiger, L., Riesenberg, R., Schauble, C., Krupar, R., Kamp, A., et al. (2012). Loss of mammal-specific tectorial membrane component carcinoembryonic antigen cell adhesion molecule 16 (CEACAM16) leads to hearing impairment at low and high frequencies. *The Journal of Biological Chemistry*, *287*, 21584–21598.

Killick, R., Malenczak, C., & Richardson, G. P. (1992). The protein composition of the avian tectorial membrane. *Hearing Research*, *64*, 21–38.

Kimura, R. S. (1966). Hairs of the cochlear sensory cells and their attachment to the tectorial membrane. *Acta Oto-Laryngologica*, *61*, 55–72.

Kronester-Frei, A. (1979). The effect of changes in endolymphatic ion concentrations on the tectorial membrane. *Hearing Research*, *1*, 81–94.
Ladhams, A., & Pickles, J. O. (1996). Morphology of the monotreme organ of Corti and macula lagena. *The Journal of Comparative Neurology*, *366*, 335–347.
Lee, P., Lin, R., Moon, J., & Lee, L. P. (2006). Microfluidic alignment of collagen fibers for in vitro cell culture. *Biomedical Microdevices*, *8*, 35–41.
Legan, P. K., Goodyear, R. J., Morin, M., Mencia, A., Pollard, H., Olavarrieta, L., et al. (2014). Three deaf mice: Mouse models for TECTA-based human hereditary deafness reveal domain-specific structural phenotypes in the tectorial membrane. *Human Molecular Genetics*, *23*, 2551–2568.
Legan, P. K., Lukashkina, V. A., Goodyear, R. J., Kossi, M., Russell, I. J., & Richardson, G. P. (2000). A targeted deletion in alpha-tectorin reveals that the tectorial membrane is required for the gain and timing of cochlear feedback. *Neuron*, *28*, 273–285.
Legan, P. K., Lukashkina, V. A., Goodyear, R. J., Lukashkin, A. N., Verhoeven, K., Van Camp, G., et al. (2005). A deafness mutation isolates a second role for the tectorial membrane in hearing. *Nature Neuroscience*, *8*, 1035–1042.
Legan, P. K., Rau, A., Keen, J. N., & Richardson, G. P. (1997). The mouse tectorins. Modular matrix proteins of the inner ear homologous to components of the sperm-egg adhesion system. *The Journal of Biological Chemistry*, *272*, 8791–8801.
Legan, P. K., & Richardson, G. P. (1997). Extracellular matrix and cell adhesion molecules in the developing inner ear. *Seminars in Cell & Developmental Biology*, *8*, 217–224.
Lenoir, M., Puel, J. L., & Pujol, R. (1987). Stereocilia and tectorial membrane development in the rat cochlea. A SEM study. *Anatomy and Embryology*, *175*, 477–487.
Lim, D. J. (1972). Fine morphology of the tectorial membrane. Its relationship to the organ of Corti. *Archives of Otolaryngology*, *96*, 199–215.
Lim, D. J. (1987). Development of the tectorial membrane. *Hearing Research*, *28*, 9–21.
Lim, D. J., & Anniko, M. (1985). Developmental morphology of the mouse inner ear. A scanning electron microscopic observation. *Acta Oto-Laryngologica*, *422*, 1–69.
Lukashkin, A. N., Legan, P. K., Weddell, T. D., Lukashkina, V. A., Goodyear, R. J., Welstead, L. J., et al. (2012). A mouse model for human deafness DFNB22 reveals that hearing impairment is due to a loss of inner hair cell stimulation. *Proceedings of the National Academy of Sciences of the United States of America*, *109*, 19351–19356.
Lukashkin, A. N., Richardson, G. P., & Russell, I. J. (2010). Multiple roles for the tectorial membrane in the active cochlea. *Hearing Research*, *266*, 26–35.
Mammano, F., & Nobili, R. (1993). Biophysics of the cochlea: Linear approximation. *The Journal of the Acoustical Society of America*, *93*, 3320–3332.
Manley, G. A. (2012). Evolutionary paths to mammalian cochleae. *Journal of the Association for Research in Otolaryngology: JARO*, *13*, 733–743.
Masaki, K., Gu, J. W., Ghaffari, R., Chan, G., Smith, R. J., Freeman, D. M., et al. (2009). Col11a2 deletion reveals the molecular basis for tectorial membrane mechanical anisotropy. *Biophysical Journal*, *96*, 4717–4724.
Morsli, H., Choo, D., Ryan, A., Johnson, R., & Wu, D. K. (1998). Development of the mouse inner ear and origin of its sensory organs. *The Journal of Neuroscience: The Official Journal of the Society for Neuroscience*, *18*, 3327–3335.
Neely, S. T., & Kim, D. O. (1986). A model for active elements in cochlear biomechanics. *The Journal of the Acoustical Society of America*, *79*, 1472–1480.
Nowotny, M., & Gummer, A. W. (2011). Vibration responses of the organ of Corti and the tectorial membrane to electrical stimulation. *The Journal of the Acoustical Society of America*, *130*, 3852–3872.
Rau, A., Legan, P. K., & Richardson, G. P. (1999). Tectorin mRNA expression is spatially and temporally restricted during mouse inner ear development. *The Journal of Comparative Neurology*, *405*, 271–280.

Richardson, G. P., Russell, I. J., Duance, V. C., & Bailey, A. J. (1987). Polypeptide composition of the mammalian tectorial membrane. *Hearing Research, 25*, 45–60.

Richter, C. P., Emadi, G., Getnick, G., Quesnel, A., & Dallos, P. (2007). Tectorial membrane stiffness gradients. *Biophysical Journal, 93*, 2265–2276.

Rueda, J., Cantos, R., & Lim, D. J. (1996). Tectorial membrane-organ of Corti relationship during cochlear development. *Anatomy and Embryology, 194*, 501–514.

Russell, I. J., Legan, P. K., Lukashkina, V. A., Lukashkin, A. N., Goodyear, R. J., & Richardson, G. P. (2007). Sharpened cochlear tuning in a mouse with a genetically modified tectorial membrane. *Nature Neuroscience, 10*, 215–223.

Saeidi, N., Karmelek, K. P., Paten, J. A., Zareian, R., DiMasi, E., & Ruberti, J. W. (2012). Molecular crowding of collagen: A pathway to produce highly-organized collagenous structures. *Biomaterials, 33*, 7366–7374.

Sathyanarayana, B. K., Hahn, Y., Patankar, M. S., Pastan, I., & Lee, B. (2009). Mesothelin, stereocilin, and otoancorin are predicted to have superhelical structures with ARM-type repeats. *BMC Structural Biology, 9*, 1.

Sellon, J. B., Ghaffari, R., Farrahi, S., Richardson, G. P., & Freeman, D. M. (2014). Porosity controls spread of excitation in tectorial membrane traveling waves. *Biophysical Journal, 106*, 1406–1413.

Shah, D. M., Freeman, D. M., & Weiss, T. F. (1995). The osmotic response of the isolated, unfixed mouse tectorial membrane to isosmotic solutions: Effect of Na +, K +, and Ca2 + concentration. *Hearing Research, 87*, 187–207.

Shiel, M. J., & Cotanche, D. A. (1990). SEM analysis of the developing tectorial membrane in the chick cochlea. *Hearing Research, 47*, 147–157.

Simmler, M. C., Cohen-Salmon, M., El-Amraoui, A., Guillaud, L., Benichou, J. C., Petit, C., et al. (2000). Targeted disruption of otog results in deafness and severe imbalance. *Nature Genetics, 24*, 139–143.

Teudt, I. U., & Richter, C. P. (2014). Basilar membrane and tectorial membrane stiffness in the CBA/CaJ mouse. *Journal of the Association for Research in Otolaryngology: JARO, 15*, 675–694.

Thalmann, I., Thallinger, G., Crouch, E. C., Comegys, T. H., Barrett, N., & Thalmann, R. (1987). Composition and supramolecular organization of the tectorial membrane. *The Laryngoscope, 97*, 357–367.

Tsuprun, V., & Santi, P. (1997). Ultrastructural organization of proteoglycans and fibrillar matrix of the tectorial membrane. *Hearing Research, 110*, 107–118.

Tsuprun, V., & Santi, P. (2002). Structure of outer hair cell stereocilia side and attachment links in the chinchilla cochlea. *The Journal of Histochemistry and Cytochemistry: Official Journal of the Histochemistry Society, 50*, 493–502.

Vader, D., Kabla, A., Weitz, D., & Mahadevan, L. (2009). Strain-induced alignment in collagen gels. *PLoS One, 4*, e5902.

Verpy, E., Leibovici, M., Michalski, N., Goodyear, R. J., Houdon, C., Weil, D., et al. (2011). Stereocilin connects outer hair cell stereocilia to one another and to the tectorial membrane. *The Journal of Comparative Neurology, 519*, 194–210.

Verpy, E., Masmoudi, S., Zwaenepoel, I., Leibovici, M., Hutchin, T. P., Del Castillo, I., et al. (2001). Mutations in a new gene encoding a protein of the hair bundle cause non-syndromic deafness at the DFNB16 locus. *Nature Genetics, 29*, 345–349.

Wang, H., Wang, X., He, C., Li, H., Qing, J., Grati, M., et al. (2015). Exome sequencing identifies a novel CEACAM16 mutation associated with autosomal dominant non-syndromic hearing loss DFNA4B in a chinese family. *Journal of Human Genetics, 60*, 119–126.

Wilmarth, P. A., Krey, J. F., Shin, J. B., Choi, D., David, L. L., & Barr-Gillespie, P. G. (2015). Hair-bundle proteomes of avian and mammalian inner-ear utricles. *Scientific Data, 2*, 150074.

Wood, A., Ashhurst, D. E., Corbett, A., & Thorogood, P. (1991). The transient expression of type II collagen at tissue interfaces during mammalian craniofacial development. *Development*, *111*, 955–968.

Xia, A., Gao, S. S., Yuan, T., Osborn, A., Bress, A., Pfister, M., et al. (2010). Deficient forward transduction and enhanced reverse transduction in the alpha tectorin C1509G human hearing loss mutation. *Disease Models & Mechanisms*, *3*, 209–223.

Xia, A., Liu, X., Raphael, P. D., Applegate, B. E., & Oghalai, J. S. (2016). Hair cell force generation does not amplify or tune vibrations within the chicken basilar papilla. *Nature Communications*, 7, 13133.

Yariz, K. O., Duman, D., Seco, C. Z., Dallman, J., Huang, M., Peters, T. A., et al. (2012). Mutations in OTOGL, encoding the inner ear protein otogelin-like, cause moderate sensorineural hearing loss. *American Journal of Human Genetics*, *91*, 872–882.

Zebhauser, R., Kammerer, R., Eisenried, A., McLellan, A., Moore, T., & Zimmermann, W. (2005). Identification of a novel group of evolutionarily conserved members within the rapidly diverging murine Cea family. *Genomics*, *86*, 566–580.

Zeng, W. Z., Grillet, N., Dewey, J. B., Trouillet, A., Krey, J. F., Barr-Gillespie, P. G., et al. (2016). Neuroplastin isoform Np55 is expressed in the stereocilia of outer hair cells and required for normal outer hair cell function. *The Journal of Neuroscience: The Official Journal of the Society for Neuroscience*, *36*, 9201–9216.

Zheng, J., Miller, K. K., Yang, T., Hildebrand, M. S., Shearer, A. E., DeLuca, A. P., et al. (2011). Carcinoembryonic antigen-related cell adhesion molecule 16 interacts with alpha-tectorin and is mutated in autosomal dominant hearing loss (DFNA4). *Proceedings of the National Academy of Sciences of the United States of America*, *108*, 4218–4223.

Zwaenepoel, I., Mustapha, M., Leibovici, M., Verpy, E., Goodyear, R., Liu, X. Z., et al. (2002). Otoancorin, an inner ear protein restricted to the interface between the apical surface of sensory epithelia and their overlying acellular gels, is defective in autosomal recessive deafness DFNB22. *Proceedings of the National Academy of Sciences of the United States of America*, *99*, 6240–6245.

Zwislocki, J. J., & Kletsky, E. J. (1979). Tectorial membrane: A possible effect on frequency analysis in the cochlea. *Science*, *204*, 639–641.

CHAPTER SEVEN

Extracellular Matrix (ECM) and the Sculpting of Embryonic Tissues

Bette J. Dzamba, Douglas W. DeSimone[1]
Department of Cell Biology, University of Virginia, School of Medicine, Charlottesville, VA, United States
[1]Corresponding author: e-mail address: dwd3m@virginia.edu

Contents

Abstract

Extracellular matrices (ECMs) are structurally and compositionally diverse networks of collagenous and noncollagenous glycoproteins, glycosaminoglycans, proteoglycans, and associated molecules that together comprise the metazoan matrisome. Proper deposition and assembly of ECM is of profound importance to cell proliferation, survival, and differentiation, and the morphogenesis of tissues and organ systems that define sequential steps in the development of all animals. Importantly, it is now clear that the instructive influence of a particular ECM at various points in development reflects more than a simple summing of component parts; cellular responses also reflect the dynamic assembly and changing topology of embryonic ECM, which in turn affect its biomechanical properties. This review highlights recent advances in understanding how biophysical features attributed to ECM, such as stiffness and viscoelasticity, play

Current Topics in Developmental Biology, Volume 130
ISSN 0070-2153
https://doi.org/10.1016/bs.ctdb.2018.03.006

important roles in the sculpting of embryonic tissues and the regulation of cell fates. Forces generated within cells and tissues are transmitted both through integrin-based adhesions to ECM, and through cadherin-dependent cell–cell adhesions; the resulting short- and long-range deformations of embryonic tissues drive morphogenesis. This coordinate regulation of cell–ECM and cell–cell adhesive machinery has emerged as a common theme in a variety of developmental processes. In this review we consider select examples in the embryo where ECM is implicated in setting up tissue barriers and boundaries, in resisting pushing or pulling forces, or in constraining or promoting cell and tissue movement. We reflect on how each of these processes contribute to morphogenesis.

1. INTRODUCTION

Embryogenesis is defined by a sequential series of dynamic processes that include cell division and growth, and the elaboration of differentiation programs leading to cell fate specification. "Work" is also required to move cells around at multiple points in development in order to manifest a body plan, sculpt tissues and assemble organs and organ systems. This elaborate choreography of molecular, cellular, and tissue-level events is responsible for the formation of the new individual and, depending on both the species and particular process observed, can occur over remarkably short periods of time. One of the most dramatic yet rapid morphogenetic events in metazoan development is the process of gastrulation, which culminates in the formation of a multilayered (i.e., diploblastic or triploblastic) embryo from a multicelled blastula. Cells of the blastula undergo collective changes in position as a result of gastrulation movements often leading to new neighbor associations. These rearrangements are required not only to effect shape changes in the gastrula but also to enable subsequent inductive interactions between groups of cells and tissues leading to progressive changes in cell fate. Repositioning of tissues in gastrulating embryos typically involves one or more distinct types of deformations or movements that include the bending, splitting, or spreading of cell sheets, and the migration of cells as individuals or collective groups. Changes in the adhesion of cells to one another and to the extracellular matrix (ECM) must be orchestrated precisely throughout gastrulation to permit this reorganization while still maintaining levels of adhesion necessary to keep the embryo intact. Many of these basic features are repeated subsequently throughout development as more complex tissues and organs emerge as a consequence of differentiation and morphogenesis.

While this review focuses primarily on the "ECM side of this equation" (i.e., cell–ECM interactions), it is important to remember that cross talk between integrin and cadherin adhesive contacts integrates cellular inputs (Mui, Chen, & Assoian, 2016; Weber, Bjerke, & DeSimone, 2011) that help guide developmentally significant processes such as cell sorting (McMillen & Holley, 2015), cell polarity, and tissue morphology (Mao & Baum, 2015). How these topologically distinct cellular inputs are sensed and coordinated to instruct cell behaviors are important questions now being addressed in multiple laboratories. In some cases, the actin cytoskeleton may provide direct connections between cadherin- and integrin-based adhesions and, thus, serve to distribute and balance mechanical forces within and between cells (Maruthamuthu, Sabass, Schwarz, & Gardel, 2011). In other instances, reciprocal regulation of cell–ECM and cell–cell adhesions is indirect and occurs over longer periods of time. In epithelia undergoing branching morphogenesis, for example, local fibronectin deposition and assembly signal a reduction in E-cadherin expression that is responsible for "loosening" contacts between cells and promoting cleft formation at these sites (Onodera et al., 2010; Sakai, Larsen, & Yamada, 2003). Alternatively, ECM may serve to stabilize or actively restrict isotropic forces and in doing so, "constrain" and shape tissues and organs (e.g., Crest, Diz-Muñoz, Chen, Fletcher, & Bilder, 2017). Cadherins can also regulate the integrin-dependent deposition and assembly of ECM during development. Noncanonical Wnt signaling leads to increased cell–cell adhesion and tension in the ectoderm which, in turn, is responsible for proper spatial deposition of fibronectin fibrils at free cell surfaces (Dzamba, Jakab, Marsden, Schwartz, & DeSimone, 2009). Cadherin adhesion is also reported to stabilize adjacent integrins in a conformationally inactive state between mesoderm cells, thereby promoting fibronectin assembly at free cell surfaces where cadherin levels are reduced and integrins are able to adopt active conformational states (Jülich et al., 2015). Mechanisms of cadherin and integrin adhesive and signaling cross talk have been the subject of several reviews (e.g., Changede & Sheetz, 2017; Weber et al., 2011) and will not be covered in detail here; however, the importance of these interactions is highlighted where relevant.

The composition of ECM varies widely throughout development and is reflected in the assembly of specialized ECMs, including basal laminae, that accompany early embryogenesis, tissue formation, and organogenesis (reviewed in Rozario & DeSimone, 2010). From a developmental standpoint, therefore, it is important to consider both how ECM composition and assembly are regulated in the embryo, and how cells sense and respond to the assembled matrix that surrounds them. For example, the supramolecular composition and density of specific ECM components will influence how a given cell or tissue responds to matrix, which in turn will depend on both the type and number of ECM receptors arrayed at cell surfaces. Viscoelastic properties of the ECM depend on the relative ratios of elastic (e.g., elastin and fibrillin) and inelastic (e.g., collagen) elements. In addition, glycosaminoglycans (GAGs) and proteoglycans (PGs) directly contribute to two key material properties of ECM: viscosity and resistance to compressive forces. GAGs and PGs are also responsible for the hydration and "swelling" of ECM, which can facilitate the passage of migratory cells.

Many ECM molecules contain specific binding sites for cell surface receptors, the most important of which are members of the integrin family of transmembrane heterodimers. The "strength" of cell–ECM adhesive contacts depends on several factors including integrin–ECM binding and affinity, receptor clustering and avidity, and associations with the actin (or, in the case of hemidesmosomes, intermediate filament) cytoskeleton. Aside from these physical linkages, integrins are also key players in the bidirectional flow of signals between the matrix and the cell interior. Integrin signaling occurs through associations with adapter proteins and kinases that accumulate at nascent focal complexes, the longevity and maturation of which depends on multiple factors including actomyosin contractility, substrate stiffness, and/or externally applied mechanical forces. Tension on integrins leads to recruitment of vinculin, talin, and other proteins, the maturation of focal adhesions, and the further accumulation of signaling molecules such as focal adhesion kinase (FAK) and c-src. Moreover, the elasticity of the ECM can be thought of as "tuning" adhesive and signaling responses. For example, increased tension on integrins due to actin retrograde flow and actomyosin contractility might lead to rupture of a nascent adhesive contact on a stiff substrate. In contrast, a softer, more elastic substrate will deform in response to cell-generated mechanical forces resulting in maintenance of the adhesion and possible further recruitment of cytoskeletal elements, scaffolding proteins, and signaling molecules. Thus, integrins are able to sense, transduce, and respond to mechanical signals arising from the material properties of the ECM, and any externally applied forces that may be propagated through matrix. The reader is referred to a number of reviews for further details on the cellular and molecular mechanisms underlying integrin mechanotransduction (e.g., Changede & Sheetz, 2017; Hoffman, Grashoff, & Schwartz, 2011; Schwartz & DeSimone, 2008; Sun, Guo, & Fässler, 2016).

Embryo development and tissue morphogenesis provide us with numerous opportunities to consider the functional significance of developmentally regulated changes in the material properties and organization of embryonic ECMs, and how this biomechanical information is sensed and interpreted by cells. There are several examples worth keeping in mind throughout this review. For instance, changes in matrix rigidity are reported to regulate stem cell differentiation (Engler, Sen, Sweeney, & Discher, 2006) and self-renewal (Gilbert et al., 2010). Simply varying the area of an ECM contact with a stem cell is sufficient to direct lineage commitment (McBeath, Pirone, Nelson, Bhadriraju, & Chen, 2004). The extent to which ECM rigidity and topology alone influence cell fate decisions in early embryos remains largely unknown and is, in part, complicated by the presence of morphogens and growth factors known to be secreted and sequestered by the ECM. The physical assembly state of a given ECM component can also serve as a developmental "checkpoint" on morphogenetic movements. During gastrulation in *Xenopus*, fibronectin expression is required for multiple tissue movements that include the convergence and extension of axial mesoderm (Davidson, Marsden, Keller, & DeSimone, 2006; Marsden & DeSimone, 2003), epibolic spreading of ectoderm (Marsden & DeSimone, 2001), and collective cell migration of mesendoderm (Davidson, Hoffstrom, Keller, & DeSimone, 2002). Fibronectin is initially deposited as a pericellular matrix just prior to the start of these movements and is further elaborated into a dense fibrillar matrix over time. By selectively blocking fibril assembly but not initial fibronectin dimer binding to cell surfaces, Rozario, Dzamba, Weber, Davidson, and DeSimone (2009) reported that in the absence of fibrillar forms of the protein, epiboly is blocked and mesendoderm migration velocity is increased, while convergence and extension movements remain unaffected. Another example of how regulated deposition and assembly of ECM can help drive morphogenesis involves the invagination of vegetal epithelium at the onset of gastrulation in sea urchins (Lane, Koehl, Wilt, & Keller, 1993). In this model, new localized secretion of chondroitin sulfate proteoglycan (CSPG) into a preexisting layer of ECM formed after fertilization results in a bilayered matrix. Once secreted, the CSPG that defines the innermost layer swells as it becomes hydrated by the surrounding seawater. The authors liken this to a "bimetallic strip" where differential expansion of the two layers leads inevitably to a bending of the cell sheet (Lane et al., 1993). Together, these observations highlight the potential importance of subtle differences in ECM assembly state for initiating and controlling morphogenesis.

In the following sections we focus largely on the biomechanical features of ECMs and how these properties help establish tissue architecture by constraining or promoting cell and tissue movements. The following sections are organized to reflect this theme and concentrate on recent examples from the literature. In Section 2, we consider the roles of ECM in resisting forces that are generated by cells and tissues. Section 3 reflects on the role of ECM in the generation and "coupling" of forces across and between tissues. In Section 4, the role of ECM in subdividing tissues is considered.

2. THE ROLE OF ECM IN RESISTING AND CONSTRAINING TISSUES

In order to generate shapes other than "spheres" mechanical imbalances must be present in tissues that are undergoing morphogenesis. Force imbalance can arise from anisotropic force generation by the cells within the tissue or by differential resistance to isotropic forces. In this section we will consider several examples of how ECM functions to channel force and generate structural anisotropies.

2.1 Drosophila Egg Chamber Elongation

As the *Drosophila* egg chamber matures its initial spherical shape elongates more than twofold along the anterior–posterior axis to form an ellipse (Fig. 1A). The follicle comprises a germ cell cluster that is surrounded by a somatic follicular epithelium. The apical surface of the epithelium faces the germ cells, and the outer basal surface is surrounded by basement membrane. The egg chamber rotates within the basement membrane leading to the alignment of actin filaments and basement membrane fibers perpendicular to the anterior–posterior axis (Cetera et al., 2014; Haigo & Bilder, 2011). These circumferentially aligned actin filaments and ECM fibers are postulated to act as a "molecular corset" (Bilder & Haigo, 2012; Cetera & Horne-Badovinac, 2015) that contributes to elongation.

The location of the basement membrane on the outside of the egg chamber makes it easily accessible for studies of its mechanical properties. A recent study (Crest et al., 2017) used atomic force microscopy to demonstrate that a gradient of anterior–posterior stiffness in the follicle basement membrane results in the center of the follicle becoming stiffer relative to the anterior and posterior ends as development proceeds. The anisotropy that leads to egg chamber elongation is not generated by asymmetric cellular forces but rather by anisotropic resistance to the force exerted by the growing egg chamber through the mechanical properties of the basement membrane. The crucial parameter regulating elongation is the existence of a stiffness gradient between the center and the ends of the chamber. Manipulations that reduce the overall stiffness of the basement membrane but maintain the regional stiffness differences do not inhibit elongation.

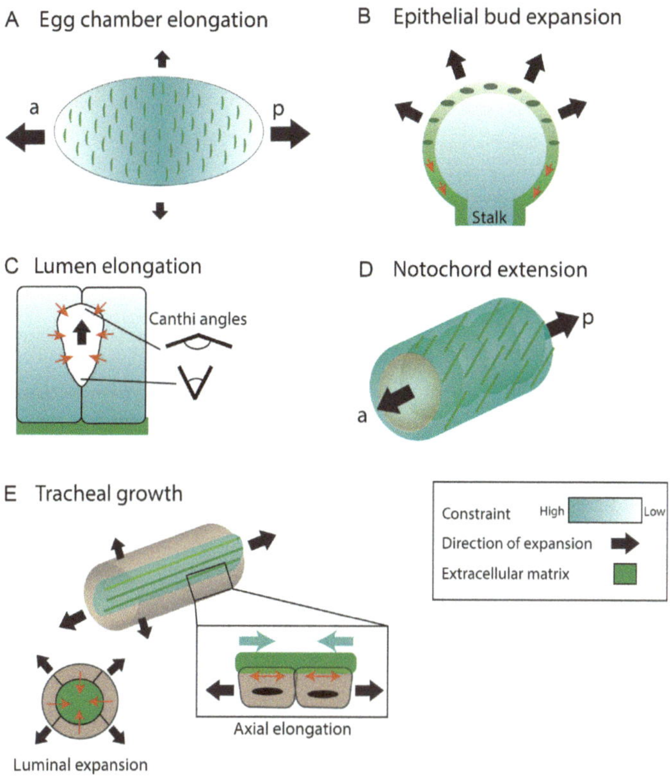

Fig. 1 The constraining and sculpting of tissues by ECM. (A) *Drosophila* egg chamber elongation. Rotation of the egg chamber around its anterior–posterior (a–p) axis results in the polarized assembly of extracellular matrix fibrils (*green*) perpendicular to the direction of elongation creating a "molecular corset." A gradient of extracellular matrix stiffness (*blue*) provides more resistance to expansion in the center of the chamber than at the ends leading to anisotropic growth (*arrows*). (B) Epithelial bud expansion. Microperforations in the basement membrane (*gray ovals*) allow the matrix (*green*) to stretch to accommodate tissue expansion at the tip of epithelial buds (*black arrows*). The increased distensibility also allows the matrix to translocate away from the tip (*red arrows*) where it accumulates around the forming duct. The denser matrix (*green*) likely constrains (*blue*) tissue expansion to sculpt the duct. (C) Lumen elongation in canaliculi. Isotropic hydrostatic pressure (*red arrows*) promotes lumen expansion between hepatocytes. The angle of the canthus nearest the fibronectin is smaller than the angle of the canthus further away, suggesting that there is more resistance to lumen expansion near the matrix and that adhesion to fibronectin (*green*) results in a gradient of intercellular stress (*blue*). The lumen elongates away from the extracellular matrix adhesion in the direction of least resistance (*arrow*). (D) Notochord extension. The outer layers of the thick perinotochordal sheath contain orthogonally oriented fibers (*green*) that constrain (*blue*) the hydrostatic swelling of large vacuoles within the notochord cells (*beige*). The constraint allows pressure to build within the notochord leading to its straightening and extension in the anterior–posterior (a–p) direction (*arrows*).

(Continued)

2.2 Epithelial Bud Expansion

It has long been recognized that the basement membrane surrounding the developing salivary gland is thinner in regions undergoing expansion such as buds and thicker in more stable regions such as stalks (Bernfield & Banerjee, 1982; Grobstein & Cohen, 1965). This observation led Grobstein and Cohen (1965) to propose that ECM acts as a "jacket" constraining morphogenesis (Fig. 1B). More recently, confocal microscopy of the basement membrane surrounding salivary gland, lung, and kidney end buds revealed that the basement membrane develops microperforations in a temporally and spatially controlled manner that correlates with active branching morphogenesis and epithelial growth. The microperforations in the salivary gland are first observed after stage 12 when clefting and expansion of the end buds begins. The number of microperforations peaks at stage 13.5, then disappears when differentiation begins at stage 15. At stage 13.5 more than 27% of the surface area of the basement membrane at the expanding bud tip is occupied by holes, compared to less than 5% of the basement membrane at the equator. Microperforations are not detected in the basement membrane at the clefts and stalks where tissue expansion is minimal (Harunaga et al., 2014).

The basement membrane is highly dynamic both locally and globally. Locally, the area of the holes in the basement membrane expands and contracts in a myosin-dependent manner. Sometimes dilation of a perforation coincides with bleb-like epithelial cell protrusions moving in and out of the hole. The perforations are more dynamic at the bud tips than at the equator. Globally, the entire basement membrane moves away from the bud tip and accumulates below the equator (Harunaga et al., 2014). Taken together these observations suggest that the basement membrane is more compliant at

---

Fig. 1—Cont'd (E). *Drosophila* tracheal expansion. The diameter and length of the *Drosophila* tracheal dorsal trunk are regulated by a chitin containing apical ECM (*green*) that transiently fills the tracheal lumen during expansion. Widening of the tube diameter (*bottom left, large arrows*) is promoted by apical secretion of ECM (*green, red arrows*) that leads to luminal expansion force. During axial elongation of the tube (*bottom right, black arrows*) appropriate length is achieved by mechanical coupling between the luminal epithelial cells and the apical ECM (*green*). Elongation, driven by apical membrane growth (*red arrows*), is constrained (*blue arrows*) by interaction with the viscoelastic apical ECM (*green*). *Panel (A): Based on Haigo and Bilder (2011); Cetera and Horne-Badovinac (2015); Crest et al. (2017). Panel (B): Based on Harunaga, Doyle, and Yamada (2014). Panel (C): Based on Li et al. (2016). Panel (D): Based on Adams, Keller, and Koehl (1990). Panel (E): Based on Hayashi and Dong (2017).*

the bud tip than at the stalk. These regional differences in basement membrane distensibility likely contribute to sculpting the forming branches.

The thinner more distensible matrix at the bud tip provides less resistance to the growing bud and allows for its expansion, whereas the thicker less compliant matrix would support the compact structure of the stalk. The regional differences in matrix compliance might also function to change the local behavior of cells. Cells in contact with thinner basement membrane at the bud tips are more motile than those in the stalks where the basement membrane is thicker (Hsu et al., 2013). Integrin-mediated sensing of the relative thickness of the ECM may regulate these regional differences in motility through mechanotransduction as has been shown in other contexts (Doyle, Carvajal, Jin, Matsumoto, & Yamada, 2015; Pelham & Wang, 1997; Plotnikov, Pasapera, Sabass, & Waterman, 2012; van Helvert, Storm, & Friedl, 2018).

2.3 Lumen Elongation

The basement membranes surrounding the egg chamber and epithelial buds undergoing branching morphogenesis are complex molecular assemblies, but ECM can break the symmetry of isotropic forces to generate an elongated structure even in a simple system, such as a minimal in vitro model of the formation of canaliculi in the liver (Li et al., 2016; Fig. 1C). The luminal cavity that gives rise to canaliculi in the mammalian liver is generated by osmotically driven apical lumen expansion. When grown in suspension, clusters of primary hepatocytes form spherical lumens. However, when groups of as few as two cultured hepatocytes are grown in microwells coated with fibronectin on their sides and bottoms, canaliculi elongate directionally away from the fibronectin matrix. Measurement of the angles of the canthi at the ends of canaliculi revealed that the angle at the end closest to the fibronectin contact is narrower than the angle farthest away, suggesting that the intercellular junction closest to fibronectin is under increased tension and thus more resistant to the isotropic hydrostatic force at one end of the lumen than the other. This differential resistance guides the anisotropic growth of the lumen away from the site of matrix contact (Li et al., 2016). It remains to be determined how contact with fibronectin locally regulates intercellular tension within the structure of the canaliculi.

2.4 Notochord Extension

The notochord functions as the primary axial skeleton of chordates during embryogenesis. In most vertebrates, the notochord is eventually replaced by ossified vertebrae, but it persists throughout the life of invertebrate chordates (Stemple, 2005; Trapani, Bonaldo, & Corallo, 2017). Notochord maturation is characterized by the development of lysosome-related vacuoles within the notochord cells (Ellis, Bagwell, & Bagnat, 2013), and the formation of a thick ECM called the perinotochordal sheath. The mechanical force responsible for stiffening and straightening the notochord is derived from the hydrostatic pressure of the expanding vacuoles being resisted by the perinotochordal sheath (Adams et al., 1990; Koehl, Quillin, & Pell, 2000; Fig. 1D). The perinotochordal sheath comprises an inner laminin-rich layer and medial and outer layers that are characterized by fibers that are oriented orthogonally to one another (Stemple, 2005; Trapani et al., 2017). In *Xenopus* the fibers of the outer sheath layer are oriented at a 54-degree angle with respect to the long axis of the notochord (Adams et al., 1990). Modeling the effects of fiber angle on mechanical performance suggests that 54 degrees is the angle at which compromise is reached between the ability to elongate and the tendency to buckle. Hydraulic cylinders wound with smaller fiber angles become shorter and wider when inflated, whereas those wound with higher fiber angles elongate but stiffen less than those wound with smaller fiber angles (Koehl et al., 2000).

Loss of function of many of the ECM components of the perinotochordal sheath results in kinked notochords (Trapani et al., 2017), suggesting that when the structure of the sheath is disrupted the hydrostatic pressure exerted by the vacuolated inner cells is not channeled into sufficient stiffening of the notochord. Loss of lysyl oxidase function via knockdown or pharmacological inhibition also results in notochord undulation and enhances the phenotype of mutations in col2a1, col8a1, or fibrillin-2 (Gansner & Gitlin, 2008; Gansner, Madsen, Mecham, & Gitlin, 2008; Gansner, Mendelsohn, Hultman, Johnson, & Gitlin, 2007). The genetic interaction between lysyl oxidases and matrix components suggests that cross-linking is one mechanism that contributes to the stiffness of the perinotochordal sheath. It is not well understood how the perinotochordal sheath with its orthogonal fiber structures is assembled.

2.5 *Drosophila* Tracheal Tube Morphogenesis

Drosophila tracheal tubes are continuous with openings in the cuticle and are required for vital gas exchange with the internal organ systems of the adult fly.

The expansion, elongation, and maintenance of these tubes depend on the apical secretion of ECM by epithelial cells lining the lumen (Fig. 1E). A major component of this apical ECM (aECM) is chitin. Diametric expansion of the tracheal lumen is inhibited in mutant embryos lacking chitin synthase, indicating the importance of chitin accumulation in this process (Devine et al., 2005; Tonning et al., 2005). In addition to undergoing luminal expansion, tracheal tubes also undergo axial elongation through epithelial membrane growth (reviewed in Hayashi & Dong, 2017). Dong, Hannezo, and Hayashi (2014) report that the tracheal aECM is a viscoelastic material mechanically coupled to the epithelial cells. The lumenal core of aECM acts as an elastic material to counter tube elongation "stretching" forces originating from epithelial apical membrane growth. This provides an interesting functional contrast to the notochord where an external sheath of ECM constrains the vacuole-driven stiffening and expansion of the tissue.

As the tracheal tubes mature aECM is removed to facilitate gas exchange once the fly emerges from its pupal case. However, a helicoid arrangement of chitin remains; these circumferential thickenings or folds are referred to as taenidia. The taenidia provide structural support that maintains patency of the tracheal lumen. How the chitin is deposited as a continuous supracellular aECM across epithelial cell boundaries is a fascinating question and involves active feedback between the aECM, cell junctions, and spirally arranged apical actin filaments (Öztürk-Çolak, Moussian, Araújo, & Casanova, 2016).

The examples presented in this section illustrate how ECM can channel the tendency of a tissue to expand isotropically due to tissue growth (egg chamber, epithelial bud) or changes in hydrostatic pressure between (canaliculi formation) or within (notochord) cells by providing anisotropic resistance to the expansion force. In the fly trachea, the ECM drives both initial diametric expansion and subsequent elastic constraint on the growth-driven elongation of the epithelial tube. Elucidating the relative compositions and organization of ECM components in these tissues will be crucial to understanding their differing mechanical properties (Hayashi & Dong, 2017; Pastor-Pareja & Xu, 2011). ECM fiber orientation contributes unique mechanical properties to the matrices surrounding the notochord (Adams et al., 1990), trachea (Öztürk-Çolak et al., 2016), and egg chamber (Haigo & Bilder, 2011), but how oriented fibers are assembled in these tissues is not well understood.

3. COORDINATION AND COUPLING OF FORCES ACROSS AND BETWEEN TISSUES

As discussed earlier, it is well established that mechanotransduction, initiated by integrin-mediated adhesion to the ECM, influences the behaviors of adherent cells (Gauthier & Roca-Cusachs, 2018; Sun et al., 2016). Here, we highlight several recent studies that explore the role of ECM in generating and propagating forces that act across and between tissues.

3.1 *Xenopus* Gastrulation

ECM can promote morphogenesis by providing a substrate for cell migration. Migration serves not only to rearrange the actively migrating cells but also to generate force that affects the behavior of neighboring cells and tissues (Fig. 2A). During *Xenopus* gastrulation the leading edge mesendoderm cells extend monopolar protrusions in the direction of travel and migrate collectively toward the animal pole of the embryo (Davidson et al., 2002). The mesendoderm migrates across a fibronectin matrix that is assembled by the blastocoel roof (Boucaut & Darribere, 1983). The axial mesoderm subsequently involutes and becomes elongated in the anterior–posterior direction as its cells orient mediolaterally and intercalate giving rise to convergence and extension of the tissue (Keller, 2002). Because these movements occur in adjacent tissues during gastrulation, it is likely that forces generated by one tissue will affect the other. Laser ablation of axial mesoderm cells in explanted dorsal marginal zone tissue plated on blastocoel roof conditioned substrates, combined with particle image velocimetry to analyze tissue displacement, revealed that axial mesoderm is under tension in the anterior–posterior direction (Hara et al., 2013). This tension is also reduced when leading edge mesendoderm is removed from the explant or when conditioned substrates are prepared using blastocoel roofs from fibronectin morphant embryos. Similar results were obtained in separate studies using antiintegrin $\alpha5\beta1$ (itgα5) monoclonal antibodies that block adhesion to fibronectin resulting in anterior–posterior "snapback" of the cell sheet as anisotropic tension is released (Davidson et al., 2002). These observations suggest that migration of leading edge mesendoderm on fibronectin generates force that is transmitted to the axial mesoderm. When fibronectin is selectively removed from the roof, but not the mesoderm, both mesendoderm migration and notochord formation are impaired. The notochords from embryos with reduced fibronectin on the blastocoel

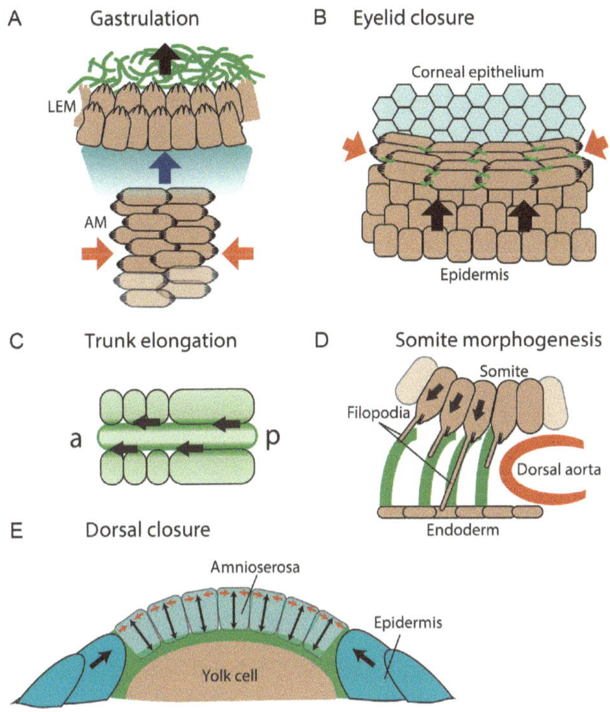

Fig. 2 Coordination of morphogenetic forces across tissues by ECM. (A) *Xenopus* gastrulation. Migration of leading edge mesendoderm cells (LEM, *black arrow*) on fibronectin (*green*) generates anteriorly directed force (*blue arrow*) that pulls on the adjacent axial mesoderm (AM) to promote mediolateral cell orientation (*red arrows*) and notochord formation. (B) Eyelid closure. Fibronectin (*green*)-dependent cell intercalation at the leading edge generates force (*red arrows*) perpendicular to the direction of eyelid closure. This force drives the movement (*black arrows*) of the epidermis over the corneal epithelium (*blue*). (C) Zebrafish tail bud trunk elongation. Fibronectin (*green*) mediates mechanical coupling between the notochord and paraxial mesoderm that results in the alignment of force in the anterior–posterior (a–p, *black arrows*) direction to drive trunk elongation. (D) Somite morphogenesis. The coordinated action of external cyclic forces generated by the pulsing of the dorsal aorta and cell-generated forces transmitted through filopodia to fibronectin results in the assembly of fibronectin pillars (*green*). The fibronectin pillars and filopodia together provide a flexible bridge between the somite and endoderm, and promote mediolateral expansion of the somite (*black arrows*). (E) *Drosophila* dorsal closure. Mechanical force generated by the apical constriction of the amnioserosa cells (*red arrows*) promotes the dorsal migration of lateral epidermal sheets to cover the embryo (*black arrows*). Integrin adhesion to laminin-containing extracellular matrix (*green*) couples the amnioserosa to the epidermis and underlying yolk cell. The apical and basal surfaces of the amnioserosa cells are mechanically coupled (*double arrows*). Matrix adhesion resists the forces generated by apical constriction (*red arrows*) and modulates force transmission between cells. *Panel (A): Based on Hara et al. (2013); Davidson et al. (2002). Panel (B): Based on Heller, Kumar, Grill, and Fuchs (2014). Panel (C): Based on Dray et al. (2013). Panel (D): Based on Sato et al. (2017). Panel (E): Based on Narasimha and Brown (2004); Goodwin et al. (2016).*

roof are shorter and wider than those from control embryos. Axial mesoderm cells in morphant embryos are significantly less elongated than controls and are not well aligned in the mediolateral direction (Hara et al., 2013). Thus, force in the anterior–posterior direction generated by fibronectin-dependent leading edge mesendoderm migration contributes to convergent extension of the adjacent axial mesoderm.

3.2 Eyelid Closure

The movement of epidermal sheets that leads to eyelid closure during mouse embryogenesis is also dependent on fibronectin–integrin interactions (Heller et al., 2014; Fig. 2B). In this case however, mediolateral cell intercalation at the leading edge generates the force that pulls the following epidermis over the cornea rather than cell migration forces promoting mediolateral intercalation in the following tissue as discussed earlier for *Xenopus* gastrulation (Hara et al., 2013; Fig. 2A). Eyelid closure results from the movement of skin epidermis over corneal epithelium. Fibronectin and itgα5 expression are enriched in the cells at the front of the epidermal sheet. The front cells are elongated mediolaterally and organize their actomyosin cytoskeleton perpendicular to the direction of sheet movement. The following cells are arranged in a more typical epithelial pattern and have cortical actin. During closure the front cells actively intercalate leading to thickening of the leading tissue. The following cells progressively elongate in the direction of travel. Laser ablation of the front cells inhibits elongation of the following cells, suggesting that the intercalation provides a "towing" force for the epidermal sheet. When itgα5, fibronectin, or myosin IIA is depleted by shRNAs eyelid closure fails. The front cells maintain their mediolateral elongation with actin filaments oriented perpendicular to the direction of closure; however, their intercalation speed is reduced. The following cells are less elongated than in control eyelids and eyelid closure is slowed, suggesting that fibronectin–integrin interactions are needed for the front cells to intercalate and generate the force that pulls the epidermal sheets over the cornea (Heller et al., 2014).

The leading edge cells driving mesendoderm migration during *Xenopus* gastrulation and epithelial sheet migration during mouse eyelid closure, both use interaction with ECM to generate force that leads to the migration of cells behind them (Fig. 2A and B). However, in the case of eyelid closure the leading edge cells are oriented perpendicular to the direction of travel, whereas leading edge mesendoderm cells orient their protrusions in the direction of travel. This raises the question of how forces become orientated. Traction forces generated by collectively migrating *Xenopus* mesendoderm leading edge cells on fibronectin are balanced by cell–cell adhesions in subsequent rows (Weber, Bjerke, & DeSimone, 2012). These leading row cells extend monopolar protrusions in the direction of travel but when the tissue is dissociated, single cells become multipolar protrusive and migrate randomly. Monopolar protrusions and directed migration can be restored in these cells by attaching a C-cadherin-coated paramagnetic bead and applying tugging forces with a magnet. These results demonstrate the importance of integrin-dependent traction stresses at the front of the cell being balanced and transmitted to following rows at sites of cadherin adhesion at the rear; this process has been termed cohesotaxis (Weber et al., 2012). Moreover, the mechanosensitive response in these cells requires keratin filaments, which accumulate at the rear of leading edge cells where they are thought to maintain the mechanical integrity of the adhesion with the following row. Recent traction force microscopy analyses of mesendoderm explants on soft fibronectin substrates confirm that traction stresses are almost entirely generated by leading edge cells and spatially coincident with a gradient of Rac1 GTPase activity that is highest in the monopolar protrusions (Sonavane et al., 2017). If keratin levels are reduced with antisense morpholinos, traction stresses are no longer limited to the leading row but are now observed throughout the explant. The Rac1 activity gradient is disrupted in these keratin-deficient cells; these and additional experiments suggest a possible mutual antagonism between Rac1 and the keratin cytoskeleton (Sonavane et al., 2017). Interestingly, the distribution of high fibronectin traction stresses at the front being balanced by forces transmitted at cell–cell adhesions at the rear is likely responsible for "sculpting" and maintaining the cleft of Brachet, an extracellular compartment defined by the space between the mesendoderm/mesoderm and the fibronectin-coated blastocoel roof. Brachet's cleft is postulated to play an important role in the formation of a gradient of chordin and other morphogens secreted by the dorsal organizer region (Plouhinec, Zakin, Moriyama, & De Robertis, 2013).

3.3 Zebrafish Trunk Elongation

Fibronectin–integrin interactions during trunk elongation in zebrafish regulate both the mechanical properties of cells and the mechanical linkage between paraxial tail bud tissue and the notochord (Fig. 2C). Morpholino-mediated knockdown of the two primary integrin receptors for fibronectin, itgα5 and itgαv, results in a truncated body axis that lacks somites. Analysis of cell motion in the tail bud reveals little change when integrins itgα5 and itgαv are knocked down, suggesting that the main function of integrin–ECM interactions in tail bud trunk elongation is not to promote cell migration. Fibronectin fibrils normally cover the surface of the paraxial mesoderm but display no obvious polarity, suggesting that any forces generated by the medial movements of cells converging toward the midline must be balanced by force in the anterior–posterior direction. The amount of fibronectin matrix on the surface of the paraxial mesoderm is reduced in the itgα5/itgαv morphants consistent with the established role of integrins in fibronectin assembly (Leiss, Beckmann, Girós, Costell, & Fässler, 2008; Schwarzbauer & DeSimone, 2011). The fibronectin fibrils that remain display a mediolateral polarity that is not observed in wild-type embryos, suggesting that anterior–posterior tension is disrupted in the morphants. Because integrin knockdown did not alter tail bud cell motion the anterior–posterior force acting on fibronectin on the presomitic mesoderm is likely not generated by cell migration. Notochords from itgα5/itgαv morphant embryos are separated from the paraxial mesoderm, suggesting that integrin–fibronectin interactions normally serve to mechanically couple the two tissues. Notochords in morphant embryos are undulated, suggesting that mechanical linkage to the paraxial mesoderm is necessary for anterior–posterior force generation that leads to proper notochord elongation and trunk extension.

In addition to promoting mechanical linkage between tissues integrin-mediated fibronectin assembly also influences the mechanical characteristics of paraxial mesoderm cells themselves. When itgα5 and itgαv are knocked down the paraxial mesoderm cells display more blebbing indicative of changes in cellular mechanics that lead to fluctuations in cytosolic pressure. Transgenic expression of itgα5 in the paraxial mesoderm is sufficient to rescue fibronectin assembly, adhesion between the notochord and paraxial mesoderm, axis elongation and to reduce blebbing (Dray et al., 2013).

3.4 Avian Somite Morphogenesis

Fibronectin "pillars" were described recently in quail embryos where they are thought to provide flexible linkage between the somitic mesoderm and endoderm (Sato et al., 2017). The organization of fibronectin into pillars is promoted by the cyclic force generated by pulsing of the dorsal aorta providing a compelling example of force generated by one tissue creating an ECM structure that, in turn, provides mechanical linkage between two other tissues (Fig. 2D). Formation of the fibronectin pillars is dependent on β1-integrin-mediated fibrillogenesis through cellular tension but also on the extrinsic force arising from the pulsing of the dorsal aorta. When dorsal aorta formation is blocked by pharmacological inhibition of VEGF signaling, scattered fibronectin fibrils, but not pillars, are observed. Pillars also fail to form when blood flow through the dorsal aorta is stopped by inhibiting heart contraction or by inducing local occlusion of the dorsal aorta.

Live-cell imaging of membrane and actin dynamics revealed that long filopodia-like structures originating from the basal side of the somites dynamically extend and retract along the fibronectin pillars and occasionally contact the endoderm. Pillar formation is reduced when filopodia formation is inhibited by sequestering Ena/VASP away from the filopodia to the mitochondrial surface using FP4-mito. Reciprocally, inhibition of fibronectin fibrillogenesis by expression of the 70 kDa aminoterminal fragment of fibronectin to disrupt fibronectin–fibronectin interaction (McKeown-Longo & Mosher, 1985) inhibits both fibronectin pillar and filopodia formation, suggesting that the two structures are mutually dependent. Inhibition of pillar formation leads to a reduction in the medioventral expansion of the somites and increases the space between the somites and endoderm.

3.5 *Drosophila* Dorsal Closure

During the process of dorsal closure in *Drosophila* mechanical force generated by the contraction of amnioserosa cells brings the lateral edges of epidermis from each side of the embryo to meet dorsally where the epidermal cells then "zipper" to create an uninterrupted epidermal covering (Fig. 2E). Integrin–ECM interactions are integral to the coordination of forces at the intracellular, intratissue, and intertissue levels. Integrin βPS and laminin are concentrated in regions of tissue interaction between the leading edge of the epidermis and amnioserosa and also between the amnioserosa and yolk cell (Narasimha & Brown, 2004). Dorsal closure fails in myospheroid mutant embryos that lack βPS integrins; the amnioserosa detaches from the yolk cell, and tears develop near the amnioserosa–epidermal interface (Narasimha & Brown, 2004).

The apical contraction of individual amnioserosa cells results in reduction of the amnioserosa tissue surface area and its ingression (Gorfinkiel, Schamberg, & Blanchard, 2011; Kiehart, Galbraith, Edwards, Rickoll, & Montague, 2000; Narasimha & Brown, 2004). Apical contraction is temporally regulated with an initial slow phase followed by a fast phase. There is also a gradient of contraction with cells at the margin contracting faster than those at the center. Integrin–ECM adhesion is necessary for both temporal and spatial coordination of contraction. In myospheroid mutants the amnioserosa cells remain in the slow phase, and the contraction gradient is inverted such that cells in the center contract more rapidly than those at the margins (Gorfinkiel, Blanchard, Adams, & Martinez Arias, 2009). Although it is unclear why myospheroid mutants would have altered contraction rates, one possible mechanism is that integrin–ECM-mediated mechanical coupling between the epidermis and amnioserosa normally results in the epidermis providing resistance to amnioserosa contraction.

Integrin–ECM interaction has recently been shown to modulate the transmission of force both within and between amnioserosa cells. Focal adhesion-like structures link amnioserosa cells to the laminin containing ECM between the amnioserosa and the yolk cell. Analysis of recoil behavior after laser ablation of apical cell membranes suggests that apical tension is inversely correlated with adhesion to the ECM. Myospheroid mutants with reduced adhesion to the ECM have higher recoil than controls. Conversely, increasing basal cell–ECM adhesion by expression of activated Talin or Rap1 decreases recoil, suggesting that forces generated apically are resisted by basal connection to the ECM. The integrin–ECM modulation of apical force is not limited to individual cells. Reduction of integrin-mediated ECM adhesion also results in more efficient transmission of force across the tissue, suggesting that in the absence of basal tethering by the ECM, forces are directed across cell–cell junctions (Goodwin et al., 2016). Disruption of cell–ECM adhesion in the amnioserosa leads to changes in the distribution and stability of cadherins at cell–cell junctions, as well as altered actin organization and changes in myosin dynamics, suggesting that the interdependence of integrin and cadherin adhesion and subsequent regulation of the cytoskeleton are essential for dorsal closure.

4. SUBDIVISION OF TISSUES BY ECM

Fibronectin plays remarkably similar roles in both the subdivision of mesenchyme into somites that generates the segmented body plan of vertebrates, and in the clefting of epithelial buds that gives rise to the branched structure of organs such as salivary gland and lung. In both cases fibronectin assembly is essential for the progression of tissue division. Both processes highlight the dynamic interplay between cell–cell and cell–ECM adhesion in morphogenesis.

4.1 Somitogenesis

The metameric structure of the vertebrate body axis is established by the sequential subdivision of the paraxial mesoderm into somites that subsequently give rise to vertebrae, muscles, and dermis (Fig. 3A). Fibronectin matrix is assembled at nascent somite boundaries and is essential for the morphogenesis and maintenance of this tissue (George, Georges-Labouesse, Patel-King, Rayburn, & Hynes, 1993; Jülich et al., 2005; Koshida et al., 2005; Marsden & DeSimone, 2003; Martins et al., 2009; Rifes et al., 2007). Spatial regulation of fibronectin assembly at the borders provides an excellent example of how patterns of transcription factor expression are "translated" into the active cell behaviors that drive morphogenesis.

The interaction of a molecular oscillator or "segmentation clock" with gradients of signaling activity along the anterior–posterior axis of the embryo results in progressive molecular segmentation that precedes the morphological division of the somites. This iterative patterning is well studied (Hubaud & Pourquié, 2014; Oates, Morelli, & Ares, 2012), but how the expression of segmentation genes leads to the morphogenesis of somites and the assembly of fibronectin that defines the somite borders has been less clear.

Segmental patterning results in the expression of EphA4 and EphrinB2 expressed on opposite sides of the nascent somite border (Barrios et al., 2003; Durbin et al., 1998). The initial clefting of the paraxial mesoderm that leads to segmentation is mediated by EphA4–EphrinB2 signaling (Watanabe, Sato, Saito, Tadokoro, & Takahashi, 2009). Eph/Ephrin signaling subsequently contributes to fibronectin assembly at the border by activation and clustering of Itgα5 (Jülich, Mould, Koper, & Holley, 2009). In *Xenopus* both FAK and Ena/VASP activities are essential for fibronectin assembly and somite formation (Kragtorp & Miller, 2006), highlighting that the spatiotemporal regulation of fibronectin fibrillogenesis involves multiple molecular interactions leading to a conducive mechanical environment that promotes fibronectin unfolding and intermolecular interactions (Baneyx, Baugh, & Vogel, 2002; Zhang, Magnusson, & Mosher, 1997; Zhong et al., 1998).

Fibronectin and Itgα5 are expressed throughout the paraxial mesoderm in zebrafish (Jülich et al., 2005; Koshida et al., 2005); however, fibrils are restricted to the surface of the paraxial mesoderm and to somite borders (Crawford, Henry, Clason, Becker, & Hille, 2003). Recent studies suggest that complex interactions between Itgα5, Cadherin 2 (Cdh2), and PAPC

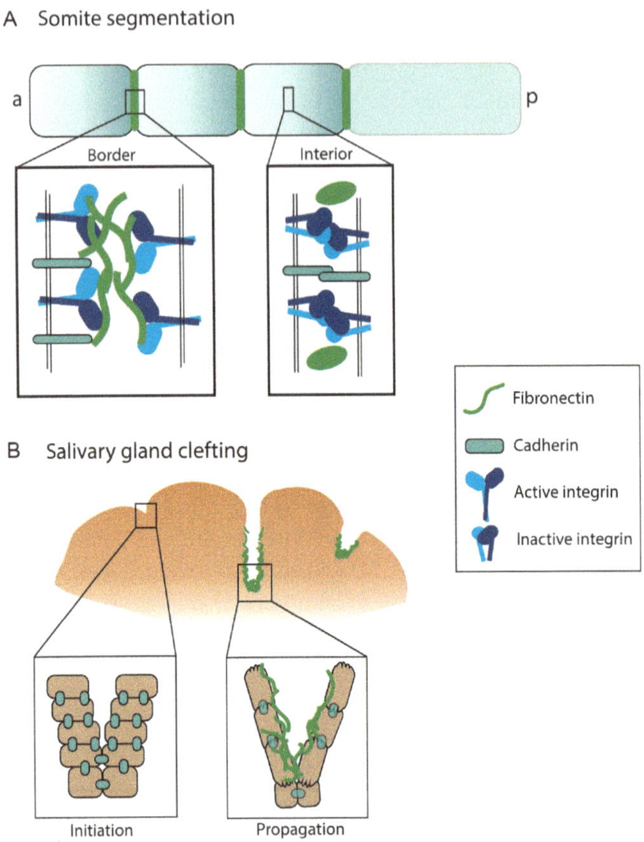

Fig. 3 Subdivision of tissues by ECM. (A) Somite segmentation. An anterior–posterior (a–p) gradient of cadherin stability within each somite results in higher levels of cadherin (*blue*) at the posterior somite border. Differential cadherin expression levels promote fibronectin (*green*) assembly at the border. On free cell surfaces at somite borders (*left box*) the extended active integrin conformation that mediates fibrillogenesis is favored (*left box*). Within somites (*right box*) cadherin adhesion promotes association between integrins on adjacent cells which stabilizes the inactive bent integrin conformation. (B) Salivary gland clefting. Transient intercellular gaps on the bud surface lead to cleft initiation (*left box*). As the cleft propogates (*right box*) cell separation is reinforced by directional fibronectin assembly (*green*) that begins at the base of the cleft and progresses as the cleft deepens. Cells in contact with fibronectin favor cell–matrix adhesion over cell–cell adhesion, decrease their cell surface levels of cadherin, and become more motile. *Panel (A): Based on Jülich et al. (2015). McMillen, Chatti, Jülich, and Holley (2016). Panel (B): Based on Sakai et al. (2003). Larsen, Wei, and Yamada et al. (2006); Onodera et al. (2010); Daley et al. (2017).*

(Chal, Guillot, & Pourquié, 2017; Jülich et al., 2015; McMillen et al., 2016) result in increased cytoskeletal and tissue tension at the posterior somite border that promotes fibronectin assembly at the boundary. Within the mesenchymal tissue Itgα5 receptors on adjacent cells physically associate and repress one another by favoring the bent inactive integrin conformation. Cdh2 that is present in adherens junctions within the tissue contributes to the repression of Itgα5 by stabilizing the intercellular Itgα5 interaction. On the surface of the paraxial mesoderm and at nascent somite borders a free cell surface allows Itgα5 to assume the extended active conformation necessary for fibronectin assembly (Jülich et al., 2015, 2009).

Although Cdh2 is not differentially expressed within the somites, stable Cdh2 adhesions form a posterior to anterior gradient within each somite resulting in higher levels of stable Cdh2 on the posterior side of the border. Mosaic analysis has demonstrated that differential Cdh2 levels are sufficient to drive the assembly of fibronectin. At the borders between high and low Cdh2 levels cortical actin belts are assembled. Expression of constitutively active myosin regulatory light chain kinase can substitute for different levels of Cdh2, suggesting that the higher Cdh2 levels at the posterior border result in higher levels of cytoskeletal tension and tissue stiffness that in turn promote fibronectin assembly (McMillen et al., 2016). Thus, Cdh2 plays context-dependent roles in regulating assembly. It inhibits fibrillogenesis within tissues by promoting intercellular Itgα5 inhibition and promotes it at tissue borders by enhancing tissue tension.

A recent study of amniote embryos sheds light on how the Cdh2 gradient is established. The expression of paraxial protocadherin (PAPC) in the anterior presomitic mesoderm is regulated by the segmentation clock. PAPC then promotes the clathrin-mediated endocytosis of Cdh2 at the anterior of the forming somite leading to the initiation of clefting and setting up the differential Cdh2 levels that drive fibronectin assembly (Chal et al., 2017). Further studies are needed to determine if a similar mechanism leads to the differential Cdh2 stability levels observed in zebrafish (McMillen et al., 2016).

Although the details of how the various signals leading to fibronectin assembly are integrated at the molecular level remain to be worked out, they ultimately cooperate to create the proper mechanical environment for integrin activation and the transmission of tension to fibronectin that leads to fibrillogenesis at the border. It also remains to be determined how fibronectin at the border functions to promote and maintain somite segmentation.

4.2 Branching Morphogenesis

Another example of reiterative tissue subdivision that requires fibronectin assembly is the clefting that subdivides epithelial buds to generate the branched structures of organs such as salivary gland, lung, and kidney (Wang, Sekiguchi, Daley, & Yamada, 2017). During submandibular salivary gland branching the interaction of the epithelial bud with its surrounding mesenchyme is thought to create increased membrane curvature which in turn leads to the generation of numerous small indentations in the basement membrane separating the two tissues (Nogawa, 1983). A subset of these indentations then progresses to form a cleft from the surface of the bud toward the interior (Fig. 3B).

The assembly of fibronectin fibrils in the nascent clefts of developing salivary gland, lung, and kidney is necessary for cleft progression and subsequent branch formation (De Langhe et al., 2005; Sakai et al., 2003). Antibodies to fibronectin or its receptor integrin alpha5 or siRNA-mediated fibronectin knockdown inhibit clefting. Conversely, addition of exogenous fibronectin to organoid cultures accelerates clefting (Sakai et al., 2003).

Fibronectin fibrillogenesis is a mechanically regulated process (Baneyx et al., 2002; Zhang et al., 1997; Zhong et al., 1998). The assembly of fibronectin at the cleft is promoted by localized actomyosin contractility controlled by ROCK kinase. Exogenous fibronectin can rescue the clefting defects caused by siRNA-mediated depletion of endogenous fibronectin, but this rescue fails when contractility and subsequent fibronectin assembly is inhibited. ROCK activity leads to the activation of β1 integrins and FAK and promotes a positive feedback loop whereby activated integrin and FAK promote fibronectin assembly which in turn leads to further activation of integrin and FAK (Daley, Gulfo, Sequeira, & Larsen, 2009; Daley, Kohn, & Larsen, 2011). It will be interesting to determine how ROCK activity is activated at the forming cleft.

Fibronectin assembly at the cleft contributes to branching via multiple mechanisms. "Pulse-chase washout" of fluorescently labeled fibronectin revealed that assembly within the submandibular gland is directional with fibronectin initially assembling at the base of the cleft to form a wedge that translocates inward as the cleft progresses. Fibronectin continues to be assembled behind the wedge creating a physical barrier that promotes cell separation (Larsen et al., 2006). Interestingly, the cleft base also contains an actin filled cytoplasmic ridge or "shelf" that is observed in electron micrographs of branching submandibular glands. This shelf may serve as a matrix attachment point and the actin within it may be responsible for generating mechanical force that drives cleft elongation (Kadoya & Yamashina, 2010).

Epithelial bud cells respond to contact with fibronectin by reducing cell–cell adhesion and increasing cell–matrix adhesion (Daley et al., 2011; Larsen et al., 2006; Sakai et al., 2003). Reduced cell–cell adhesion between the outer bud cells provides the transient plasticity required for large-scale tissue rearrangement. Fibronectin contributes to the downregulation of E-cadherin in the outer cells through upregulation of the transcription factor Btbd7. Similar to loss of fibronectin function, knockdown of Btbd7 by siRNA in organoid cultures or by homologous recombination in mouse leads to defects in branching morphogenesis of salivary gland, lung, and kidney (Daley et al., 2017; Onodera et al., 2010). Btbd7 expression in MDCK cells leads to a reduction of E-cadherin localization to cell–cell junctions and an overall decrease in E-cadherin protein levels (Onodera et al., 2010). In vivo there is less E-cadherin localization in outer bud cells when compared to inner bud cells, but this differential localization is lost in Btbd7 knockout salivary glands with outer cells displaying increased E-cadherin similar to inner bud cells (Daley et al., 2017). Loss of Btbd7 does not lead to increased Cadherin 1 (Cdh1) mRNA expression in the outer cells, suggesting that the increase in outer cell E-cadherin protein levels is likely mediated posttranscriptionally. In addition to its role as a transcription factor Btbd7 is known to interact with E3 ubiquitin ligases (Metzger, Hristova, & Weissman, 2012). In MDCK cells, overexpression of Btbd7 leads to enhanced ubiquitylation, endocytosis, and degradation of E-cadherin. In cultured salivary glands, inhibitors of ubiquitin ligase and dynamin lead to increased E-cadherin in the outer cells and inhibit branching morphogenesis, phenocopying the Btbd7 knockout.

Fibronectin regulation of Btbd7 and the subsequent downregulation of cell–cell adhesion contributes to the enhanced motility of outer bud cells relative to inner bud cells (Kadoya & Yamashina, 2010; Larsen et al., 2006). This motility is also dependent on integrin-mediated interaction with the basement membrane and myosin II (Hsu et al., 2013). In the pancreas, E-cadherin is also downregulated in the outer bud cells but in this case Btbd7 is not involved (Daley et al., 2017); instead, integrin signaling through Src appears to be critical (Shih, Panlasigui, Cirulli, & Sander, 2016).

Lung branching morphogenesis requires both fibronectin assembly (Sakai et al., 2003) and spatiotemporally regulated smooth muscle cell differentiation at the nascent cleft (Kim et al., 2015). Contraction of smooth muscle at the cleft likely provides mechanical force to promote clefting. Studies of canonical Wnt signaling in embryonic lung organ cultures suggest that fibronectin is involved in regulating smooth muscle differentiation at the nascent cleft. Inhibition of Wnt using Dickkopf-1 demonstrates that Wnt is required for clefting, smooth muscle actin expression, and fibronectin matrix assembly. The defects in clefting and smooth muscle actin expression can be rescued by addition of exogenous fibronectin to the cultures. Conversely, antibody-mediated interference with fibronectin function inhibits smooth muscle actin expression (De Langhe et al., 2005).

In both somitogenesis and branching morphogenesis cross talk between integrin and cadherin adhesion is the nexus for tissue division. The reciprocity between the two adhesion systems and fibronectin is complex and remains to be understood. In the case of somitogenesis, spatial regulation of cadherin stability appears to be upstream of fibronectin assembly, whereas during branching morphogenesis assembly of fibronectin at the cleft results in a loss of cadherins from the surface.

In this review we have highlighted a selected number of recent studies that illustrate the importance of ECM materials in a diverse array of morphogenetic processes. Continued investigation of ECM functions in the context of embryonic development across multiple species is highly likely to reveal additional, remarkable biological insights. Future progress in this area will require not only a better appreciation of the physical properties of ECM and its role in mechanotransduction and other cell signaling pathways but also how embryonic matrices are assembled and remodeled over time. Deciphering the inherent compositional and biomechanical complexity of ECMs will be key to understanding morphogenesis.

ACKNOWLEDGMENTS

We thank Dr. Mary Kate Worden for a critical reading of this manuscript and for helpful suggestions. The authors are supported by USPHS Grant number GM094793.

REFERENCES

Adams, D. S., Keller, R., & Koehl, M. A. (1990). The mechanics of notochord elongation, straightening and stiffening in the embryo of Xenopus laevis. *Development*, *110*, 115–130. https://doi.org/10.1093/icb/40.1.53.

Baneyx, G., Baugh, L., & Vogel, V. (2002). Fibronectin extension and unfolding within cell matrix fibrils controlled by cytoskeletal tension. *Proceedings of the National Academy of Sciences of the United States of America*, *99*, 5139–5143. https://doi.org/10.1073/pnas.072650799.

Barrios, A., Poole, R. J., Durbin, L., Brennan, C., Holder, N., & Wilson, S. W. (2003). Eph/Ephrin signaling regulates the mesenchymal to epithelial transition of the paraxial mesoderm during somite morphogenesis. *Current Biology*, *13*, 1571–1582. https://doi.org/10.1016/j.cub.2003.08.030.

Bernfield, M., & Banerjee, S. D. (1982). The turnover of basal lamina glycosaminoglycan correlates with epithelial morphogenesis. *Developmental Biology*, *90*, 291–305. https://doi.org/10.1016/0012-1606(82)90378-5.

Bilder, D., & Haigo, S. L. (2012). Expanding the morphogenetic repertoire: Perspectives from the Drosophila egg. *Developmental Cell*, *22*, 12–23. https://doi.org/10.1016/j.devcel.2011.12.003.

Boucaut, J., & Darribere, T. (1983). Fibronectin in early amphibian embryos. Migrating mesodermal cells contact fibronectin established prior to gastrulation. *Cell Tissue Research*, *234*, 135–145.

Cetera, M., & Horne-Badovinac, S. (2015). Round and round gets you somewhere: Collective cell migration and planar polarity in elongating Drosophila egg chambers. *Current Opinion in Genetics & Development*, *32*, 10–15. https://doi.org/10.1016/j.gde.2015.01.003.

Cetera, M., Ramirez-San Juan, G. R., Oakes, P. W., Lewellyn, L., Fairchild, M. J., Tanentzapf, G., et al. (2014). Epithelial rotation promotes the global alignment of contractile actin bundles during Drosophila egg chamber elongation. *Nature Communications*, *5*, 1–12. https://doi.org/10.1038/ncomms6511.

Chal, J., Guillot, C., & Pourquié, O. (2017). *PAPC* couples the segmentation clock to somite morphogenesis by regulating N-cadherin-dependent adhesion. *Development*, *144*, 664–676. https://doi.org/10.1242/dev.143974.

Changede, R., & Sheetz, M. (2017). Integrin and cadherin clusters: A robust way to organize adhesions for cell mechanics. *BioEssays*, *39*, 1–12. https://doi.org/10.1002/bies.201600123.

Crawford, B. D., Henry, C. A., Clason, T. A., Becker, A. L., & Hille, M. B. (2003). Activity and distribution of paxillin, focal adhesion kinase, and cadherin indicate cooperative roles during zebrafish morphogenesis. *Molecular Biology of the Cell*, *14*, 3065–3081. https://doi.org/10.1091/mbc.E02-08-0537.

Crest, J., Diz-Muñoz, A., Chen, D.-Y., Fletcher, D. A., & Bilder, D. (2017). Organ sculpting by patterned extracellular matrix stiffness. *eLife*, *6*, e24958.

Daley, W. P., Gulfo, K. M., Sequeira, S. J., & Larsen, M. (2009). Identification of a mechanochemical checkpoint and negative feedback loop regulating branching morphogenesis. *Developmental Biology*, *336*, 169–182. https://doi.org/10.1016/j.ydbio.2009.09.037.

Daley, W. P., Kohn, J. M., & Larsen, M. (2011). A focal adhesion protein-based mechanochemical checkpoint regulates cleft progression during branching morphogenesis. *Developmental Dynamics*, *240*, 2069–2083. https://doi.org/10.1002/dvdy.22714.

Daley, W. P., Matsumoto, K., Doyle, A. D., Wang, S., DuChez, B. J., Holmbeck, K., et al. (2017). Btbd7 is essential for region-specific epithelial cell dynamics and branching morphogenesis in vivo. *Development*, *144*, 2200–2211.

Davidson, L. A., Hoffstrom, B. G., Keller, R., & DeSimone, D. W. (2002). Mesendoderm extension and mantle closure in Xenopus laevis gastrulation: Combined roles for integrin $\alpha5\beta1$, fibronectin, and tissue geometry. *Developmental Biology*, *242*, 109–129. https://doi.org/10.1006/dbio.2002.0537.

Davidson, L. A., Marsden, M., Keller, R., & DeSimone, D. W. (2006). Integrin $\alpha5\beta1$ and fibronectin regulate polarized cell protrusions required for Xenopus convergence and extension. *Current Biology*, *16*, 833–844.

De Langhe, S. P., Sala, F. G., Del Moral, P. M., Fairbanks, T. J., Yamada, K. M., Warburton, D., et al. (2005). Dickkopf-1 (DKK1) reveals that fibronectin is a major target of Wnt signaling in branching morphogenesis of the mouse embryonic lung. *Developmental Biology*, *277*, 316–331. https://doi.org/10.1016/j.ydbio.2004.09.023.

Devine, W. P., Lubarsky, B., Shaw, K., Luschnig, S., Messina, L., & Krasnow, M. A. (2005). Requirement for chitin biosynthesis in epithelial tube morphogenesis. *Proceedings of the National Academy of Sciences of the United States of America*, *102*, 17014–17019. https://doi.org/10.1073/pnas.0506676102.

Dong, B., Hannezo, E., & Hayashi, S. (2014). Balance between apical membrane growth and luminal matrix resistance determines epithelial tubule shape. *Cell Reports*, 7, 941–950.

Doyle, A. D., Carvajal, N., Jin, A., Matsumoto, K., & Yamada, K. M. (2015). Local 3D matrix microenvironment regulates cell migration through spatiotemporal dynamics of contractility-dependent adhesions. *Nature Communications*, *6*, 1–15. https://doi.org/10.1038/ncomms9720.

Dray, N., Lawton, A., Nandi, A., Jülich, D., Emonet, T., & Holley, S. A. (2013). Cell-fibronectin interactions propel vertebrate trunk elongation via tissue mechanics. *Current Biology*, *23*, 1335–1341.

Durbin, L., Brennan, C., Shiomi, K., Cooke, J., Barrios, A., Shanmugalingam, S., et al. (1998). Eph signaling is required for segmentation and differentiation of the somites. *Genes & Development*, *12*, 3096–3109. https://doi.org/10.1101/gad.12.19.3096.

Dzamba, B. J., Jakab, K. R., Marsden, M., Schwartz, M. A., & DeSimone, D. W. (2009). Cadherin adhesion, tissue tension, and noncanonical Wnt signaling regulate fibronectin matrix organization. *Developmental Cell*, *16*, 421–432. https://doi.org/10.1016/j.devcel.2009.01.008.

Ellis, K., Bagwell, J., & Bagnat, M. (2013). Notochord vacuoles are lysosome-related organelles that function in axis and spine morphogenesis. *The Journal of Cell Biology*, *200*, 667–679. https://doi.org/10.1083/jcb.201212095.

Engler, A. J., Sen, S., Sweeney, H. L., & Discher, D. E. (2006). Matrix elasticity directs stem cell lineage specification. *Cell*, *126*, 677–689. https://doi.org/10.1016/j.cell.2006.06.044.

Gansner, J. M., & Gitlin, A. J. D. (2008). Essential role for the alpha 1 chain of type VIII collagen in zebrafish notochord formation. *Developmental Dynamics*, *237*, 3715–3726. https://doi.org/10.1002/dvdy.21779.

Gansner, J. M., Madsen, E. C., Mecham, R. P., & Gitlin, J. D. (2008). Essential role for fibrillin-2 in zebrafish notochord and vascular morphogenesis. *Developmental Dynamics*, *237*, 2844–2861. https://doi.org/10.1002/dvdy.21705.

Gansner, J. M., Mendelsohn, B. A., Hultman, K. A., Johnson, S. L., & Gitlin, J. D. (2007). Essential role of lysyl oxidases in notochord development. *Developmental Biology*, *307*, 202–213. https://doi.org/10.1016/j.ydbio.2007.04.029.

Gauthier, N. C., & Roca-Cusachs, P. (2018). Mechanosensing at integrin-mediated cell–matrix adhesions: From molecular to integrated mechanisms. *Current Opinion in Cell Biology*, *50*, 20–26. https://doi.org/10.1016/j.ceb.2017.12.014.

George, E. L., Georges-Labouesse, E. N., Patel-King, R. S., Rayburn, H., & Hynes, R. O. (1993). Defects in mesoderm, neural tube and vascular development in mouse embryos lacking fibronectin. *Development*, *119*, 1079–1091. https://doi.org/10.1083/jcb.200107107.

Gilbert, P. M., Havenstrite, K. L., Magnusson, K. E. G., Sacco, A., Leonardi, N. A., Kraft, P., et al. (2010). Substrate elasticity regulates skeletal muscle stem cell. *Science*, *329*, 1078–1081. https://doi.org/10.1126/science.1191035.

Goodwin, K., Ellis, S. J., Lostchuck, E., Zulueta-Coarasa, T., Fernandez-Gonzalez, R., & Tanentzapf, G. (2016). Basal cell-extracellular matrix adhesion regulates force transmission during tissue morphogenesis. *Developmental Cell*, *39*, 611–625.

Gorfinkiel, N., Blanchard, G. B., Adams, R. J., & Martinez Arias, A. (2009). Mechanical control of global cell behaviour during dorsal closure in Drosophila. *Development*, *136*, 1889–1898. https://doi.org/10.1242/dev.030866.

Gorfinkiel, N., Schamberg, S., & Blanchard, G. B. (2011). Integrative approaches to morphogenesis: Lessons from dorsal closure. *Genesis*, *49*, 522–533. https://doi.org/10.1002/dvg.20704.

Grobstein, C., & Cohen, J. (1965). Collagenase: Effect on the morphogenesis of embryonic salivary epithelium in vitro. *Science*, *150*, 626–628.

Haigo, S. L., & Bilder, D. (2011). Global tissue revolutions in a morphogenetic movement controlling elongation. *Science*, *331*, 1071–1074. https://doi.org/10.1126/science.1199424.

Hara, Y., Nagayama, K., Yamamoto, T. S., Matsumoto, T., Suzuki, M., & Ueno, N. (2013). Directional migration of leading-edge mesoderm generates physical forces: Implication in Xenopus notochord formation during gastrulation. *Developmental Biology*, *382*, 482–495.

Harunaga, J. S., Doyle, A. D., & Yamada, K. M. (2014). Local and global dynamics of the basement membrane during branching morphogenesis require protease activity and actomyosin contractility. *Developmental Biology*, *394*, 197–205. https://doi.org/10.1016/j.ydbio.2014.08.014.

Hayashi, S., & Dong, B. (2017). Shape and geometry control of the Drosophila tracheal tubule. *Development, Growth & Differentiation*, *59*, 4–11. https://doi.org/10.1111/dgd.12336.

Heller, E., Kumar, K. V., Grill, S. W., & Fuchs, E. (2014). Forces generated by cell intercalation tow epidermal sheets in mammalian tissue morphogenesis. *Developmental Cell*, *28*, 617–632. https://doi.org/10.1016/j.devcel.2014.02.011.

Hoffman, B. D., Grashoff, C., & Schwartz, M. A. (2011). Dynamic molecular processes mediate cellular mechanotransduction. *Nature*, *475*, 316–323. https://doi.org/10.1038/nature10316.

Hsu, J. C., Koo, H., Harunaga, J. S., Matsumoto, K., Doyle, A. D., & Yamada, K. M. (2013). Region-specific epithelial cell dynamics during branching morphogenesis. *Developmental Dynamics*, *242*, 1066–1077.

Hubaud, A., & Pourquié, O. (2014). Signalling dynamics in vertebrate segmentation. *Nature Reviews. Molecular Cell Biology*, *15*, 709–721. https://doi.org/10.1038/nrm3891.

Jülich, D., Cobb, G., Melo, A. M., McMillen, P., Lawton, A. K., Mochrie, S. G. J., et al. (2015). Cross-scale integrin regulation organizes ECM and tissue topology. *Developmental Cell*, *34*, 33–44.

Jülich, D., Geisler, R., Holley, S. A., Van Bebber, F., Busch-Nentwich, E., Dahm, R., et al. (2005). Integrinα5 and delta/notch signaling have complementary spatiotemporal requirements during zebrafish somitogenesis. *Developmental Cell*, *8*, 575–586. https://doi.org/10.1016/j.devcel.2005.01.016.

Jülich, D., Mould, A. P., Koper, E., & Holley, S. A. (2009). Control of extracellular matrix assembly along tissue boundaries via integrin and Eph/Ephrin signaling. *Development*, *136*, 2913–2921.

Kadoya, Y., & Yamashina, S. (2010). Cellular dynamics of epithelial clefting during branching morphogenesis of the mouse submandibular gland. *Developmental Dynamics*, *239*, 1739–1747. https://doi.org/10.1002/dvdy.22312.

Keller, R. (2002). Shaping the vertebrate body plan by polarized embryonic cell movements. *Science*, *298*, 1950–1954. https://doi.org/10.1126/science.1079478.

Kiehart, D. P., Galbraith, C. G., Edwards, K. A., Rickoll, W. L., & Montague, R. A. (2000). Multiple forces contribute to cell sheet morphogenesis for dorsal closure in Drosophila. *The Journal of Cell Biology*, *149*, 471–490. https://doi.org/10.1083/jcb.149.2.471.

Kim, H. Y., Pang, M. F., Varner, V. D., Kojima, L., Miller, E., Radisky, D. C., et al. (2015). Localized smooth muscle differentiation is essential for epithelial bifurcation during branching morphogenesis of the mammalian lung. *Developmental Cell*, *34*, 719–726. https://doi.org/10.1016/j.devcel.2015.08.012.

Koehl, M. A. R., Quillin, K. J., & Pell, C. A. (2000). Mechanical design of fiber-wound hydraulic skeletons: The stiffening and straightening of embryonic notochords1. *Integrative and Comparative Biology*, *40*, 28–41.

Koshida, S., Kishimoto, Y., Ustumi, H., Shimizu, T., Furutani-Seiki, M., Kondoh, H., et al. (2005). Integrin α5-dependent fibronectin accumulation for maintenance of somite boundaries in zebrafish embryos. *Developmental Cell*, *8*, 587–598. https://doi.org/10.1016/j.devcel.2005.03.006.

Kragtorp, K. A., & Miller, J. R. (2006). Regulation of somitogenesis by Ena/VASP proteins and FAK during Xenopus development. *Development*, *133*, 685–695. https://doi.org/10.1242/dev.02230.

Lane, M. C., Koehl, M. A., Wilt, F., & Keller, R. (1993). A role for regulated secretion of apical extracellular matrix during epithelial invagination in the sea urchin. *Development*, *117*, 1049–1060.

Larsen, M., Wei, C., & Yamada, K. M. (2006). Cell and fibronectin dynamics during branching morphogenesis. *Journal of Cell Science*, *119*, 3376–3384. https://doi.org/10.1242/jcs.03079.

Leiss, M., Beckmann, K., Girós, A., Costell, M., & Fässler, R. (2008). The role of integrin binding sites in fibronectin matrix assembly in vivo. *Current Opinion in Cell Biology*, *20*, 502–507. https://doi.org/10.1016/j.ceb.2008.06.001.

Li, Q., Zhang, Y., Pluchon, P., Robens, J., Herr, K., Mercade, M., et al. (2016). Extracellular matrix scaffolding guides lumen elongation by inducing anisotropic intercellular mechanical tension. *Nature Cell Biology*, *18*, 311–318. https://doi.org/10.1038/ncb3310.

Mao, Y., & Baum, B. (2015). Tug of war—The influence of opposing physical forces on epithelial cell morphology. *Developmental Biology*, *401*, 92–102. https://doi.org/10.1016/j.ydbio.2014.12.030.

Marsden, M., & DeSimone, D. W. (2001). Regulation of cell polarity, radial intercalation and epiboly in Xenopus: Novel roles for integrin and fibronectin. *Development*, *128*, 3635–3647.

Marsden, M., & DeSimone, D. W. (2003). Integrin-ECM interactions regulate cadherin-dependent cell adhesion and are required for convergent extension in Xenopus. *Current Biology*, *13*, 1182–1191. https://doi.org/10.1016/S0960-9822(03)00433-0.

Martins, G. G., Rifes, P., Amândio, R., Rodrigues, G., Palmeirim, I., & Thorsteinsdóttir, S. (2009). Dynamic 3D cell rearrangements guided by a fibronectin matrix underlie somitogenesis. *PLoS One*, *4*, e7429–14.

Maruthamuthu, V., Sabass, B., Schwarz, U. S., & Gardel, M. L. (2011). Cell-ECM traction force modulates endogenous tension at cell-cell contacts. *Proceedings of the National Academy of Sciences*, *108*, 4708–4713. https://doi.org/10.1073/pnas.1011123108.

McBeath, R., Pirone, D. M., Nelson, C. M., Bhadriraju, K., & Chen, C. S. (2004). Cell shape, cytoskeletal tension, and RhoA regulate stem cell lineage commitment. *Developmental Cell*, *6*, 483–495. https://doi.org/10.1016/S1534-5807(04)00075-9.

McKeown-Longo, P. J., & Mosher, D. F. (1985). Interaction of the 70,000-mol-wt amino-terminal fragment of fibronectin with the matrix-assembly receptor of fibroblasts. *The Journal of Cell Biology*, *100*, 364–374. https://doi.org/10.1083/jcb.100.2.364.

McMillen, P., Chatti, V., Jülich, D., & Holley, S. A. (2016). A sawtooth pattern of cadherin 2 stability mechanically regulates somite morphogenesis. *Current Biology*, *26*, 542–549. https://doi.org/10.1016/j.cub.2015.12.055.

McMillen, P., & Holley, S. A. (2015). Integration of cell-cell and cell-ECM adhesion in vertebrate morphogenesis. *Current Opinion in Cell Biology*, *36*, 48–53.

Metzger, M. B., Hristova, V. A., & Weissman, A. M. (2012). HECT and RING finger families of E3 ubiquitin ligases at a glance. *Journal of Cell Science*, *125*, 531–537. https://doi.org/10.1242/jcs.091777.

Mui, K. L., Chen, C. S., & Assoian, R. K. (2016). The mechanical regulation of integrin-cadherin crosstalk organizes cells, signaling and forces. *Journal of Cell Science, 129*, 1093–1100. https://doi.org/10.1242/jcs.183699.

Narasimha, M., & Brown, N. H. (2004). Novel functions for integrins in epithelial morphogenesis. *Current Biology, 14*, 381–385. https://doi.org/10.1016/j.cub.2004.02.033.

Nogawa, B. H. (1983). Determination of the curvature of epithelial cell mass by mesenchyme in branching morphogenesis of mouse salivary gland. *Journal of Embryology and Experimental Morphology, 73*, 221–232.

Oates, A. C., Morelli, L. G., & Ares, S. (2012). Patterning embryos with oscillations: Structure, function and dynamics of the vertebrate segmentation clock. *Development, 139*, 625–639. https://doi.org/10.1242/dev.063735.

Onodera, T., Sakai, T., Hsu, J. C., Matsumoto, K., Chiorini, J. A., & Yamada, K. M. (2010). Btbd7 regulates epithelial cell dynamics and branching morphogenesis. *Science, 329*, 562–565. https://doi.org/10.1126/science.1191880.

Öztürk-Çolak, A., Moussian, B., Araújo, S. J., & Casanova, J. (2016). A feedback mechanism converts individual cell features into a supracellular ECM structure in Drosophila trachea. *eLife, 5*, e09373.

Pastor-Pareja, J. C., & Xu, T. (2011). Shaping cells and organs in Drosophila by opposing roles of fat body-secreted collagen IV and Perlecan. *Developmental Cell, 21*, 245–256.

Pelham, R. J., & Wang, Y.-l. (1997). Cell locomotion and focal adhesions are regulated by substrate flexibility. *Proceedings of the National Academy of Sciences of the United States of America, 94*, 13661–13665. https://doi.org/10.1073/pnas.94.25.13661.

Plotnikov, S. V., Pasapera, A. M., Sabass, B., & Waterman, C. M. (2012). Force fluctuations within focal adhesions mediate ECM-rigidity sensing to guide directed cell migration. *Cell, 151*, 1513–1527. https://doi.org/10.1016/j.cell.2012.11.034.

Plouhinec, J.-L., Zakin, L., Moriyama, Y., & De Robertis, E. M. (2013). Chordin forms a self-organizing morphogen gradient in the extracellular space between ectoderm and mesoderm in the Xenopus embryo. *Proceedings of the National Academy of Sciences of the United States of America, 110*, 20372–20379. https://doi.org/10.1073/pnas.1319745110.

Rifes, P., Carvalho, L., Lopes, C., Andrade, R. P., Rodrigues, G., Palmeirim, I., et al. (2007). Redefining the role of ectoderm in somitogenesis: A player in the formation of the fibronectin matrix of presomitic mesoderm. *Development, 134*, 3155–3165. https://doi.org/10.1242/dev.003665.

Rozario, T., & DeSimone, D. W. (2010). The extracellular matrix in development and morphogenesis: A dynamic view. *Developmental Biology, 341*, 126–140.

Rozario, T., Dzamba, B., Weber, G. F., Davidson, L. A., & DeSimone, D. W. (2009). The physical state of fibronectin matrix differentially regulates morphogenetic movements in vivo. *Developmental Biology, 327*, 386–398.

Sakai, T., Larsen, M., & Yamada, K. M. (2003). Fibronectin requirement in branching morphogenesis. *Nature, 423*, 876–881.

Sato, Y., Nagatoshi, K., Hamano, A., Imamura, Y., Huss, D., Uchida, S., et al. (2017). Basal filopodia and vascular mechanical stress organize fibronectin into pillars bridging the mesoderm-endoderm gap. *Development, 144*, 281–291.

Schwartz, M. A., & DeSimone, D. W. (2008). Cell adhesion receptors in mechanotransduction. *Current Opinion in Cell Biology, 20*, 551–556. https://doi.org/10.1016/j.ceb.2008.05.005.

Schwarzbauer, J. E., & DeSimone, D. W. (2011). Fibronectins, their fibrillogenesis, and in vivo functions. *Cold Spring Harbor Perspectives in Biology, 3*, a005041.

Shih, H. P., Panlasigui, D., Cirulli, V., & Sander, M. (2016). ECM signaling regulates collective cellular dynamics to control pancreas branching morphogenesis. *Cell Reports, 14*, 169–179.

Sonavane, P. R., Wang, C., Dzamba, B., Weber, G. F., Periasamy, A., & DeSimone, D. W. (2017). Mechanical and signaling roles for keratin intermediate filaments in the assembly and morphogenesis of mesendoderm tissue at gastrulation. *Development, 144*, 4363–4376. https://doi.org/10.1242/dev.155200.

Stemple, D. L. (2005). Structure and function of the notochord: An essential organ for chordate development. *Development, 132*, 2503–2512. https://doi.org/10.1242/dev.01812.

Sun, Z., Guo, S. S., & Fässler, R. (2016). Integrin-mediated mechanotransduction. *The Journal of Cell Biology, 215*, 445–456. https://doi.org/10.1083/jcb.201609037.

Tonning, A., Hemphälä, J., Tång, E., Nannmark, U., Samakovlis, C., & Uv, A. (2005). A transient luminal chitinous matrix is required to model epithelial tube diameter in the Drosophila trachea. *Developmental Cell, 9*, 423–430. https://doi.org/10.1016/j.devcel.2005.07.012.

Trapani, V., Bonaldo, P., & Corallo, D. (2017). Role of the ECM in notochord formation, function and disease. *Journal of Cell Science, 130*, 3203–3211.

van Helvert, S., Storm, C., & Friedl, P. (2018). Mechanoreciprocity in cell migration. *Nature Cell Biology, 20*, 8–20.

Wang, S., Sekiguchi, R., Daley, W. P., & Yamada, K. M. (2017). Patterned cell and matrix dynamics in branching morphogenesis. *The Journal of Cell Biology, 216*, 559–570.

Watanabe, T., Sato, Y., Saito, D., Tadokoro, R., & Takahashi, Y. (2009). EphrinB2 coordinates the formation of a morphological boundary and cell epithelialization during somite segmentation. *Proceedings of the National Academy of Sciences of the United States of America, 106*, 7467–7472. https://doi.org/10.1073/pnas.0902859106.

Weber, G. F., Bjerke, M. A., & DeSimone, D. W. (2011). Integrins and cadherins join forces to form adhesive networks. *Journal of Cell Science, 124*, 1183–1193. https://doi.org/10.1242/jcs.064618.

Weber, G. F., Bjerke, M. A., & DeSimone, D. W. (2012). A mechanoresponsive cadherin-keratin complex directs polarized protrusive behavior and collective cell migration. *Developmental Cell, 22*, 104–115.

Zhang, Q., Magnusson, M. K., & Mosher, D. F. (1997). Lysophosphatidic acid and microtubule-destabilizing agents stimulate fibronectin matrix assembly through rho-dependent actin stress fiber formation and cell contraction. *Molecular Biology of the Cell, 8*, 1415–1425. https://doi.org/10.1091/mbc.8.8.1415.

Zhong, C., Chrzanowska-Wodnicka, M., Brown, J., Shaub, A., Belkin, A. M., & Burridge, K. (1998). Rho-mediated contractility exposes a cryptic site in fibronectin and induces fibronectin matrix assembly. *The Journal of Cell Biology, 141*, 539–551. https://doi.org/10.1083/jcb.141.2.539.

CHAPTER EIGHT

The Fish Egg's Zona Pellucida

Eveline S. Litscher[1], Paul M. Wassarman
Department of Cell, Developmental, and Regenerative Biology, Icahn School of Medicine at Mount Sinai, New York, NY, United States
[1]Corresponding author: e-mail address: eveline.litscher@mssm.edu

Contents

Abstract

All fish eggs are surrounded by an envelope, called the zona pellucida (ZP), that plays various roles during oogenesis, egg deposition, fertilization, and embryogenesis. The fish egg ZP consists of only a few proteins that are homologs of mammalian ZP proteins ZP1, ZP3, and ZP4. Unlike the situation in mammals, in fishes there are often multiple copies of ZP genes, perhaps a consequence of ancient polyploidization, gene amplification, and mutation. Like mammalian ZP proteins, fish egg ZP1-like proteins exhibit conserved organization with distinct domains and motifs, but unlike mammalian ZP1 and ZP4 have a glutamine (Q)- and/or proline (P)-rich stretch as an N-terminal extension. Such extensions may play a role in assembly of ZP fibrils and/or account for certain properties of the fish egg ZP, such as elasticity. Recent proposals suggest that fish egg ZP proteins can adopt amyloid-like structures, serve as antifreeze proteins in Antarctic icefishes, and protect eggs subjected to desiccating conditions in small shallow pools. In this chapter, these and other aspects of fish egg ZP proteins are presented.

Current Topics in Developmental Biology, Volume 130
ISSN 0070-2153
https://doi.org/10.1016/bs.ctdb.2018.01.002

ABBREVIATIONS

aa	amino acids
AFGP	antifreeze glycoproteins
C/c/cys	cysteine
C-terminal	carboxy-terminal
CFCS	consensus furin cleavage site
CT	cytoplasmic tail
CTP	carboxy-terminal propeptide
ECM	extracellular matrix
EHP	external hydrophobic patch
IHP	internal hydrophobic patch
IMP	ice melting-promoting
MW	molecular weight
N-terminal	amino-terminal
polyQ/P	sequence rich in Q and/or P
PPII	left-handed polyP helical structure, left-handed polyproline II helix, polyproline II
PRR	proline-rich region
Q/P	glutamine/proline
SS	signal sequence
TMD	transmembrane domain
VE	vitelline envelope
ZP	zona pellucida
ZP-C	C-terminal subdomain
ZP-N	N-terminal subdomain
ZPD	zona pellucida domain

COLOR OF AMINO ACID SEQUENCES

Cys	red
CFCS	orange and underlined
IHP/EHP	yellow highlight
polyQ/P	underlined
SS	orange
TMD/hydrophobic sequence	dark blue
trefoil domain	light blue
ZPD	green

1. INTRODUCTION TO THE FISH EGG'S ZONA PELLUCIDA

All fish eggs are surrounded by a relatively thick, fibrillar, and elastic extracellular matrix (ECM), variously called a vitelline envelope (VE), chorion, egg envelope, or zona pellucida (ZP). The ECM appears and increases in thickness concomitant with oocyte growth and with increasing numbers of surrounding follicle cells. The ECM, here called the ZP, envelopes the egg and plays significant roles during oogenesis, egg deposition, fertilization, and embryogenesis (Litscher & Wassarman, 2014, 2015). The 32 fish species addressed in this chapter are listed alphabetically in Table 1.

Typically, the fish egg envelope consists of only 2–4 ZP proteins, homologs of mammalian ZP proteins, that are expressed either in the liver under the control of estrogen or in ovaries or, in some species, in both organs (Conner & Hughes, 2003; Spargo & Hope, 2003). All ZP proteins possess a common motif, the ZP domain (ZPD), that participates in polymerization of ZP proteins into higher order structures, such as fibrils and matrices (Jovine, Darie, Litscher, & Wassarman, 2005). The ZPD is a conserved module with a bipartite structure consisting of N-terminal subdomain (ZP-N) and C-terminal subdomain (ZP-C) that adopt an immunoglobulin (Ig)-like fold (Han et al., 2010; Monné & Jovine, 2011) and may have dual functions. For example, the ZP-N domain can serve as an independent structural domain in the absence of ZP-C and as a folding unit that is responsible for polymerization of individual ZP proteins into fibrils. On the other hand, the ZP-C subdomain is always associated with ZP-N and may have other important function(s) (Jovine, Janssen, Litscher, & Wassarman, 2006; Jovine, Qi, Williams, Litscher, & Wassarman, 2002, 2004).

Table 1 List of Fish Species

Catfish—*Anarhichas lupus*
Cod—*Gadus morhua*
Flounder—*Platichthys flesus*
Hagfish—*Eptatretus burgeri, cirrhatus, stoutii*
Halibut—*Hippoglossus hippoglossus*
Icefish—*Champsocephalus gunnari; Chaenocephalus aceratus*
Killifish—*Austrofundulus limnaeus; Fundulus heteroclitus*
Lamprey—*Lampetra fluviatilis, japonica, tridentata; Petromyzon marinus*
Lumpsucker—*Cyclopterus lumpus*
Medaka—*Oryzias latipes*
Ray—*Torpedo marmorata*
Rockcod—*Notothenia neglecta, rossii; Nototheniops larseni, nudifrons*
Salmon—*Salmo salar*
Shark—*Callorhinchus milii; Centrophorus uyato; Mustelus canis, schmitti*
Sturgeon—*Acipenser sinensis, transmontanus*
Toothfish—*Dissostichus mawsoni*
Rainbow trout—*Oncorhynchus mykiss*
Zebrafish—*Brachydanio rerio*

ZP1-like proteins often have an N-terminal Q/P-rich region followed by a trefoil domain and a ZPD with 10 or 12 Cys residues. Homologs are called VEα, VEβ, ChgHα/β, ZI-1,2, ZP1, ZPB, and ZP4, based on sequence similarity and subdivision of these proteins into α/β, or 1/2 reflects a division into subfamilies of 2–3 closely related proteins. Generally, ZP1-like proteins have a higher molecular weight (MW) (50–75 kDa) than ZP3 proteins (40–50 kDa) (Conner & Hughes, 2003; Litscher & Wassarman, 2015). ZP3 proteins, often the smallest of the ZP proteins, consist almost entirely of a ZPD that has 8 Cys residues. Homologs of ZP3 are variously called VEγ, ChgLγ, ZI-3, or ZPC (Del Giacco, Diani, & Cotelli, 2000; Hyllner, Westerlund, Olsson, & Schopen, 2001; Sugiyama, Murata, Iuchi, Nomura, & Yamagami, 1999). Like ZP1-like proteins, some ZP3 proteins are divided into subfamilies of 2–5 or more closely related species. For example, in the zebrafish genome tandem repeats of 5 or more *zp3* genes are found (Liu, Wang, & Gong, 2006). ZP3 proteins sometimes possess a relatively short stretch of repetitive Q/P residues.

ZPd, a protein similar to ZP3, has been found only in a shark species (*Callorhinchus milii*) and consists mainly of a ZPD and an N-terminal EGF domain (Venkatesh et al., 2014). Occasionally homologs of ZP2, also called ZPx and ZPax, are found within a fish genome. They are characterized by multiple copies of the ZP-N subdomain at the N-terminus followed by a ZPD that has 10 Cys residues (Callebaut, Mornon, & Monget, 2007).

It is likely that all ZP proteins were derived from a common ancestral gene that underwent an initial duplication event hundreds of millions of years ago. This event gave rise to "modern ZP3" and to 4 other gene families called *zp1/zp4, zp2, zpax*, and *zpd* (Claw & Swanson, 2012). *zp1/zp4* is an example of gene duplication since adjacent regions of these two genes are located on the same chromosome in the mouse (Smith, Paton, Hughes, & Burt, 2005; Spargo & Hope, 2003). Sequences of fish ZP proteins have been compared with orthologs from mice and humans and, although amino acid (aa) sequences between species vary in length, there is a high degree of identity in overlapping sequences (Litscher & Wassarman, 2007, 2015). For example, sequence comparisons of trout ZP proteins (ZP1α, ZP1β, and ZP3) with orthologs from mice or humans (ZP1 and ZP3) have average identities of 31%–36% (Table 2). It has been suggested that proteins with sequence identities of 40% or more usually have the same function and those with 25%–40% possibly perform similar functions; fish and mammalian ZP1/ZP4 and ZP3 would be in the latter category (Chothia, Gough, Vogel, & Teichmann, 2003). In mammals ZP proteins are structural and sperm-binding proteins,

Table 2 Sequence Identity of ZP Proteins

Identity (%)	Trout ZP1α	Trout ZP1β	Trout ZP3
Mouse ZP1/ZP3	33	32	35
Human ZP1/ZP3	33	31	36

Whole sequences of trout (*O. mykiss*) ZP1α/β and ZP3 are compared with sequences of mouse and human ZP1 and ZP3. The numbers represent % identity of trout ZP1α/β and ZP3 with the corresponding mouse and human orthologs (it should be noted that the aa sequences vary in length).

whereas in many fish ZP proteins are solely structural proteins since the egg envelope has a micropyle for sperm entry.

ZP proteins are synthesized as precursor polypeptides with a signal sequence (SS) at the N-terminus and a C-terminal propeptide (CTP) containing a transmembrane domain (TMD); however, some fishes have no recognizable TMD in their CTP. They have a ZPD with 8, 10, or 12 conserved Cys residues present as intramolecular disulfides, followed by a consensus furin cleavage site (CFCS) located close to the C-terminus, and, if present, a TMD or hydrophobic sequence downstream of the CFCS. ZP1-like proteins always have a trefoil domain upstream of the ZPD (Darie, Biniossek, Jovine, Litscher, & Wassarman, 2004; Wassarman, 2008). It has been demonstrated that elements within the CTP are required for secretion and assembly of ZP proteins. In addition to a CFCS, the CTP contains a hydrophobic peptide, called an external hydrophobic patch (EHP) and, with some exceptions, a TMD followed by a short hydrophilic C-terminus. Another hydrophobic peptide, called an internal hydrophobic patch (IHP), is present within the ZPD. Sequence alignments of vertebrate ZP1–4 orthologs, including fish species, revealed that the IHP and EHP are well-conserved units within vertebrate ZP proteins (Jovine et al., 2004; Litscher & Wassarman, 2014).

2. HAGFISH AND LAMPREY EGGS

Vertebrates fall into two major clades, cyclostomes (jawless) and gnathostomes (jawed). Most living vertebrates have jaws, like sharks, bony fishes, or mammals, but two small fish groups, hagfishes and lampreys, are without jaws. They are eel-shaped and scale-less and do not have vertebrae, but overall their anatomy is similar to that of other fishes although more simple than jawed vertebrates. Today, hagfishes and lampreys are the only living representatives of jawless vertebrates (Janvier, 2010; Shimeld & Donoghue, 2012).

Despite recent advances, hagfishes and lampreys are not easy systems to work with compared to other vertebrate models, but are interesting because of their phylogenetic background. They are the only surviving members from an ancient and diverse group of jawless fishes and, consequently, are informative of the genomic makeup and developmental biology of ancestral vertebrates (Heimberg, Cowper-Sal-lari, Sémon, Donoghue, & Peterson, 2010). The first lamprey whole-genome sequence was published in 2013 and provided insight into the origin and evolution of the vertebrate lineage (Smith et al., 2013). Hagfishes and lampreys are thought to belong to the oldest living group of vertebrates, more than 350 million years old, and are good candidates with which to address questions concerning the evolutionary origin and duplication events of ancestral ZP genes.

Hagfishes have a worldwide marine distribution. They live on soft bottoms and sediments in burrows to depths of 1300 m and more than 80 species have been described. They are attractive to biologists in the context of vertebrate evolution but are also of interest to commercial fisheries as a source of food (Ota & Kuratani, 2006).

Female hagfishes tend to have a small number of eggs (20–40), and the eggs are notably larger than those of many oceanic vertebrates (Fig. 1). The eggs are ellipsoid in shape with a long axis (e.g., *Eptatretus cirrhatus*, 11 mm × 29 mm) and have anchoring filaments (2–4 mm in length) on both ends of the axis (Martini & Beulig, 2013). The laid eggs stick together at their ends with the anchor-filament-tufts to form a cluster. A micropyle with a funnel for sperm entry is located at the animal pole, but no one has reported observing their reproductive behavior or deposition of fertilized eggs. An opercular ring opens when the young hagfish hatches.

The hagfish egg is surrounded by a relatively thick envelope (e.g., *Eptatretus stoutii*, 200–250 μm at the operculum end) and is described as a fibrous structure covering the entire egg including the bottom of the micropylar region. This is unlike the situation in teleost eggs in which the bottom of the micropyle adjoins directly the egg's plasma membrane (Koch, Spitzer, Pithawalla, Castillos, & Wilson, 1993). The overall thickness of the egg envelope gives it substantial mechanical strength, possibly a requirement for large eggs in a benthic (e.g., ocean floor) environment (Ota & Kuratani, 2006).

Lampreys live in coastal areas, rivers, and lakes with more than 40 species described. Some species (e.g., *Petromyzon marinus*, sea lamprey) have a parasitic lifestyle in which they feed on blood and fluids of other fishes. Like hagfishes they are also appreciated as food for humans. The ovary of the female *Lampetra fluviatilis* (river lamprey) fills the body cavity and

Fig. 1 Ovulated New Zealand hagfish egg (*E. cirrhatus*) showing an opercular ring (Or) and anchoring filaments (Af). *Reproduced with permission from Oxford University Press: Zintzen, V., Roberts, C. D., Shepherd, L., Stewart, A. L., Struthers, C. D., Anderson, M. J., et al. (2015). Review and phylogeny of the New Zealand hagfishes (Myxiniformes: Myxinidae), with a description of three new species.* Zoological Journal of the Linnean Society, 174, *363–393.*

contains 20,000–40,000 eggs. Ovulated eggs are ovoid in shape, 1 mm × 1.5 mm, with a peculiar tuft at the animal pole, the only place through which sperm can enter the egg, and the egg envelope is 6 μm (vegetative pole) to 11 μm (animal pole) in thickness (Dziewulska & Domagała, 2009). In *Lampetra japonica* (Arctic lamprey), the egg envelope is a double-layered fibrous matrix with a radial pattern (Kobayashi & Yamamoto, 1994) (Fig. 2). The embryo develops in its translucent egg envelope for about 3 weeks until hatching (*Lampetra tridentata*, Pacific lamprey) (Fig. 3). Sea lamprey ZP genes have been predicted based on their similarity to other fish

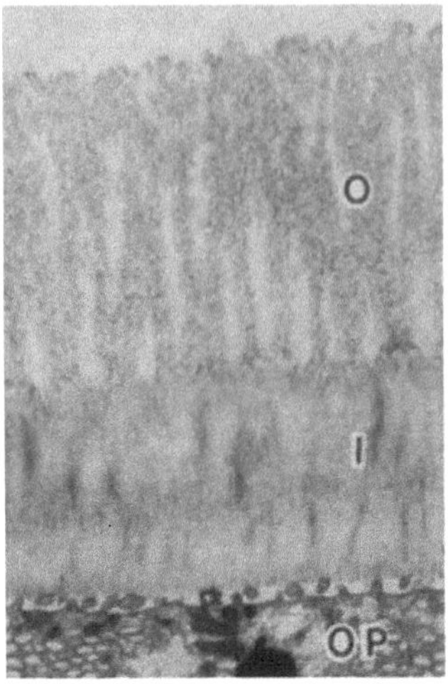

Fig. 2 Transmission electron micrograph of a section through the animal pole region of an unfertilized Arctic lamprey egg (*L. japonica*) showing the outer (O) and inner (I) layer of the egg envelope, about 300 μm in width. *OP*, ooplasm. *Reproduced with permission from John Wiley and Sons: Kobayashi, W., Yamamoto, T. S. (1994). Fertilization of the Lamprey (*Lampetra japonica*) eggs: Implication of the presence of fast and permanent blocks against polyspermy.* Journal of Experimental Zoology Part A: Ecological Genetics and Physiology, 269, *166–176.*

ZP genes and classified as ZP homologs (e.g., *zp2, zp4,* and *zpax*). However, the predicted protein sequences are not full-length due to incomplete coverage of the genome (Xu et al., 2012).

3. CARTILAGINOUS FISH EGGS

Cartilaginous fishes (*chondrichthyes*) represent the oldest surviving jawed vertebrates and, as the name suggests, have a skeleton made out of cartilage. They include sharks, rays, and skates (*elasmobranchii*) and chimeras (*holocephali*).

Chondrichthyes have highly diversificated reproductive strategies and are characterized by internal fertilization and genital tracts that are specialized for uterine gestation. More than 50% of *chondrichthyes* give birth to live young as compared to 2%–3% for *ostheichthyes*. Egg laying *chondrichthyes* typically produce relatively large oocytes within often elaborately designed egg capsules made primarily of collagen and fertilization must occur before egg capsule formation is completed. The capsule provides protection and anchorage throughout development of the embryo. Some of the largest oocytes

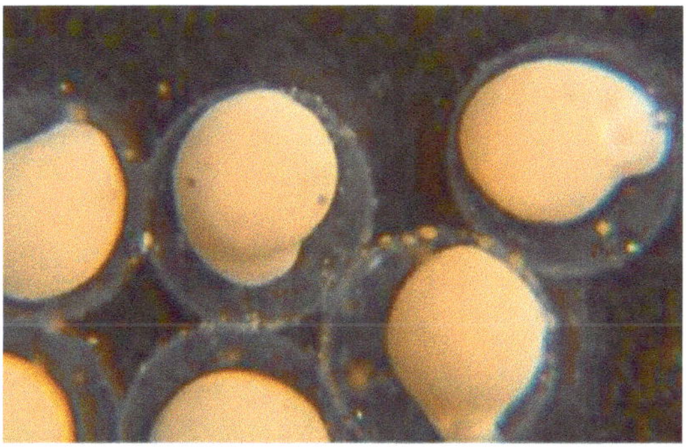

Fig. 3 Developing Pacific lamprey embryos (*L. tridentata*) within transparent egg envelopes. *Reproduced with permission from NOAA Northwest Fisheries Science Center.*

Table 3 Approximate Egg Diameter and ZP Width in *Chondrichthyes*

Species	Egg Diameter (cm)	ZP Width (μm)
Torpedo marmorata	3	25
Mustelus canis	2	50–70
Mustelus schmitti	3	30
Centrophorus uyato	≤10	50–70[a]

[a]A thick ZP is seen early in oogenesis; oocyte diameter 2–4 mm. As the oocyte grows, the width of the ZP narrows. Maximal thickness of ZP early in oogenesis has also been observed in *Mustelus* species.

known, ≥10 cm in diameter, are produced in *chondrichthyes* (Davenport, Weaver, & Wourms, 2011; Galíndez, Díaz Andrade, & Estecondo, 2014; Prisco, Loredana, & Piero, 2002) (Table 3). The ZP has no surface ornamentation or micropyle as in teleosts and appears as a relatively thick fibrous matrix surrounding the oocyte (Davenport et al., 2011) (Fig. 4).

The relatively small genome of *Callorhinchus milii* (elephant shark), a member of the subclass *holocephali*, makes it an attractive cartilaginous fish model for whole-genome sequencing and comparative analysis (Venkatesh, Tay, Dandona, Patil, & Brenner, 2005). The elephant shark is oviparous with each egg encased in a spindle-shaped capsule, 25 cm in length by 10 cm in width. The entire genome of the elephant shark has been sequenced, and, interestingly, the degree of conserved synteny and similarity of sequences compared to the human genome is higher than that between human and teleost fishes (Venkatesh et al., 2014). Molecular analysis of egg envelope proteins identified 3, perhaps 4, ZP encoding genes, *zp2*, *zp3*, *zp4*, and possibly *zpd*; a fifth gene, *zpax*, was found to be identical to *zp2*. Conceptual translations

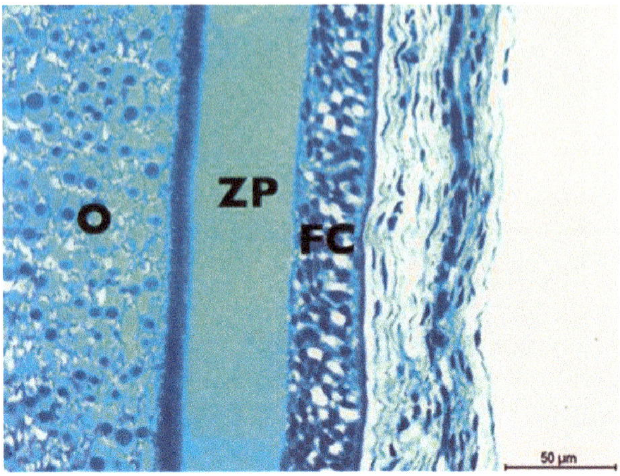

Fig. 4 Photomicrograph of a section through a little gulper shark oocyte (*C. uyato*). The oocyte is about 6 mm in diameter and the ZP 50–60 µm in width. *FC*, follicle cells; *O*, oocyte. *Reproduced with permission from I. Davenport, Xavier University of Louisiana, New Orleans, LA 70125.*

Table 4 Comparison of ZP Sequence Homologies

	O. latipes	*H. sapiens*	Sequence Overlap (aa)
C. milii	Identity/Similarity (%)	Identity/Similarity (%)	*O. latipes/H. sapiens*
ZP2	35/67	28/60	851/477
ZP3	45/71	42/70	309/390
ZP4		39/69	—/400

C. milii (elephant shark) ZP2, ZP3, and ZP4 are compared to orthologs in *O. latipes* (medaka) and *H. sapiens* (human). *C. milii* ZP2 is compared to *O. latipes* ZPax, a homolog of ZP2.

of *zp2–zp4* and comparison of their aa sequences to ZP orthologs in fishes and mammals revealed their homology (Table 4).

Elephant shark ZP2-4 and ZPd (only partial sequences available) have a ZPD, 8 or 10 Cys residues, a CFCS, and a C-terminal TMD. ZP4 has, in addition, a trefoil domain (Table 5). Interestingly, elephant shark ZP4 has a P-rich motif N-terminal to the trefoil domain (68 P out of 201 residues; p_3–p_{203}), a sequence which may contain helical elements (e.g., a left-handed polyproline II helix [PPII]) (Sequence 1). Also, comparison with ZP4 orthologs of other fish species shows that elephant shark ZP4 is structurally more closely related to sturgeon (see Section 5) and mammalian ZP1 with respect to the number of Cys residues in its ZPD. For example, elephant

Table 5 Characteristics of *C. milii* ZP Proteins

ZP Protein	Length (aa)	Trefoil (aa)	ZPD (aa)	Cys (no.)	CFCS	TMD (aa)
ZP2	954	—	619–882	10	$rmgr_{890}$	928–950
ZP3	451	—	68–325	8	$rgkr_{362}$	417–439
ZP4	616	207–251	254–523	10	kr_{532}	578–600
ZPd (partial)	278	—	2–176	8	$rsgr_{186}$	230–252

SEQUENCE 1 *C. milii* (elephant shark) ZP4 (NCBI: XP_007898913).

```
1   vlptgavlsp lprptlspvp sgtlpsrpvp lplpspslps rsvpspaspf pspalsstps
61  pvlptgavls plprptlspv psgtlpsrsv plpipwpshp srsvpfpspa lssipspvlp
121 travplpsps lfprptlspv psgsfpprpt plpspspsll prpvttpahs plprptlspv
181 glprpiaspa pspalaprpt pspdegdtcd vdaqhriacg npaisevach lkgccfdaps
241 artpcfygrt erveCtrdlf sivipkdvtv pplnlssvsl angraseCsp kvrtaafivf
301 rfpvsaCgtt qvfdgqtity etdviaerei ldgpsgvitr dstyrlhveC rfagdaelpf
361 qvlvntpapp slvvkagvln memriakeql ystwyakady pvlkflrqpv fvevrvlnrt
421 dahinlmlyd Cwatpspdsd hrtqwrilvn gCpyygdnyl trlhrvnrss evpfpthrkr
481 felktfgfwd skiqnllhgk ayfhCsavvC htsrmesCsl pCaddkhngf kreaevdvqs
541 epvqrrlvtl kspiifapet aaktvskssq iheslsgvil lgvavgslli ilmsgvtmcw
601 kcrelqqhks tgsrsp
```

Table 6 Comparison of Elephant Shark ZP4 With Trout ZP1α, Sturgeon ZP4, and Mammalian ZP1

Elephant shark ZP4	$_{373}$vv-kagvlnmemriak-----------eqly
Trout ZP1α	$_{366}$vvsvdavlhvelrlangrClskgCdemqeay
Sturgeon ZP4	$_{333}$va-eqgplrielriar-----------dkay
Mouse ZP1	$_{391}$vt-qsgplrlelriat-----------dktf
Human ZP1	$_{399}$mt-qpgplrlelriat-----------detf

Aligned, partial sequences of elephant shark (*C. milii*) ZP4, trout (*O. mykiss*) ZP1α, sturgeon (*A. sinensis*) ZP4, mouse and human ZP1 and positions of two additional Cys residues in trout ZP1α. Cys, *red*; IHP, *yellow highlight*; partial ZP-C subdomain, *green*.

shark, Chinese sturgeon (*Acipenser sinensis*), and mouse/human ZP1 have 10 Cys residues, and rainbow trout (*Oncorhynchus mykiss*) ZP1α (and ZP1β) has 12 Cys residues in the ZPD; the additional 2 Cys residues are located between IHP and ZP-C (Table 6).

4. "MODERN" FISH EGGS

Bony fishes (*osteichthyes*) are a diverse and large group that have an ossified endoskeleton. They include lobe- (*sarcopterygii*) and ray-finned (*actinopterygii*) fishes. *Sarcopterigyii* include coelacanths and lungfishes whose ancestors gave rise to four-limbed tetrapods. *Actinopterygii* are divided into 3 groups: *chondrostei* (e.g., sturgeon), *holostei* (e.g., gar), and *teleostei* (e.g., carp, trout, and halibut). *Teleostei* are a very diverse group that includes more than 30,000 known species and most of today's living fishes (Litscher & Wassarman, 2015).

Osteichthyes usually produce a large number of small eggs with little yolk that are fertilized externally. Development of embryos is within the protective egg envelope, but without the "stiff" egg capsules that surround chondrichthyan eggs. Eggs are spherical and vary in size from 0.2 to 5 mm in diameter. Benthic eggs are on average larger than pelagic eggs and have, in general, a much thicker egg envelope than pelagic eggs. For example, *Salmo salar* (Atlantic salmon) deposit their eggs in a nest of pebbles, and the egg envelope is 30–60 μm thick (Reid & Chaput, 2012; Songe et al., 2016). Often, these nonbuoyant benthic eggs have an egg envelope with a "sticky" outer surface which adheres to substrates (Berois, Arezo, & Papa, 2011). On the other hand, *Gadus morhua* (Atlantic cod) has small pelagic eggs and their egg envelope is 7.5 μm thick; the highly hydrated eggs float, and the egg envelope has a "smooth" outer surface (Lonning, Kjorsvik, & Falk-Petersen, 1988) (Table 7). Rainbow trout egg ZPs, manually collected and purified, exhibit a one-layer envelope with fibers radially oriented (Brivio, Bassi, & Cotelli, 1991) (Fig. 5). The fibers are less than 1 μm in diameter and have a wave-shaped fibrillary appearance.

In some fish species, the egg envelope is made up of 2–3 layers, which differ in ultrastructure, and an adhesive layer (jelly layer) outside of the egg envelope may also be present. Sperm do not bind to the egg envelope, but enter the egg via an opening in the egg envelope, called the micropyle, to reach the egg surface. In some cases (e.g., in sturgeon), the egg envelope has several micropyles (Cherr & Clark, 1985).

In early reports the egg envelope has sometimes been called a "zona radiata" because of a radial pattern of pore channels in the envelope as seen by light microscopy. The pore channels are remnants from retracted microvilli on the oocyte surface and retracted processes from follicle cells that traversed the ZP during oocyte growth, but retreated shortly before ovulation (Laale, 1980).

Table 7 Approximate Egg Diameter and ZP Width in Bony Fishes

Species	Egg Diameter (mm)	ZP Width (μm)
Pelagic Eggs[a]		
Gadus morhua (Atlantic cod) (liver[b])	1.3–1.6	7.5
Brachydanio rerio (Zebrafish) (ovary)	0.5	1.3
Platichthys flesus (Flounder)	0.8	2.5
Hippoglossus hippoglossus (Halibut) (liver)	3.0	9.0
Benthic Eggs[c]		
Salmo salar (Atlantic salmon)	5–7	30–60
Oncorynchus mykiss (Rainbow trout) (liver)	5.5	15
Oryzias latipes (Medaka) (liver/ovary)	1.35	9.1
Anarhichas lupus (Atlantic catfish)	5.4	38
Cyclopterus lumpus (Lumpsucker)	2.3	60
Acipenser transmontanus (White sturgeon)	3.5–4.0	100
Acipenser sinensis (Chinese sturgeon)	3.7–4.9	—

[a]E.g., Floating eggs.
[b]Site of synthesis, if known.
[c]E.g., Nonbuoyant eggs.

Fig. 5 Photomicrograph of isolated and purified rainbow trout ZP (*O. mykiss*). The ZP is about 20 μm in width. *Reproduced with permission from F. Cotelli, University of Milano, 20122 Milano, Italy.*

Characterization of "modern" fish ZP proteins has been carried out for many species, using a variety of methods. The published data comprise a vast literature on composition and characterization of ZPs and ZP proteins, as well as analysis and expression of ZP genes. ZP genes are expressed in either the liver under the control of estrogen or locally in oocytes, or in both tissues

(Sano et al., 2013). Often, a female fish releases several tens of thousands of relatively small eggs at a time, and in order to synthesize large amounts of protein in a relatively short time it may be necessary to use an additional organ (besides the ovary) and, perhaps, make use of amplified ZP genes (Conner & Hughes, 2003; Glasauer & Neuhauss, 2014; Litscher & Wassarman, 2014; Sano et al., 2017). For example, ZP precursor proteins are synthesized in the liver and then transported via circulating blood to the oocyte for uptake; this is similar to the synthesis of egg yolk precursor, vitellogenin, in the liver of amphibians and birds (Darie et al., 2005). Synthesis of individual ZP proteins must be coordinated in a timely manner to ensure that they are present simultaneously so they can assemble first into fibrils and then into a matrix. It is possible that ZP precursors are transported in the bloodstream in small vesicles, perhaps bound to specific receptors, that dock on to the oocyte's plasma membrane for uptake into oocytes. Precursor proteins would then follow the biosynthetic pathway described earlier. On the other hand, some fish species (e.g., zebrafish) release relatively small clutches of eggs (hundreds of eggs) and do not express ZP proteins in the liver, only in the ovary (Mold, Dinitz, & Sambandan, 2009).

5. STURGEON EGGS

Sturgeons (*acipenseridae*) are large, freshwater, or anadromous fishes of north temperate regions. They are modern relicts that were dominant during Paleozoic times and are represented today by about two dozen species in four genera (also referred to as "primitive" fish) (Billard & Lecointre, 2001). They are excellent food fishes, and their eggs are commercially important as caviar.

Sturgeons lay a large number of eggs and spawn in the spring, migrating upstream to deposit their eggs over hard substrates. The eggs are adhesive and attach to rocks and pebbles. Interestingly, sturgeons are characterized by a high capacity for hybridization; e.g., many species will hybridize if they live within the same geographical area. Usually, hybrids $2n \times 2n$ or $4n \times 4n$ are fertile, but crosses $2n \times 4n$ (triploid) are sterile such that no oocytes will form in females.

In *Acipenser transmontanus* (white sturgeon) the ZP is laid down during oocyte growth. A thin matrix appears concomitantly with proliferation of granulosa cells and reaches a maximum thickness when the oocyte diameter is greater than 3 mm and consists of 3 main layers. The 3-layer ZP matrix, 100 μm in width, consists of at least 7 components, 3 of which are

Table 8 Characteristics of *A. sinensis* ZP Proteins

Protein	Length (aa)	Trefoil (aa)	ZPD (aa)	Cys (no.)	CFCS	TMD (aa)
ZP3.1	440	—	69–326	8	$rkrr_{360}$	409–431
ZP3.2	394	—	27–270	8	$rkrr_{301}$	348–370
ZP3.3	397	—	27–273	8	$rrkr_{304}$	351–373
ZP4	592	167–210	213–485	10	$rfvr_{491}$	561–583
ZPax	946	—	595–876	10	$rsar_{882}$	921–943

considered to be main components with MWs of 86–100 and 47 kDa. They have been immunolocalized to the ovary and liver of females (Murata, Conte, McInnis, Fong, & Cherr, 2014). The liver-synthesized ZP proteins are released into the blood and transferred to the ovary and oocytes to be incorporated into the ZP. It is possible that follicle cells also participate in ZP protein synthesis.

In *A. sinensis* (Chinese sturgeon), 5 ZP genes have been identified, cloned, and sequenced; these include 3 variants of *zp3* (*zp3.1, zp3.2*, and *zp3.3*), *zp4*, and *zpax*. The transcribed sequences of the ZP proteins range in size from 394 to 946 aa residues. All 5 ZP proteins have a ZPD, a CFCS and a TMD, and ZP4 has a trefoil domain (Li et al., 2011) (Table 8). ZP3.1, ZP3.2, and ZP3.3 have a ZPD with 8 Cys residues, ZP4 and ZPax have a ZPD with 10 Cys residues, and, in the case of ZP4, the ZPD sequence is similar to elephant shark ZP4 and to the mouse and human ZP1 which have 10 Cys residues (see also Section 3; Table 6).

6. RAINBOW TROUT ZONA PELLUCIDA PROTEINS

The ZP of rainbow trout eggs consists of 3 proteins, ZP1α, ZP1β, and ZP3, with MWs of 58–60, 52, and 47 kDa, respectively, under denaturing and reducing conditions (Darie et al., 2004; Hyllner et al., 2001). Each protein possesses a SS, a ZPD, and a CFCS followed by a hydrophobic C-terminus. ZP1α and ZP1β also have a trefoil domain just upstream of the ZPD and a long stretch of polyQ/P between the SS and trefoil domain. ZP3 has a short polyQ/P motif between the SS and ZPD. All 3 proteins possess a relatively large number of Cys residues (ZP1α, 18; ZP1β, 18; and ZP3, 12) of which 12 are present in the ZPD of ZP1α and ZP1β, 8 in the ZPD of ZP3, and 6 in the trefoil domain of ZP1α and ZP1β.

ZP1α contains a long stretch of polyQ/P (41 Q, 61 P) N-terminal of the trefoil domain, encompassing 181 aa residues (q_{23}–q_{203}), an IHP (d_{370}–e_{376}) between the two ZPD subdomains (ZP-N and ZP-C), and 2 extra Cys residues (c_{384}lskgc_{389}) followed by a ZP-N. The C-terminal sequence has a CFCS (rqrr_{533}) followed by an EHP (s_{551}–i_{557}) that is located within the hydrophobic cytoplasmic tail (CT) (Sequence 2).

ZP1β contains also a long stretch of polyQ/P (35 Q, 54 P) N-terminal of the trefoil domain, encompassing 144 aa residues (q_{21}–q_{164}), an IHP (p_{330}–e_{336}) between the two ZPD subdomains (ZP-N and ZP-C), and 2 extra Cys residues (c_{344}ltkgc_{349}) followed by a ZP-N. The C-terminal sequence has a CFCS (rkrr_{492}) followed by an EHP (s_{511}–t_{517}) which is located within the hydrophobic CT (Sequence 3).

ZP3 contains a shorter stretch of polyQ/P (14 Q, 19 P) encompassing 76 aa residues (q_{23}–q_{98}), an IHP (y_{231}–t_{238}) between the two ZPD

SEQUENCE 2 ***O. mykiss*** **(rainbow trout) ZP1α (NCBI: NP_001117745).**

```
1   malqwsvvcl vavamlgclc daqlkwpyqp pqnpaqplpp rpaeplpqwp tqplpqrpae
61  plpqrpaqpl pqrpaqplpq rpaqplpqwp aqplpqrpaq plpqrpaqpl pqwptqplpq
121 rpaeplpqrp aqplpqrpaq plpqwpaqpl tqwpaqpfpq rpaqplpqrp aqtlpqrpaq
181 pflqkpaqni pqqipytkgd tkqtcevvdk dkvlcglsgi naaqcqaisc cfdgrmcfyg
241 ktvtvqCtkd gqfvvvvsrd atlpklelds isllgangah Ctpvgttsaf aiyqfkvteC
301 gtvvteepdt ivyenrmsss yvvgngpfgd itrdshydll fqCrytgtsv etliievrty
361 pnpnpvvsvd avlhvelrla ngrClskgCd emqeaytsyy tvadypvtkv lrdpvyaevr
421 ilgmtdpnvv lileqCwant sptgerlprw dllvngCpyq ddryltvpir sdssyfppge
481 ffshykrflf kmftfvdpts mvplqenvyi hCsatvChal agsCeqmCnr qrrdlsaqgq
541 kktkgdvvvs sqkvimidpr fya
```

SEQUENCE 3 ***O. mykiss*** **(rainbow trout) ZP1β (NCBI: AAF71259).**

```
1   mkwsavclva vatlgwlcda qiylekpgwp piqtppswpa qppqkpiqpp qrpaqppqwp
61  vqppqrpaqp pqrpaqppqr paqpqqwpaq ppqrpaqppq wpaqppqrpa qppqrpaqpp
121 qrpaqppqrp aqppqwpvhp pqwpvqpqtp lqrpkfpsdt qskqscdvdg qhkvqcglpd
181 itaahcdain ccfdgrmcfy gksvtvqCtk dgqfvvvvar datlpsleld sisllgtnga
241 hChpigttsv faiyqfkvte Cgtvmteetd tiiyenrmss syqvgvgpfg sitrdsqydl
301 tfqCrykgst ivavvidvkp vpppnsdiap gplivelrlg sggCltkgCn eeevaytsyy
361 teadypvtkv lrdpvytevr ilartdpniv ltlgrCwatt tpnplslpqw dllidgCpyq
421 ddrylttpit vgpssglsyp thyrrfvlkm ftfvdpmsma plretvfihC ntavClpslg
481 dsCeprCyrk rrdipaavqk tariksnlvs sgeliltdpr eltn
```

subdomains (ZP-N and ZP-C) followed by a ZP-N. The C-terminal sequence has a CFCS ($rkgr_{412}$) followed by an EHP (w_{424}–s_{436}) which is located within the hydrophobic CT. ZP3 has 4 extra Cys residues (C9–12) between ZPD and CFCS, linked C_9–C_{11} and C_{10}–C_{12} (Sequence 4).

Rainbow trout ZP1α- and ZP1β-IHP sequences are found in similar positions compared to mouse and human. The 2 additional Cys residues, found only in teleostean fish ZP1-like proteins, not in sturgeons and sharks, are covalently linked as disulfides and positioned between the IHP and the ZP-C (Tables 6 and 9). ZP1α- and ZP1β-EHP sequences are located 17 or 18 residues, respectively, downstream of the CFCS within a hydrophobic

SEQUENCE 4 *O. mykiss* (rainbow trout) ZP3 (NCBI: NP_001117746).

```
1   mamkwsvvcl vavamlgclc vagnwppfsk pvqqpfrpnr qppqqpqqpq qppyqkprip
61  pkdqtqakqk fetpldwtyp ldpkpepkii qsseartpva ansvraeCre nmvhveakhd
121 llgigqliql edltlgdCpm tgfdninqvl ifesplqsCg sqlrmttnsl iyiftlyykp
181 kplantplir tndaminieC hyprkhnvss laliptwtpf saakyaeell yfsmrlmtad
241 wqyeragnmy vlgdmvniea svmqyfhvpl rifvdsCvat lepninanpr yafienhgCl
301 idakmtgshs qfmprsadyk lyfqveafrf qiqkgsdpin pqktkipfqa asdypatldm
361 ifitChlkat tiafpidfey kaCsyintwr eaggndgvcg ccdstcsnrk grdttkhqkl
421 vniwegdvql gpifisekva q
```

Table 9 Comparisons of IHP and EHP Locations of Trout ZP1α/β With Mouse and Human Orthologs

IHP	
ZP1α	341fqcrytg.ts v.etliievr typnpnpvvs vdavlhvelr langrClskg Cdemqeay
ZP1β	302fqcrykg..s tivavvidvk pvpppnsdia .pgplivelr lgsggCltkg Cneeevay
mZP1	366vrcifna..s dflpiqasif s.pqppapvt qsgplrlelr iat....... ....dktf
hZP1	374vrcvfna..s dflpiqasif p.ppspapmt qpgplrlelr iak....... ....detf
EHP	
ZP1α	526qmcn....rq rrdlsaqgqk ktkg.dvvvs sqkvimidpr fya563
ZP1β	485prcy....rk rrdipaavqk tariksnlvs sgeliltdpr eltn524
mZP1	537ttcdsgiarr rr.ssghhni tlraldivss pgavgfedaa klepsgss.. ..rnsssr
hZP1	544tacstgttrq rr.ssghrnd tarpqdivss pgpvgfedsy gqeptlgptd sngnsslr

Aligned, partial sequences of trout (*O. mykiss*) ZP1α/β with orthologs in mouse (mZP1) and human ZP1 (hZP1). Note that trout ZP1α/β has 2 additional Cys residues between the IHP and ZP-C subdomain. Cys, *red*; IHP, EHP *yellow highlight*; partial ZP-N, ZP-C subdomains, *green*.

stretch similar to mouse and human ZP1. However, unlike mouse and human that possess a TMD downstream of the EHP, trout ZP1α and ZP1β do not have a defined TMD to anchor them in the plasma membrane, but they do have a hydrophobic sequence (Sequences 2 and 3; Table 9).

ZP1α, ZP1β, and ZP3 are synthesized in hepatocyte cells in the liver under control of estradiol-17β and are present as precursor proteins in the blood. They are transported as monomers via circulating blood to oocytes for incorporation and assembly into a ZP. Although ZP1α, ZP1β, and ZP3 do not have a TMD they do have an IHP and EHP which prevent their premature polymerization into fibrils in the bloodstream. However, it is not known how these proteins are transported to their site of assembly or assembled into fibrils and matrix. Are they transported in small vesicles bound to specific receptors that dock on to the oocyte's plasma membrane for uptake? Perhaps they are processed at the apical surface of the oocyte's membrane by a membrane-anchored furin-like protease before assembly. On the other hand, there is evidence that ZP proteins with a deleted TMD can be secreted from cells and assembled into a matrix as long as the two hydrophobic patches, IHP and EHP, are present (Darie et al., 2005; Jovine et al., 2004).

Mass spectrometric evidence suggests that ZP1α, ZP1β, and ZP3 undergo cleavage at CFCS after arrival in the ovary, followed by polymerization and assembly of a matrix. For example, soluble ZP precursors at the site of assembly are converted into insoluble proteins and fibrils by proteolysis of the hydrophobic C-terminal tail and dissociation of the EHP from the IHP. In this manner the proteins are activated for assembly into dimers, tetramers, and higher order structures. It is likely that heterodimers serve as building blocks of fibrils (Darie, Janssen, Litscher, & Wassarman, 2008).

Small amounts of unprocessed ZP1α + ZP3 and ZP1β + ZP3 heterodimers, with their CFCS intact, have been detected in oocyte preparations, suggesting that covalent cross-linking and heterodimerization happen prior to cutting at the CFCS. Heterodimerization of fish ZP1-like proteins via covalent bonds can occur through their N-terminal Q-rich region catalyzed by transglutaminase; however, under physiological conditions, this event occurs after fertilization and is described as "hardening of the envelope" (Chang, Wang, & Huang, 2002). It is possible that the heterodimers detected in samples subjected to mass spectrometry represent an artifact and are due to transglutaminase release during sample preparation.

In vitro experiments have shown that ZP1β and ZP3 can polymerize into dimers and oligomers of variable size depending on experimental

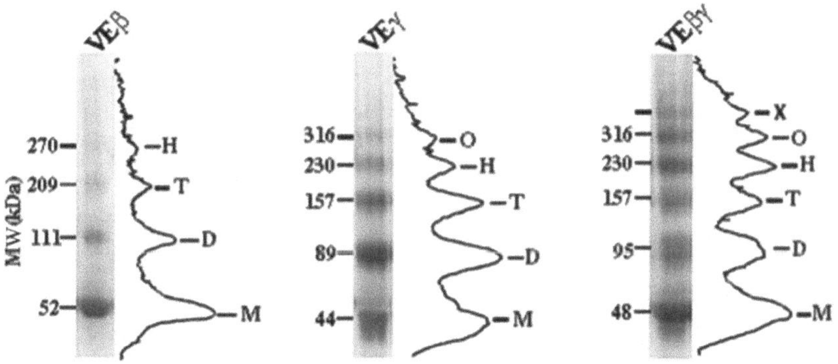

Fig. 6 Densitometric plots of rainbow trout ZP1β (VEβ), ZP3 (VEγ), and ZP1β+ ZP3 (VEβγ) subjected to Blue Native-PAGE in 6 M urea. The positions of the monomer (M), dimer (D), tetramer (T), hexamer (H), and octomer (O) are indicated on the *right*. An additional high MW band in the lane containing ZP1β+ZP3 (VEβγ) is labeled X. Apparent MWs of the bands are indicated on the *left*.

conditions. They are present primarily as monomers in 6 M urea, but dimerize and polymerize into large oligomers in Tris-buffer or water and are held together by noncovalent bonds. For example, when purified proteins are analyzed by gel electrophoresis under native conditions (Blue Native-PAGE; Schagger & von Jagow, 1991) (Fig. 6), ZP1β and ZP3 are resolved as a series of discrete bands with MWs from 52 and 44 kDa (monomers), respectively, to higher MW bands of noncovalently linked homodimers, homotetramers, and higher order homopolymers. In addition, a mixture of ZP1β and ZP3 displays a ladder of monomers, dimers, oligomers, etc., but it is not clear if these dimers and oligomers are heteropolymers of ZP1β and ZP3. Estimates of the MWs of the oligomers strongly suggest that ZP1β and ZP3 are present as monomers, dimers, and tetramers. Trimers are not detected, and it is likely that the formation of oligomers larger than tetramers occurs by polymerization of ZP1β and ZP3 dimers. ZP1α is present in relatively small amounts and was not characterized further. Size exclusion chromatography showed that ZP1β does not form higher oligomers on its own, but preferably associates with ZP3 to form higher oligomers. ZP3, on the other hand, is mostly found in high MW homopolymers. In samples prepared for transmission electron microscopy (Tris-buffered solutions or water), purified ZP1β and ZP3 appear as relatively large aggregates of long fibrils with contiguous beads located periodically along the fibrils. Measurements of bead sizes reveal that they have a diameter of 151 ± 23 Å for ZP1β and 135 ± 16 Å for ZP3 and indicate that each bead consists of a homodimer (Darie et al., 2008; Litscher & Wassarman, 2015) (Fig. 7).

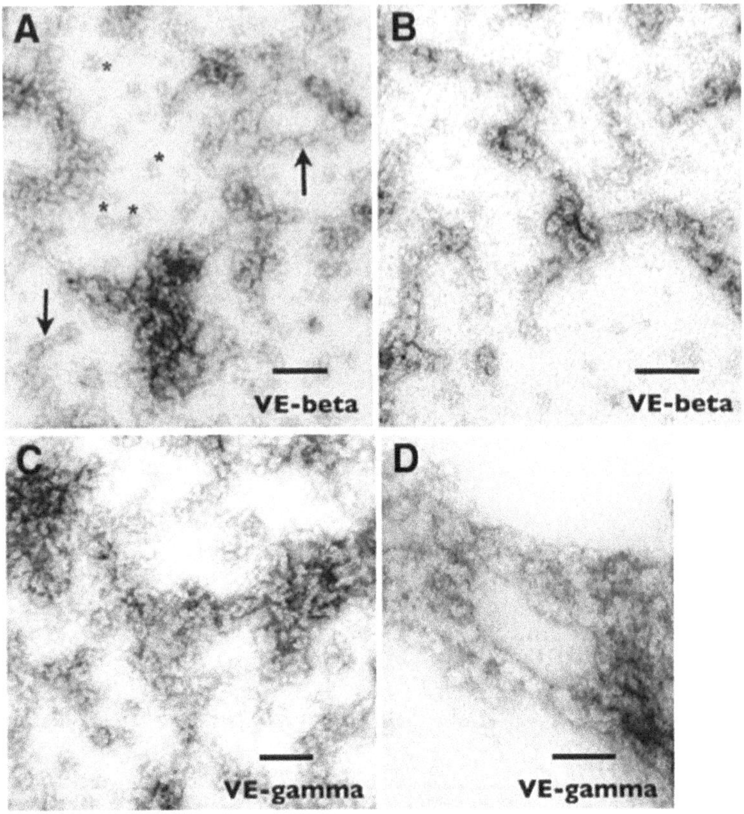

Fig. 7 Transmission electron micrographs of ZP1β (VEβ) and ZP3 (VEγ) in buffer. Micrographs showing ZP1β (VEβ) (A, 200,000X; B, 265,000X) and ZP3 (VEγ) (C, 222,000X; D, 370,000X) fibrils. In panel (A) *arrows* indicate contiguous beads along a fibril, and *asterisks* indicate individual beads.

In summary, these results suggest that in the absence of urea, ZP proteins first dimerize, then polymerize, and finally form large aggregates of long fibrils that are unable to enter native gels. In distilled water or buffer they form very long fibrils composed of contiguous beads and probably consist of ZP1 or ZP3 as homodimers. On the other hand, in the presence of urea, ZP1β and ZP3 give rise to a ladder of smaller oligomers that can be detected by Blue Native-PAGE; a useful method to isolate and characterize protein complexes under native conditions. Coomassie Brilliant Blue G-250 binds to proteins and provides a negative charge for electrophoretic separation without denaturing the proteins, thus allowing analysis of molecular mass, oligomeric state, and composition.

As discussed earlier, the bipartite structure of the ZPD has been implicated in the assembly of the ZP. It was suggested that the N-terminal half of the ZPD (ZP-N) is a conserved module responsible for polymerization through noncovalent interactions into long fibrils (Jovine et al., 2006, 2002; Litscher, Janssen, Darie, & Wassarman, 2008). Adjoining regions, such as long stretches of polyQ/P, the trefoil domain, the ZP-C subdomain, and the protease-sensitive interdomain linking ZP-N with ZP-C, are very likely to be involved in additional functions (e.g., N-terminal cross-linking of ZP1 proteins, formation of ZP1 or ZP3 homodimers, ZP1/ZP3 heterodimers).

7. POSTFERTILIZATION CHANGES IN THE ZONA PELLUCIDA

After fertilization, the ZP hardens as a result of cross-linking between proteins of the ZP. The conversion from a soft to hardened ZP is thought to be due, at least in part, to transglutaminase activity that catalyzes formation of ϵ-(γ-glutamyl) lysine cross-links between ZP proteins (Chang et al., 2002). The hardened egg envelope protects the developing embryo until hatching.

At the time of hatching, polymerized and cross-linked ZP proteins are digested by hatching enzymes. Two hatching enzymes, a high choriolytic enzyme (HCE) and a low choriolytic enzyme (LCE), have been identified and characterized in the killifish *Fundulus heteroclitus* (Atlantic killifish) (Kawaguchi et al., 2010). It is suggested that HCE cuts proline-rich (P-X-Y) repeat sequences in the N-terminal region of ZP1 and ZP3 into small peptide fragments of various lengths (3–12 aa). HCE itself does not cleave the ϵ-(γ-glutamyl) lysine bonds responsible for the chorion hardening. Subsequently, HCE-digested envelopes swell, and the structure of the ZP matrix is loosened, consisting now of ZP fibrils with "shortened" ZP1 and ZP3. The swelling might be caused by exposure of hydrophilic regions which in turn would expose previously sequestered sequences that are then accessible to LCE. LCE cleaves the loosened matrix, possibly within the protease-sensitive region of ZP-N and ZP-C of ZP1 and ZP3, resulting in the complete solubilization of the ZP matrix (Kawaguchi et al., 2010) (Fig. 8).

A similar hatching mechanism is also found in medaka (*Oryzias latipes*) and rainbow trout, and it is possible that the mechanism of egg envelope digestion postfertilization is well conserved in "modern" fish eggs.

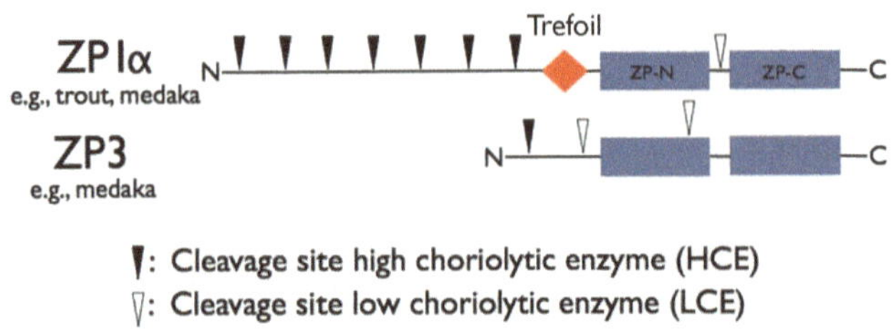

Fig. 8 Schematic representation of ZP1α and ZP3 with approximate HCE and LCE cleavage sites. For example, there are sites within the N-terminal proline-rich region and the protease-sensitive linker region between the ZP-N and ZP-C subdomains (ZP1α), within ZP-N subdomain, and upstream of ZP-N (ZP3). *Modified after Kawaguchi, M., Yasumasu, S., Shimizu, A., Sano, K., Iuchi, I., & Nishida, M. (2010). Conservation of the egg envelope digestion mechanism of hatching enzyme in euteleostean fishes.* The FEBS Journal. 277, *4973–4987.*

8. ANTIFREEZE PROTECTION BY ZONA PELLUCIDA PROTEINS

Antarctic icefishes live in cold antarctic and subantarctic waters, and they adapt to freezing temperature by synthesizing antifreeze glycoproteins (AFGPs) to lower the freezing point of body fluids. AFGPs are distributed throughout the body via circulation. However, AFGP genes do not seem to be expressed in eggs and so the question arises, how do the eggs prevent themselves from freezing? Is it possible that extensive duplication of ZP genes, under the selective pressure of freezing temperatures, contributes to antifreeze protection? In this context, it has recently been proposed that ZP proteins can be used as AFGPs and amplification of ZP copy numbers thereby enhances freeze prevention (Cao et al., 2016). In addition, the thickness of the ZP can vary according to the environment, and its morphology may be an indication of the ecological conditions for spawning and egg development (Riehl & Kock, 1989) (Table 10).

Amplification of ZP genes is found in several species of Antarctic fishes. For example, the Antarctic toothfish (*Dissostichus mawsoni*) has multiple numbers of ZP proteins, among them at least 10 isoforms of ZP3 and multiple copies of each isoform. There seems to be a positive correlation between ZP gene expansion and the severity of freezing water temperatures

Table 10 Approximate Egg Diameter and ZP Width in Antarctic Icefishes

Species	Egg Diameter (mm)	ZP Width (μm)
Champsocephalus gunnari (Mackerel icefish)	3.5–4.1	22–25
Chaenocephalus aceratus (Blackfin icefish)	4.4–4.7	20–22
Notothenia rossii (Marbled rockcod)	5.0	20–25
Notothenia neglecta (Yellowbelly rockcod)	4.6	23–28
Nototheniops nudifrons	2.4–2.5	25
Nototheniops larseni	1.7	12–15

since subantarctic species and species that live in warmer waters (e.g., medaka) have a lower copy number of these genes.

What is required for ZP proteins of icefishes to exert an "ice melting-promoting" (IMP) activity? IMP activity is conformation dependent, is correlated with the amount of ZP protein monomers in solution, and involves an electrostatical potential through charged aa in "acidic patches" on the surface of ZP proteins (Cao et al., 2016). For example, a tropical fish like zebrafish has multiple copies of *zp3* genes, but zebrafish ZP3 exhibits weaker electrostatic surface potentials and does not exhibit IMP activity found in its Antarctic toothfish counterpart. Several *notothenioid* species that live in freezing continental shelf waters have an expanded copy number of ZP genes, and perhaps release of a surplus of unpolymerized ZP proteins from the oocyte into the perivitelline space results in a lowering of the freezing point of perivitelline fluid and thereby freeze resistance of oocytes.

9. ZONA PELLUCIDA FIBRILS AND DESICCATION

Populations of the annual killifish (*Austrofundulus limnaeus*) live in small bodies of water in arid parts of Venezuela and their embryos survive in these temporary pools of water. The embryos are drought tolerant and survive under desiccating conditions by reducing evaporative water loss. It has been suggested that the egg envelope and the structure of the ZP proteins are important for the survival of the embryo (Podrabsky, Carpenter, & Hand, 2001).

ZPs were isolated from killifishes, solubilized under varying conditions, and examined by spectroscopy and microscopy. Long and unbranched fibrils with a diameter of 4–6 nm and indeterminate length were found in these preparations, and staining with Congo red suggested that they exhibited typical characteristics of amyloid fibrils (see also Section 10). Generally, amyloid fibrils are enriched in β-sheet secondary structure in which the peptide strands are aligned orthogonal to the direction of fibril growth, a pattern called "cross-β motif" with a typical 3–4 nm width.

The unfertilized ZP is 80% soluble in 8 M Guanidine-HCl and consists of two major proteins with MWs of 34 and 24 kDa. The secondary structure of these protein mixtures indicates that they contain β-sheets (~47%), turns (~30%), α-helices (~10%), and intermolecular β-sheets (~14%). Dehydration caused an increase in intermolecular β-sheet content of the envelope proteins that was reversible on rehydration. These structural changes may help decrease water permeability and increase resistance of embryos to water stress. Interestingly, the data from unfertilized and fertilized egg envelopes were very similar, except for the solubility assays in which fertilized egg envelopes were much less soluble (i.e., only 10% soluble). It is likely that the reduced solubility of the fertilized killifish envelopes is due to the expanded Q-repeats (see below) by which transglutaminase links Q- to K-residues through isopeptide bonds (Chang et al., 2002).

Recently published aa sequences of killifish egg envelope proteins suggest there are 2 major types of ZP proteins, called ZP1/4-like and ZP3 (partial sequences and isoforms included). For example, a predicted ZP4 sequence includes a polyQ/P repeat motif, called a "low complexity" sequence, followed by a trefoil domain and ZPD (Sequence 5).

In general, polyQ is conformationally adaptable and can adopt either a β-sheet or a polyproline II (PPII) helix conformation; a polyQ tract interrupted by polyP is pushed toward a PPII-like conformation and away from β-sheet conformation (Adzhubei, Sternberg, & Makarov, 2013; Darnell, Derryberry, Kurutz, & Meredith, 2009; Darnell, Orgel, Pahl, & Meredith, 2007; Rath, Davidson, & Deber, 2005).

Considering motifs rich in polyQ/P in fish ZP1-like proteins, it is possible that an N-terminal repeat sequence of polyQ/P has a propensity toward PPII helix conformation (see also Section 11). On the other hand, fibril formation often involves β-sheet structures, and a PPII helix conformation of N-terminal-polyQ/P motifs in ZP1/4 proteins is probably not involved in polymerization and fibril formation (Jahn et al., 2010). Hence, as an in vivo example of functional amyloid fibrils, it is likely that the β-sheet structures of the ZP-N subdomain is responsible for polymerization and assembly into fibrillar structures (as found in solubilized killifish ZP preparations) with the amyloid characteristics mentioned earlier.

SEQUENCE 5 ***A. limnaeus*** **(annual killifish) ZP4 (NCBI: XP_013867760).**

```
1   mkliycclla aalysylaaa qsgkpqypqi psypprqppp qtppqkpanp ppqpppqtpp
61  qvpkypqkpa yppppppplt pplvpqypqk payppppqppp qpppltpplv pqypqkpayp
121 ppppppqtpp qvphypskpa npppppppqt pphvpqypqk panpppppa npppqepphp
181 qfppppqvpq pdsfhiceva enykiqcgpp sttahdcevl nccydgrmcy ygkyvtlqCt
241 kdgefivvva rdstlpnidl esinflgtgd hChpvgttsa faiyqfpvtd Cgtvmteepg
301 viiyqnkmts lfevavgphg sitrdslfev qvqCkyvgta vqalvidvgl lppplpvaal
361 gplrvelrig ngqCvtkgCv edqvaytsyy tdadypitki lrdpvyvevh mldrtdania
421 ltlgrCwant npiadylpqw dllihgCpnr ddryltrlip vdassgvyyp syykrfefqm
481 ftfvsggasd ytkktpsdqp ftplsehvfv hCevsiCqit atdnCeprCf rkrreisgst
541 kklyrdettl isskeliitq pise
```

10. ZP1 AND AMYLOID FIBRILS

A structural model of the fish ZP-N subdomain (ZP1) consensus sequence has been created based on the structure of the mouse ZP3 ZP-N homolog (Louros et al., 2014). Similar to the mammalian fold, the derived model contained an Ig-like β-sandwich fold with 2 antiparallel β-sheets, each composed of 4 β-strands, A, B, D, E and C, E′, F, G, and linked by 2 disulfide bonds.

Two segments of the ZP-N sequence, β-strands A and G, were predicted to interact for polymerization of ZP1/ZP3 fibrils in the fish egg envelope. The synthesized peptides (A = vtvqa*t, G = fellfqa*; *cysteine substituted with alanine) each self-assemble (A/A and G/G) into long fibrils of indeterminate length, but with a defined width of 10–12 nm. Peptide A fibrils appeared to be long and unbranched and tended to interact laterally to form ribbons and aggregates. Peptide G fibrils sometimes form double helical structures (e.g., 2 fibrils of 5–7 nm width, twisted around each other). These observations led to a proposal that A-G β-sheets interact with adjacent ZP-N A-G β-sheets to form dimers and polymerize into fibrils, possibly ZP1–ZP3 heterodimer fibrils. The A/G fibrils formed in vitro showed the typical physical characteristics of amyloid fibers. They stained with Congo red and had a characteristic green birefringence under polarized light, showed uniform antiparallel β-sheets, and produced a typical cross-β diffraction pattern (Fandrich, 2007; Jahn et al., 2010; Louros et al., 2014). The analogy between amyloid fibers and ZP fibrils suggests that ZP matrices have properties of a functional amyloid-like structure built by assembling ZP fibrils into the meshwork or matrix of a functional ZP (Fowler, Koulov, Balch, & Kelly, 2007) (see also Section 9).

11. ZP1 AMINO-TERMINAL DOMAINS

In many fish species, ZP1 and ZP1-like (ZP4) proteins contain N-terminal motifs that are rich in Q and/or P residues. For example, rainbow trout ZP1α has 41 Q and 61 P residues in 181 aa residues upstream of the trefoil domain, and elephant shark ZP4 has 68 P residues in 201 aa residues upstream of the trefoil domain. Sequences rich in Q and/or P residues are also called polyQ, polyP, and polyQ–polyP tracts or regions and are often referred to as "low complexity" domains, but predicting their structures is difficult (Rath et al., 2005). However, it has been shown that a polyP tract adopts a type of left-handed helical structure (PPII), and that a polyQ tract can readily adopt either a β-sheet or a PPII-like conformation (Adzhubei et al., 2013). The presence of a polyP region adjacent to polyQ can shift the balance of polyQ–polyP toward a PPII conformation (Darnell et al., 2007). Due to the conformational properties of P, it is assumed that PPII is also a dominant conformation of P-rich regions (PRRs), i.e., in longer spans of multiple, but nonconsecutive P residues. It has been proposed that a major structural feature of PPII is its flexibility and involvement in structural elasticity and self-assembly processes. PPII helices are often found immediately preceding α-helices in loops and interdomain linker regions, and in N- and C-terminal regions.

Q residues are also involved in modification of the ZP during fertilization. For example, transglutaminidase cross-linking of Q to K residues results in the formation of insoluble protein polymers in a modified ZP, called the fertilization envelope (Chang et al., 2002) (see also Section 7). Interestingly, ZP4 of *C. milii* has no stretch of Q residues in its N-terminal domain, but does have a long PRR extension, perhaps because the egg is encased by a newly made stiff collagenous egg capsule following fertilization.

Q/P-rich regions in ZP1 and ZP4 N-terminal domains are present in many fish species and are also present in birds and reptiles, but are not found in mammalian ZP1/ZP4. For example, chicken ZP1 has an additional ZP-N domain followed by a Q/P-rich region and mouse ZP1 has an additional ZP-N domain, but no Q/P-rich region (Table 11; Fig. 9). These findings are consistent with the fact that birds are positioned between fish and mammals phylogenetically. Note that for all ZP proteins the ZP-N

Table 11 ZP1/ZP4 N-Terminal Domains Upstream of the ZPD

Species	ZP-N[a]	Q/P-Rich Regions (aa)[b]	Trefoil[c]	Polypeptide (aa)[d]
Elephant shark				
ZP4	–	68 P/1 Q (201)	+	616 (P: 3–203)
Sturgeon[e]				
ZP4	–	8 Q/9 P	+	592
Trout[f]				
ZP1α	–	41 Q/61 P (181)	+	563 (Q/P: 23–203)
ZP1β	–	35 Q/54 P (144)	+	524 (Q/P: 21–164)
Medaka				
ZP1	–	42 Q/72 P (208)	+	591 (Q/P: 23–230)
ZP1 *minor*	–	88 Q/75 P (239)	+	634 (Q/P: 24–262)
Annual killifish				
ZP4	–	25 Q/90 P (171)	+	564 (Q/P:21–191)
Zebrafish				
ZP1	–	5 Q/8 P	+	431
Chicken				
ZP1	+	70 Q/59 P (454)	+	934 (Q/P: 126–579)
Cobra[g]				
ZP1	+	79 Q/103 P (542)	+	1035 (Q/P: 133–674)
Mouse				
ZP1	+	3 Q/20 P	+	623
Human				
ZP1	+	5 Q/16 P	+	638

[a]ZP-N, ~90–95 aa, single copy of ZP-N subdomain in N-terminal position.
[b]Number of Q/P residues; length of polyQ/P tract.
[c]Trefoil, ~40–45 aa.
[d]Length of polypeptide; position of Q/P tract.
[e]Sturgeon, *A. sinensis*.
[f]Rainbow trout, *O. mykiss*.
[g]King cobra, *Ophiophagus hannah*.

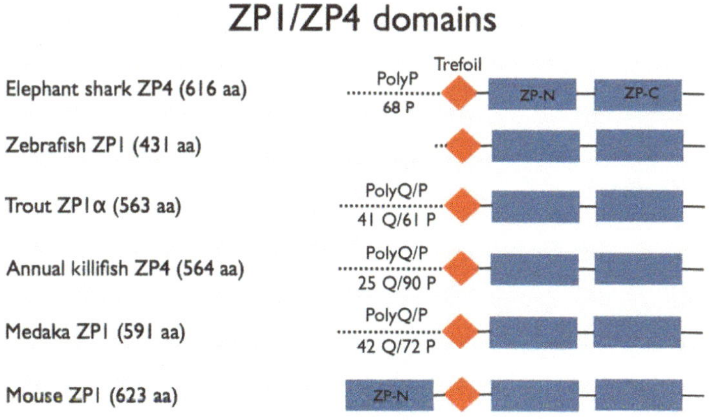

Fig. 9 Schematic representation of domains in ZP1/ZP4 with emphasis on the N-terminal regions that are polyP-rich and polyQ/P-rich in fishes. Such regions are not found in mammalian ZP1/4 that have an additional ZP-N subdomain instead.

subdomain is responsible for polymerization of ZP fibrils. On the other hand, PPII conformations at the N-terminus of ZP1/ZP4 in fish species may be responsible for elasticity and/or self-assembly of ZP fibrils (Bochicchio & Tamburro, 2002).

ACKNOWLEDGMENTS

Many thanks to Drs. Luca Jovine and Costel Darie who contributed to our research on rainbow trout egg ZP proteins carried out at the Icahn School of Medicine at Mount Sinai. This research was supported in part by the National Institutes of Health (NICHD).

REFERENCES

Adzhubei, A. A., Sternberg, M. J. E., & Makarov, A. A. (2013). Polyproline-II helix in proteins: Structure and function. *Journal of Molecular Biology*, *425*, 2100–2132.

Berois, N., Arezo, M. J., & Papa, N. G. (2011). Gamete interactions in teleost fish: The egg envelope. Basic studies and perspectives as environmental biomonitor. *Biological Research*, *44*, 119–124.

Billard, R., & Lecointre, G. (2001). Biology and conservation of sturgeon and paddlefish. *Reviews in Fish Biology and Fisheries*, *10*, 355–392.

Bochicchio, B., & Tamburro, A. M. (2002). Polyproline II structure in proteins: Identification by chiroptical spectroscopies, stability, and functions. *Chirality*, *14*, 782–792.

Brivio, M. F., Bassi, R., & Cotelli, F. (1991). Identification and characterization of the major components of the *Oncorhynchus mykiss* egg chorion. *Molecular Reproduction and Development*, *28*, 85–93.

Callebaut, I., Mornon, J. P., & Monget, P. (2007). Isolated ZP-N domains constitute the N-terminal extensions of zona pellucida proteins. *Bioinformatics*, *23*, 1871–1874.

Cao, L., Huang, Q., Wu, Z., Cao, D. D., Ma, Z., et al. (2016). Neofunctionalization of zona pellucida proteins enhances freeze-prevention in the eggs of Antarctic notothenioids. *Nature Communications*, 7, 12987.

Chang, Y. S., Wang, Y. W., & Huang, F. L. (2002). Cross-linking of ZP2 and ZP3 by transglutaminase is required for the formation of the outer layer of fertilization envelope of carp egg. *Molecular Reproduction and Development*, *63*, 237–244.

Cherr, G. N., & Clark, W. R., Jr. (1985). Gamete interaction in the white sturgeon *Acipenser transmontanus*: A morphological and physiological review. *Environmental Biology of Fishes*, *14*, 11–22.

Chothia, C., Gough, J., Vogel, C., & Teichmann, S. A. (2003). Evolution of the protein repertoire. *Science*, *300*, 1701–1703.

Claw, K. G., & Swanson, W. J. (2012). Evolution of the egg: New findings and challenges. *Annual Review of Genomics and Human Genetics*, *13*, 109–125.

Conner, S. J., & Hughes, D. C. (2003). Analysis of fish ZP1/ZPB homologous genes—Evidence for both genome duplication and species-specific amplification models of evolution. *Reproduction*, *126*, 347–352.

Darie, C. C., Biniossek, M. L., Gawinowicz, M. A., Milgrom, Y., Thumfart, Y. O., et al. (2005). Mass spectrometric evidence that proteolytic processing of rainbow trout egg vitelline envelope proteins takes place on the egg. *The Journal of Biological Chemistry*, *280*, 37585–37598.

Darie, C. C., Biniossek, M. L., Jovine, L., Litscher, E. S., & Wassarman, P. M. (2004). Structural characterization of fish egg vitelline envelope proteins by mass spectrometry. *Biochemistry*, *43*, 7459–7478.

Darie, C. C., Janssen, W. G., Litscher, E. S., & Wassarman, P. M. (2008). Purified trout egg vitelline envelope proteins VEβ and VEγ polymerize into homomeric fibrils from dimers *in vitro*. *Biochimica et Biophysica Acta*, *1784*, 385–392.
Darnell, G. D., Derryberry, J. M., Kurutz, J. W., & Meredith, S. C. (2009). Mechanism of cis-inhibition of polyQ fibrillation by polyP: PPII oligomers and the hydrophobic effect. *Biophysical Journal*, *97*, 2295–2305.
Darnell, G., Orgel, J. P. R. O., Pahl, R., & Meredith, S. C. (2007). Flanking polyproline sequences inhibit β-sheet structure in polyglutamine segments by inducing PPII-like helix structure. *Journal of Molecular Biology*, *374*, 688–704.
Davenport, I. R., Weaver, A. L., & Wourms, J. P. (2011). A novel set of structures within the elasmobranch, ovarian follicle. *Journal of Morphology*, *272*, 557–565.
Del Giacco, L., Diani, S., & Cotelli, F. (2000). Identification and spatial distribution of the mRNA encoding an egg envelope component of the cyprinid zebrafish, *Danio rerio*, homologous to the mammalian ZP3 (ZPC). *Development Genes and Evolution*, *210*, 41–46.
Dziewulska, K., & Domagała, J. (2009). Ripening of the oocyte of the river lamprey (*Lampetra fluviatilis* L.) after river entry. *Journal of Applied Ichthyology*, *25*, 752–756.
Fandrich, M. (2007). On the structural definition of amyloid fibrils and other polypeptide aggregates. *Cellular and Molecular Life Sciences*, *64*, 2066–2078.
Fowler, D. M., Koulov, A. V., Balch, W. E., & Kelly, J. W. (2007). Functional amyloid—From bacteria to humans. *Trends in Biochemical Sciences*, *32*, 217–224.
Galíndez, E. J., Díaz Andrade, M., & Estecondo, S. (2014). Morphological indicators of initial reproductive commitment in *Mustelus schmitti* (springer 1939) (*Chondrichthyes, Triakidae*): Folliculogenesis and ovarian structure over the life cycle. *Brazilian Journal of Biology*, *74*(Suppl), S154–S163.
Glasauer, M. K., & Neuhauss, C. F. (2014). Whole-genome duplication in teleost fishes and its evolutionary consequences. *Molecular Genetics and Genomics*, *289*, 1045–1060.
Han, L., Monné, M., Okumura, H., Schwend, T., Cherry, A. L., et al. (2010). Insights into egg coat assembly and egg-sperm interaction from the X-ray structure of full-length ZP3. *Cell*, *143*, 404–415.
Heimberg, A. M., Cowper-Sal-lari, R., Sémon, M., Donoghue, P. C. J., & Peterson, K. J. (2010). microRNAs reveal the interrelationships of hagfish, lampreys, and gnathostomes and the nature of the ancestral vertebrate. *Proceedings of the National Academy of Sciences of the United States of America*, *107*, 19379–19383.
Hyllner, S. J., Westerlund, L., Olsson, P. E., & Schopen, A. (2001). Cloning of rainbow trout egg envelope proteins: Members of a unique group of structural proteins. *Biology of Reproduction*, *64*, 805–811.
Jahn, T. R., Makin, O. S., Morris, K. L., Marshall, K. E., Tian, P., et al. (2010). The common architecture of cross-β amyloid. *Journal of Molecular Biology*, *395*, 717–727.
Janvier, P. (2010). MicroRNAs revive old views about jawless vertebrate divergence and evolution. *Proceedings of the National Academy of Sciences of the United States of America*, *107*, 19137–19138.
Jovine, L., Darie, C. C., Litscher, E. S., & Wassarman, P. M. (2005). Zona pellucida domain proteins. *Annual Review of Biochemistry*, *74*, 83–114.
Jovine, L., Janssen, W. G., Litscher, E. S., & Wassarman, P. M. (2006). The PLAC1-homology region of the ZP domain is sufficient for protein polymerisation. *BMC Biochemistry*, *7*, 11.
Jovine, L., Qi, H., Williams, Z., Litscher, E., & Wassarman, P. M. (2002). The ZP domain is a conserved module for polymerization of extracellular proteins. *Nature Cell Biology*, *4*, 457–461.
Jovine, L., Qi, H., Williams, Z., Litscher, E. S., & Wassarman, P. M. (2004). A duplicated motif controls assembly of zona pellucida domain proteins. *Proceedings of the National Academy of Sciences of the United States of America*, *101*, 5922–5927.

Kawaguchi, M., Yasumasu, S., Shimizu, A., Sano, K., Iuchi, I., & Nishida, M. (2010). Conservation of the egg envelope digestion mechanism of hatching enzyme in euteleostean fishes. *The FEBS Journal, 277*, 4973–4987.

Kobayashi, W., & Yamamoto, T. S. (1994). Fertilization of the lamprey (*Lampetra japonica*) eggs: Implications of the presence of fast and permanent blocks against polyspermy. *The Journal of Experimental Zoology, 269*, 166–176.

Koch, E. A., Spitzer, R. H., Pithawalla, R. B., Castillos, F. A., III, & Wilson, L. J. (1993). The hagfish oocyte at late stages of oogenesis: Structural and metabolic events at the micropylar region. *Tissue and Cell, 25*, 259–273.

Laale, H. W. (1980). The perivitelline space and egg envelopes of bony fishes: A review. *Copeia, 1980*, 210–226.

Li, C. J., Wei, Q. W., Chen, X. H., Zhou, L., Cao, H., et al. (2011). Molecular characterization and expression pattern of three zona pellucida 3 genes in the Chinese sturgeon, *Acipenser sinensis. Fish Physiology and Biochemistry, 37*, 471–484.

Litscher, E. S., Janssen, W. G., Darie, C. C., & Wassarman, P. M. (2008). Purified mouse egg zona pellucida glycoproteins polymerize into homomeric fibrils under non-denaturing conditions. *Journal of Cellular Physiology, 214*, 153–157.

Litscher, E. S., & Wassarman, P. M. (2007). Egg extracellular coat proteins: From fish to mammals. *Histology and Histopathology, 22*, 337–347.

Litscher, E. S., & Wassarman, P. M. (2014). Evolution, structure, and synthesis of vertebrate egg-coat proteins. *Trends in Developmental Biology, 8*, 65–76.

Litscher, E. S., & Wassarman, P. M. (2015). *A guide to zona pellucida domain proteins*. Hoboken, NJ: John Wiley and Sons. 172 pp.

Liu, X., Wang, H., & Gong, Z. (2006). Tandem-repeated zebrafish *zp3* genes possess oocyte-specific promoters and are insensitive to estrogen induction. *Biology of Reproduction, 74*, 1016–1025.

Lonning, S., Kjorsvik, E., & Falk-Petersen, I. B. (1988). A comparative study of pelagic and demersal eggs from common marine fishes in northern Norway. *Sarsia, 73*, 49–60.

Louros, N. N., Petronikolou, N., Karamanos, T., Cordopatis, P., Iconomidou, V. A., & Hamodrakas, S. J. (2014). Structural studies of "aggregation-prone" peptide-analogues of teleostean egg chorion ZPB proteins. *Biopolymers, 102*, 427–436.

Martini, F. H., & Beulig, A. (2013). Morphometrics and gonadal development of the hagfish *Eptatretus cirrhatus* in New Zealand. *PLoS One, 8*, e78740.

Mold, D. E., Dinitz, A. E., & Sambandan, D. R. (2009). Regulation of zebrafish zona pellucida gene activity in developing oocytes. *Biology of Reproduction, 81*, 101–110.

Monné, M., & Jovine, L. (2011). A structural view of egg coat architecture and function in fertilization. *Biology of Reproduction, 85*, 661–669.

Murata, K., Conte, F. S., McInnis, E., Fong, T. H., & Cherr, G. N. (2014). Identification of the origin and localization of chorion (egg envelope) proteins in an ancient fish, the white sturgeon, *Acipenser transmontanus. Biology of Reproduction, 90*, 1–12.

Ota, K. G., & Kuratani, S. (2006). The history of scientific endeavors towards understanding hagfish embryology. *Zoological Science, 23*, 403–418.

Podrabsky, J. E., Carpenter, J. F., & Hand, S. C. (2001). Survival of water stress in annual fish embryos: Dehydration avoidance and egg envelope amyloid fibers. *American Journal of Physiology. Regulatory, Integrative and Comparative Physiology, 280*, R123–R131.

Prisco, M., Loredana, R., & Piero, A. (2002). Ultrastructural studies on developing follicles of the spotted ray *Torpedo marmorata. Molecular Reproduction and Development, 61*, 78–86.

Rath, A., Davidson, A. R., & Deber, C. M. (2005). The structure of "unstructured" regions in peptides and proteins: Role of the polyproline II helix in protein folding and recognition. *Biopolymers, 80*, 179–185.

Reid, J. E., & Chaput, G. (2012). Spawning history influence on fecundity, egg size, and egg survival of Atlantic salmon (*Salmo salar*) from the Miramichi River, New Brunswick, Canada. *ICES Journal of Marine Science, 69*, 1678–1685.

Riehl, R., & Kock, K. H. (1989). The surface structure of Antarctic fish eggs and its use in identifying fish eggs from the Southern Ocean. *Polar Biology*, *9*, 197–203.

Sano, K., Kawaguchi, M., Katano, K., Tomita, K., Inokuchi, M., et al. (2017). Comparison of egg envelope thickness in teleosts and its relationship to the sites of ZP protein synthesis. *Journal of Experimental Zoology. Part B, Molecular and Developmental Evolution*, *328*, 240–258.

Sano, K., Kawaguchi, M., Watanabe, S., Nagakura, Y., Hiraki, T., & Yasumasu, S. (2013). Inferring the evolution of teleostean zp genes based on their sites of expression. *Journal of Experimental Zoology. Part B, Molecular and Developmental Evolution*, *320*, 332–343.

Schagger, H., & von Jagow, G. (1991). Blue native electrophoresis for isolation of membrane protein complexes in enzymatically active form. *Analytical Biochemistry*, *199*, 223–231.

Shimeld, S. M., & Donoghue, P. C. J. (2012). Evolutionary crossroads in developmental biology: Cyclostomes (lamprey and hagfish). *Development*, *139*, 2091–2099.

Smith, J. J., Kuraku, S., Holt, C., Sauka-Spengler, T., Jiang, N., et al. (2013). Sequencing of the sea lamprey (*Petromyzon marinus*) genome provides insights into vertebrate evolution. *Nature Genetics*, *45*, 415–421.

Smith, J., Paton, I. R., Hughes, D. C., & Burt, D. W. (2005). Isolation and mapping the chicken zona pellucida genes: An insight into the evolution of orthologous genes in different species. *Molecular Reproduction and Development*, *70*, 133–145.

Songe, M. M., Willems, A., Sarowar, M. N., Rajan, K., Evensen, O., et al. (2016). A thicker chorion gives ova of Atlantic salmon (*Salmo salar* L.) the upper hand against *Saprolegnia* infections. *Journal of Fish Diseases*, *39*, 879–888.

Spargo, S. C., & Hope, R. M. (2003). Evolution and nomenclature of the zona pellucida gene family. *Biology of Reproduction*, *68*, 358–362.

Sugiyama, H., Murata, K., Iuchi, I., Nomura, K., & Yamagami, K. (1999). Formation of mature egg envelope subunit proteins from their precursors (choriogenins) in the fish, *Oryzias latipes:* Loss of partial C-terminal sequences of the choriogenins. *Journal of Biochemistry*, *125*, 469–475.

Venkatesh, B., Lee, A. P., Ravi, V., Maurya, A. K., Lian, M. M., et al. (2014). Elephant shark genome provides unique insights into gnathostome evolution. *Nature*, *505*, 174–179.

Venkatesh, B., Tay, A., Dandona, N., Patil, J. G., & Brenner, S. (2005). A compact cartilaginous fish model genome. *Current Biology*, *15*, R82–83.

Wassarman, P. M. (2008). Zona pellucida glycoproteins. *The Journal of Biological Chemistry*, *283*, 24285–24289.

Xu, Q., Li, G., Cao, L., Wang, Z., Ye, H., et al. (2012). Proteomic characterization and evolutionary analyses of zona pellucida domain-containing proteins in the egg coat of the cephalochordate, *Branchiostoma belcheri*. *BMC Evolutionary Biology*, *12*, 239.

CHAPTER NINE

Egg-Coat and *Zona Pellucida* Proteins of Chicken as a Typical Species of Aves

Shunsuke Nishio*, Hiroki Okumura†, Tsukasa Matsuda*,[1]
*Graduate School of Bioagricultural Sciences, Nagoya University, Nagoya, Japan
†Faculty of Agriculture, Meijo University, Nagoya, Japan
[1]Corresponding author: e-mail address: tmatsuda@agr.nagoya-u.ac.jp

Contents

Abstract

Birds are oviparous vertebrates in terrestrial animals. Birds' eggs accumulate mass of egg yolk during the egg development and are accordingly much larger than the eggs of viviparous vertebrates. Despite such difference in size and contents, the birds' eggs are surrounded with the egg-coat morphologically and compositionally resembling the mammalian egg-coat, *zona pellucida*. On the other hand, there are some differences in part between the two egg-coats, though relationships of such structural differences to any biological roles specific for the extracellular matrix of birds' eggs are not fully understood. In birds, unlike mammals, ZP proteins constituting the egg-coat are highly conserved and therefore those of chicken are described as a representative of birds. The egg-coat ZP proteins, ZP1, ZP3, and ZPD as the majors, accumulate and form the matrix by self-assembly around the egg rapidly growing in the ovarian follicle, in which ZP1 is from liver and both ZP3 and ZPD are from follicular granulosa cells. Although details of the egg-coat–sperm interaction on fertilization remain to be investigated, the lytic degradation process of egg coat matrix for the sperm penetration has become to be clarified gradually. ZP1 is the primary target of sperm acrosin, and the limited cleavage in the specific region leading to the loss of intermolecular cross-linkages is crucial for the

Current Topics in Developmental Biology, Volume 130
ISSN 0070-2153
https://doi.org/10.1016/bs.ctdb.2018.02.008

lysis of egg-coat matrix. Possible roles of the ZP1 with the additional sequence characteristic to birds are discussed from a viewpoint of giving both robustness and elastomeric nature to the egg-coat matrix for the birds' eggs.

1. EGG DEVELOPMENT AND THE EGG-COAT OF BIRDS

Eggs of birds, familiar as a material for cooking in case of domestic fowl, are usually unfertilized and laid eggs, in which the egg (the true egg to be embryo) is surrounded with egg white, thin shell membrane, and hard but brittle eggshell. These layers surrounding the egg are synthesized by oviduct epithelial cells and given to the ovulated eggs, except for the egg-coat, termed the inner vitelline membrane or perivitelline layer in birds (Fig. 1). In birds as an oviparous vertebrate, mass of egg yolk to nourish growing embryo until hatching is synthesized by the liver of laying female birds and accumulated in the egg of ovary during oocyte development and maturation.

In chicken, the oocytes grow to the sizes of ~1 mm in diameter slowly over time for months and years, and then egg yolk lipoproteins as well as some others are gradually accumulated in a part of such small oocytes in a few months, resulting in the increase in the oocyte size to 7–8 mm in diameter. After that and from 7 to 10 days before ovulation, the egg yolk components become to be accumulated rapidly in one of the growing oocytes with growth and ovulation hierarchy (Gilbert, 1979; Johnson, 1986; Fig. 1). During such a rapid growth stage of the maturing egg shortly before ovulation, egg mass increases dramatically to the size of 35–45 mm in diameter (5–6 times), i.e., increase about 30 times in the surface area, which is covered with the egg-coat.

Even in the immature oocytes, the egg-coat is detected around the oocytes by immunocytological analysis and the constituent ZP proteins are also detectable immunochemically and chemically by LC-MS analysis (Nishio et al., 2014 and unpublished results). In harmony with the rapid growth of maturing eggs, the constituent egg-coat proteins are synthesized by the ovary cells of specific follicles including the rapidly growing oocyte and an exceptional component protein is transported specifically to the surface of such oocytes via blood circulation from liver as described in detail in Section 4. The egg-coat as an extracellular matrix surrounding the rapidly growing and matured large eggs including mass of egg yolk must be permeable enough for yolk lipoproteins to diffuse easily across the egg-coat matrix

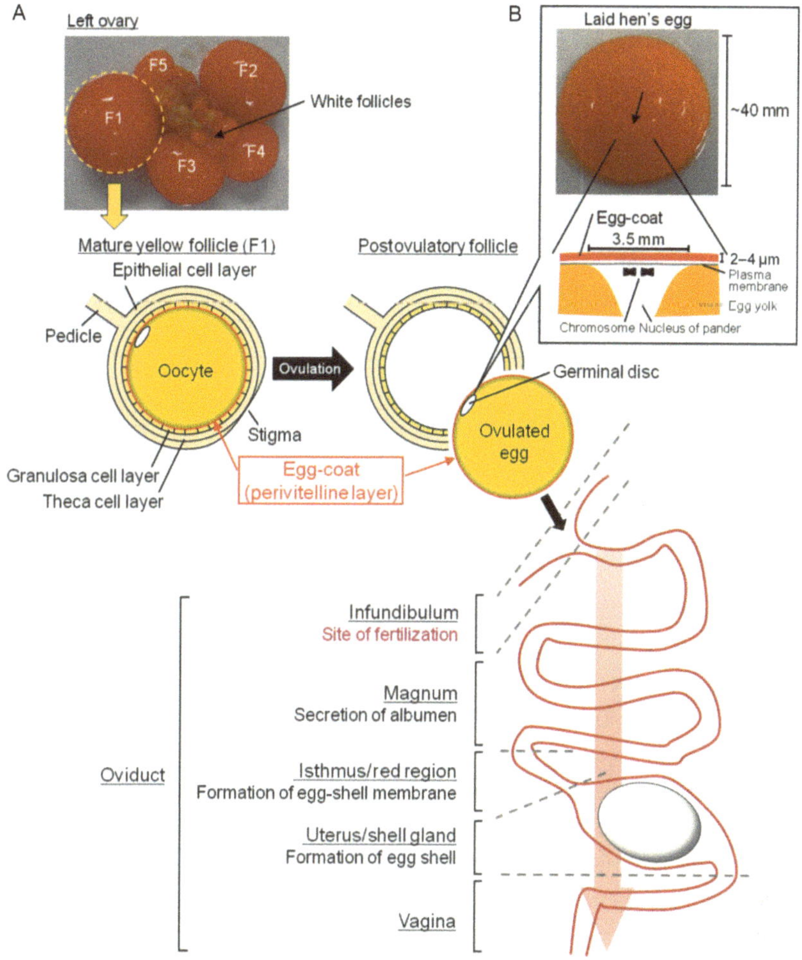

Fig. 1 A series of events for oocyte development and egg formation in chicken. (A) In laying hens, about 12,000 oocytes are present in a mature ovary and are covered with several cell layers. Each oocyte surrounded by cell layers becomes a follicle. About 2000 follicles in the ovary accumulate white yolk and grow slowly over 2 months. Those follicles accumulating white yolk are called white follicles. About 10 days before ovulation, the white follicle rapidly accumulates yellow yolk in the oocyte. Those massive follicles are called yellow follicles and largest one, termed F1 follicle, is ovulated at each day. Ovulated egg is only surrounded the egg-coat, perivitelline layer, and dropped in infundibulum region of oviduct. Infundibulum is the anterior end of oviduct, and ovulated egg is retained here about 15–30 min. During this time, ovulated egg encounters the sperm from male for fertilization. Next, egg transfers magnum where egg albumen is secreted. After magnum, egg surrounded egg albumen transfers isthmus/red region. In this region, egg with egg albumen is wrapped with egg-shell membrane. Finally, calcium is deposited on the egg-shell membrane to form eggshell. (B) Ovulated eggs and laid eggs have a visible whitish area termed germinal disc on the surface. The female pronucleus, cell organelles, and cytoplasm are localized in the germinal disc region, and ejaculated sperm preferentially bind to and penetrate the egg-coat around this area at the fertilization.

(Bujo et al., 1994), and at the same time must be robust enough to protect the large egg including mass of egg yolk physically in the gravity field.

The eggs of birds are a extremely telolecithal egg, having the yolk localized at one end, and the female pronucleus, cell organelles, and cytoplasm situate at the other end termed germinal disc, which can be seen as a whitish disc of a few mm in diameter on the surface of laid eggs (Fig. 1). Morphology of the egg-coat around the germinal disc differs from that of the other region of the egg in that egg plasma membrane forms microvilli structure, which protrudes from the egg to the perivitelline space across mesh of the egg-coat matrix. At the egg region other than germinal disc, the egg-coat appears to make contact with the egg, where the egg plasma membrane is not clearly visible under microscopes (Bakst & Howarth, 1977; Fig. 1).

The egg is ovulated from the largest follicle with maturation hierarchy and taken in by infundibulum of the oviduct (Johnson, 1986). On ovulation, the egg is pushed out in a body cavity from the rupturing follicular capsule, where the egg-coat surrounding the egg is separated from the granulosa cell layer as a part of the follicular capsule tissue (Romanoff, 1960). By contrast to mammalian egg ovulation, the bird egg is ovulated without any accessory cells such as cumulus oophorus in mammals (Tanghe, Van Soom, Nauwynck, Coryn, & de Kruif, 2002), and therefore, the egg-coat is the sole structure surrounding and protecting large eggs with mass of egg yolk from accidental and mechanical rupture in the intraperitoneal space of hens. On fertilization, the egg-coat plays roles in the egg–sperm interaction and somehow allows sperm to bind to and penetrate the egg-coat matrix without making any damages to the egg. Unlike mammalian egg-coat (*zona pellucida*), the egg-coat matrix of bird eggs is degraded by lytic degradation of the constituent ZP proteins by specific proteases released from sperm acrosome (Takeuchi et al., 2001; S. Nishio & T. Matsuda, to be published elsewhere). Independently of fertilized or unfertilized eggs, the egg captured by the oviduct is further surrounded with a few layers consisting of specific glycoproteins, distinct from the egg-coat constituent ZP proteins, secreted from the infundibulum epithelial cells (Bellairs, Harkness, & Harkness, 1963).

2. STRUCTURE OF CHICKEN EGG-COAT AND THE CONSTITUTING ZP PROTEINS

The chicken egg-coat is a thin layer of matrix with 2–4 μm thickness formed with mesh of connected filaments with 0.4–0.7 μm in (Bakst &

Howarth, 1977). These filaments of the mesh are formed by the self-assembly of ZP proteins. The ZP proteins constituting chicken egg-coat have thus far been identified to be ZP1, ZP2, ZP3, ZP4, ZPD, and ZPAX2 (Bausek, Waclawek, Schneider, & Wohlrab, 2000; Nishio et al., 2014; Nishio & Matsuda, to be published elsewhere; Okumura et al., 2004; Takeuchi et al., 1999; Waclawek, Foisner, Nimpf, & Schneider, 1998), though two ZPAX genes are present in the chicken genome (Fig. 2). Among these, ZP1, ZP3, and ZPD are the major constituents of the egg-coat involved in the matrix formation. The primary structures of the pre-pro form for these ZP proteins are deduced from their longest open reading frame of the cDNA with reference of their genomic DNA sequences, and the domain structures with the consensus motif for *N*-glycosylation (NXS/T) or furin cleavage (RXRR) are shown schematically in Fig. 3. These ZP proteins constituting the egg-coat and their primary structure are highly conserved among many species, such as chicken, turkey, and Japanese quail in the class of Aves (Litscher & Wassarman, 2014). This class-specific conserved nature of the egg-coat and the ZP proteins seems to be a feature of Aves, because considerable diversity in the egg-coat ZP proteins is observed among the species of mammals and reptiles (Litscher & Wassarman, 2014).

All of these six ZP proteins contain the conserved ZP module, consisting of ZP-N and ZP-C domains, at the polypeptide C-terminal side. ZP2, ZP3, ZP4, and ZPD have the hydrophobic transmembrane region, near the C-terminus, whereas ZP1 and ZPAXs have no such hydrophobic region predicted to be a transmembrane region. Independently of the presence or absence the hydrophobic region, all but ZPAXs of the ZP proteins conserve the consensus furin cleavage site (CFCS), adjacently at the C-terminal side of ZP-C domain. At the C-terminally adjacent region, a short hydrophobic sequence, termed external hydrophobic patch (EHP), is present in ZP3, which makes hydrophobic interaction with another hydrophobic short sequence, termed internal hydrophobic patch (IHP), at the N-terminal region of ZP-C domain (Han et al., 2010). Such short hydrophobic sequences as EHP and IHP are conserved in part among ZP proteins except for ZPAXs (Fig. 4).

Both of ZP1 and ZP4 have a trefoil domain at the N-terminal side of ZP module and the N-terminal domain homologous to the ZP-N domain. Only one difference in the domain structure between ZP1 and ZP4 is the presence of Pro-rich repeated sequence region, described in detail later (Section 8) between the N-terminal domain and the trefoil domains in ZP1 but absence in ZP4. Like ZP1, ZP2 also has a long N-terminal region but

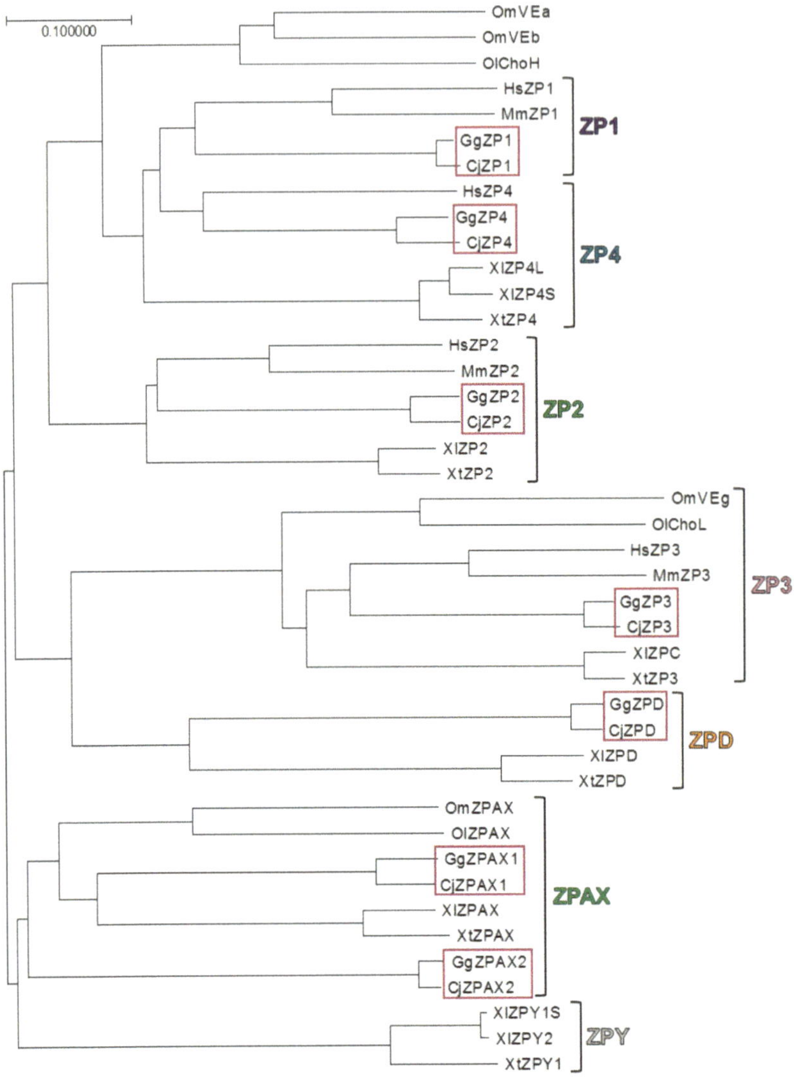

Fig. 2 Phylogenetic tree of ZP modules in ZP glycoproteins. *Om, Oncorhynchus mykiss; Ol, Oryzias latipes; Hs, Homo sapiens; Mm, Mus musculus; Gg, Gallus gallus; Cj, Coturnix japonica; Xl, Xenopus laevis; Xt, Xenopus tropicalis.*

the sequence of this region differs from that of ZP1. This N-terminal side of ZP2 is predicted to be a tandem repeat of five sequence units similar to the ZP-N domain. Among these ZP proteins, ZP3 has the shortest and the longest flanking regions of ZP module, respectively, at the N-terminal and the

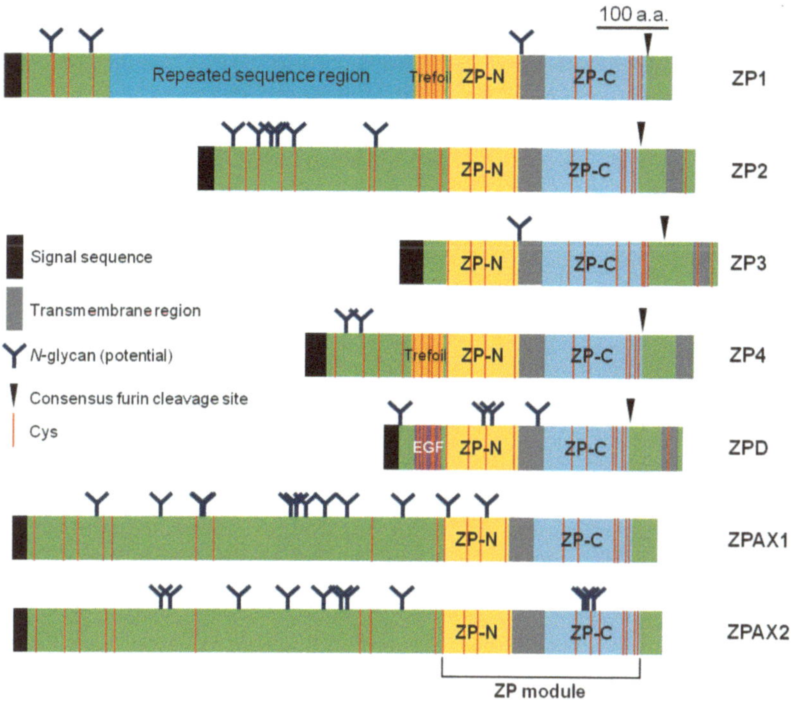

Fig. 3 Schematic drawing of chicken ZP glycoproteins. Accession numbers of chicken ZP glycoproteins are following: ZP1 (CAC16087.1), ZP2 (NP_001034187), ZP3 (NP_989720), ZP4 (NP_990210), ZPD (BAD13713), ZPAX1 (XP_015140443), and ZPAX2 (XP_419968). All domains but the repeated sequence region are predicted by PROSITE (http://prosite.expasy.org/). Signal sequences and transmembrane (TM) region are predicted by SignalP server (http://www.cbs.dtu.dk/services/SignalP/) and TMHMM Server (http://www.cbs.dtu.dk/services/TMHMM/), respectively.

C-terminal sides of ZP module. Only ZP3 has an additional ZP-C subdomain in the ZP-C domain.

All of the seven ZP proteins have the *N*-glycosylation motif, Asn-X-Ser/Thr, and the presence of *N*-glycans sensitive to peptide-*N*-glycanase F have experimentally shown for ZP1, ZP2, ZP3, and ZPAX2. Only ZP2 gives a smear band with lower electrophoretic mobility resembling mouse ZP1, ZP2, and ZP3, whereas the other ZP1, ZP3, ZPD, and ZPAX are not (Nishio et al., 2014). ZP2 expressed and secreted by Chinese hamster ovary (CHO) cells is *N*-glycosylated but not detected as a smear, suggesting that unidentified *N*-glycan(s) characteristic to chicken follicular granulosa cells is added to ZP2 (Nishio & Matsuda, to be published elsewhere).

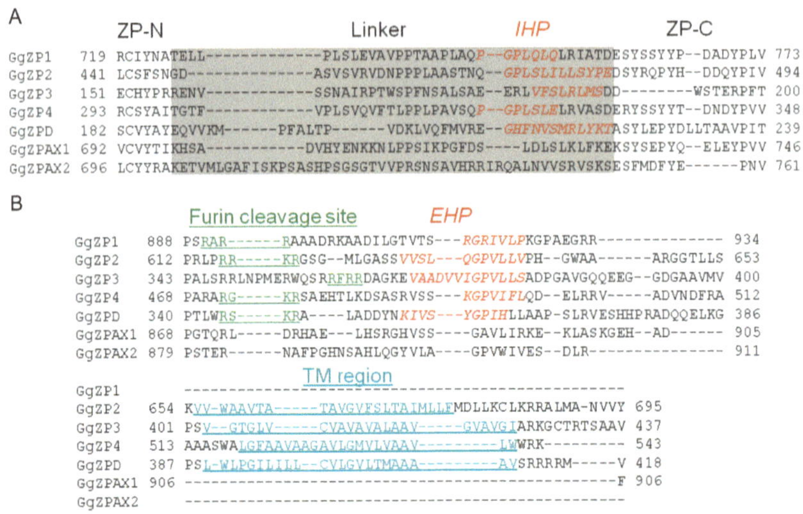

Fig. 4 Hydrophobic regions termed IHP, EHP, and TM and conserved furin cleavage site (CFCS) of chicken ZP glycoproteins. Sequence alignments of linker region between ZP-N and ZP-C domains in ZP module (A) and carboxy-terminal peptide (B). Chicken ZP-glycoprotein sequences were aligned using MAFFT (https://mafft.cbrc.jp/alignment/software/). IHP and EHP of GgZP1 and GgZP4 were referred to Litscher and Wassarman (2014). Those of GgZP2 and GgZPD were predicted based on the crystal structures of mouse ZP2 and human uromodulin (Bokhove et al., 2016), while those of GgZP3 were referred to the crystal structure of GgZP3 (Han et al., 2010).

3. THE CRYSTAL STRUCTURE OF ZP3 HOMODIMER AND FURIN CLEAVAGE-INDUCED DIMERIZATION

In 1980, it was demonstrated for the first time that ZP3 is involved in the interaction of sperm with egg-coat (Bleil & Wassarman, 1980). Since then, it had been one of the most important issues for the investigation of fertilization, or the egg–sperm interactions, to determine three-dimensional structure of ZP3 molecule by X-ray crystallography. However, it was extremely difficult to crystallize mammalian ZP3 because they are heavily glycosylated (~50% of the molecular mass in mouse ZP3) and easily aggregated in concentration or enzymatic deglycosylation (Han et al., 2010). In contrast, chicken ZP3 has only one putative *N*-glycosylation site (Takeuchi et al., 1999; Waclawek et al., 1998) and exhibits relatively high homology to mammalian one (~53% to human ZP3), and additionally, it was reported that avian ZP3 being overexpressed in CHO cells is secreted into the medium with correct processing (Sasanami, Toriyama, & Mori, 2003).

Hence, chicken ZP3 was expected to be more suitable for crystallization. In CHO cells, the recombinant chicken ZP3 containing (1) substitution of the C-terminal transmembrane region with a hexahistidine tag, (2) removal of the putative *N*-glycosylation site by replacement of Asn159 with Gln, (3) inactivation of the CFCS by replacement of Arg359, Arg361, and Arg362 with Ala, and (4) deletion of the N-terminal region showing higher species diversity of amino acid sequences was expressed. The recombinant ZP3 that was secreted into the medium was purified and successfully crystallized after limited proteolysis by trypsin. Finally, crystal structure of chicken ZP3 was solved at 2.0 and 2.6 Å resolutions (PDB ID: 3NK4 and 3NK3, respectively) (Han et al., 2010).

In the determined crystal structures, chicken ZP3 appeared to be a homodimer in which the ZP-N domain of one molecule associated non-covalently with the ZP-C domain of another one each other. Whereas the majority of ZP3 were secreted from the expressing cells into the medium as a homodimer, recombinant ZP3s with amino acid mutations in the predicted binding regions between ZP-N and ZP-C domains did not secreted at all. This suggests that the dimerization of ZP3 is necessary for the biogenesis or trafficking of ZP3 in the cells (Han et al., 2010) including granulosa cells of birds and oocytes of mammals.

Interestingly, the C-terminal propeptide containing the EHP and located downstream of the CFCS was remained and observed between the ZP-N and ZP-C domains in the crystal structure, although the propeptide was cleaved after limited trypsinization at Arg358 that reside immediately upstream of the mutated CFCS of recombinant ZP3. The structure-based mutational analyses confirmed that the propeptide mediates the inter-molecular binding of ZP-N to ZP-C domains with hydrophobic interactions (Han et al., 2010). More surprisingly, the propeptide was spontaneously released from ~40% of the recombinant ZP3 during incubation at 39°C for 30 h, and the recombinant ZP3 formed homopolymers similar to native ZP3 after removal of the propeptide. These results suggest the mechanisms of ZP3 secretion from the cells and incorporation into egg-coat as follows: ZP3 molecules are synthesized in the cells as homodimeric precursors that are anchored on the plasma membrane via the C-terminal propeptides. The propeptides are cleaved at the CFCS during or just after the process of secretion, and dissociated ZP3 is released slowly into the extracellular space to be incorporated into the egg-coat matrix, although it is unexplained whether the ZP3 being released from the cells maintain dimeric form or not (Han et al., 2010).

4. EXPRESSION OF ZP GENES AT DEVELOPMENTAL STAGES OF GROWING OOCYTE

One of the most characteristic features of avian egg-coat is being assembled from six subfamilies of ZP proteins (ZP1–4, ZPD, and ZPAX) that are expressed either in the ovary (granulosa cells or oocytes) or the liver (Fig. 5). Furthermore, these ZP protein subfamilies exhibit distinct

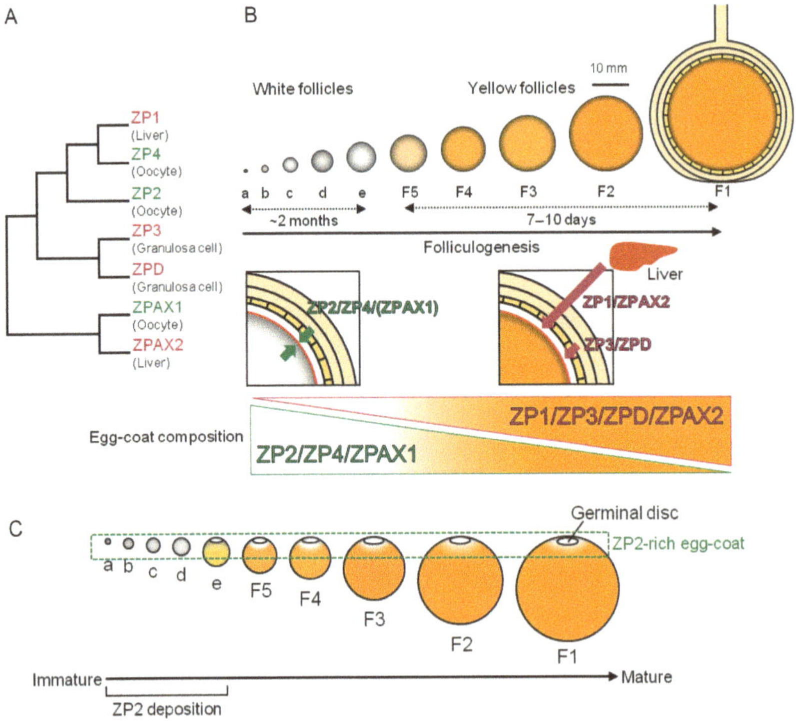

Fig. 5 Stage- and area-specific ZP-glycoprotein composition of the egg-coat at developmental stages of oocyte. (A) Phylogenetic tree of chicken ZP glycoproteins. Expression tissues or cells are shown under each name of ZP glycoproteins. (B) Changes in the composition of chicken egg-coat during oocyte development. During the developmental stage as white follicles, the major components of egg-coat are ZP2, ZP4, and ZPAX1, which are produced from both or either of oocyte and granulosa cells. When rapid accumulation of yellow yolk begins, the major components of egg-coat are changed to ZP1 and ZPAX2 synthesized from liver, and ZP3 and ZPD synthesized from granulosa cells. (C) A speculative fate of the ZP2-rich egg-coat constructed during the immature white follicle stages. Although four ZP glycoproteins, ZP1, ZP3, ZPD, and ZPAX2, are deposited in the egg-coat at rapidly growing stages, it seems that the ZP2-rich egg-coat is still retained as it is around the germinal disc.

expression patterns during follicular development. Recent studies in chicken and quail revealed that both ZP2 and ZP4 are expressed in the immature oocytes of the white follicles, and the expression levels of them gradually decrease as the follicles increased in size with oocyte development (Kinoshita et al., 2010; Nishio et al., 2014; Rodler, Sasanami, & Sinowatz, 2012; Serizawa et al., 2011). Interestingly, ZP2 being expressed in the immature white follicles accumulates in the egg-coat of the germinal disc region in matured yellow follicle (Fig. 5; Nishio et al., 2014). Considering that ZP2 is involved in taxon-specific binding of the acrosome-intact spermatozoa and the secondary binding of the acrosome-reacted spermatozoa during the penetration of egg-coat in mammals (Avella, Baibakov, & Dean, 2014; Bleil, Greve, & Wassarman, 1988), the accumulation of avian ZP2 in the germinal disc region of egg-coat might be associated with the preferential binding to and penetration of chicken spermatozoa through the egg-coat on the germinal disc region (Bramwell & Howarth, 1992). ZP4 may replace the function of ZP1 in formation of egg-coat matrix (Greve & Wassarman, 1985) in chicken, because ZP4 is regarded as more closely related to ZP1 based on structural conservation (Nishio et al., 2014). In contrast, both ZP3 and ZPD are expressed in the granulosa cells surrounding matured oocytes of the yellow follicles (Benson, Christensen, Fairchild, & Davis, 2009; Okumura et al., 2004; Pan, Sasanami, Kono, Matsuda, & Mori, 2001; Sasanami et al., 2002; Sato et al., 2009; Takeuchi et al., 1999; Waclawek et al., 1998), and the expression levels of them increase during the follicular development (Nishio et al., 2014; Pan et al., 2001; Rodler et al., 2012; Sato et al., 2009). The production of ZP3 is stimulated by testosterone (Pan et al., 2001). More interestingly, ZP1 is expressed in the liver in an estrogen-dependent manner to be secreted into the bloodstream and incorporated in the egg-coat of maturing oocyte (Bausek et al., 2000; Benson et al., 2009; Hanafy, Sasanami, & Mori, 2007; Nishio et al., 2014; Sasanami, Pan, & Mori, 2003). Considering that the diameter of avian oocyte extremely increases during follicular development, for example, in chicken, from ~7 to 40 mm during the final 8–9 days before ovulation (Wyburn, Aitken, & Johnston, 1965), and the site of ZP protein synthesis (the liver or the ovary) is related to the egg-coat morphology (thickness) of teleost fish (Sano et al., 2017), the liver-specific synthesis of avian ZP1 might contribute to the quick expansion of the egg-coat throughout the rapid growth of oocyte. ZPAX2 is identified in the egg-coat surrounding ovulated chicken ovum by mass spectrometry (Mann, 2008) and a recent genomic study suggested that some genes including *ZPAX2* may be important for abdominal fat deposition in chicken (Zhang et al., 2014). Two chicken *ZPAX* genes (*ZPAX1* and *ZPAX2*) present in genome are expressed differently. *ZPAX1* gene is expressed highly in the white follicles and weakly in the granulosa cells of yellow follicles, while *ZPAX2* gene is expressed specifically in the liver. Similar to ZP1, ZPAX2 protein is incorporated into the egg-coat matrix during rapid growth of yellow follicles (Nishio & Matsuda, to be published elsewhere). As a result of characteristic expression patterns of ZP proteins, major components of avian egg-coat surrounding matured oocyte just before ovulation are ZP1, ZP3, ZPD, and a minor component of that is ZPAX2.

5. SECRETION, TRANSPORT, AND ASSEMBLY FOR MATRIX FORMATION AROUND THE OOCYTE

Processing and secretion mechanisms of ZP proteins are studied well in ZP3 of chicken and quail. In the granulosa cell, ZP3 being synthesized into the endoplasmic reticulum as a transmembrane protein is transported to the Golgi apparatus and proteolytically cleaved at a CFCS prior to the secretion to dissociate the C-terminal region containing the EHP and transmembrane region (Han et al., 2010; Sasanami, Hanafy, Toriyama, & Mori, 2003; Sasanami et al., 2002). Premature ZP3 possessing the C-terminal region probably forms dimers through hydrophobic interactions, and it is suggested that the dimer formation is a prerequisite for the ZP3 synthesis (Han et al., 2010). Furthermore, the dissociation of EHP by furin cleavage enhances polymerization of ZP3 and probably incorporation of them into the egg-coat (Han et al., 2010). An N-linked oligosaccharide occupying the highly conserved *N*-glycosylation site on the linker region between ZP-N and ZP-C domains is not required for the intracellular processing of ZP3 (Han et al., 2010; Sasanami, Toriyama, et al., 2003) but might play important roles in the efficient transport of ZP3 to the apical surface of polarized epithelial cells including granulosa cells (Sasanami et al., 2005).

Assembly of avian ZP1, ZP3, and ZPD to form the egg-coat matrix is studied well also in chicken and quail. ZP1 being expressed in the liver and secreted into the bloodstream is transported to the maturing follicles as above. It is suggested that interaction between ZP1 and ZP3 induces egg-coat matrix formation and the intermolecular disulfide formation between ZP1 molecules (Ohtsuki, Hanafy, Mori, & Sasanami, 2004; Okumura, Aoki, Sato, Nadano, & Matsuda, 2007; Okumura, Okajima, Nadano, & Matsuda, 2007; Sasanami et al., 2006). The formation of intermolecular disulfide-linked ZP1 dimer, which is found in the egg-coat (Takeuchi et al., 2001), is inhibited in chicken serum (Okumura, Aoki, et al., 2007; Okumura, Okajima, et al., 2007), and ZP1 purified from the serum but not ZP1 from the egg-coat is incorporated into the egg-coat after injection into the female quail (Kinoshita et al., 2008), implying that interaction with ZP3 on the surface of maturing oocyte induces some conformational changes in ZP1 that arrived from the liver, although the molecular details are still remain elusive. The third major component in matured avian egg-coat, ZPD, is suggested to form homopolymers and attaches to the ZP1–ZP3 matrix to form the ZPD aggregates (Okumura et al., 2015), which are easily detached from the egg-coat matrix by mechanical treatment (Okumura et al., 2004). These findings, together with mapping of the intermolecular contacts among ZP glycoproteins being involved in the egg-coat matrix formation by using the individual recombinant domains of chicken ZP proteins and the antibodies being raised against them as probes, a model for the architecture of chicken egg-coat matrix, are proposed (Okumura et al., 2015). This model is compatible with the model of mouse egg-coat matrix (Greve & Wassarman, 1985). Therefore, the chicken model might be applicable to the egg-coat of all other vertebrates by replacing the ZP-glycoprotein components, although the matrix formation mechanisms and detailed domain configurations of the ZP proteins in the matrix remain to be elucidated.

6. INTERACTION OF EGG-COAT WITH SPERM AT FERTILIZATION

In the infundibulum following the upperpart oviduct, the ovulated and engulfed egg encounters sperm, which are transported from the distal oviduct, and fertilized (Olsen, 1942, 1952). The initial steps of fertilization are interaction or binding of sperm to, and afterword or simultaneous penetration of, the egg-coat surrounding the egg. In birds including chicken, the soluble enzymes are released from the acrosome of interacting sperm, and the lytic dissociation of egg-coat matrix is visible under microscopes, resulting in the formation of holes on the matrix (Fig. 6; Howarth & Digby, 1973; Okamura & Nishiyama, 1978; Takeuchi et al., 1999). The sperm keep moving forward without detaching the degrading matrix and penetrate the matrix through the hole. Such lytic degradation of the egg-coat by interacting sperm is not observed in fertilization of mammals (Bedford, 1998), though excessive concentrations of the proteases were

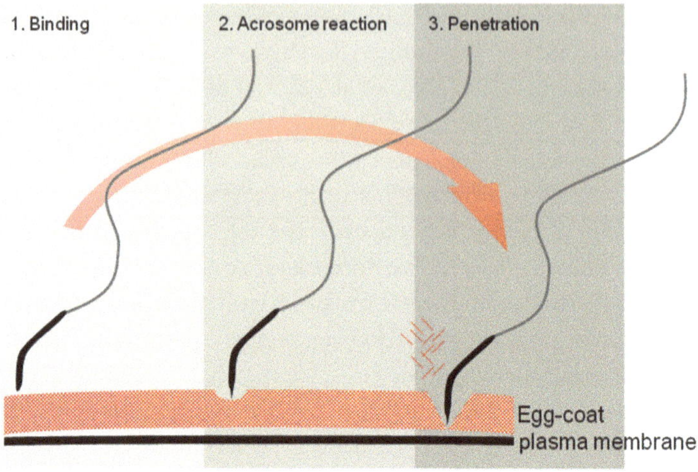

Fig. 6 A canonical scheme of sperm–egg-coat interaction in birds. When ejaculated sperm meets the egg at the infundibulum, sperm head binds to the surface of egg-coat and the acrosome reaction triggered by the ZP glycoproteins occurs. Sperm proteases released from acrosome degrade the egg-coat to make a hole for sperm penetration. A series of these events is limited for a short time (~15 min) until the fertilized egg is surrounded with another layer of protein matrix in the oviduct. This scheme of sperm–egg-coat interaction is based on the previous electron microscopic images of chicken sperm penetration through the egg-coat (Okamura & Nishiyama, 1978; Takeuchi et al., 2001).

reported to induce lytic degradation of the *zona pellucida* matrix in vitro in earlier studies using rabbit sperm extracts and eggs (Stambaugh & Buckley, 1968).

Some components of the egg-coat proteins have been suggested to be the sperm receptor(s) of the egg indirectly based on some observations that hole formation on the egg-coat by sperm is inhibited by antibodies specific for ZP proteins and that purified egg-coat components bind to sperm protein components or the surface of sperm. Preincubation of small pieces of isolated egg-coat with either of two monoclonal antibodies 8E1 and 2D4 specific for ZP1, which had been raised against the chicken egg-coat suspension, inhibited the hole formation on the egg-coat, whereas the polyclonal antiserum specific for another major component ZP3 did not, suggesting important roles of ZP1 in the egg-coat recognition by sperm (Takeuchi et al., 2001). Polyclonal antiserum specific for ZP1, but not ZP3, also showed such an inhibitory effect on the sperm-induced hole formation on the egg-coat (Bausek, Ruckenbauer, Pfeifer, Schneider, &

Wohlrab, 2004). All of these antibodies well immunostained the isolated egg-coat matrix, indicating that such antibody molecules bound to ZP3 in the egg-coat matrix do not affect the interaction of sperm with the egg-coat. The epitope recognized by the monoclonal antibody 8E1 was later identified to be Pro891-Arg900, the decapeptide, which locates at the C-terminus of ZP domain and is C-terminally flanked by the furin cleavage consensus sequence (RARR) (Okumura, Aoki, et al., 2007; Okumura, Okajima, et al., 2007). Assuming that ZP1 is processed at this cleavage site by cell membrane furin upon secretion in the liver or transmission across the granulosa cells layer in ovary, this decapeptide epitope should be situated at the C-terminus of the ZP domain in ZP1. Irrespective of the furin cleavage at this site, this C-terminal end region of ZP domain in ZP1 must be exposed to the egg-coat matrix surface sufficiently enough for the mAB 8E1 to access it.

Both of ZP1 purified from laying hen's serum and ZP3 purified from the culture medium of granulosa cells isolated from the largest follicles were shown to bind to the very tip of the sperm head of live sperm using immunofluorescence microscopy, suggesting that both of ZP1 and ZP3 can interact individually with some membrane molecules of sperm head (Bausek et al., 2004). The N159Q mutant of ZP3 lacking the N-linked sugar chain, which had been expressed and secreted by transfected CHO cells, still retained the binding ability to the tip of sperm head, but another mutation from T to A at the *O*-glycosylation site T168 markedly reduced the sperm-binding ability of ZP3 (Han et al., 2010). Sperm-binding activity of these native and mutated ZP3 molecules indicates that the *O*-glycan attached to T168 of ZP3 contributes to the ZP3 binding to the sperm head, but that glycan is not necessarily chicken-type ones. An earlier study using lectins and deglycosylation enzymes also showed possible contribution of carbohydrate chains of the egg-coat to the interaction of sperm with egg-coat (Robertson, Wishart, & Horrocks, 2000). Wheat germ agglutinin (WGA) specific for terminal GlcNAc (b1–4) GlcNAc residues and Con A specific for Man-Glc residues well bind to chicken egg-coat. Preincubation of the egg-coat with WGA, but not Con A, markedly inhibited the sperm-induced hole formation on the egg-coat, and moreover, free D-GlcNAc also inhibited the hole formation, i.e., the number of holes were reduced to 27% at 75 mM GlcNAc in the sperm–egg-coat interaction assay. The pre-treatment of egg-coat with peptide-*N*-glycosidase F (PNGsase F), which had released some carbohydrates from the egg-coat pieces, remarkably reduced the sperm-induced hole formation on the egg-coat.

7. LYTIC DEGRADATION OF THE EGG-COAT AND ZP PROTEIN FRAGMENTATION DURING SPERM PENETRATION

An earlier study on the in vitro fertilization using ovulated eggs recovered from infundibulum of laying hens showed a clearly visible result that the eggs incubated with sperm for a time longer than that of in vivo are ruptured, and the egg yolk leaks out from the egg locally in the region of the germinal disc (Howarth & Digby, 1973). This simple but very laborious experiment clearly demonstrated that the sperm penetration of the egg-coat in birds is achieved by a lytic degradation of the egg-coat matrix around the egg-coat–sperm interaction sites. This is in contrast to the sperm penetration of the egg-coat, *zona pellucida*, in mammals, in which the egg-coat matrix appears to be split on the sperm penetration and a penetration path rather than the hole is observed in the thicker *zona pellucida* matrix (Bedford, 1998), though the involvement of proteolytic events in such nonlytic sperm penetration of mammalian egg-coat is not necessarily denied. Several studies using electron microscopy clearly demonstrated the lytic degradation of the egg-coat matrix locally around the site of sperm binding to the egg-coat (Okamura & Nishiyama, 1978; Takeuchi et al., 2001). In the transmittance electron microscopic images, some sperm contact with the egg-coat and the filamentous mesh structure of egg-coat matrix around the tip of sperm head is indistinct and appears to be lysed, and some other sperm are penetrating through the hole formed by the lytic degradation of the egg-coat on the way to the perivitelline space of the egg. Such lytic degradation of the egg-coat matrix on the sperm interaction is markedly inhibited with serine protease inhibitors, such as SBTI (Ho & Meizel, 1975) and TLCK (Howarth & Digby, 1973), indicating that the sperm-induced local degradation of the egg-coat, leading to the hole formation, is proteolytic hydrolysis of the egg-coat proteins.

The egg-coat is an extracellular matrix, and the constituent ZP proteins are insoluble naturally. Lytic degradation of egg-coat matrix could be reworded as degradative solubilization of ZP proteins. Sperm-induced proteolytic degradation, as well as solubilization, of the major components, ZP1 and ZP3, of chicken egg-coat was shown by an in vitro assay using the egg-coat suspension suitable for quantitative analysis and ejaculated live sperm or sperm extracts containing acrosomal enzymes (Takeuchi et al., 2001). The incubation of the egg-coat with a smaller number of sperm did not induce detectable levels of ZP protein solubilization. With the increase in the sperm number incubated with the egg-coat, ZP1 (97 kDa) became detectable in the soluble fraction as proteolytic fragments with various sizes of 40–50 kDa. ZP3 (42 kDa) also was solubilized as an N-terminally truncated fragment of 38 kDa only when the egg-coat was incubated with a larger number of sperm. Interestingly, by the preincubation of the egg-coat with the monoclonal antibodies specific for ZP1, such sperm-induced degradation and solubilization of not only ZP1 but also ZP3 was inhibited remarkably. Such degradation inhibition of both ZP1 and ZP3 by the ZP1-specific monoclonal antibodies was observed similarly when the egg-coat was incubated with the sperm extract containing acrosome proteases. Incubation of the egg-coat with the sperm extract results in the degradation of both ZP1 and ZP3 into smaller fragments including the major 34 and 20 kDa fragments, and almost the same degradation profiles are observed when the egg-coat is incubated with bovine trypsin under the same condition. However, the degradation of ZP1 and ZP3 by the trypsin is not at all inhibited by the ZP1-specific monoclonal antibodies, suggesting that the sperm protease(s) have characteristic substrate specificity distinct from general serine proteases like trypsin. Thus, at the initial stage of sperm interaction with and penetration of the egg-coat, ZP1 in the egg-coat matrix seems to be the primary target of sperm proteases, and the degradation and solubilization of ZP1 may result in the exposure of ZP3 and the subsequent solubilization together with the truncation (Takeuchi et al., 2001).

Peptide fragments released from the egg-coat matrix during the incubation of egg-coat suspension with sperm proteases have recently been identified using N-terminal peptide sequencing and LC-MS/MS analysis (Nishio & Matsuda, to be published elsewhere). According to the identified sequences and peptides, the sperm acrosin first cleaves the Pro-rich repeated sequence region, connecting the N-terminal ZP-N domain and the C-terminal trefoil domain followed by ZP module. Since the N-terminal ZP-N domain is cross-linked to another ZP1 through intermolecular disulfide bridge, the cleavage at this connecting region results in the loss of inter-ZP1 cross-linkage. Following this ZP1 cleavage by sperm acrosin, complexes of ZP module-containing fragments from ZP1 and ZP3 become detectable in the soluble fraction of the reaction mixture, indicating the release of not only ZP1 but also ZP3 from the matrix by the cleavage of ZP1. Interestingly, the specific cleavage sites in the ZP1 repeated sequence region are mostly at the peptide bonds between Arg and Pro, known to be the peptide bond resistant to the proteolysis by general serine proteases like trypsin (known as "Keil rule"; Keil, 1992; Rodriguez, Gupta, Smith, & Pevzner, 2008). This unusual cleavage site as well as the unique substrate-specificity of sperm acrosin suggests that the chicken egg-coat matrix is resistant and protective to general serine proteases such as those from infectious bacteria, but susceptible to sperm acrosin for sperm acceptance by penetration of the matrix. This may indicate that the repeated sequence region with the Arg-Pro sequence functions as both of a protective matrix against bacterial infection and a receptive matrix for sperm penetration on fertilization. Such repeated sequence region with the Arg-Pro sites is not present in mouse and human ZP1s. Further studies are required to clarify relationship between the absence of the repeated sequence region in ZP1 and the apparently nonlytic event of sperm penetration of *zona pellucida* in mouse fertilization.

8. THE ROBUST AND ELASTOMERIC EGG-COAT FOR BIRDS' EGGS WITH MASS OF EGG YOLK

Vertebrate eggs and the egg-coat surrounding it are quite different in size, i.e., those of oviparous vertebrates are larger than those of the viviparous mammals, which do not any more need a large mass of egg yolk for nourishing the growing embryo. Especially, the terrestrial vertebrates such as birds as well as reptiles have eggs much larger than mammalian eggs. Chicken eggs are about 40 mm in diameter, whereas mouse eggs are about 80 μm in diameter, indicating that the surface area of a chicken egg is about 10^{5-6} times as large as that of a mouse egg. The egg-coat matrix surrounding such a huge egg of chicken must be robust enough to physically protect the oocyte, egg, and embryo from rupture due to gravity in the terrestrial environment. Moreover, the chicken egg-coat must be elastomeric and the matrix fiber and mesh must be expansive in response to rapid increase in size with the growth of oocyte during several days before ovulation. There are some differences in morphology of the egg-coat between chicken and mice. The chicken egg-coat is 2–3 μm thickness, which is no more than 1/2–1/3 of mouse one (Litscher & Wassarman, 2014), and appears to be constructed with thick fibers interconnected in places and to be a net an opening rather than a continuous layer of mouse *zona pellucida* (Bedford, 1998). Such a net-like structure of chicken egg-coat might make it possible for the egg-coat matrix to extend elastically and flexibly during the rapid growth of oocytes. In addition, the thinner egg-coat of chicken might result from such horizontal stretching of the egg-coat on the surface of rapidly growing oocytes.

The egg-coat matrixes of both chicken and mouse mainly constitute of the three common components, ZP1, ZP2, and ZP3, though chicken ZP2 is limited at the germinal disc region. Chicken ZPD is associated with egg-coat matrix, but not involved in the matrix formation. Chicken ZP4 is a very minor component in the egg-coat of mature eggs, while mouse ZP4 is not expressed because of the pseudogene. The domain structure based on the primary structure is well conserved in all of ZP1, ZP2, and ZP3, except for the region connecting the N-terminal ZP-N domain with the trefoil domain of the ZP module N-terminal of ZP1. The size of chicken ZP1 polypeptide precursor is about 1.5 times (934/623) as long as that of mouse ZP1, whereas there are no large differences in the size of ZP2 (695 and 713) and ZP3 (446 and 423) between chicken and mouse. The longer connecting region of chicken ZP1 consists of a tandem repeat of 20 Pro-rich segments with about 24-amino acid residues, termed the repeated sequence region (Bausek et al., 2000). Pro residues are rich in some extracellular proteins forming fibrous aggregates, such as collagen and elastin, and some relationship of high Pro content with elastomeric fibril formation was suggested on some natural elastic proteins and model peptides (Rauscher, Baud, Miao, Keeley, & Pomès, 2006). Such a Pro-rich repeated sequence region is conserved in the ZP1 sequences of birds, but not in mammals. This may indicate that the Pro-rich repeated sequence region of ZP1 conserved in birds as well as reptiles, i.e., the oviparous terrestrial vertebrates, plays a role in giving the robust and elastic nature to the birds' egg-coat matrix. A single ancestral gene encoding the ZP module is suggested to evolve into the various egg-coat proteins with several gene duplication events as well as addition to the ZP module N-terminus of some domains and sequences, including ZP-N, trefoil, EGF domains, and the repeated sequence region (Litscher & Wassarman, 2014). Although the repeated sequence region in ZP1 is conserved among vertebrate, except for mammals, the length and the sequence of the repeated unit are highly diverse among Aves (birds) and three reptiles, Crocodilia (crocodiles and alligators); Lepidosauria (lizards and snakes); Testudines (turtles) (Fig. 7). It seems likely that ZP1 of birds' egg-coat has evolved together with the change in egg size to adapt to changes in their habitat environment of birds, terrestrial, and oviparous vertebrates.

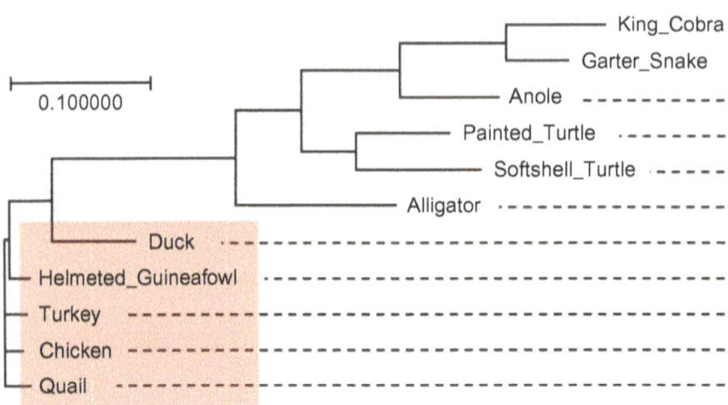

Repeated sequence region	
Length (a.a.)	No. of units
5–18	32
18–21	22
9–13	50
9–11	5
7–11	15
7–19	23
20–28	19
20–24	18
20–24	19
20–24	19 or 20
20–24	19

Fig. 7 Comparison of ZP module sequences and the repeated sequence regions among Aves and reptiles ZP1 proteins. Several ZP1 sequences of Aves and reptiles were retrieved from NCBI database. Phylogenetic tree was generated using the amino acid sequences of ZP module of ZP1 (*left*). Number of units in the repeated sequence region and the length (the number of residues) of one repeated shown on the *right* were analyzed manually (Nishio & Matsuda, to be published elsewhere).

REFERENCES

Avella, M. A., Baibakov, B., & Dean, J. (2014). A single domain of the ZP2 zona pellucida protein mediates gamete recognition in mice and humans. *The Journal of Cell Biology*, *23*, 801–809.

Bakst, M. R., & Howarth, B., Jr. (1977). The fine structure of the hen's ovum at ovulation. *Biology of Reproduction*, *17*, 361–369.

Bausek, N., Ruckenbauer, H. H., Pfeifer, S., Schneider, W. J., & Wohlrab, F. (2004). Interaction of sperm with purified native chicken ZP1 and ZPC proteins. *Biology of Reproduction*, *71*, 684–690.

Bausek, N., Waclawek, M., Schneider, W. J., & Wohlrab, F. (2000). The major chicken egg envelope protein ZP1 is different from ZPB and is synthesized in the liver. *The Journal of Biological Chemistry*, *275*, 28866–28872.

Bedford, J. M. (1998). Mammalian fertilization misread? Sperm penetration of the eutherian zona pellucida is unlikely to be a lytic event. *Biology of Reproduction*, *59*, 1275–1287.

Bellairs, R., Harkness, M. L., & Harkness, R. D. (1963). The vitelline membrane of the hen's egg: A chemical and ultrastructural study. *Journal of Ultrastructure Research*, *8*, 339–359.

Benson, A. P., Christensen, V. L., Fairchild, B. D., & Davis, A. J. (2009). The mRNA for zona pellucida proteins B1, C and D in two genetic lines of turkey hens that differ in fertility. *Animal Reproduction Science*, *111*, 149–159.

Bleil, J. D., Greve, J. M., & Wassarman, P. M. (1988). Identification of a secondary sperm receptor in the mouse egg zona pellucida: Role in maintenance of binding of acrosome-reacted sperm to eggs. *Developmental Biology*, *128*, 376–385.

Bleil, J. D., & Wassarman, P. M. (1980). Mammalian sperm-egg interaction: Identification of a glycoprotein in mouse egg zonae pellucidae possessing receptor activity for sperm. *Cell*, *20*, 873–882.

Bokhove, M., Nishimura, K., Brunati, M., Han, L., de Sanctis, D., Rampoldi, L., et al. (2016). A structured interdomain linker directs self-polymerization of human uromodulin. *Proceedings of the National Academy of Sciences of the United States of America*, *113*, 1552–1557.

Bramwell, R. K., & Howarth, B., Jr. (1992). Preferential attachment of cock spermatozoa to the perivitelline layer directly over the germinal disc of the hen's ovum. *Biology of Reproduction, 47*, 1113–1117.

Bujo, H., Hermann, M., Kaderli, M. O., Sugawara, S., Nimpf, J., Yamamoto, T., et al. (1994). Chicken oocyte growth is mediated by an eight ligand binding repeat member of the LDL receptor family. *EMBO Journal, 13*, 5165–5175.

Gilbert, A. B. (1979). Female genital organs. In A. S. King & J. McLelland (Eds.), *Form and function in birds: Vol. 1* (pp. 237–360). London: Academic Press.

Greve, J. M., & Wassarman, P. M. (1985). Mouse egg extracellular coat is a matrix of interconnected filaments possessing a structural repeat. *Journal of Molecular Biology, 181*, 253–264.

Han, L., Monné, M., Okumura, H., Schwend, T., Cherry, A. L., Flot, D., et al. (2010). Insights into egg-coat assembly and egg-sperm interaction from the X-ray structure of full-length ZP3. *Cell, 143*, 404–415.

Hanafy, A. M., Sasanami, T., & Mori, M. (2007). Sensitivity of expression of perivitelline membrane glycoprotein ZP1 mRNA in the liver of Japanese quail (*Coturnix japonica*) to estrogenic compounds. *Comparative Biochemistry and Physiology. Toxicology & Pharmacology: CBP, 144*, 356–362.

Ho, J. J., & Meizel, S. (1975). Hydrolysis of the hen egg vitelline membrane by cock sperm acrosin and other enzymes. *The Journal of Experimental Zoology, 194*, 429–437.

Howarth, B., Jr., & Digby, S. T. (1973). Evidence for the penetration of the vitelline membrane of the hen's egg by a trypsin-like acrosomal enzyme. *Journal of Reproduction and Fertility, 33*, 123–125.

Johnson, A. L. (1986). Reproduction in the female. In P. D. Sturkie (Ed.), *Avian physiology* (4th ed., pp. 403–431). New York: Springer-Verlag.

Keil, B. (1992). *Specificity of proteolysis*. New York: Springer-Verlag Berlin Heidelberg.

Kinoshita, M., Mizui, K., Ishiguro, T., Ohtsuki, M., Kansaku, N., Ogawa, H., et al. (2008). Incorporation of ZP1 into perivitelline membrane after in vivo treatment with exogenous ZP1 in Japanese quail (*Coturnix japonica*). *The FEBS Journal, 275*, 3580–3589.

Kinoshita, M., Rodler, D., Sugiura, K., Matsushima, K., Kansaku, N., Tahara, K., et al. (2010). Zona pellucida protein ZP2 is expressed in the oocyte of Japanese quail (*Coturnix japonica*). *Reproduction (Cambridge, England), 139*, 359–371.

Litscher, E. S., & Wassarman, P. M. (2014). Evolution, structure, and synthesis of vertebrate egg-coat proteins. *Trends in Developmental Biology, 8*, 65–76.

Mann, K. (2008). Proteomic analysis of the chicken egg vitelline membrane. *Proteomics, 8*, 2322–2332.

Nishio, S., Kohno, Y., Iwata, Y., Arai, M., Okumura, H., Oshima, K., et al. (2014). Glycosylated chicken ZP2 accumulates in the egg-coat of immature oocytes and remains localized to the germinal disc region of mature eggs. *Biology of Reproduction, 91*, 107 (1–10).

Ohtsuki, M., Hanafy, A. M., Mori, M., & Sasanami, T. (2004). Involvement of interaction of ZP1 and ZPC in the formation of quail perivitelline membrane. *Cell and Tissue Research, 318*, 565–570.

Okamura, F., & Nishiyama, H. (1978). The passage of spermatozoa through the vitelline membrane in the domestic fowl, *Gallus gallus*. *Cell and Tissue Research, 188*, 497–508.

Okumura, H., Aoki, N., Sato, C., Nadano, D., & Matsuda, T. (2007). Heterocomplex formation and cell-surface accumulation of hen's serum zona pellucida B1 (ZPB1) with ZPC expressed by a mammalian cell line (COS-7): A possible initiating step of egg-envelope matrix construction. *Biology of Reproduction, 76*, 9–18.

Okumura, H., Kohno, Y., Iwata, Y., Mori, H., Aoki, N., Sato, C., et al. (2004). A newly identified zona pellucida glycoprotein, ZPD, and dimeric ZP1 of chicken egg envelope are involved in sperm activation on sperm-egg interaction. *The Biochemical Journal, 384*, 191–199.

Okumura, H., Okajima, T., Nadano, D., & Matsuda, T. (2007). Association of chicken zona pellucida glycoprotein (ZP) B1 with ZPC induces formation of ZPB1-ZPC fibrous aggregates containing disulfide-bridged ZPB1 dimer. *Biochemical and Biophysical Research Communications, 364*, 682–688.

Okumura, H., Sato, T., Sakuma, R., Fukushima, H., Matsuda, T., & Ujita, M. (2015). Identification of distinctive interdomain interactions among ZP-N, ZP-C and other domains of zona pellucida glycoproteins underlying association of chicken egg-coat matrix. *FEBS Open Bio, 5*, 454–465.

Olsen, M. W. (1942). Maturation, fertilization, and early cleavage in the hen's egg. *Journal of Morphology, 70*, 513–533.

Olsen, M. W. (1952). Intra-ovarian insemination in the domestic fowl. *The Journal of Experimental Zoology, 119*, 461–481.

Pan, J., Sasanami, T., Kono, Y., Matsuda, T., & Mori, M. (2001). Effects of testosterone on production of perivitelline membrane glycoprotein ZPC by granulosa cells of Japanese quail (*Coturnix japonica*). *Biology of Reproduction, 64*, 310–316.

Rauscher, S., Baud, S., Miao, M., Keeley, F. W., & Pomès, R. (2006). Proline and glycine control protein self-organization into elastomeric or amyloid fibrils. *Structure, 14*, 1667–1676.

Robertson, L., Wishart, G. J., & Horrocks, A. J. (2000). Identification of perivitelline N-linked glycans as mediators of sperm-egg interaction in chickens. *Journal of Reproduction and Fertility, 120*, 397–403.

Rodler, D., Sasanami, T., & Sinowatz, F. (2012). Assembly of the inner perivitelline layer, a homolog of the mammalian zona pellucida: An immunohistochemical and ultrastructural study. *Cells, Tissues, Organs, 195*, 330–339.

Rodriguez, J., Gupta, N., Smith, R. D., & Pevzner, P. A. (2008). Does trypsin cut before proline? *Journal of Proteome Research*, 7, 300–305.

Romanoff, A. L. (1960). *Fertilization and fertility. In the avian embryo: Structural and functional development.* New York: The Macmillan Company.

Sano, K., Kawaguchi, M., Katano, K., Tomita, K., Inokuchi, M., Nagasawa, T., et al. (2017). Comparison of egg envelope thickness in teleosts and its relationship to the sites of ZP protein synthesis. *Journal of Experimental Zoology. Part B, Molecular and Developmental Evolution, 328*, 240–258.

Sasanami, T., Hanafy, A. M., Toriyama, M., & Mori, M. (2003). Variant of perivitelline membrane glycoprotein ZPC of Japanese quail (*Coturnix japonica*) lacking its cytoplasmic tail exhibits the retention in the endoplasmic reticulum of Chinese hamster ovary (CHO-K1) cells. *Biology of Reproduction, 69*, 1401–1407.

Sasanami, T., Matsushima, K., Ohtsuki, M., Kansaku, N., Hiyama, G., & Mori, M. (2005). Vectorial secretion of perivitelline membrane glycoprotein ZPC of Japanese quail (*Coturnix japonica*) in polarized Madin-Darby canine kidney cells. *Cells, Tissues, Organs, 180*, 169–177.

Sasanami, T., Ohtsuki, M., Ishiguro, T., Matsushima, K., Hiyama, G., Kansaku, N., et al. (2006). Zona pellucida domain of ZPB1 controls specific binding of ZPB1 and ZPC in Japanese quail (*Coturnix japonica*). *Cells, Tissues, Organs, 183*, 41–52.

Sasanami, T., Pan, J., Doi, Y., Hisada, M., Kohsaka, T., & Toriyama, M. (2002). Secretion of egg envelope protein ZPC after C-terminal proteolytic processing in quail granulosa cells. *European Journal of Biochemistry, 269*, 2223–2231.

Sasanami, T., Pan, J., & Mori, M. (2003). Expression of perivitelline membrane glycoprotein ZP1 in the liver of Japanese quail (*Coturnix japonica*) after in vivo treatment with diethylstilbestrol. *The Journal of Steroid Biochemistry and Molecular Biology, 84*, 109–116.

Sasanami, T., Toriyama, M., & Mori, M. (2003). Carboxy-terminal proteolytic processing at a consensus furin cleavage site is a prerequisite event for quail ZPC secretion. *Biology of Reproduction, 65*, 1613–1619.

Sato, T., Kinoshita, M., Kansaku, N., Tahara, K., Tsukada, A., Ono, H., et al. (2009). Molecular characterization of egg envelope glycoprotein ZPD in the ovary of Japanese quail (*Coturnix japonica*). *Reproduction, 137*, 333–343.

Serizawa, M., Kinoshita, M., Rodler, D., Tsukada, A., Ono, H., Yoshimura, T., et al. (2011). Oocytic expression of zona pellucida protein ZP4 in Japanese quail (*Coturnix japonica*). *Animal Science Journal, 82*, 227–235.

Stambaugh, R., & Buckley, J. (1968). Zona pellucida dissolution enzymes of the rabbit sperm head. *Science, 161*, 585–586.

Takeuchi, Y., Nishimura, K., Aoki, N., Adachi, T., Sato, C., Kitajima, K., et al. (1999). A 42-kDa glycoprotein from chicken egg-envelope, an avian homolog of the ZPC family glycoproteins in mammalian zona pellucida. Its first identification, cDNA cloning and granulosa cell-specific expression. *European Journal of Biochemistry, 260*, 736–742.

Takeuchi, Y., Cho, R., Iwata, Y., Nishimura, K., Kato, T., et al. (2001). Morphological and biochemical changes of isolated chicken egg-envelope during sperm penetration: degradation of the 97-kilodalton glycoprotein is involved in sperm-driven hole formation on the egg-envelope.. *Biology of Reproduction, 64*, 822–830.

Tanghe, S., Van Soom, A., Nauwynck, H., Coryn, M., & de Kruif, A. (2002). Minireview: Functions of the cumulus oophorus during oocyte maturation, ovulation, and fertilization. *Molecular Reproduction and Development, 61*, 414–424.

Waclawek, M., Foisner, R., Nimpf, J., & Schneider, W. J. (1998). The chicken homologue of zona pellucid protein-3 is synthesized by granulosa cells. *Biology of Reproduction, 59*, 1230–1239.

Wyburn, G. M., Aitken, R. N., & Johnston, H. S. (1965). The ultrastructure of the zona radiata of the ovarian follicle of the domestic fowl. *Journal of Anatomy, 99*, 469–484.

Zhang, H., Du, Z. Q., Dong, J. Q., Wang, H. X., Shi, H. Y., Wang, N., et al. (2014). Detection of genome-wide copy number variations in two chicken lines divergently selected for abdominal fat content. *BMC Genetics, 15*, 517.

CHAPTER TEN

The Mouse Egg's Zona Pellucida

Paul M. Wassarman[1], Eveline S. Litscher
Department of Cell, Developmental, and Regenerative Biology, Icahn School of Medicine at Mount Sinai, New York, NY, United States
[1]Corresponding author: e-mail address: paul.wassarman@mssm.edu

Contents

Abstract

All mammalian eggs are surrounded by a highly specialized extracellular matrix (ECM), called the zona pellucida (ZP), that functions before, during, and after fertilization. Unlike somatic cell ECM the mouse ZP is composed of three different proteins, ZP1–3, that are synthesized and secreted by growing oocytes and assembled into long interconnected fibrils. ECM or vitelline envelope (VE) that surrounds fish, reptilian, amphibian, and avian eggs also consists of a limited number of proteins all closely related to ZP1–3. Messenger RNAs encoding ZP1–3 are expressed only by growing oocytes at very high levels from single-copy genes present on different chromosomes. Processing at the amino- and carboxy-termini of nascent ZP1–3 permits secretion of mature proteins into the extracellular space and assembly into fibrils and matrix. Structural features of nascent ZP proteins prevent assembly within secretory vesicles of growing oocytes. Homozygous knockout female mice that fail to synthesize either ZP2 or ZP3 are unable to construct a ZP, ovulate few if any eggs, and are infertile. ZP1–3 have a common structural feature, the ZP domain (ZPD), that has been conserved through 600 million years of evolution and is essential for ZP protein assembly into fibrils. The ZPD consists of two subdomains, each with four conserved cysteine residues present as two intramolecular disulfides, and resembles an immunoglobulin (Ig) domain found in a wide variety of proteins that have diverse functions, from receptors to mechanical transducers. ZP2 and ZP3 function as receptors for acrosome-reacted and acrosome-intact sperm, respectively, during fertilization of ovulated eggs, but are inactivated as sperm receptors as a result of fertilization.

Current Topics in Developmental Biology, Volume 130
ISSN 0070-2153
https://doi.org/10.1016/bs.ctdb.2018.01.003

ABBREVIATIONS

aa amino acids
AI acrosome-intact
AR acrosome-reacted
ARx acrosome reaction
CFCS consensus furin cleavage site
CT cytoplasmic tail
C-terminal carboxy-terminal
CTP carboxy-terminal propeptide
Cys cysteine
ECM extracellular matrix
EHP external hydrophobic patch
ER endoplasmic reticulum
GPI glycosyl phosphatidylinositol
IHP internal hydrophobic patch
MW molecular weight
N-terminal amino-terminal
PGCs primordial germ cells
SCS sperm combining-site
SS signal sequence
SV secretory vesicles
TMD transmembrane domain
VE vitelline envelope
ZP zona pellucida
ZP-C zona pellucida C-terminal subdomain
ZPD zona pellucida domain
ZP-N zona pellucida N-terminal subdomain

1. INTRODUCTION TO THE ZONA PELLUCIDA

Nearly all animal cells are surrounded by an acellular component called the extracellular matrix (ECM) that participates in a myriad of biological activities (Alberts et al., 2014). ECM affects gene expression, differentiation, morphogenesis, cellular adhesion, cellular migration, and intercellular communication. Integrins are transmembrane receptors that mediate interactions between ECM and intracellular actin cytoskeleton, permitting outside-in and inside-out signal transduction. Most cells are anchored to ECM and loss of anchoring can cause cells to become cancerous.

ECM of most animal cells is made up of proteoglycans, such as hyaluronic acid, heparin-, chondroitin-, and keratin-sulfate, and fibrous proteins, such as collagens, elastins, fibronectins, and laminins (Franz, Stewart, & Weaver, 2010;

Kreiss & Vale, 1999). On the other hand, eggs from monotremes, marsupials, and placental mammals have an ECM, called the zona pellucida (ZP), constructed of three or four proteins called ZP1–4 (Litscher & Wassarman, 2015). The ZP supports oocyte growth and follicle cell proliferation during oogenesis, regulates fertilization of eggs, and protects ovulated eggs and preimplantation embryos in the reproductive tract prior to implantation in the uterus. ECM surrounding fish, reptilian, amphibian, and avian eggs, called the vitelline envelope (VE), consists of a small number of proteins related to ZP1–4. It has been suggested that ZP proteins have properties similar to amyloids that self-aggregate and form cross-β-sheet fibrillar structures (Egge, Muthusubramanian, & Cornwall, 2015; Hewetson et al., 2017; Louros, Iconomidou, Giannelou, & Hamodrakas, 2013). Although ZP1–4 are different from proteins that constitute somatic cell ECM, the ZP has some properties quite reminiscent of those of somatic cell ECM.

2. OOGENESIS IN MICE

The ZP is laid down during the process of oogenesis in females that results in production of unfertilized eggs (Gosden, 2013; Rodrigues, Limback, McGinnis, Plancha, & Albertini, 2008; Wassarman & Albertini, 1994; Zuccotti, Merico, Cecconi, Redi, & Garagna, 2011) (Fig. 1A). In mice oogenesis begins early in fetal development with appearance of 15–100 primordial germ cells (PGCs) in yolk sac endoderm and in the region of the allantois arising from the primitive streak. PGCs migrate into the endodermal epithelium of the hindgut where 170–350 PGCs are found by day-9 of fetal development, and then along the dorsal mesentery of the genital ridges found in the roof of the coelom, the site of gonad development. There are about 5000 PGCs in 11–12-day embryos and more than 20,000 PGCs by the time the genital ridges are fully colonized at day 13–14. These PGCs are the sole source of adult germ cells. The origin and migration of PGCs to the genital ridges are the same in both males and females. Gonadal sex differentiation, to either testis or ovary, occurs in the 12–13-day embryo. The 13-day female embryo possesses a differentiated ovary with all PGCs converted to actively dividing oogonia.

As early as day-12 of embryogenesis, a few oogonia enter first meiotic prophase, stop dividing, and become oocytes. In day-14 embryos germ cells are about equally divided between oogonia and oocytes. By day-15 the ovary contains only oocytes at various stages of first meiotic prophase. It takes about 4 days for oocytes to complete nuclear progression from

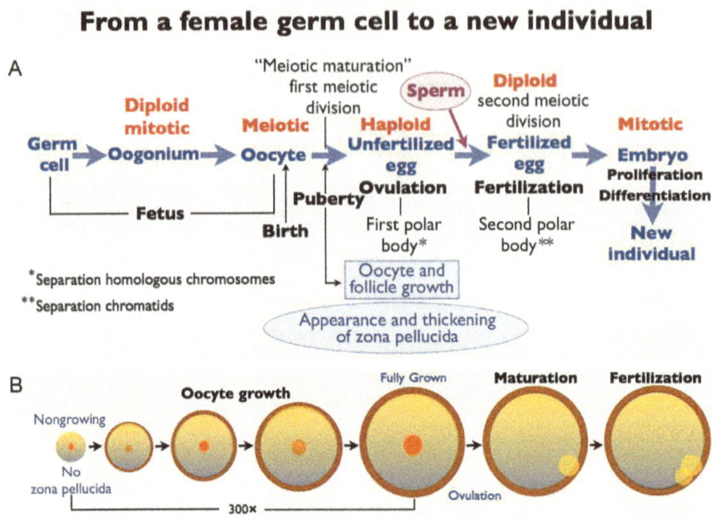

Fig. 1 (A) Schematic representation of steps involved in the conversion of female germ cells in the fetus to fertilized eggs in the adult. In the fetus *primordial germ cells* convert to mitotic *oogonia* and then to meiotic *oocytes* with crossing-over and recombination. Soon after birth all oocytes are arrested in meiosis at diplotene of the first meiotic prophase. At puberty during each reproductive cycle some oocytes grow about 300-fold over 2–3 weeks, their surrounding follicle cells proliferate and differentiate, and fully grown oocytes are ovulated. At about the time of ovulation oocytes undergo *meiotic maturation* with emission of a first polar body following separation of homologous chromosomes (*1st meiotic division*). In this manner fully grown oocytes become haploid *unfertilized eggs*. Upon fusion with a single sperm, *fertilized eggs* emit a second polar body following separation of chromatids (*2nd meiotic division*), but are restored to a diploid state by the haploid sperm genome. (B) Schematic representation of ZP production during oocyte growth in mice. Nongrowing oocytes lack a ZP, but as soon as oocytes begin to grow they lay down a ZP that continues to thicken throughout the growth phase (2–3 weeks; about a 300-fold increase in volume) and results in a 6.2 ± 1.9 μm thick ZP around fully grown oocytes and ovulated eggs.

leptotene through pachytene, with crossing-over and recombination, and by day-18 some oocytes have reached diplotene of the first meiotic prophase. By day-5 *postpartum*, nearly all oocytes are arrested in late diplotene or the dictyate stage of meiosis, where they remain until stimulated to resume meiosis at the time of ovulation. This population of small (12 μm diameter) nongrowing oocytes is the sole source of unfertilized eggs in the sexually mature adult.

Nearly half the population of mouse oocytes is lost during the first 2–3 weeks after birth (Fig. 2); a similar loss of oocytes occurs after human

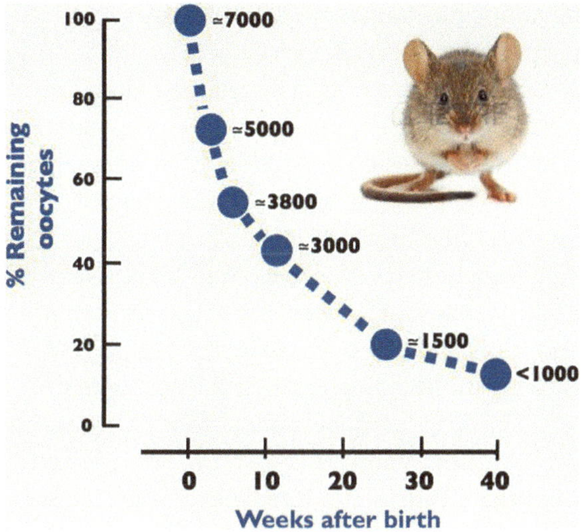

Fig. 2 A graph depicting the dramatic loss of female germ cells during fetal development of mice followed by a progressive loss during the first 40 weeks after birth. More than 70% of germ cells are lost during fetal development in mice such that a newborn mouse has about 7000 and 40 weeks later less than 1000 remain. A human female has about 1,500,000 germ cells at birth, by puberty there are about 300,000, and at menopause (age 45–55) less than 1000 remain.

births. However, during this period more oocytes begin to grow than at any other time in the life of the mouse. Oocytes grow from about 12 to 80 µm in diameter over 2–3 weeks (more than a 300-fold increase in oocyte volume, from about 0.9 to 300 pL). Nongrowing oocytes are contained within primordial follicles that begin as single layers of 4–6 flattened epithelial-like cells around oocytes and become three layers of cuboidal granulosa cells by the time oocytes have completed their growth phase. Primordial follicles produced in the fetus are very stable and have a half-life of about 10 months during adulthood (Lei & Spradling, 2013).

Over several days, while oocytes remain a constant size, the number of follicle cells increases to >50,000 in Graafian follicles. The developing follicle is a syncytium of cells connected by intercellular cytoplasmic channels called gap junctions (Kidder & Mahwi, 2002; Wassarman, 2002). Gap junctions form between oocytes and cells of the innermost cumulus cells, the corona radiata, between other cumulus cells, between cumulus cells and follicle cells not surrounding oocytes, and between all other follicle cells. Communication between the oocyte and follicle cells is bidirectional; follicle cells regulate oocyte growth and oocytes regulate follicular development.

Follicles exhibit an incipient antrum when they are several layers thick and, as the antrum expands, oocytes take up an acentric position surrounded by two or more layers of cumulus cells. The corona radiata becomes columnar is shape and these cells form gap junctions with the oolemma. A fully developed Graafian follicle consists of one fully grown oocyte positioned at the end of a stalk of follicle cells and surrounded by several layers of cumulus cells and a large, fluid-filled antrum.

Small, nongrowing oocytes do not have a ZP; however, a very thin ZP appears around oocytes as soon as they enter their growth phase and it continues to thicken throughout oocyte growth (Fig. 1B). As oocytes grow from about 40 to 80 μm in diameter, their ZP increases from about 1.6 to 6.2 μm in width. The ZP is a viscoelastic, porous ECM penetrable by antibodies, enzymes, and small viruses (Gwatkin, 1977; Kim & Kim, 2013). During oogenesis the ZP stabilizes intercellular connections, or gap junctions, between the corona radiata and oocytes ensuring proper transfer of nutrients, metabolites, and other molecules to growing oocytes from the external milieu.

In sexually mature mice, fully grown oocytes respond to a hormonal stimulus in vivo or to release from follicles into a suitable medium in vitro, and resume meiosis just prior to ovulation. Oocytes undergo meiotic maturation, a process characterized by dissolution of the nuclear envelope, chromatin condensation into bivalents, separation of homologous chromosomes, emission of a first polar body, and arrest of meiosis at metaphase II. Oocytes acquire the ability to undergo meiotic maturation or "meiotic competence" during oocyte growth (Sorensen & Wassarman, 1976). Upon completion of meiotic maturation fully grown oocytes become unfertilized eggs. Ovulated eggs complete meiosis, with separation of chromatids to become haploid and emission of a second polar body, upon fertilization by a single sperm. Simultaneously, the group of follicle cells left behind in the ovary becomes an endocrine gland, the corpus luteum, that supports pregnancy by secretion of progesterone after implantation of expanded blastocysts into the uterus.

3. ZONA PELLUCIDA PROTEINS

The mouse egg's ZP is a gel-like structure that is composed of three proteins, called ZP1–3, that have apparent MWs of about 200, 120, and 83 kDa, respectively, and account for virtually all proteins (3.5 ng) in the ZP (Bleil & Wassarman, 1980a; Litscher & Wassarman, 2015; Wassarman, 1988a) (Fig. 3). ZP2 and ZP3 are present in roughly equimolar amounts

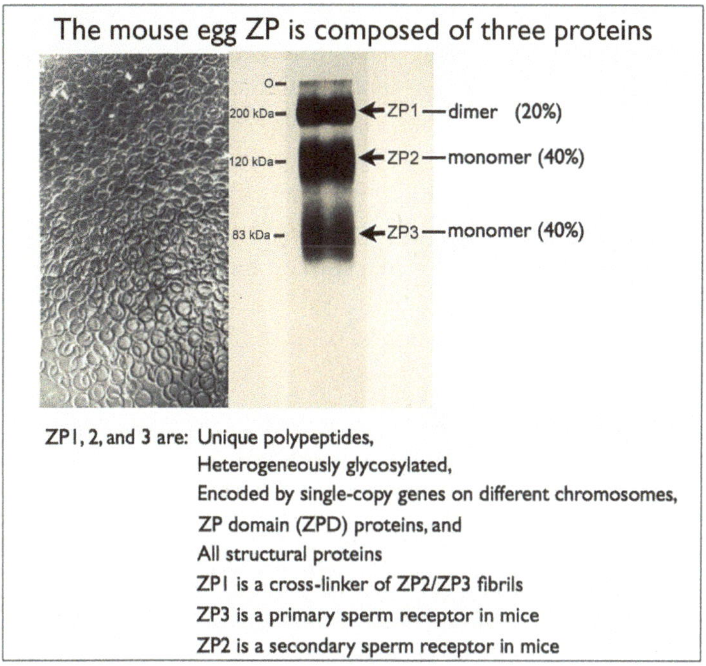

Fig. 3 The mouse egg ZP is composed of three proteins, ZP1–3. Shown is a light micrograph of mouse ZP isolated from fully grown oocytes, a one-dimensional gel of mouse ZP proteins separated under denaturing conditions, and a list of some of the properties of mouse ZP proteins. ZP1, ZP2, and ZP3 have MWs of about 200, 120, and 83 kDa, respectively, and represent about 20%, 40%, and 40% of total ZP protein (about 3.5 ng total), respectively.

in the ZP and are monomers, whereas ZP1 is the least abundant protein and is a dimer of identical polypeptide chains held together by intermolecular disulfides. The primary structures of ZP2- and ZP3-related proteins from different mammals are well conserved having about 65%–98% identity, whereas ZP1-related proteins are conserved to a lesser degree having about 40% identity. Considerable evidence suggests that ZP1–3 are structural proteins and that ZP2 and ZP3 are receptors for sperm that are inactivated as receptors following fertilization (Wassarman & Litscher, 2016). It has been proposed that all ZP proteins are derived from a common ancestral gene, possibly ZP3, and that ancestral ZP3 functioned as an egg ECM protein (Claw & Swanson, 2012; Spargo & Hope, 2003).

Each ZP protein consists of a unique polypeptide that is often heterogeneously glycosylated with asparagine- (N-) and serine/threonine- (O-) linked oligosaccharides that in some cases are sialylated and sulfated.

ZP1–3 possess 4, 6, and 5 N-linked oligosaccharides, respectively. The ZP of human eggs and eggs of many other mammals contains a fourth protein, ZP4, that has a trefoil domain and is homologous with ZP1 (Conner, Lefievre, Hughes, & Barratt, 2005; Hughes & Barratt, 1999). In some mammals, such as cat, cow, dog, and pig, ZP1 is replaced completely by ZP4. In mice, ZP4 is not present since it is encoded by a pseudo-gene that is not expressed during oogenesis (Goudet, Mugnier, Callebaut, & Monget, 2008). ZP1 and ZP4 may be an example of duplication of a single gene since adjacent regions of these two genes in the mouse are located on the same chromosome (Smith, Paton, Hughes, & Burt, 2005).

4. ZONA PELLUCIDA DOMAIN PROTEINS

All three mouse ZP proteins possess a zona pellucida domain (ZPD)—ZP1, amino acids (aa) 271–540; ZP2, aa 364–628; and ZP3, aa 45–302. ZP1–3 are prototypical ZPD proteins (Bork & Sander, 1992) that are present in virtually all vertebrate egg ZP and VE, as well as in a wide variety of tissues and organs in both vertebrates and invertebrates (Jovine, Darie, Litscher, & Wassarman, 2005; Litscher & Wassarman, 2015; Plaza, Chanut-Delalande, Fernandes, Wassarman, & Payre, 2010). The ZPD arose more than 600 million years ago, and hundreds of ZPD proteins with diverse functions, from receptors to mechanical transducers, have been identified in a wide variety of organisms, from jellyfish, flies, and nematodes to humans. In mammals they are found in brain, heart, intestine, kidney, liver, nose, ovary, pancreas, spleen, tongue, and other locations. ZPD proteins are present at the apical surface of many epithelia and participate in functioning of the senses, including taste and smell. In addition to a ZPD, many of these proteins possess other domains, such as epidermal growth factor (EGF), scavenger receptor Cys-rich (SRCR), von Willebrand factor (vWF), plasminogen apple N-terminal (PAN), complement c1r/c1s, Uegf, Bmp1 (CUB), and whey acidic protein (WAP) domains. ZPD proteins include betaglycan, cuticlin, dumpy, endoglin, glycoprotein-2, hensin, mesoglein, nompA, oikosin, tectorin, uromodulin, and vomeroglandin. Precursors of ZPD proteins almost invariably contain a single transmembrane domain (TMD) composed of either hydrophobic aa or a glycosyl phosphatidylinositol (GPI) anchor that are missing from the secreted proteins. Mutations in genes encoding ZPD proteins can result in severe human pathologies, including deafness, vascular and renal disease, cancer, and possibly infertility.

A ZPD consists of about 260 aa, has 8 conserved cysteine (Cys) residues, and has a bipartite structure with 2 subdomains, an amino (N)-terminal subdomain (ZP-N) and a carboxy (C)-terminal subdomain (ZP-C). Each subdomain has four conserved Cys residues present as intramolecular disulfides. The ZP-N disulfides are linked 1 → 4 and 2 → 3 and the ZP-C disulfides are linked 5 → 7 and 6 → 8. The subdomains are separated from each other by a short protease-sensitive region. ZP3 has only eight Cys residues in its ZPD, but ZP1 and ZP2 have two additional Cys residues, called a and b, in their ZP-C that form an a → b linked disulfide. A comparison of ZPD sequences for ZP1–4 from platypus to humans reveals an average %-identity of about 73% and %-similarity of about 90%.

Considering the two ZPD subdomains as separate units, the sequential order is always ZP-N → ZP-C and never ZP-C → ZP-N. However, a ZP-N can be present in the absence of a ZP-C. For example, ZP1 and ZP2 possess 1 and 3 ZP-Ns, respectively, at their N-termini (Callebaut, Mornon, & Monget, 2007), and a ZP-N is present in the absence of a ZP-C in several proteins, including Plac1, Oosp1, and papillote (Bökel, Prokop, & Brown, 2005; Cocchia et al., 2000; Yan et al., 2001). No protein has been found to possess only a ZP-C which suggests that its role is dependent on its partner ZP-N. There is evidence that ZP-N is a conserved module used for polymerization of extracellular proteins into fibrils (Jovine et al., 2005; Janssen, Litscher, & Wassarman, 2006; Jovine, Qi, Williams, Litscher, & Wassarman, 2002), consistent with the suggestion that the ZPD might play a role in polymerization of proteins (Killick, Legan, Malenczak, & Richardson, 1995; Legan, Rau, Keen, & Richardson, 1997). The strong structural similarity between the ZP-N and ZP-C suggests that the ZPD may have arisen by duplication of an ancestral gene encoding a protein containing a single ZP-N. It is likely that the bipartite structure of the ZPD endows it with dual functions. Generally, the ZPD module is a conserved evolutionary unit essential for polymerization of proteins, whereas adjoining regions contribute to functional diversification that may be caused by selective pressure related to species-specific functions.

The three-dimensional structure of the ZPD has been determined by X-ray crystallographic analysis of the ZP-N of mouse ZP3 at 2.3 Å resolution (Monné, Han, Schwend, Burendahl, & Jovine, 2008), full-length chicken ZP3 at 2.0 Å resolution (Han et al., 2010; Monné & Jovine, 2011), the protease-resistant core (aa 295–610) of human uromodulin at 3.2 Å resolution (Bokhove et al., 2016), and the ZP-C (aa 463–664) of mouse ZP2 at 2.25 Å resolution (Bokhove et al., 2016). These analyses revealed that both the ZP-N and ZP-C adopt an immunoglobulin (Ig)-like fold consisting of an antiparallel sandwich of two β-sheets made up of eight strands of polypeptide (A, B, D, and E and C, E′, F, and G) that enclose a hydrophobic core, with two buried disulfides that clamp both sides of the sandwich. The ZP-N of mouse ZP3 has approximate dimensions of 52 × 29 × 28 Å (Monné et al., 2008). In ZP3 crystals, two ZP3 molecules are arranged as homodimers in antiparallel orientation to form an asymmetric structure. The two ZPDs are bonded by electrostatic interactions between the ZP-N and ZP-C of opposing molecules: ZP-N1:ZP-C2 and ZP-N2: ZP-C1. On the other hand, there are no ZP-N1:ZP-N2 or ZP-C1:ZP-C2 contacts within the homodimer. However, certain structural features make the ZPD distinct from other known Ig-like domains such that it represents a new Ig superfamily subtype structure. The ZP-N and ZP-C of uromodulin are similar to those of ZP3 having the same fold, disulfide connectivities, and a conserved tyrosine residue in the ZP-N; mutation of this residue in tectorin is associated with hearing loss (Legan et al., 2005).

ZP1/ZP2-like proteins have two extra Cys residues, called a and b, compared to ZP3-like proteins. For some time it was thought that ZP-Cs of different ZPD proteins had different conserved disulfide linkages (5 → 7 and 6 → 8 linkages in ZP3-like proteins vs 5 → 6, 7 → a, and 8 → b linkages in ZP1/ZP2-like proteins) and different overall structures, and could account for differences in polymerization specificity. However, it is now appears that ZP-Ns and ZP-Cs from different ZPD proteins share the same conserved disulfide linkages (5 → 7 and 6 → 8) and the same overall architecture as ZP3 (Bokhove et al., 2016). Consequently, other molecular features must regulate whether ZPD proteins assemble into homopolymers like uromodulin fibrils or heteropolymers like ZP fibrils. In this context, of interest is the observation that for different ZPD proteins the interdomain region or linker between the ZP-N and ZP-C can be either flexible, as found in ZP2 and ZP3, or rigid, as found in uromodulin (Bokhove et al., 2016). In at least two ZPD proteins, β-glycan and endoglin, the linker region is extremely short. It has been postulated that differences in plasticity of the interdomain regions modulate polymerization of structurally similar ZPD proteins.

5. ZONA PELLUCIDA PROTEIN SYNTHESIS

Are oocytes, follicle cells, or both the site of synthesis of mouse ZP proteins? This question was resolved in the early 1980s using growing oocytes, with or without surrounding follicle cells, cultured in vitro in the presence of radiolabeled aa or sugars. It was concluded that ZP proteins are synthesized only by growing mouse oocytes, not by follicle cells (Bleil & Wassarman, 1980b; Greve, Salzmann, Roller, & Wassarman, 1982; Salzmann, Greve, Roller, & Wassarman, 1983; Shimizu, Tsuji, & Dean, 1983). Subsequently, in situ hybridization and ribonuclease protection experiments revealed that messenger RNA (mRNA) encoding ZP proteins is undetectable in nongrowing oocytes, but is present in small and midstage growing oocytes, and in fully grown oocytes. For example, ZP3 mRNA is undetectable in nongrowing oocytes (<1000 copies/oocyte), but increases to about 300,000 copies/oocyte in midstage growing oocytes, declines to about 240,000 copies/oocyte in fully grown oocytes, and decreases to undetectable levels in fertilized eggs (Roller, Kinloch, Hiraoka, Li, & Wassarman, 1989).

In mice, ZP1–3 are encoded by single-copy genes located on different chromosomes. Genes encoding ZP1–3 are present on chromosomes #19 (12 exons), #7 (18 exons), and #5 (8 exons), respectively (Epifano, Liang, & Dean, 1995; Liang & Dean, 1993). Genomic organization of ZP loci is identical in oocytes and somatic cells, but ZP genes are hypomethylated in oocytes where they are expressed as compared to somatic cells where they are not expressed (Ringuette, Sobieski, Chamow, & Dean, 1986). At least two ovary-specific DNA-binding proteins, ZAP-1 (Millar, Lader, & Dean, 1993) and OSP-1 (Schickler, Lira, Kinloch, & Wassarman, 1992), bind to promoters of ZP2 and ZP3. As little as 153 nucleotides of mZP3 5′-flanking sequence is sufficient to target expression of ZP3 and foreign proteins (e.g., firefly luciferase) to growing oocytes, but not to follicle cells (Lira, Kinloch, Mortillo, & Wassarman, 1990; Lira, Schickler, & Wassarman, 1993).

Mouse ZP proteins are synthesized by growing oocytes as precursor polypeptides. ZP1–3 precursors have MWs of 69 kDa (623 aa; monomer), 79 kDa (713 aa), and 47 kDa (424 aa), respectively (Litscher & Wassarman, 2015) (Fig. 4). All precursors possess an N-terminal signal sequence (SS; 20–30 aa) and a C-terminal propeptide (CTP; 70–80 aa) that includes a consensus furin cleavage site (CFCS; 4 aa), TMD (20 aa), and short cytoplasmic tail (CT; 10–15 aa). ZP1 also possesses a trefoil domain (42 aa) with 6 Cys residues that form 3 intramolecular disulfides. The SS targets nascent ZP proteins to the secretory pathway where it is removed. The TMD is a sequence of hydrophobic aa that anchors the proteins in secretory vesicles (SV) and on plasma membrane. The CFCS has the sequence R-X-X-R or R-X-R/K-R that is recognized and cleaved by a member of the

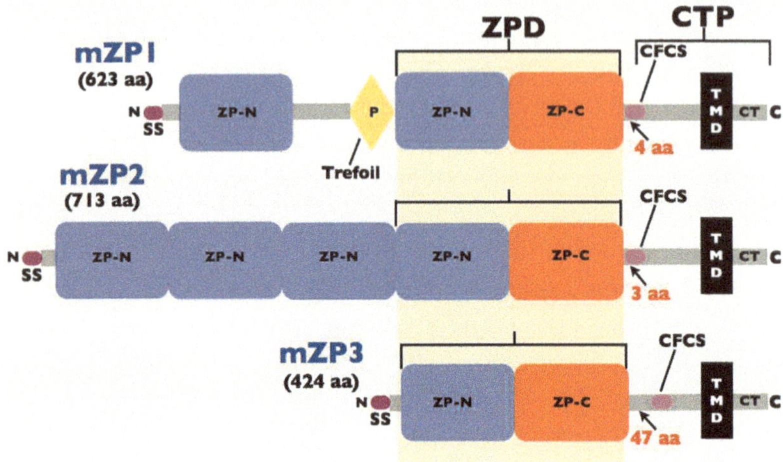

Fig. 4 Schematic representation of the organization of mouse ZP proteins mZP1–3. In each case, the polypeptide contains an SS (signal sequence) at the N-terminus, a ZPD (zona pellucida domain), and a CFCS (consensus furin cleavage site), TMD (transmembrane domain), and CT(cytoplasmic tail) at the CTP (C-terminal propeptide). mZP1 (623 aa) also has a trefoil (P) domain adjacent to the ZPD and an extra ZP-N subdomain close to the N-terminus of the polypeptide. mZP2 (713 aa) has three extra ZP-N subdomains between the ZPD and N-terminus of the polypeptides. mZP3 (424 aa), the smallest of the three proteins, consists primarily of a ZPD. It should be noted that for ZP1 and ZP2 there are only three or four amino acids between the ZPD and CFCS, but for ZP3 there is a stretch of 47 aa. The latter includes the proposed SCS that is a site of positive Darwinian selection during evolution.

furin-like family of proprotein convertases and releases nascent ZP proteins into the extracellular space to form fibrils.

The oocyte's Golgi, an important site of processing of secretory proteins like ZP1–3, expands dramatically in growing oocytes, from flattened stacked lamellae in nongrowing oocytes to swollen stacked lamellae associated with numerous vacuoles in fully grown oocytes (Wassarman & Josefowicz, 1978). These changes, as well as those of the endoplasmic reticulum (ER), suggest that growing oocytes are actively engaged in trafficking of many proteins, including ZP proteins. As oocytes begin to grow, they secrete nascent ZP proteins into localized "pockets" between the oocyte surface and surrounding follicle cells, and as growth continues the pockets of ZP fibrils coalesce, form a uniform ZP, and then the ZP continues to thicken throughout oocyte growth (Wassarman & Mortillo, 1991) (Fig. 5). Nascent ZP proteins assemble solely on the innermost layer of the thickening ZP, suggesting that the ZP thickens solely from the inside during oocyte growth (Qi, Williams, & Wassarman, 2002).

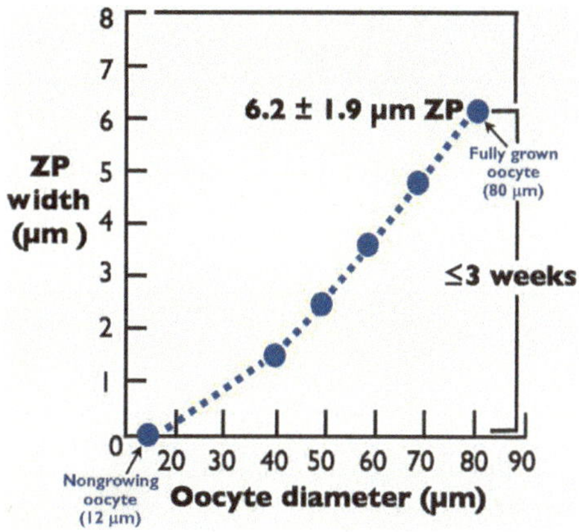

Fig. 5 A graph depicting changes in the width of the mouse oocyte's ZP during growth of the oocyte. Mouse oocytes grow from about 12 μm (nongrowing oocyte) to about 80 μm in diameter (fully grown oocyte) over 2–3 weeks. Concomitant with oocyte growth, the ZP increases from zero (nongrowing oocyte) to about 6.2 ± 1.9 μm (fully grown oocyte) in width.

6. ZONA PELLUCIDA-KNOCKOUT MICE

With the advent of gene targeting procedures mouse lines have been established in which ZP genes have been inactivated by homologous recombination and insertional mutagenesis. Homozygous null males are unaffected by the mutations. Oocytes and eggs from mutant females carrying a single ZP3 allele (heterozygous nulls; $ZP3^{+/-}$) possess a ZP and the females reproduce normally (Table 1). However, the ZP is less than one-half the thickness (2.7 ± 1.2 μm) of wild-type ZP (6.2 ± 1.9 μm) and contains about one-half the amount of ZP2 and ZP3 present in wild-type ZP (Wassarman, Qi, & Litscher, 1997). On the other hand, oocytes and eggs recovered from homozygous null females that are devoid of either ZP2 ($ZP2^{-/-}$) or ZP3 ($ZP3^{-/-}$) do not have a ZP and are infertile (Liu et al., 1996; Rankin et al., 1996, 2001) (Table 1). This is due to a paucity of growing oocytes and follicles in ovaries that leads to a scarcity of ovulated eggs in oviducts. Both ZP2 and ZP3 must be present for incorporation into fibrils and establishment of a matrix.

The paucity of growing oocytes and follicles in $ZP3^{-/-}$ mice is reflected in differences in the size of ovaries from wild-type and $ZP3^{-/-}$ females (Wassarman & Litscher, 2012). Ovaries from wild-type females are always larger than ovaries from homozygous null females, with the largest difference (about fourfold) seen in juvenile mice (<21 days of age) that possess

Table 1 Phenotypes of ZPI-3 Null Females

	Presence of ZP			
	Early-Growing Oocyte	**Late-Growing Oocyte**	**Fully Grown Oocyte**	**Fertility**
Wild Type	+	+	+	+
$ZP1^{-/-}$	+[a]	+	+	+/−[b]
$ZP2^{-/-}$	+[c]	−	−	−[d]
$ZP3^{-/-}$	−	−	−	−[d]
$ZP3^{+/-}$	+[e]	+	+	+

[a]A loosely organized, abnormal ZP (Rankin et al., 1999).
[b]Fewer than normal offspring (Rankin et al., 1999).
[c]A very thin, abnormal ZP (Rankin et al., 2001).
[d]No offspring (Liu et al., 1996; Rankin et al., 1996).
[e]A thinner than normal ZP (Wassarman et al., 1997).

an unusually large number of developing follicles (Table 2). This is most demonstrable in juvenile females since the ovarian stroma and vasculature are incompletely developed and do not mask the effect of a decreased number of developing follicles on ovarian weight. In the absence of a ZP the formation of gap junctions between oocytes and surrounding follicle cells is reduced, thereby compromising transfer of nutrients, metabolites, and other molecules to growing oocytes and resulting in fewer oocytes and ovulated eggs. Presumably the ZP helps to stabilize gap junctions under normal conditions of oocyte growth and follicle expansion. It is noteworthy that female mice that are homozygous null for proteins associated with follicle cell gap junctions, such as connexin-37 and -43, are deficient in both multilayered follicles and normal growing oocytes, and are infertile (Ackert, Gittens, O'Brien, Eppig, & Kidder, 2001; Simon et al., 1997).

Oocytes and eggs from homozygous null females devoid of ZP1 ($ZP1^{-/-}$) have a ZP, but the females are not as fertile as wild-type mice (Table 1). The ZP is quite fragile and exhibits large pores penetrable by follicle cells, presumably because of insufficient cross-linking of ZP fibrils by ZP1 (Rankin, Talbot, Lee, & Dean, 1999). The poorly organized ZP permits fertilization to occur, but it is vulnerable during passage of cleavage-stage embryos down the oviduct and leads to early embryonic loss and fewer offspring. The presence of ZP2 and ZP3 supports formation of heterodimers and fibrils, but in the absence of ZP1 the fibrils are insufficiently packed together. The appearance and properties of the ZP from $ZP1^{-/-}$ females are entirely consistent with a lack of cross-linking of fibrils.

Table 2 Null Mutation and Weight of Ovary

	Weight of Ovary (mg)		
Age	**Wild Type**	**mZP3$^{-/-}$**	**Wild Type / mZP3$^{-/-}$**
20 days	1.0 ± 0.17	0.26 ± 0.1	4:1
35 days	1.7 ± 0.17	1.0 ± 0.34	1.7:1
70 days	3.6 ± 0.43	2.8 ± 0.65	1.3:1

7. ZONA PELLUCIDA PROTEIN SECRETION AND ASSEMBLY

Certain elements involved in secretion of ZP proteins by growing oocytes are located in the CTP. One of these elements is the CFCS that must be cleaved by a proprotein convertase for secretion of ZP proteins to take place (Litscher & Wassarman, 1999; Williams & Wassarman, 2001). When the CFCS is mutated, it abolishes cleavage and results in accumulation of unprocessed ZP proteins in the ER. Under these conditions, secretion of ZP proteins is reduced as much as 20-fold, but the low level of secreted protein is cleaved by either a convertase at a significantly reduced rate or some other protease.

Another element is an external hydrophobic patch (EHP) located in the CTP that interacts with an internal hydrophobic patch (IHP) located in the ZPD (Han et al., 2010; Jovine, Qi, Williams, Litscher, & Wassarman, 2004, 2007). An EHP:IHP interaction is also present in another ZPD protein, the uromodulin precursor (Schaeffer, Santambrogio, Perucca, Casari, & Rampoldi, 2009). The EHP:IHP interaction locks the ZPD in a conformation that is incompatible with polymerization, thereby preventing formation of ZP fibrils within growing oocytes prior to excision of the CTP (Fig. 6). This mechanism for preventing polymerization applies to many, if not most, ZPD proteins because it relies on sequence elements, EHP and IHP, and coupling between proteolytic processing and polymerization. Cleavage of inhibitory sequences from protein precursors, with concomitant exposure of polymerization elements, regulates assembly of several other proteins, including fibrinogen, fibrillin, tau, C9, and Cks1, among others. Evidence from mutation/deletion experiments suggest that the EHP and IHP are functionally related to each other and, together with the CFCS and TMD, control incorporation of nascent ZP proteins into the ZP. Although secretion is severely inhibited when either the EHP or IHP is mutated in the

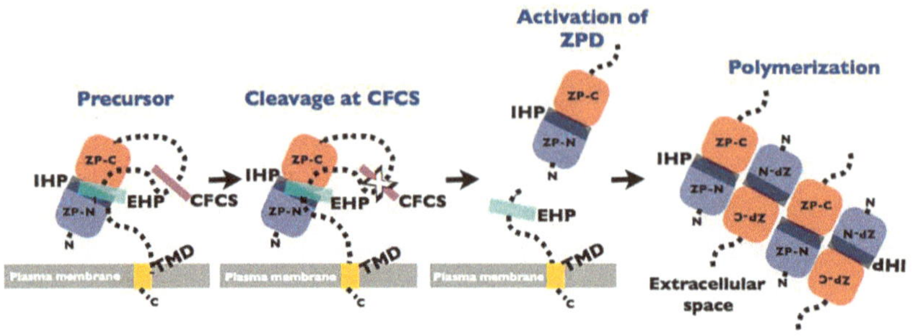

Fig. 6 Schematic representation of a general mechanism for assembly of nascent ZP proteins. In all ZPD precursor proteins (Precursor), the ZPD is followed by a C-terminal propeptide (CTP) that contains a basic cleavage site, such as a CFCS (consensus furin cleavage site), and EHP (external hydrophobic patch), and a TMD (transmembrane domain) or glycosylphosphatidylinositol (GPI) anchor site. Precursors do not polymerize within the cell either as a result of direct interaction between the EHP and IHP (internal hydrophobic patch) or because they adopt a conformation dependent on the presence of both hydrophobic patches. C-terminal processing at the CFCS by a proprotein convertase (Cleaved at CFCS) leads to dissociation of mature proteins from the EHP and activation of the ZPD (Activated ZPD) for assembly (Polymerized) into fibrils and matrices.

absence of a TMD, when the TMD is present nascent protein is secreted, but not incorporated into the thickening ZP (Jovine et al., 2004, 2007).

Within growing oocytes ZP proteins are localized to SV that originate from the trans-Golgi. The vast majority of SV are doughnut-shaped with an empty central cavity or lumen (Qi et al., 2002). This is consistent with nascent ZP proteins associating with SV membrane via the protein's TMD, leaving the central portion of the SV empty. The oocyte's SV are large, about $2.3 \pm 0.3\,\mu m$ in diameter, as compared to those of somatic cells that are about $0.1–0.2\,\mu m$ in diameter. However, for oocytes from ZP3 homozygous null mice in which ZP3 is absent and the amount of ZP2 is about half that of wild-type oocytes, SV are only $0.7 \pm 0.2\,\mu m$ in diameter (Qi & Wassarman, unpublished results). The size of the oocyte's SV is clearly dependent on the amount of cargo available. The SV fuse with the oocyte's plasma membrane and release nascent ZP proteins into the extracellular space to be incorporated into the thickening ZP. Incorporation of nascent protein is exclusively at the ZP's innermost surface, indicating that the ZP thickens from the inside to the outside (Qi et al., 2002).

Once on the external surface of oocytes ZP proteins that have had their CTP removed assemble into long interconnected fibrils and matrix. A ZP2–3 heterodimer is located every 14–15 nm along the fibrils and the fibrils are interconnected by ZP1. The ZP1–3 assemblage is held together by noncovalent bonds. The structural periodicity of fibrils can be visualized in electron micrographs of ZP solubilized by proteases or reducing agents, as well as in micrographs of solubilized ZP preparations decorated with monoclonal antibodies directed against either ZP2 or ZP3 (Greve & Wassarman, 1985; Wassarman & Mortillo, 1991).

8. ZONA PELLUCIDA AND FERTILIZATION

For sperm to fertilize ovulated eggs they must bind to and penetrate the ZP, reach the egg's plasma membrane, and fuse with it. To penetrate the ZP, acrosome-intact (AI) sperm must first bind to the ZP and undergo the acrosome reaction (ARx) (Buffone, Hirohashi, & Gerton, 2014), a process that involves multiple membrane fusions (Wassarman & Litscher, 2008). The ARx is a Ca^{2+}-dependent event that leads to signaling through sperm G proteins to produce intracellular ionic changes and second messengers (Lackey & Gray, 2015). Completion of the ARx results in exposure of the sperm's inner acrosomal membrane and associated proteases. Proteases, together with the forward thrust of sperm, facilitate movement of bound sperm through the ZP in a relatively short time. Both AI and acrosome-reacted (AR) sperm can bind to the ZP of ovulated eggs, but only AR sperm can penetrate the ZP and fuse with plasma membrane. Izumo, an AR sperm protein, binds to Juno, an egg plasma membrane protein, prior to fusion of the two gametes (Aydin, Sultana, Li, Thavalingam, & Lee, 2016; Bianchi, Doe, Goulding, & Wright, 2014; Ellerman et al., 2009; Grayson, 2015; Han et al., 2016; Inoue, Ikawa, Isotani, & Okabe, 2005; Nishimura et al., 2016; Ohto et al., 2016). Izumo is an Ig superfamily sperm protein and Juno is a folate receptor-like egg protein. Both proteins are essential, but not sufficient for sperm–egg fusion to occur.

The notion that the ZP harbors species-restricted receptors for sperm dates back more than a century. Results of experiments in the 1970s strongly suggested that solubilized ZP preparations from unfertilized hamster and mouse eggs contained sperm receptors (Gwatkin, 1977). This suggestion was extended in the 1980s and 1990s by evidence demonstrating that AI sperm bind to ZP3 and AR sperm bind to ZP2 of ovulated eggs (Bleil & Wassarman, 1980c, 1983, 1986; Bleil, Greve, & Wassarman, 1988; Mortillo & Wassarman, 1991; Vazquez et al., 1989) (Fig. 7). This led to the conclusion that both ZP3 and ZP2 are receptors for sperm and that ZP3 induces the ARx (Wassarman, 1987, 1988b, 1999). However,

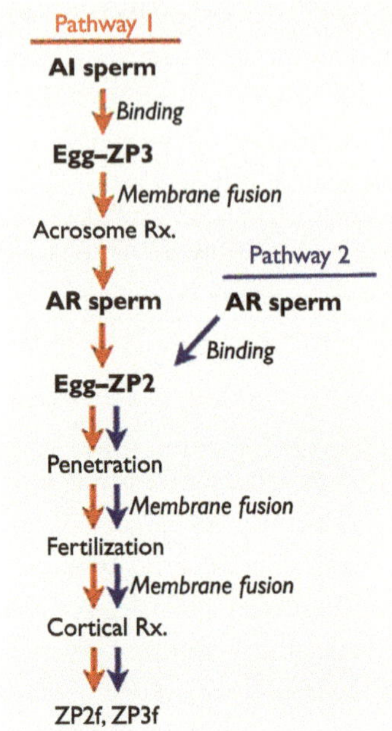

Fig. 7 Description of two pathways to fertilization in mice. In *Pathway 1* acrosome-intact (AI) sperm bind to egg–ZP3 (primary receptor) in the zona pellucida (ZP). As a result of binding to egg–ZP3, AI sperm undergo the acrosome reaction (ARx; fusion between outer and inner acrosomal membranes) and acrosome-reacted sperm (AR sperm) then bind to egg–ZP2 (secondary receptor). In *Pathway 2* sperm that undergo the acrosome reaction (AR sperm) *prior* to reaching the egg bind directly to egg–ZP2, not to egg–ZP3. In both Pathways 1 and 2 AR sperm penetrate the ZP and fertilize the egg (fusion between sperm and egg plasma membranes). Fusion of sperm and egg induces the cortical reaction in eggs (fusion between cortical granule membrane and egg plasma membrane) which results in exocytosis of cortical granule components, including enzymes, into the ZP. This leads to inactivation of ZP2 (ZP2 → ZP2f) and ZP3 (ZP3 → ZP3f) as sperm receptors and in a decrease in the solubility, or "hardening," of the ZP. AI and AR sperm cannot bind to the ZP of fertilized eggs.

in the early 2000s it was proposed that binding of sperm to the ZP was not due to specific ZP proteins, but to the supramolecular structure of the ZP (Dean, 2004; Rankin et al., 2003). More recent experiments have refuted this proposal and once again confirmed that AR sperm bind directly to ZP2 of ovulated eggs (Avella, Baibakov, & Dean, 2014; Baibakov, Boggs, Yauger, Baibakov, & Dean, 2012).

Binding of AR sperm to ZP2 is due to a sequence of about 100 aa near the N-terminus of the protein and this sequence provides some species specificity for sperm binding (Avella et al., 2014). On the other hand, binding of

AI sperm to ZP3 is dependent on a portion of the ZP3 polypeptide encoded by exon-7 at its C-terminus, called the sperm combining-site (SCS), and on serine-/threonine- (O-) linked carbohydrate (Bleil & Wassarman, 1988; Florman, Bechtol, & Wassarman, 1984; Florman & Wassarman, 1985; Han et al., 2010; Kinloch, Sakai, & Wassarman, 1995; Litscher et al., 1995; Litscher & Wassarman, 1996; Rosiere & Wassarman, 1992; Wassarman & Litscher, 2016; Williams, Litscher, Jovine, & Wassarman, 2006). The involvement of polypeptide and carbohydrate in binding is akin to binding of sperm to eggs in marine animals and amphibia, sexual agglutination in budding yeast, binding of pollen to the plant stigma, and binding of bacteria, animal viruses, parasites, and other pathogens to their cellular hosts. Following fertilization both ZP2 and ZP3 are modified such that AI and AR sperm can no longer bind to the ZP, thereby assisting in prevention of polyspermy. In addition, the solubility of the ZP decreases considerably due in part to limited proteolysis of ZP2 (Bleil, Beall, & Wassarman, 1981; Moller & Wassarman, 1989). Cleavage of ZP2, without release of (a) peptide(s), apparently increases interactions between ZP fibrils and results in a loss of water from and a thinning of the ZP.

ZP3 from different mammalian eggs exhibits a high degree of divergence in its SCS compared to other regions of the polypeptide. This divergence is attributed to positive Darwinian selection and perhaps ensures some species specificity during fertilization (Claw & Swanson, 2012; Swanson, Yang, Wolfner, & Aquadro, 2001; Wassarman & Litscher, 1995; Williams et al., 2006) (Table 3). This would be analogous to the accelerated evolution of

Table 3 Divergent ZP3 Sequences in Mammals

	↓ ↓ ↓↓ ↓ ↓ ↓
Mouse	$_{331}$SSSSQFQIHGPRQWSKLV
Rat	$_{331}$SSSSEFETHEPAQWSTLV
Cow	$_{329}$SGRSMRLSH--RE-GRPV
Pig	$_{328}$PSLSR-KLSMPKRQS--A
Dog	$_{325}$PGRSRRLSHLERGWRRSV
Macaca	$_{330}$PSHSRRQPHVVSQWSRSA
Chimpanzee	$_{330}$PSHSRRQPHVVSQWSRSA
Human	$_{330}$PSHSRRQPHVMSQWSRSA

Highly divergent aa's (↓) in mouse ZP3 as compared to homologous C-terminal sequences of rat, cow, pig, dog, macaca, chimpanzee, and human. Note that mouse S_{334} (*underlined*) is conserved in all eight species.

coat proteins of mollusk eggs, accessory gland proteins of flies, and agglutinin proteins of yeast that are thought to be responsible for facilitating speciation. An interesting feature of the SCS of ZP3 is the presence of four closely spaced Cys residues at its N-terminus. For ZP3 from mammals, amphibia, and birds this region has the signature sequence Cys-X-Cys-Cys-X-X-X-X-Cys, with the first Cys residue preceded by an acidic and a hydrophobic residue. Since these Cys residues are present as intramolecular disulfides, they could form a disulfide knot that might have a significant effect on the presentation of the SCS to sperm.

9. SUMMARY

(1) Mouse and all other mammalian eggs are surrounded by a relatively thick, gel-like ECM called the ZP. (2) The mouse egg's ZP is composed of three unique proteins, ZP1–3, encoded by single-copy genes on different chromosomes. (3) ZP genes are expressed solely by growing mouse oocytes and, therefore, only by female mice. (4) ZP1–3 are prototype ZPD proteins that are synthesized and processed only by growing oocytes and secreted for self-assembly into fibrils and matrix. (5) ZP2 and ZP3 polymerize into long fibrils that have a structural repeat, probably a ZP2:ZP3 dimer, and are cross-linked by ZP1. (6) Failure to synthesize either ZP2 or ZP3 results in eggs that lack a ZP and in females that are infertile. (7) During fertilization of ovulated mouse eggs ZP2 and ZP3 serve as receptors for AR and AI sperm, respectively. (8) Following fertilization of mouse eggs ZP2 and ZP3 are inactivated as sperm receptors and the solubility of the ZP decreases considerably.

ACKNOWLEDGMENTS

Many thanks to the postdoctoral fellows, graduate students, and research assistants who contributed to our research on ZP proteins and mammalian fertilization at Harvard Medical School, Roche Institute of Molecular Biology, and Icahn School of Medicine at Mount Sinai. Our research was supported in part by the National Institutes of Health (NICHD) and by Hoffmann-La Roche Inc.

REFERENCES

Ackert, C. L., Gittens, J. E., O'Brien, M. J., Eppig, J. J., & Kidder, G. M. (2001). Intercellular communication via connexin 43 gap junctions is required for ovarian folliculogenesis in the mouse. *Developmental Biology*, *233*, 258–270.

Alberts, B., Johnson, A., Lewis, J., Morgan, D., Raff, M., Roberts, K., et al. (2014). *Molecular biology of the cell* (6th ed.). New York: Garland Science.

Avella, M. A., Baibakov, B., & Dean, J. (2014). A single domain of the ZP2 zona pellucida protein mediates gamete recognition in mice and humans. *Journal of Cell Biology, 205*, 801–809.

Aydin, H., Sultana, A., Li, S., Thavalingam, A., & Lee, J. E. (2016). Molecular architecture of the human sperm IZUMO1 and egg JUNO fertilization complex. *Nature, 534*, 562–565.

Baibakov, B., Boggs, N. A., Yauger, B., Baibakov, G., & Dean, J. (2012). Human sperm bind to the N-terminal domain of ZP2 in humanized zonae pellucidae in transgenic mice. *Journal of Cell Biology, 197*, 897–905.

Bianchi, E., Doe, B., Goulding, D., & Wright, G. J. (2014). Juno is the egg Izumo receptor and is essential for mammalian fertilization. *Nature, 508*, 483–487.

Bleil, J. D., Beall, C. F., & Wassarman, P. M. (1981). Mammalian sperm-egg interaction: Fertilization of mouse eggs triggers modification of the major zona pellucida glycoprotein, ZP2. *Developmental Biology, 86*, 189–197.

Bleil, J. D., Greve, J. M., & Wassarman, P. M. (1988). Identification of a secondary sperm receptor in the mouse egg zona pellucida: Role in maintenance of binding of acrosome-reacted sperm to eggs. *Developmental Biology, 128*, 376–385.

Bleil, J. D., & Wassarman, P. M. (1980a). Structure and function of the zona pellucida: Identification and characterization of the proteins of the mouse oocyte's zona pellucida. *Developmental Biology, 76*, 185–202.

Bleil, J. D., & Wassarman, P. M. (1980b). Synthesis of zona pellucida proteins by denuded and follicle-enclosed mouse oocytes during culture *in vitro*. *Proceedings of the National Academy of Sciences of the United States of America, 77*, 1029–1033.

Bleil, J. D., & Wassarman, P. M. (1980c). Mammalian sperm-egg interaction: Identification of a glycoprotein in mouse egg zonae pellucidae possessing receptor activity for sperm. *Cell, 20*, 873–882.

Bleil, J. D., & Wassarman, P. M. (1983). Mammalian sperm-egg interaction: Sequence of events and induction of the acrosome reaction by a zona pellucida glycoprotein. *Developmental Biology, 95*, 317–324.

Bleil, J. D., & Wassarman, P. M. (1986). Autoradiographic visualization of the mouse egg's sperm receptor bound to sperm. *Journal of Cell Biology, 102*, 1363–1371.

Bleil, J. D., & Wassarman, P. M. (1988). Galactose at the nonreducing terminus of O-linked oligosaccharides of mouse egg zona pellucida glycoprotein ZP3 is essential for the glycoprotein's sperm receptor activity. *Proceedings of the National Academy of Sciences of the United States of America, 85*, 6778–6782.

Bökel, C., Prokop, A., & Brown, N. H. (2005). Papillote and Piopio: Drosophila ZP-domain proteins required for cell adhesion to the apical extracellular matrix and microtubule organization. *Journal of Cell Science, 118*, 633–642.

Bokhove, M., Nishimura, K., Brunati, M., Han, L., de Sanctis, D., Rampoldi, L., et al. (2016). A structured interdomain linker directs self-polymerization of human uromodulin. *Proceedings of the National Academy Sciences of the United States of America, 113*, 1552–1557.

Bork, P., & Sander, C. (1992). A large domain common to sperm receptors (Zp2 and Zp3) and TGF-beta type III receptor. *FEBS Letters, 300*, 237–240.

Buffone, M. G., Hirohashi, N., & Gerton, G. L. (2014). Unresolved questions concerning mammalian sperm acrosomal exocytosis. *Biology of Reproduction, 90*, 1–8.

Callebaut, I., Mornon, J. P., & Monget, P. (2007). Isolated ZP-N domains constitute the N-terminal extensions of zona pellucida proteins. *Bioinformatics, 23*, 1871–1874.

Claw, K. G., & Swanson, W. J. (2012). Evolution of the egg: New findings and challenges. *Annual Review of Genomics and Human Genetics, 13*, 109–125.

Cocchia, M., Huber, R., Pantano, S., Chen, E. Y., Ma, P., Forabosco, A., et al. (2000). PLAC1, and Xq26 gene with placenta specific expression. *Genomics, 68*, 305 312.

Conner, S. J., Lefievre, L., Hughes, D. C., & Barratt, C. L. (2005). Cracking the egg: Increased complexity in the zona pellucida. *Human Reproduction, 20*, 1148–1152.

Dean, J. (2004). Reassessing the molecular biology of sperm-egg recognition with mouse genetics. *Bioessays*, *26*, 29–38.

Egge, N., Muthusubramanian, A., & Cornwall, G. A. (2015). Amyloid properties of the mouse egg zona pellucida. *PLoS One*, *10*, e0129907.

Ellerman, D. A., Pei, J., Gupta, S., Snell, W. J., Myles, D., & Primakoff, P. (2009). Izumo is part of a multiprotein family whose members form large complexes on mammalian sperm. *Molecular Reproduction and Development*, *76*, 1188–1199.

Epifano, O., Liang, L.-F., & Dean, J. (1995). Mouse Zp1 encodes a zona pellucida protein homologous to egg envelope proteins in mammals and fish. *Journal of Biological Chemistry*, *270*, 27254–27258.

Florman, H. M., Bechtol, K. B., & Wassarman, P. M. (1984). Enzymatic dissection of the functions of the mouse egg's receptor for sperm. *Developmental Biology*, *106*, 243–255.

Florman, H. M., & Wassarman, P. M. (1985). O-linked oligosaccharides of mouse egg ZP3 account for its sperm receptor activity. *Cell*, *41*, 313–324.

Franz, C., Stewart, K. M., & Weaver, V. M. (2010). The extracellular matrix at a glance. *Journal of Cell Science*, *123*, 4195–4200.

Gosden, R. G. (2013). Oocyte development and loss. *Seminars in Reproductive Medicine*, *31*, 393–398.

Goudet, G., Mugnier, S., Callebaut, I., & Monget, P. (2008). Phylogenetic analysis and identification of pseudogenes reveal a progressive loss of zona pellucida genes during evolution of vertebrates. *Biology of Reproduction*, *78*, 796–806.

Grayson, P. (2015). Izumo1 and Juno: The evolutionary origins and coevolution of essential sperm-egg binding partners. *Royal Society Open Science*, *2*, 150296.

Greve, J. M., Salzmann, G. S., Roller, R. J., & Wassarman, P. M. (1982). Biosynthesis of the major zona pellucida glycoprotein secreted by oocytes during mammalian oogenesis. *Cell*, *31*, 749–759.

Greve, J. M., & Wassarman, P. M. (1985). Mouse egg extracellular coat is a matrix of interconnected filaments possessing a structural repeat. *Journal of Molecular Biology*, *181*, 253–264.

Gwatkin, R. B. L. (1977). *Fertilization mechanisms in man and mammals*. New York: Plenum, 161 pp.

Han, L., Monné, M., Okumura, H., Schwend, T., Cherry, A. L., Flot, D., et al. (2010). Insights into egg-coat assembly and egg-sperm interaction from the X-ray structure of full-length ZP3. *Cell*, *143*, 1–12.

Han, L., Nishimura, K., Sedat Al Hosseini, H., Bianchi, E., Wright, G. J., & Jovine, L. (2016). Divergent evolution of vitamin B9 binding underlies Juno-mediated adhesion of mammalian gametes. *Current Biology*, *26*, R100–101.

Hewetson, A., Do, H. Q., Myers, C., Muthusubramanian, A., Sutton, R. B., Wylie, B., et al. (2017). Functional amyloids in reproduction. *Biomolecules*, *7*, 46.

Hughes, D. C., & Barratt, C. L. R. (1999). Identification of the true orthologue of the mouse ZP1 gene: Evidence for greater complexity in the mammalian zona pellucida. *Biochimica et Biophysica Acta*, *1447*, 303–306.

Inoue, N., Ikawa, M., Isotani, A., & Okabe, M. (2005). The immunoglobulin superfamily protein Izumo is required for sperm to fuse with eggs. *Nature*, *434*, 234–238.

Jovine, L., Darie, C. C., Litscher, E. S., & Wassarman, P. M. (2005). Zona pellucida domain proteins. *Annual Review of Biochemistry*, *74*, 83–114.

Jovine, L., Janssen, W. G., Litscher, E. S., & Wassarman, P. M. (2006). The PLAC1-homology region of the ZP domain is sufficient for protein polymerization. *BMC Biochemistry*, *7*, 11.

Jovine, L., Qi, H., Williams, Z., Litscher, E., & Wassarman, P. M. (2002). The ZP domain is a conserved module for protein polymerization. *Nature Cell Biology*, *4*, 457–461.

Jovine, L., Qi, H., Williams, Z., Litscher, E. S., & Wassarman, P. M. (2004). A duplicated motif controls assembly of zona pellucida domain proteins. *Proceedings of the National Academy of Sciences of the United States of America, 101*, 5922–5927.

Jovine, L., Qi, H., Williams, Z., Litscher, E. S., & Wassarman, P. M. (2007). Features that affect secretion and assembly of zona pellucida glycoproteins during mammalian oogenesis. In S. K. Gupta, K. Koyama, & J. F. Murray (Eds.), *Gamete biology* (pp. 187–201). UK: Nottingham University Press.

Kidder, G. M., & Mhawi, A. A. (2002). Gap junctions and ovarian folliculogenesis. *Reproduction, 123*, 613–620.

Killick, R., Legan, P. K., Malenczak, C., & Richardson, G. P. (1995). Molecular cloning of chick beta-tectorin, an extracellular matrix molecule of the inner ear. *Journal of Cell Biology, 129*, 535–547.

Kim, J., & Kim, J. (2013). Viscoelastic characterization of mouse zona pellucida. *IEEE Transactions of Biomedical Engineering, 60*, 569–575.

Kinloch, R. A., Sakai, Y., & Wassarman, P. M. (1995). Mapping the mouse ZP3 combining site for sperm by exon swapping and site-directed mutagenesis. *Proceedings of the National Academy of Sciences of the United States of America, 92*, 263–267.

Kreiss, T., & Vale, R. (1999). *Guidebook to the extracellular matrix, anchor, and adhesion proteins* (2nd ed.). . Oxford, UK: Oxford University Press.

Lackey, B. R., & Gray, S. L. (2015). Second messengers, steroids, and signaling cascades: Crosstalk in sperm development and function. *General and Comparative Endocrinology, 224*, 294–302.

Legan, P. K., Lukashkina, V. A., Goodyear, R. J., Lukashkin, A. N., Verhoeven, K., Van Camp, G., et al. (2005). A deafness mutation isolates a second role for the tectorial membrane in hearing. *Nature Neuroscience, 8*, 1035–1042.

Legan, P. K., Rau, A., Keen, J. N., & Richardson, G. P. (1997). The mouse tectorins. Modular matrix proteins of the inner ear homologous to components of the sperm-egg adhesion system. *Journal of Biological Chemistry, 272*, 8791–8801.

Lei, L., & Spradling, A. C. (2013). Female mice lack adult germ-line stem cells but sustain oogenesis using stable primordial follicles. *Proceedings of the National Academy of Sciences of the United States of America, 110*, 8585–8590.

Liang, L.-F., & Dean, J. (1993). Conservation of mammalian secondary sperm receptor genes enables the promoter of the human gene to function in mouse oocytes. *Developmental Biology, 156*, 399–408.

Lira, S. A., Kinloch, R. A., Mortillo, S., & Wassarman, P. M. (1990). An upstream region of the mouse ZP3 gene directs expression of firefly luciferase specifically to growing oocytes in transgenic mice. *Proceedings of the National Academy of Sciences of the United States of America, 87*, 7215–7219.

Lira, S. A., Schickler, M., & Wassarman, P. M. (1993). Cis-acting DNA elements involved in oocyte-specific expression of mouse sperm receptor gene mZP3 are located close to the gene's transcription start site. *Molecular Reproduction and Development, 36*, 494–499.

Litscher, E. S., Juntunen, K., Seppo, A., Niemalä, R., Penttilä, L., Renkonen, O., et al. (1995). Oligosaccharide constructs with defined structures that inhibit binding of mouse sperm to unfertilized eggs *in vitro*. *Biochemistry, 34*, 4662–4669.

Litscher, E. S., & Wassarman, P. M. (1996). Characterization of a mouse ZP3-derived glycopeptide, gp55, that exhibits sperm receptor and acrosome reaction-inducing activity *in vitro*. *Biochemistry, 35*, 3980–3986.

Litscher, E. S., & Wassarman, P. M. (1999). Mouse zona pellucida glycoproteins mZP2 and mZP3 undergo carboxy-terminal proteolytic processing in growing oocytes. *Biochemistry, 38*, 12280–12287.

Litscher, E. S., & Wassarman, P. M. (2015). *A guide to zona pellucida domain proteins*. Hoboken, NJ: John Wiley & Sons, 208 pp.

Liu, C., Litscher, E. S., Mortillo, S., Sakai, Y., Kinloch, R. A., Stewart, C. L., et al. (1996). Targeted disruption of the mZP3 gene results in production of eggs lacking a zona pellucida and infertility in female mice. *Proceedings of the National Academy of Sciences of the United States of America*, *93*, 5431–5436.

Louros, N. N., Iconomidou, V. A., Giannelou, P., & Hamodrakas, S. J. (2013). Structural analysis of peptide-analogues of human zona pellucida ZP1 protein with amyloidogenic properties: Insights into mammalian zona pellucida formation. *PLoS One*, *8*, e73258.

Millar, S. E., Lader, E., & Dean, J. (1993). ZAP-1 DNA binding activity is first detected at the onset of zona pellucida gene expression in embryonic mouse oocytes. *Developmental Biology*, *158*, 410–413.

Moller, C. C., & Wassarman, P. M. (1989). Characterization of a proteinase that cleaves zona pellucida glycoprotein ZP2 following activation of mouse eggs. *Developmental Biology*, *132*, 103–112.

Monné, M., Han, L., Schwend, T., Burendahl, S., & Jovine, L. (2008). Crystal structure of the ZP-N domain of ZP3 reveals the core fold of animal egg coats. *Nature*, *456*, 653–657.

Monné, M., & Jovine, L. (2011). A structural view of egg coat architecture and function in fertilization. *Biology of Reproduction*, *85*, 661–669.

Mortillo, S., & Wassarman, P. M. (1991). Differential binding of gold-labeled zona pellucida glycoproteins mZP2 and mZP3 to mouse sperm membrane compartments. *Development*, *113*, 141–149.

Nishimura, K., Han, L., Bianchi, E., Wright, G. J., de Sanctis, D., & Jovine, L. (2016). The structure of sperm Izumo1 reveals unexpected similarities with Plasmodium invasion proteins. *Current Biology*, *26*, R661–662.

Ohto, U., Ishida, H., Krayukhina, E., Uchiyama, S. S., Inoue, N., & Shimizu, T. (2016). Structure of IZUMO1-JUNO reveals sperm-oocyte recognition during mammalian fertilization. *Nature*, *534*, 566–569.

Plaza, S., Chanut-Delalande, H., Fernandes, I., Wassarman, P. M., & Payre, F. (2010). From A to Z: Apical structures and zona pellucida-domain proteins. *Trends in Cell Biology*, *20*, 524–532.

Qi, H., Williams, Z., & Wassarman, P. M. (2002). Secretion and assembly of zona pellucida glycoproteins by growing mouse oocytes microinjected with epitope-tagged cDNAs for mZP2 and mZP3. *Molecular Biology of the Cell*, *13*, 530–541.

Rankin, T. L., Coleman, J. S., Epifano, O., Hoodbhoy, T., Turner, S. G., Castle, P. E., et al. (2003). Fertility and taxon-specific sperm binding persists after replacement of mouse sperm receptors with human homologs. *Developmental Cell*, *5*, 33–43.

Rankin, T., Familiari, M., Lee, E., Ginsburg, A., Dwyer, N., Blanchette-Mackie, J., et al. (1996). Mice homozygous for an insertional mutation in the ZP3 gene lack a zona pellucida and are infertile. *Development*, *122*, 2903–2910.

Rankin, T., O'Brien, M., Lee, E., Wigglesworth, K., Eppig, J., & Dean, J. (2001). Defective zonae pellucidae in Zp2-null mice disrupt folliculogenesis, fertility, and development. *Development*, *128*, 1119–1126.

Rankin, T. L., Talbot, P., Lee, E., & Dean, J. (1999). Abnormal zonae pellucidae in mice lacking ZP1 result in early embryonic loss. *Development*, *126*, 3847–3855.

Ringuette, M. J., Sobieski, D. A., Chamow, S. M., & Dean, J. (1986). Oocyte-specific gene expression: Molecular characterization of a cDNA coding for ZP-3, the sperm receptor of the mouse zona pellucida. *Proceedings of the National Academy Sciences of the United States of America*, *83*, 4341–4345.

Rodrigues, P., Limback, D., McGinnis, L. K., Plancha, C. E., & Albertini, D. F. (2008). Oogenesis: Prospects and challenges for the future. *Journal of Cellular Physiology*, *216*, 355–365.

Roller, R. J., Kinloch, R. A., Hiraoka, B. Y., Li, S. S.-L., & Wassarman, P. M. (1989). Gene expression during mammalian oogenesis and early embryogenesis: Quantification of three messenger-RNAs abundant in fully grown mouse oocytes. *Development*, *106*, 251–261.

Rosiere, T. K., & Wassarman, P. M. (1992). Identification of a region of mouse zona pellucida glycoprotein mZP3 that possesses sperm receptor activity. *Developmental Biology, 154*, 309–317.
Salzmann, G. S., Greve, J. M., Roller, R. J., & Wassarman, P. M. (1983). Biosynthesis of the sperm receptor during oogenesis in the mouse. *EMBO Journal, 2*, 1451–1456.
Schaeffer, C., Santambrogio, S., Perucca, S., Casari, G., & Rampoldi, L. (2009). Analysis of uromodulin polymerization provides new insights into the mechanisms regulating ZP domain-mediated protein assembly. *Molecular Biology of the Cell, 20*, 589–599.
Schickler, M., Lira, S. A., Kinloch, R. A., & Wassarman, P. M. (1992). A mouse oocyte-specific protein that binds to a region of mZP3 promoter involved in regulating oocyte-specific expression of the mZP3 gene. *Molecular and Cellular Biology, 12*, 120–127.
Shimizu, S., Tsuji, M., & Dean, J. (1983). *In vivo* biosynthesis of three sulfated glycoproteins of murine zonae pellucidae by oocytes grown in follicle culture. *Journal of Biological Chemistry, 258*, 5858–5863.
Simon, A. M., Goodenough, D. A., Li, E., & Paul, D. L. (1997). Female infertility in mice lacking connexin 37. *Nature, 385*, 525–529.
Smith, J., Paton, I. R., Hughes, D. C., & Burt, D. W. (2005). Isolation and mapping the chicken zona pellucida genes: An insight into the evolution of orthologous genes in different species. *Molecular Reproduction and Development, 70*, 133–145.
Sorensen, R. A., & Wassarman, P. M. (1976). Relationship between growth and meiotic maturation of the mouse oocyte. *Developmental Biology, 50*, 531–536.
Spargo, S. C., & Hope, R. M. (2003). Evolution and nomenclature of the zona pellucida gene family. *Biology of Reproduction, 68*, 358–362.
Swanson, W. J., Yang, Z., Wolfner, M. F., & Aquadro, C. F. (2001). Positive Darwinian selection drives the evolution of several female reproductive proteins in mammals. *Proceedings of the National Academy of Sciences of the United States of America, 98*, 2509–2514.
Vazquez, M. H., Phillips, D. M., & Wassarman, P. M. (1989). Interaction of mouse sperm with purified sperm receptors covalently-linked to silica beads. *Journal of Cell Science, 92*, 713–722.
Wassarman, P. M. (1987). The biology and chemistry of fertilization. *Science, 235*, 553–560.
Wassarman, P. M. (1988a). Zona pellucida glycoproteins. *Annual Review of Biochemistry, 57*, 415–442.
Wassarman, P. M. (1988b). Fertilization in mammals. *Scientific American, 255*, 78–84.
Wassarman, P. M. (1999). Mammalian fertilization: Molecular aspects of gamete adhesion. *Cell, 96*, 175–183.
Wassarman, P. M. (2002). Channels of communication in the ovary. *Nature Cell Biology, 4*, s7–9.
Wassarman, P. M., & Albertini, D. F. (1994). Cellular and molecular biology of the mammalian ovum. In E. Knobil & J. D. Neill (Eds.), *The physiology of reproduction: Vol. 1* (2nd ed.). New York: Raven Press.
Wassarman, P. M., & Josefowicz, W. J. (1978). Oocyte development in the mouse: An ultrastructural comparison of oocytes isolated at various stages of growth and meiotic competence. *Journal of Morphology, 156*, 209–236.
Wassarman, P. M., & Litscher, E. S. (1995). Sperm-egg recognition mechanisms in mammals. *Current Topics in Developmental Biology, 30*, 1–19.
Wassarman, P. M., & Litscher, E. S. (2008). Mammalian fertilization is dependent on multiple membrane fusion events. In E. H. Chen (Ed.), *Cell fusion: Overviews and methods*. Totowa, NJ: Humana Press.
Wassarman, P. M., & Litscher, E. S. (2012). Influence of the zona pellucida of the mouse egg on folliculogenesis and fertility. *International Journal of Developmental Biology, 56*, 833–839.
Wassarman, P. M., & Litscher, E. S. (2016). A bespoke coat for eggs: Getting ready for fertilization. *Current Topics in Developmental Biology, 117*, 539–552.

Wassarman, P. M., & Mortillo, S. (1991). Structure of the mouse egg extracellular coat, the zona pellucida. *International Review of Cytology*, *130*, 85–110.

Wassarman, P. M., Qi, H., & Litscher, E. S. (1997). Mutant female mice carrying a single mZP3 allele produce eggs with a thin zona pellucida, but reproduce normally. *Proceedings of the Royal Society, Biological Sciences*, *264*, 323–328.

Williams, Z., Litscher, E. S., Jovine, L., & Wassarman, P. M. (2006). Polypeptide encoded by mouse ZP3 exon-7 is necessary and sufficient for sperm binding *in vitro*. *Journal of Cellular Physiology*, *207*, 30–39.

Williams, Z., & Wassarman, P. M. (2001). Secretion of mouse ZP3, the sperm receptor, requires cleavage of its polypeptide at a consensus furin cleavage-site. *Biochemistry*, *40*, 929–937.

Yan, C., Pendola, F. L., Jacob, R., Lau, A. L., Eppig, J. J., & Matzuk, M. M. (2001). Oosp1 encodes a novel mouse oocyte-secreted protein. *Genesis*, *31*, 105–110.

Zuccotti, M., Merico, V., Cecconi, S., Redi, C. A., & Garagna, S. (2011). What does it take to make a developmentally competent mammalian egg? *Human Reproduction Update*, *17*, 525–540.

CHAPTER ELEVEN

Conceptus Coats of Marsupials and Monotremes

Stephen Frankenberg, Marilyn B. Renfree[1]
School of BioSciences, University of Melbourne, Parkville, VIC, Australia
[1]Corresponding author: e-mail address: m.renfree@unimelb.edu.au

Contents

Abstract

Mammals evolved from oviparous reptiles that laid eggs in a dry, terrestrial environment, thus requiring large amounts of yolk to support development and tough, outer coats to protect them. Eutherian mammals such as humans and mice exhibit an "extreme" form of viviparity in which yolk and conceptus coats have become largely redundant. However, the "other" mammals—monotremes and marsupials—have retained and modified some features of reptilian development that provide valuable insights into the evolution of viviparity in mammals. Most striking of these are the conceptus coats, which include the zona pellucida, the mucoid coat, and the shell coat. We discuss current knowledge of these coats in monotremes and marsupials, their possible roles, and recently identified components such as the zona pellucida protein ZPAX, conceptus coat mucin (CCM), and nephronectin (NPNT).

Current Topics in Developmental Biology, Volume 130
ISSN 0070-2153
https://doi.org/10.1016/bs.ctdb.2018.03.004

1. INTRODUCTION

Vertebrates exhibit enormous diversity in their modes of reproduction, ranging between the extremes of oviparity and viviparity, as well as adaptations to either aquatic or dry terrestrial environments. It is therefore hardly surprising that they also exhibit enormous variety in the structure of their eggs, ranging from the over-sized egg of the kiwi (equating to about a third of the mother's body size at laying) to the miniscule ova of mammals, measuring as little as 100 μm in diameter, even in large species such as elephants. The two properties that most overtly characterize egg structure are their size, which directly correlates with yolk or deutoplasmic content, and the extracellular coats that invest the ovum and subsequently the conceptus. Of all vertebrate lineages, the relationship between the conceptus coats and mode of reproduction is perhaps best exemplified in mammals, whose modes range from oviparity in monotremes to "extreme viviparity" in eutherians. Although monotremes and marsupials have been evolving independently from eutherians for millions of years (Fig. 1), many features of their reproduction and early development are likely to be similar to those in their most recent common mammalian and therian ancestors, respectively; thus, monotremes and marsupials provide convenient intermediates between ancestral reptiles and eutherians and therefore ideal models for understanding the evolution of reproduction and development toward humans and mice. This evolutionary progression was associated with decreasing complexity of the conceptus coats due to increasing redundancy in their roles.

A unique feature of mammalian development is blastocyst expansion, which results in a dramatic increase in total conceptus volume relative to the size of the zygote (Fig. 2) (reviewed by Frankenberg, de Barros, Rossant, & Renfree, 2016). This compensates for the greatly reduced contribution of yolk to development (due to reliance on a placenta) that results in much smaller ova, since a large conceptus volume is still required to accommodate the extensive system of fetal membranes required later for nutrient transfer. Marsupial oocytes are about two to three times the size of eutherian oocytes (~250–300 μm cf. 70–100 μm in eutherian mammals) (Lutwak-Mann, 1971; Renfree, 1994; Selwood & Johnson, 2006). Monotreme zygotes are considerably larger—around 4 mm in diameter—though still much smaller than those of birds and reptiles, whereas the laid egg measures around 16–18 mm in diameter due to blastocyst expansion (Flynn & Hill, 1939; Griffiths, 1978).

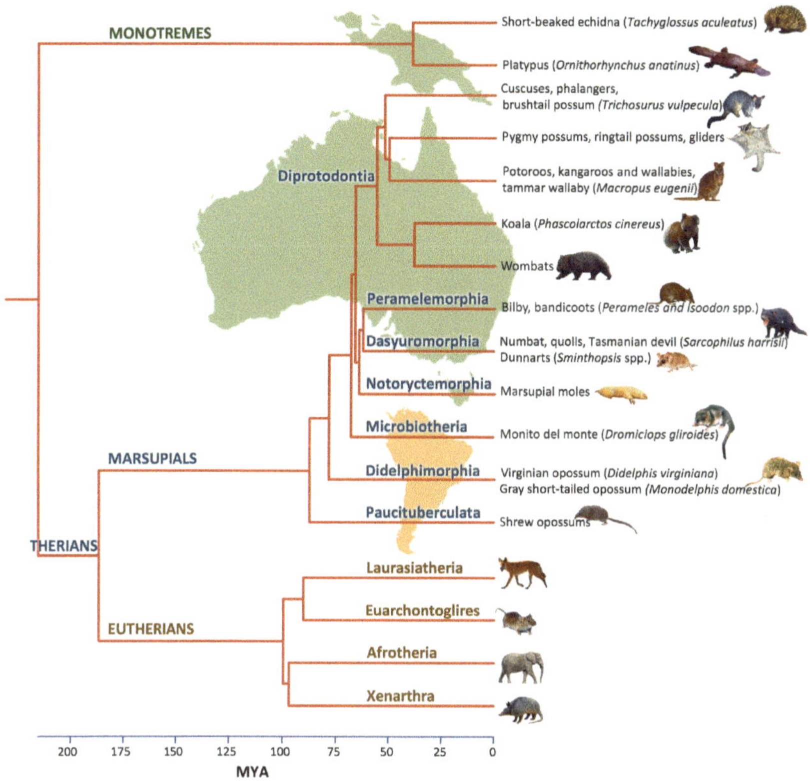

Fig. 1 Phylogeny of mammals, with some representative species indicated. *Adapted from Meredith, R. W., Janecka, J. E., Gatesy, J., Ryder, O. A., Fisher, C. A., Teeling, E. C., et al. (2011). Impacts of the cretaceous terrestrial revolution and KPg extinction on mammal diversification.* Science, 334*(6055), 521–524. doi:10.1126/science.1211028.*

Three conceptus coats surround the early monotreme conceptus: an inner zona pellucida, a mucoid coat (often called albumen layer), and a thin, outer shell coat. The early shell coat is unlike the rigid shell of a laid bird or reptile egg in that it is able to stretch to accommodate the expanding conceptus, while continuing to thicken slightly. Once the conceptus has reached around 10 mm in diameter, the shell begins to thicken dramatically and becomes much more rigid. This is the shell that protects the fetus during the ~10 day period of incubation between laying and hatching (Griffiths, 1978).

Early marsupial conceptuses are also surrounded by a zona pellucida, a mucoid coat, and a shell coat. A key difference with monotremes, however, is that instead of a rigid shell coat layer being deposited late in the gestation, the early nonrigid shell breaks down to allow placental attachment to the endometrium. Thus, placental adaptations have been important influences

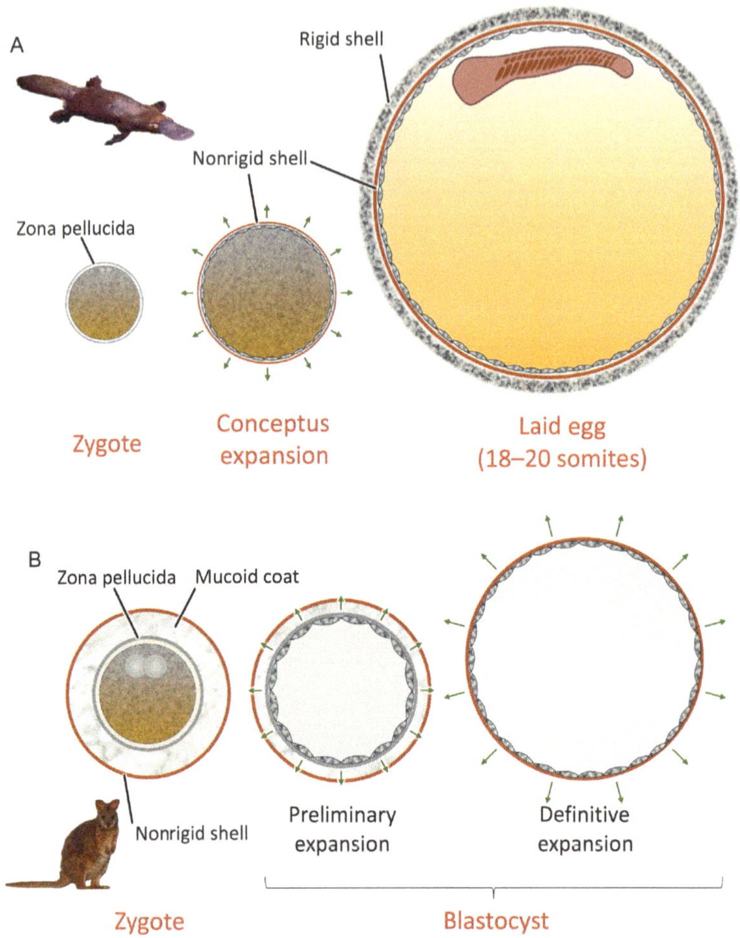

Fig. 2 Comparison between (A) monotreme and (B) marsupial early development. Unlike in birds and reptiles, mammals undergo a period of conceptus expansion after blastocyst formation so that the total conceptus volume is considerably larger than the original ovum volume. In both monotremes and marsupials, a nonrigid shell stretches to accommodate conceptus expansion. Subsequently in monotremes, a thick, rigid shell is laid down to prepare for oviposition, while in marsupials the nonrigid shell breaks down to allow placental attachment.

in the evolution of conceptus coat composition and structure during mammalian evolution.

The functions of the conceptus coats of marsupials are many. The mucoid coat is thought to provide a block to polyspermy after fertilization. The shell coat is likely to play a role in fetomaternal tolerance, as well as immunity, nutrition, morphogenesis (physical scaffold), and physical protection. Much less is known of these stages in the monotreme, but they have a well-formed zona pellucida and mucoid coat before the three-layered shell is laid down.

In this review, we give an overview of our current knowledge of the composition and structure of conceptus coats in marsupials and monotremes and relate these to the evolution of viviparity in mammals. We focus on three conceptus coats: the zona pellucida, the mucoid coat, and the shell coat. The zona pellucida is a preovulatory coat, laid down in the ovary as folliculogenesis proceeds. The mucoid layer and the shell coat are postovulatory coats that are laid down after fertilization and derived from secretions from the reproductive tract.

2. PREOVULATORY COATS

2.1 The Zona Pellucida

The zona pellucida (or "zona") is the most ancient of the coats that envelope vertebrate conceptuses. As a "preovulatory coat," the zona is deposited around the oocyte prior to ovulation by secretions from the surrounding somatic cells in the ovary. In mammals, deposition begins in the primary follicle stage and thus its diameter increases as the oocyte grows. However, the marsupial zona is very thin in marsupials (ranging from 2 to 6 μm depending on the species) compared to those of eutherian mammals (Bedford, 1996). In monotremes, the zona is even thinner—about 0.5 μm around the time of ovulation—although it appears thicker during oogenesis (Hughes & Carrick, 1978). At later stages of development, it appears to merge with the mucoid layer and is difficult to distinguish from the latter; this combined layer is often referred to as "the zona-albumen layer" (Hughes & Carrick, 1978).

2.2 Glycoproteins Constituting the Zona Pellucida

The human genome contains four genes that, respectively, encode the related glycoproteins ZP1, ZP2, ZP3, and ZP4 that constitute the zona pellucida (Wassarman & Litscher, 2016). Until recently, orthologues of these four genes were the only known representatives of the ZP gene family known in mammals. However, a recent study by Moros-Nicolas et al. (2017) identified three *ZP3*-related genes in marsupials, which they named *ZP3-A*, *ZP3-B*, and *ZP3-C*. The phylogenetic trees published by these authors indicate that *ZP3-C* is in fact a direct orthologue of eutherian *ZP3*, while *ZP3-A* and *ZP3-B* are paralogues, and we have confirmed this by analysis of synteny (unpublished data). In order to establish a more rational nomenclature for the ZP3 subfamily so that orthologues across all mammalian taxa share the same name, we propose that marsupial *ZP3-C* retain the name *ZP3*, while *ZP3-A* and *ZP3-B* be renamed *ZP3LA* (*ZP3-like A*) and *ZP3LB* (*ZP3-like B*), respectively. Apparently functional orthologues of *ZP3* (*ZP3-C*) and *ZP3LB* are present in all marsupial genomes examined, whereas *ZP3LA* is a disabled pseudogene in the gray short-tailed opossum (*Monodelphis domestica*). *ZP3* and *ZP3LA* are more similar to each other than either is to *ZP3LB*, suggesting that *ZP3* and *ZP3LA* originated from a more recent duplication, whereas *ZP3LB* diverged earlier. However, our own unpublished analyses (summarized in Fig. 3) show that direct orthologues of *ZP3LA*, but not *ZP3LB*, are present in the genomes of sauropsids (birds and reptiles). Thus, both *ZP3LA* and *ZP3LB* were probably present in the genome of the most recent amniote ancestor, while only *ZP3LA* persisted in sauropsids. *ZP3* may have arisen (by duplication of *ZP3LA*) much more recently, since *bona fide* orthologues of *ZP3* are only found in therian (marsupial and eutherian) genomes, whereas the multiple *ZP3*-like genes in birds and reptiles actually represent either orthologues or more recent, lineage-specific duplications of *ZP3LA* (S. Frankenberg, unpublished data).

Moros-Nicolas et al. (2017) showed that the genomes of Australian marsupials have the highest number of identifiable ZP genes (at least six) of any mammal, suggesting that marsupials have unusually high complexity in zona pellucida structure. In addition, we have identified (unpublished data) in the genome of the tammar wallaby an apparently functional orthologue of *ZPAX*, a gene previously identified in *Xenopus* and chicken. *ZPAX* appears to be a disabled pseudogene in other marsupials whose genomes have been sequenced, although an apparently functional orthologue is also present in the platypus genome. Its expression in the tammar wallaby ovary (unpublished data) further supports its functionality in that species. Thus, macropod marsupials appear to contain no fewer than seven different glycoproteins in their zona pellucida.

The ZP3 subfamily was suggested to represent the most ancient branch of the ZP glycoprotein family (Litscher & Wassarman, 2007, 2014), and our own sequence analyses support this. ZP3-related proteins are also simpler in structure than other ZP proteins, comprising an N-terminal signal peptide, a zona pellucida domain (ZPD; shared by all ZP proteins), a consensus furin cleavage site, and a C-terminal transmembrane domain. All these features are also variably present in other ZP proteins.

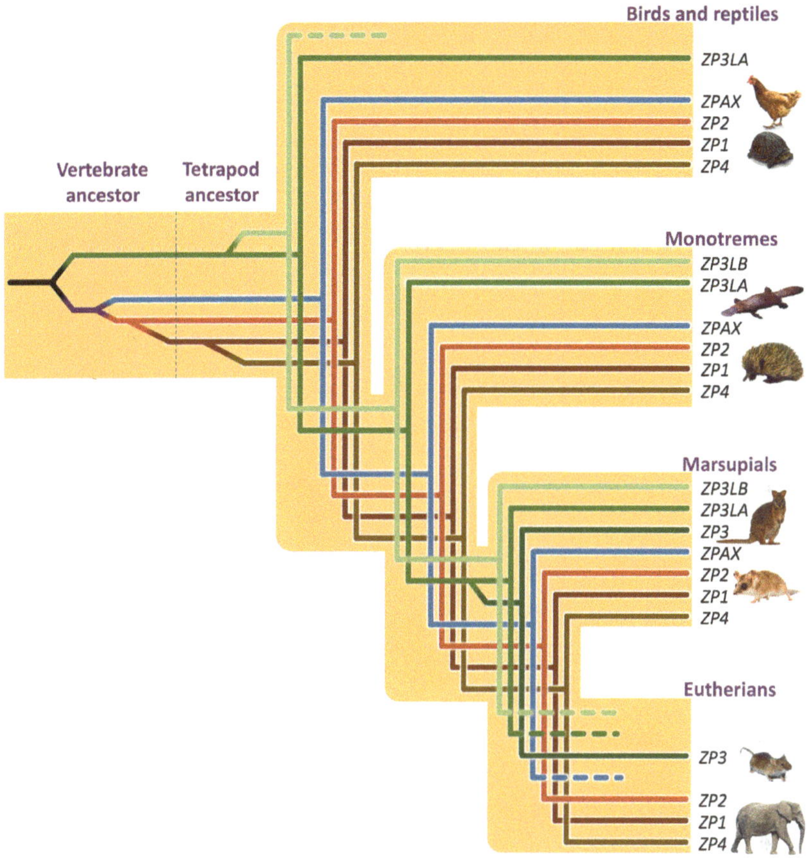

Fig. 3 Model for the phylogeny of ZP genes in amniotes. The most recent common vertebrate ancestor is presumed to have had at least four ZP genes: *ZP3-like*, *ZPAX*, *ZP2*, and *ZP1/4*. *ZP1/4* duplicated to produce *ZP1* and *ZP4* probably in an ancestor of tetrapods, while *ZP3-like* genes underwent two duplications: the first (in a tetrapod ancestor) gave rise to *ZP3LB*; the second (in a therian ancestor) gave rise to *ZP3* and *ZP3LA*. *Broken lines* represent gene extinctions. The indicated persistence of a gene in a given lineage does not necessarily mean that all species of that lineage have a functional orthologue (e.g., absence of *ZPAX* in most marsupials and of *ZP4* in some eutherians). Note that in birds and reptiles, independent duplications of some genes have occurred but are not shown.

The ZPD comprises two subdomains: an N-terminal part (ZP-N) and a C-terminal part (ZP-C). Other ZP proteins typically contain up to three extra repeats of the ZP-N domain within their N-terminal region. ZP1 and ZP4 (which are closely related) each contain only one additional ZP-N domain, but also contain a trefoil domain on the N-terminal side of the ZPD. ZP2 contains three ZP-N repeats outside the ZPD, while ZPAX was reported to contain five repeats. However, our analysis suggests while sequences corresponding to the reported first, second, and fifth ZP-N repeats in ZPAX are, respectively, homologous to the first, second, and third ZP-N repeats of ZP2, we were unable to identify homology with ZP-N within the intervening sequence corresponding to the reported third and fourth ZP-N repeats. Protein fold modeling instead identified this region (in tammar, chicken, and *Xenopus* ZPAX) as resembling *endo*-alpha-sialidase, an enzyme that catalyzes endohydrolysis of polysialic acids. This strongly suggests that ZPAX might be involved in removing glycosyl groups from ZP proteins that are involved in sperm binding, thus acting as a block to polyspermy. Why ZPAX is conserved in at least in macropods but lost in other marsupials is unclear, but could relate to differences in selection pressure for sperm competition.

3. POSTOVULATORY COATS

3.1 Monotreme Mucoid Coat (Albumen Layer)

During the passage of the monotreme conceptus down the oviduct, it acquires a two-layered coating named the "dense" and "fluid" albumen layer by Caldwell (1887) and Hill (1933). Subsequently, Hughes, Carrick, and Shorey (1975) concluded it is a mucoid coat of sulfated acidic glycoprotein, based only on their observations of marsupial (brushtail possum) conceptuses. Flynn and Hill (1947) observed that this albumen layer traps spermatozoa as it passes down the oviduct (Griffiths, 1968, 1978). It thus appears to be analogous to the marsupial mucoid coat. The mucoid coat persists to term (Hughes, 1977).

3.2 Monotreme Shell Coat

The shell is laid down as a thin transparent membrane during passage of the conceptus in the tubal region of the oviduct (Caldwell, 1887). The shell membrane thickens, and a second layer is deposited on its outer surface. In the echidna, the basal layer is 1.6 μm thick, and peripheral to this is the rodlet layer which reaches about 15–20 μm in thickness, but then compresses slightly (Fig. 4) (Caldwell, 1887; Griffiths, 1968; Hill, 1933; Semon, 1894). As the conceptus expands, the shell increases in its surface area and thickness, but most increase in thickness occurs after the conceptus reaches a diameter of around 10 mm. This is due to the deposition of a third layer (Fig. 4) (Hill, 1933). The third layer has two zones: the inner, coarse granules overlying the rodlet layer and the outer zone with larger and more irregular granules.

Fig. 4 Development of the shell coats of (A) platypus and (B) echidna. Images are scaled proportionally to each other, according to the approximate total shell coat thicknesses measured by Hill and as indicated. The *right-hand-most image* for both species depicts the shell coat of a newly laid egg. *Adapted from the figures of Hill, J. P. (1933). V. The development of the Monotremata. Part II. The structure of the egg-shell.* The Transactions of the Zoological Society of London, 21*(6), 443–456.*

There are pore canals that reach from the inner zone to open on the surface. Uterine secretion is absorbed into the conceptus before and during the laying down of the third layer of the shell; thus, the shell pores may facilitate this absorption. By way of comparison, Hill (1933) concluded that the marsupial shell coat does not advance beyond the basal layer of the monotreme shell. The conceptus grows by accumulation of the nutritive uterine fluid and reaches its full size of 15–17 mm diameter (Fig. 2), in which the embryo at laying has about 18–20 somites (Hughes, 1993). Hughes et al. (1975) implied that the monotreme shell is composed of an ovokeratin, but this has not been tested as yet.

Ultrastructurally, the monotreme shell has an open lattice structure, and there are presumptive shell membrane precursors in the cytoplasm of non-ciliated secretory cells in the endometrial glands of platypus (Griffiths, 1978; Hughes, 1977).

3.3 Marsupial Mucoid Coat

In all marsupials, transport down the oviduct is rapid and takes less than 24 h after fertilization (Tyndale-Biscoe & Renfree, 1987). A mucoid coat (or "mucin layer") is deposited within a few hours of fertilization and thickens as it passes down the oviduct. The mucoid coat proteins are secreted by epithelial cells of the oviduct isthmus, ampulla, and crypts and is an acid glycoprotein produced by nonciliated oviducal epithelial cells (Arnold & Shorey, 1985; Casey, Martinus, & Selwood, 2002; Roberts & Breed, 1996b; Roberts, Breed, & Mayrhofer, 1994). It is thought to be homologous to the mucoid coats of other amniotes (Hughes, 1977). The mucoid coat is of variable thickness, from 6.7 μm in the bandicoot (*Perameles nasuta*) (Lyne & Hollis, 1976) to 24.3 μm in the tammar (Renfree & Tyndale-Biscoe, 1978), but is apparently much thicker in the Virginia opossum (*Didelphis virginiana*) at 230 μm (McCrady, 1938). Numerous sperm are trapped in the mucoid coat (Renfree & Lewis, 1996) (Fig. 5C), a block to polyspermy likely being its primary function, although some have suggested that it also provides nutrients to the developing embryo.

3.4 Marsupial Shell Coat

The shell coat of marsupials is a proteinaceous disulfide-rich keratin and is a strongly acidic mucoprotein (Hughes, 1974; Hughes & Shorey, 1973). It is laid down around the conceptus as it traverses the ampulla and isthmus of the oviduct (Hughes & Shorey, 1973). Shell deposition occurs primarily from the epithelial cells at the uterotubal junction (Hartman, 1916; McCrady, 1938; Roberts & Breed, 1996b) but continues even after the conceptus enters the uterus (within 24 h after ovulation) where it thickens to 7–10 μm by the early blastocyst stage in the brushtail possum. The shell coat surrounds the embryo for the first two-thirds of pregnancy (Shaw, 1996) and in the diapausing tammar is 8.5 μm thick (Renfree, 1994). In the tammar, as the blastocyst expands, the shell attenuates from about 6 μm to around 0.5 μm (Hughes, 1974; Renfree, 1973). Despite this thinning, the actual volume of the shell coat material increases dramatically from 0.001 mm^3 during embryonic diapause to >0.250 mm^3 at day 18 of gestation in the tammar (Shaw, 1996). Similarly, the shell coat of the brushtail possum thickens after reaching the uterus (Hughes, 1974). Continued deposition has also been recorded in the brushtail possum, the dunnart, and the gray short-tailed opossum (Baggott & Moore, 1990; Casey et al., 2002; Roberts & Breed, 1996a; Roberts et al., 1994). Roberts and Breed (1996a) suggested that in

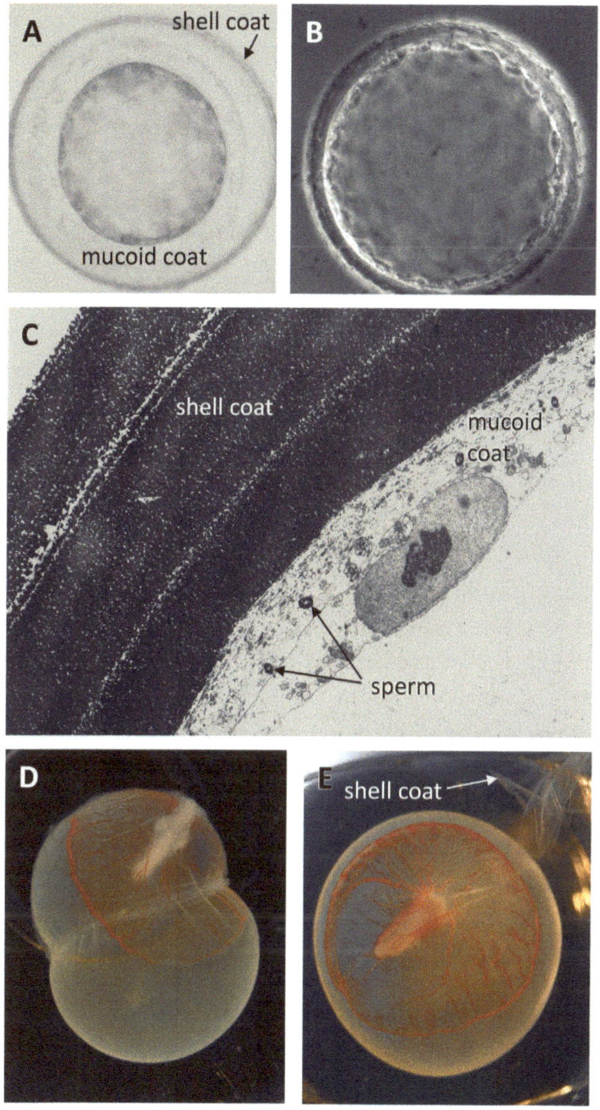

Fig. 5 Conceptus coats during early development of the tammar wallaby. (A) A newly formed blastocyst with a thick mucoid coat and a shell coat. (B) Partly expanded diapausing blastocyst, in which the mucoid coat is compressed. (C) Ultrastructure of a diapausing blastocyst, showing a thickened shell coat and compressed mucoid coat containing several trapped sperm tails. (D) A freshly flushed conceptus that is just beginning to "hatch" from its shell coat. (E) The same conceptus as in (D) after it has completely hatched. The remnant of the shell coat lies adjacent as indicated.

the fat-tailed dunnart, the shell width increases little during cleavage, but is widest at the primitive streak stage when uterine glands show evidence of shell secretion. In the stripe-faced dunnart, the shell is sticky when first deposited (Selwood & Woolley, 1991), but this may in fact be due to incomplete coverage of the mucoid coat which is extremely sticky until the shell coat is laid down and difficult to manipulate in vitro (Renfree & Lewis, 1996 and unpublished observations). Cleavage without shell results in swollen blastomeres in the opossum (Moore & Taggart, 1993), and Selwood (1989) suggested that the shell is required for normal cleavage, but this was in embryos treated enzymatically to remove the shell and the mucoid coat. In the tammar there is no support for this suggestion, because cleavage can proceed in vitro in single-cell embryos to the 8-cell stage in the absence of the shell coat in the tammar (Renfree & Lewis, 1996).

In the brushtail possum and the dunnart, electron-lucent vesicles in nonciliated cells at the uterotubal junction appear to be shell precursors (Casey et al., 2002; Roberts & Breed, 1996a; Roberts et al., 1994). They are also found in the luminal and glandular epithelium of the cranial end of the uterus in both species and in the opossum (Hughes & Hall, 1984; Phillips & Fadem, 1987; Roberts et al., 1994). Although not specifically identified, it is assumed there are similar cells in the tammar, given that the shell thickens after it enters the uterus (Shaw, 1996). The ultrastructure of the shell transforms from a granular structure during cleavage and unilaminar blastocyst stages to a fibrous structure during the later bilaminar and trilaminar blastocyst stages (Roberts & Breed, 1996a). Ultramicrographs of the brushtail possum (Hughes & Hall, 1984), bandicoots, and the Virginian opossum (Krause & Cutts, 1983) suggest that similar transformations occur in those species. The site of the uterine secretions led Roberts et al. (1994) to conclude that the marsupial shell coat is homologous to the rodlet layer (Layer 2) in monotremes.

The shell coat acts as a barrier, and molecules greater than 40 kDa will not penetrate it (Hughes & Shorey, 1973). Small molecules can cross the shell coat, mucoid coat, and zona pellucida. The fetal fluids of the tammar contain glucose, amino acids, and protein. However, once the conceptus has escaped the shell coat, the composition of the fetal fluids changes dramatically (Renfree, 1973). Glucose, amino acids, and protein all increase in concentration and are presumed to have crossed the bilaminar yolk sac membrane (Renfree, 1973). In the brushtail possum, Hughes and Shorey (1973) concluded that the embryo would have virtually unrestricted access to almost all enzymes, antibodies, and protein hormones.

The marsupial conceptus "hatches" from the shell coat at around two-thirds of the gestation period, at an embryonic stage between somite formation and early limb bud stages. The shell is sensitive to proteases in some species (Hughes, 1977), but in others (Selwood, 1989) requires long exposure and relatively high concentrations to be disrupted, suggesting that it is highly covalently bound. In two dasyurid species—agile antechinus and stripe-faced dunnart—hatching occurs in vivo and in vitro (Yousef & Selwood, 1993) and is initiated immediately above the midline of the neural plate and primitive streak in the anterior region of the embryo. In the tammar, hatching occurs around day 18 of the 26.5-day active gestation, and the embryo within its yolk sac membrane squeezes from the shell (Menzies, Pask, & Renfree, 2011) (Fig. 5). In the tammar, the site of perforation of the shell is variable, and is not consistently above the midline of the neural plate. The period of shell loss is associated with activity of an alkaline proteinase that is extruded between the interface of the trophoblast and the uterine epithelium in the bilaminar yolk sac region between 18.5 and 19 days (Denker & Tyndale-Biscoe, 1986). Similar proteolytic activity occurs in both embryonic and uterine cells, reaching maximal activity around day 18.

3.5 Identification of Components of Marsupial Postovulatory Coats

Only one published study (Casey et al., 2002) has attempted to identify specific proteins contained within marsupial postovulatory coats. The authors electrophoresed pooled mucoid and shell coats from both stripe-faced dunnart and brushtail possum, excised bands, and subjected them to N-terminal sequencing. Due to inadequate genomic resources at the time, none of the short sequences (~12–15 amino acids) could be positively identified. However, with improved resources it has become possible to identify many of them, one of which (USM1 (uterine-secreted microprotein 1)) we have extensively characterized (Hollis & Lyne, 1977; Lyne & Hollis, 1976, 1977). Other identified proteins from the possum coat sequences include ESM1 (endothelial cell-specific molecule 1), SDF4 (stromal cell-derived factor 4), and NPNT (nephronectin) (S. Frankenberg, unpublished data) (Fig. 6). ESM1 corresponds to a sequence that was previously identified as an isoform of Enolase 1 and named CP4 (Cui & Selwood, 2003); however, comparison of the published *CP4* coding sequence with genomic databases is inconsistent with this conclusion, which appears to have resulted from PCR-derived sequencing errors (S. Frankenberg, unpublished data).

Fig. 6 Identification of coat proteins from *Trichosurus vulpecula* and *Sminthopsis macroura* after N-terminal sequencing of excised gel bands by Casey et al. (2002). The short N-terminal sequences of excised bands are aligned with the best matched sequence from tBLASTn searches, with identically matched residues highlighted in *bold*. In all cases, the N-terminus of the gel band sequence aligns closely with the predicted signal peptide cleavage site (*blue arrows*) of its matched sequence. Observed molecular weights of excised bands and the predicted molecular weights of signal peptide-cleaved protein matches are as indicated. Where there are large discrepancies between observed and predicted molecular weights, this is likely due to other posttranslational modifications.

From the dunnart coat sequences from Casey et al. (2002), we identified two further proteins: lysozyme G1 (LYG1) and a novel mucin protein related to FCGBP, which we hereafter refer to as conceptus coat mucin (CCM) (Fig. 6). Each of these newly identified coat proteins is discussed later.

3.5.1 Uterine-Secreted Microprotein 1

USM1 belongs to a large family of small, secreted microproteins that are mostly found in deuterostomes but also present in the genomes of some protostomes such as molluscs (Frankenberg, Fenelon, Dopheide, Shaw, & Renfree, 2011). Among vertebrates, the family can be divided into three branches: (i) MSMB-related; (ii) MSMP-related; and (iii) USM1-related. In humans, MSMB is one of the most abundant proteins in seminal fluid. The human genome encodes only one *MSMB* gene, whereas many other vertebrate genomes contain multiple copies. By contrast, a single *MSMP* orthologue is present in the vast majority of vertebrate genomes. *USM1*-related genes are present (sometimes in multiple copies) in all vertebrate genomes except eutherian mammals. It appears likely that USM1 became redundant in eutherians due to a reduced reliance on postovulatory coats for intrauterine development.

The precise role of USM1 in marsupial conceptus coats is still unclear. Secretion of the MSMB/MSMP/USM-family (MMU family) proteins among vertebrates is mostly associated with mucosal tissues, especially in reproductive organs, while other pieces of evidence also suggest a role in innate immunity. Intriguingly, an MSMB-like protein in the egg capsule of the longfin inshore squid (*Loligo pealeii*) acts as a pheromone for conspecific male squid (Cummins et al., 2011). The amino acid sequences of MSMB- and USM-related proteins (but not MSMP) evolve rapidly, suggesting they may have important roles in species specificity of fertilization (Frankenberg et al., 2011). More studies are needed to resolve these questions.

3.5.2 Stromal Cell-Derived Factor 4 (CAB45)

SDF4 has roles in secretory cargo sorting at the trans-Golgi network (von Blume et al., 2012). It seems probable that it has no specific role in the conceptus coats, and that its presence in the conceptus coats is incidental to it being highly abundant in the uterine lumen.

3.5.3 Endothelial Cell-Specific Molecule 1 (endocan)

Endocan is an endothelial cell-associated proteoglycan that is upregulated by proangiogenic molecules and proinflammatory cytokines (Kechagia, Ezra, Burton, & Bailly, 2016). Its roles include inhibition of leukocyte binding to the vascular endothelium. Like SD4F, it seems probable that its presence in the conceptus coats is incidental to being highly abundant in the uterine lumen.

3.5.4 Nephronectin

NPNT is an extracellular matrix integrin-binding protein related to fibronectin. Like fibronectin, it binds integrins via the sequence motif RGD (reviewed by Sun et al., 2018). RGD-based integrin binding of embryonic cells to conceptus coats has previously been characterized in the chick (Lash, Gosfield, Ostrovsky, & Bellairs, 1990). As extraembryonic cells of the area opaca migrate to surround the yolk, they adhere to the inner surface of the zona pellucida (also called vitelline envelope). Marsupial embryonic cells also adhere to the inner surface of the zona pellucida (unlike in eutherians), and dissociated migratory cells from later bilaminar blastocysts (after the zona has disappeared) adhere to the shell coat in culture (Yousef & Selwood, 1996), suggesting that NPNT might be the ECM protein that facilitates this process. If so, the source of NPNT secretion would be expected to be the conceptus rather than the uterus. Since the coats isolated by Casey et al. (2002) were not actively separated from zonae pellucidae, it is likely that zona proteins were included in the gel bands. Although we have detected NPNT expression in both the conceptus and the endometrium by transcriptome sequencing (S. Frankenberg & M. B. Renfree, unpublished data), its more plausible role in cell adhesion suggests that it is derived from the the conceptus. Notably, we have also found that the genes for specific integrin subunits (α5, αv, and β1) that are likely to be involved in NPNT- and/or fibronectin-mediated adhesion (Sato et al., 2009) are dynamically and differentially expressed within embryonic-vs-extraembryonic regions of the early tammar conceptus (S. Frankenberg & M. B. Renfree, unpublished data).

3.5.5 Lysozyme G1 (LYG1)

LYG1 belongs to the g-type (originally "goose-type") lysozyme family, which are conserved in fishes, sauropsids, and mammals and have antibacterial roles (Irwin & Gong, 2003). The mammalian *LYG1* and *LYG2* paralogous genes lie in a cluster that has conserved synteny with avian *LYG* genes. Avian LYG homologues are the predominant lysozymes in the egg white of several bird species (Irwin & Gong, 2003). Thus, it appears likely that marsupial LYG1 has a similar function and represents at least one homologous feature shared between avian egg white and the marsupial mucoid coat.

3.5.6 Conceptus Coat Mucin

By sequence homology, we deduced that CCM is related to Fc Fragment of IgG Binding Protein (FCGBP), although the latter's capacity to bind IgG is strongly disputed. FCGBP is a nongel-forming mucin that has been suggested to cross link MUC2 (a gel-forming mucin) in intestinal mucus (Johansson, Thomsson, & Hansson, 2009), and it seems likely that CCM may have a similar function. We identified orthologues of CCM in the genomes of frogs, reptiles, birds, and mammals, as evidenced by conservation of synteny (Fig. 7). Among eutherians, functional orthologues appear to

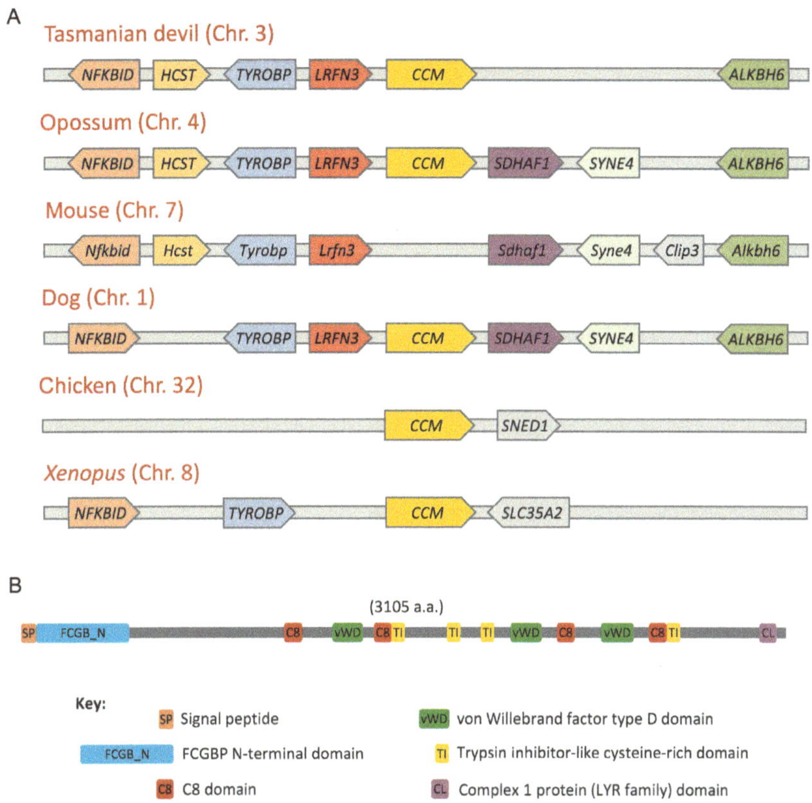

Fig. 7 Analysis of CCM conservation and structure. (A) Analysis of conserved synteny of genes flanking *CCM* homologues among vertebrates. In all vertebrate genomes in which a *CCM* orthologue could be identified, it flanked the genes *LRFN3* and/or *TYROBP*. The only eutherian genomes in which a *CCM* orthologue was identified were from the order Carnivora (exemplified here by the dog). The conserved synteny of genes flanking dog *CCM* clearly identify it as a *bona fide* orthologue of other vertebrate *CCM* genes. (For clarity, intergenic distances are not drawn to scale.) (B) Conserved domain structure of tammar wallaby CCM, as determined from GENSCAN transcript prediction.

be conserved only in Carnivora. In the rabbit, which forms a mucoid coat around its conceptus, traces of CCM sequence are identifiable at the orthologous locus but do not encode a functional ORF.

The predicted protein sequence of tammar CCM has a length of 3105 amino acids and shares a number of conserved domains with other mucins, such as repeated units of von Willebrand factor type D domain, C8 domain, and Trypsin inhibitor-like cysteine rich domain (VWD-C8-TIL) (Lang et al., 2016). However, it lacks repeats of the conserved WxxW motif characterized in MUC2 and other gel-forming mucins (MUC6, MUC5AC, MUC5AB). CMM instead appears to be more similar to FCGBP, a nongel-forming mucin that covalently binds to other mucins such as MUC2 (Johansson et al., 2009). It therefore seems likely that CCM contributes to the structure of the mucoid coat, but is not the principle component.

4. SUMMARY AND CONCLUSIONS

The extracellular conceptus coats of monotremes and marsupials have several distinct functions. The zona pellucida is laid down in the ovary, as in all other mammals, but its role, if any, in blocking polyspermy is unknown. Marsupials appear to have the greatest complexity in zona pellucida structure of any tetrapod, with the tammar wallaby genome encoding no fewer than seven different ZP proteins. One marsupial-specific function is zona-blastomere adhesion, which could involve NPNT-integrin binding.

In marsupials, the mucoid coat clearly traps supernumerary sperm before the shell coat is laid down. A recently identified mucin—CCM—is the first identified protein to be a probable structural component of the marsupial mucoid coat.

The permeability of the shell coat allows accumulation of nutrients in the yolk sac of the monotreme, and also in marsupials, although these are very different structures. Whether the shell has any role in protecting the embryo from immune rejection remains to be discovered, but after shell loss in marsupials, the duration of organogenesis and formation of the full-term fetus is remarkably brief.

Recent technological developments have greatly aided the study of conceptus coat composition in marsupials, with several proteins now identified. However, much work still remains to be done to identify new proteins and to relate their function to the hypothesized roles of the conceptus coats and to the evolution of viviparity in mammals.

REFERENCES

Arnold, R., & Shorey, C. D. (1985). Structure of the uterine luminal epithelium of the brush-tailed possum (Trichosurus vulpecula). *Journal of Reproduction and Fertility*, *74*(2), 565–573.

Baggott, L. M., & Moore, H. D. M. (1990). Early embryonic-development of the gray short-tailed opossum, monodelphis-domestica, invivo and invitro. *Journal of Zoology*, *222*, 623–639.

Bedford, J. M. (1996). What marsupial gametes disclose about gamete function in eutherian mammals. *Reproduction, Fertility, and Development*, *8*(4), 569–580.

Caldwell, W. H. (1887). The embryology of monotremata and marsupialia. Part I. *Proceedings of the Royal Society of London, 42*, 177–180.
Casey, N. P., Martinus, R., & Selwood, L. (2002). Outer egg coats of the marsupial conceptus: Secretion and protein composition. *Molecular Reproduction and Development, 62*(2), 181–194. https://doi.org/10.1002/mrd.10073.
Cui, S., & Selwood, L. (2003). Cloning and expression of a novel cDNA encoding shell coat protein, cp4, from the brushtail possum (Trichosurus vulpecula). *Molecular Reproduction and Development, 65*(2), 141–147. https://doi.org/10.1002/mrd.10296.
Cummins, S. F., Boal, J. G., Buresch, K. C., Kuanpradit, C., Sobhon, P., Holm, J. B., et al. (2011). Extreme aggression in male squid induced by a beta-MSP-like pheromone. *Current Biology, 21*(4), 322–327. S0960-9822(11)00061-3 [pii]. https://doi.org/10.1016/j.cub.2011.01.038.
Denker, H. W., & Tyndale-Biscoe, C. H. (1986). Embryo implantation and proteinase activities in a marsupial (*Macropus eugenii*). Histochemical patterns of proteinases in various gestational stages. *Cell and Tissue Research, 246*(2), 279–291.
Flynn, T. T., & Hill, J. P. (1939). The development of the Monotremata. Part IV. Growth of the ovarian ovum, maturation, fertilisation, and early cleavage. *The Transactions of the Zoological Society of London, 24*(6), 445–622.
Flynn, T. T., & Hill, J. P. (1947). The development of the monotremata. Part VI. The later stages of cleavage and the formation of the primary germ layers. *The Transactions of the Zoological Society of London, 26*(1–2), 1–151.
Frankenberg, S. R., de Barros, F. R., Rossant, J., & Renfree, M. B. (2016). The mammalian blastocyst. *Wiley Interdisciplinary Reviews: Developmental Biology, 5*(2), 210–232. https://doi.org/10.1002/wdev.220.
Frankenberg, S., Fenelon, J., Dopheide, B., Shaw, G., & Renfree, M. B. (2011). A novel MSMB-related microprotein in the postovulatory egg coats of marsupials. *BMC Evolutionary Biology, 11*, 373. https://doi.org/10.1186/1471-2148-11-373.
Griffiths, M. (1968). *Echidnas*. Pergamon Press.
Griffiths, M. (1978). *The biology of the monotremes*. Elsevier.
Hartman, C. G. (1916). Studies in the development of the opossum didelphys Virginiana L. I. History of the early cleavage. II. Formation of the blastocyst. *Journal of Morphology, 27*(1), 1–83.
Hill, J. P. (1933). V. The development of the Monotremata. Part II. The structure of the egg-shell. *The Transactions of the Zoological Society of London, 21*(6), 443–456.
Hollis, D. E., & Lyne, A. G. (1977). Endodermal formation in blastocytes of the marsupials Isoodon macrousus and Perameles nasuta. *Australian Journal of Zoology, 25*(2), 207–223.
Hughes, R. L. (1974). Morphological studies on implantation in marsupials. *Journal of Reproduction and Fertility, 39*(1), 173–186.
Hughes, R. L. (1977). Egg membranes and ovarian function during pregnancy in monotremes and marsupials. In J. H. Calaby & C. H. Tyndale-Biscoe (Eds.), *Reproduction and evolution* (pp. 281–291). Canberra: Australian Academy of Sciences.
Hughes, R. L. (1993). Monotreme development with particular reference to the extraembryonic membranes. *The Journal of Experimental Zoology, 266*(6), 480–494. https://doi.org/10.1002/jez.1402660602.
Hughes, R., & Carrick, F. (1978). Reproduction in female monotremes. *Australian Zoologist, 20*, 233–253.
Hughes, R. L., Carrick, F. N., & Shorey, C. D. (1975). Proceedings: Reproduction in the platypus, Ornithorhynchus anatinus, with particular reference to the evolution of viviparity. *Journal of Reproduction and Fertility, 43*(2), 374–375.
Hughes, R., & Hall, L. (1984). *Embryonic development in the common brushtail possum* Trichosurus vulpecula. *Possums and gliders* (pp. 197–212).

Hughes, R. L., & Shorey, C. D. (1973). Observations on the permeability properties of the egg membranes of the marsupial, Trichosurus vulpecula. *Journal of Reproduction and Fertility*, *32*(1), 25–32.

Irwin, D. M., & Gong, Z. (2003). Molecular evolution of vertebrate goose-type lysozyme genes. *Journal of Molecular Evolution*, *56*(2), 234–242. https://doi.org/10.1007/s00239-002-2396-z.

Johansson, M. E., Thomsson, K. A., & Hansson, G. C. (2009). Proteomic analyses of the two mucus layers of the colon barrier reveal that their main component, the Muc2 mucin, is strongly bound to the Fcgbp protein. *Journal of Proteome Research*, *8*(7), 3549–3557. https://doi.org/10.1021/pr9002504.

Kechagia, J. Z., Ezra, D. G., Burton, M. J., & Bailly, M. (2016). Fibroblasts profiling in scarring trachoma identifies IL-6 as a functional component of a fibroblast-macrophage pro-fibrotic and pro-inflammatory feedback loop. *Scientific Reports*, *6*, 28261. https://doi.org/10.1038/srep28261.

Krause, W. J., & Cutts, J. H. (1983). Ultrastructural observations on the shell membrane of the North American opossum (Didelphis virginiana). *The Anatomical Record*, *207*(2), 335–338. https://doi.org/10.1002/ar.1092070212.

Lang, T., Klasson, S., Larsson, E., Johansson, M. E., Hansson, G. C., & Samuelsson, T. (2016). Searching the evolutionary origin of epithelial mucus protein components-mucins and FCGBP. *Molecular Biology and Evolution*, *33*(8), 1921–1936. https://doi.org/10.1093/molbev/msw066.

Lash, J. W., Gosfield, E., 3rd., Ostrovsky, D., & Bellairs, R. (1990). Migration of chick blastoderm under the vitelline membrane: The role of fibronectin. *Developmental Biology*, *139*(2), 407–416.

Litscher, E. S., & Wassarman, P. M. (2007). Egg extracellular coat proteins: From fish to mammals. *Histology and Histopathology*, *22*(3), 337–347. https://doi.org/10.14670/HH-22.337.

Litscher, E. S., & Wassarman, P. M. (2014). Evolution, structure, and synthesis of vertebrate egg-coat proteins. *Trends in Developmental Biology*, *8*, 65–76.

Lutwak-Mann, C. (1971). The rabbit blastocyst and its environment: Physiological and biochemical aspects. In R. J. Blandau (Ed.), *The biology of the blastocyst* (pp. 243–260). Chicago: University of Chicago Press.

Lyne, A. G., & Hollis, D. E. (1976). Early embryology af the marsupials *Isoodon macrourus* and *Perameles nasuta*. *Australian Journal of Zoology*, *24*(3), 361–382.

Lyne, A. G., & Hollis, D. E. (1977). The early development of marsupials, with special reference to bandicoots. *Reproduction and Evolution*, 293–302.

McCrady, E., Jr. (1938). The embryology of the opossum (Vol. 16): Wistar Institute Press.

Menzies, B. R., Pask, A. J., & Renfree, M. B. (2011). Placental expression of pituitary hormones is an ancestral feature of therian mammals. *EvoDevo*, *2*, 16. https://doi.org/10.1186/2041-9139-2-16.

Moore, H. D., & Taggart, D. A. (1993). In vitro fertilization and embryo culture in the grey short-tailed opossum, Monodelphis domestica. *Journal of Reproduction and Fertility*, *98*(1), 267–274.

Moros-Nicolas, C., Chevret, P., Izquierdo-Rico, M. J., Holt, W. V., Esteban-Diaz, D., Lopez-Bejar, M., et al. (2017). Composition of marsupial zona pellucida: A molecular and phylogenetic approach. *Reproduction, Fertility, and Development*. https://doi.org/10.1071/RD16519.

Phillips, D. M., & Fadem, B. H. (1987). The oocyte of a new world marsupial, Monodelphis domestica: Structure, formation, and function of the enveloping mucoid layer. *The Journal of Experimental Zoology*, *242*(3), 363–371. https://doi.org/10.1002/jez.1402420316.

Renfree, M. B. (1973). The composition of fetal fluids of the marsupial *Macropus eugenii*. *Developmental Biology*, *33*(1), 62–79. doi:0012-1606(73)90165-6 [pii].

Renfree, M. B. (1994). Endocrinology of pregnancy, parturition and lactation of marsupials. In G. E. Lamming (Ed.), *Vol. 3. Pregnancy and lactation* (4th ed., pp. 677–766). London: Chapman & Hall.

Renfree, M. B., & Lewis, A. M. (1996). Cleavage in vivo and in vitro in the marsupial *Macropus eugenii*. *Reproduction, Fertility, and Development, 8*(4), 725–742.

Renfree, M. B., & Tyndale-Biscoe, C. H. (1978). Manipulation of marsupial embryos and pouch young. In J. C. Daniel (Ed.), *Methods in mammalian embryology* (pp. 307–331). New York, San Francisco, London: Academic Press.

Roberts, C. T., & Breed, W. G. (1996a). The marsupial shell membrane: An ultrastructural and immunogold localization study. *Cell and Tissue Research, 284*(1), 99–110.

Roberts, C. T., & Breed, W. G. (1996b). Variation in ultrastructure of mucoid coat and shell membrane secretion of a dasyurid marsupial. *Reproduction, Fertility, and Development, 8*(4), 645–648.

Roberts, C. T., Breed, W. G., & Mayrhofer, G. (1994). Origin of the oocyte shell membrane of a dasyurid marsupial: An immunohistochemical study. *The Journal of Experimental Zoology, 270*(3), 321–331. https://doi.org/10.1002/jez.1402700311.

Sato, Y., Uemura, T., Morimitsu, K., Sato-Nishiuchi, R., Manabe, R., Takagi, J., et al. (2009). Molecular basis of the recognition of nephronectin by integrin alpha8beta1. *The Journal of Biological Chemistry, 284*(21), 14524–14536. https://doi.org/10.1074/jbc.M900200200.

Selwood, L. (1989). Development in vitro of investment-free marsupial embryos during cleavage and early blastocyst formation. *Gamete Research, 23*(4), 399–413. https://doi.org/10.1002/mrd.1120230405.

Selwood, L., & Johnson, M. H. (2006). Trophoblast and hypoblast in the monotreme, marsupial and eutherian mammal: Evolution and origins. *BioEssays, 28*(2), 128–145. https://doi.org/10.1002/bies.20360.

Selwood, L., & Woolley, P. A. (1991). A timetable of embryonic development, and ovarian and uterine changes during pregnancy, in the stripe-faced dunnart, Sminthopsis macroura (Marsupialia: Dasyuridae). *Journal of Reproduction and Fertility, 91*(1), 213–227.

Semon, R. (1894). Zur Entwickelungsgeschichte der Monotremen. *Denkschriften der Medicinisch-Naturwissenschaftlichen Gesellschaft zu Jena, 5*, 61–74.

Shaw, G. (1996). The uterine environment in early pregnancy in the tammar wallaby. *Reproduction, Fertility, and Development, 8*(4), 811–818.

Sun, Y., Kuek, V., Qiu, H., Tickner, J., Chen, L., Wang, H., et al. (2018). The emerging role of NPNT in tissue injury repair and bone homeostasis. *Journal of Cellular Physiology, 233*(3), 1887–1894. https://doi.org/10.1002/jcp.26013.

Tyndale-Biscoe, C. H., & Renfree, M. B. (1987). *Reproductive physiology of marsupials*. Cambridge: Cambridge University Press.

von Blume, J., Alleaume, A. M., Kienzle, C., Carreras-Sureda, A., Valverde, M., & Malhotra, V. (2012). Cab45 is required for Ca(2+)-dependent secretory cargo sorting at the trans-Golgi network. *The Journal of Cell Biology, 199*(7), 1057–1066. https://doi.org/10.1083/jcb.201207180.

Wassarman, P. M., & Litscher, E. S. (2016). A bespoke coat for eggs: Getting ready for fertilization. *Current Topics in Developmental Biology, 117*, 539–552. https://doi.org/10.1016/bs.ctdb.2015.10.018.

Yousef, A., & Selwood, L. (1993). Embryonic development in culture of the marsupials Antechinus stuartii (Macleay) and Sminthopsis macroura (Spencer) during preimplantation stages. *Reproduction, Fertility, and Development, 5*(4), 445–458.

Yousef, A., & Selwood, L. (1996). The type and differentiation of cells in vitro from unilaminar and bilaminar blastocysts of two marsupials, Antechinus stuartii and Sminthopsis macroura. *Reproduction, Fertility, and Development, 8*(4), 743–752.

CHAPTER TWELVE

The Human Egg's Zona Pellucida

Satish K. Gupta[1]
Reproductive Cell Biology Laboratory, National Institute of Immunology, New Delhi, India
[1]Corresponding author: e-mail address: skgupta@nii.ac.in

Contents

Abstract

Human zona pellucida (ZP) matrix, a delicate network of thin interconnected filaments, is primarily composed of four glycoproteins, namely, ZP1, ZP2, ZP3, and ZP4. All four zona proteins share common structural elements such as signal peptide,

Current Topics in Developmental Biology, Volume 130
ISSN 0070-2153
https://doi.org/10.1016/bs.ctdb.2018.01.001

"ZP domain," consensus furin cleavage site, transmembrane-like domain, and short cytoplasmic tail. In addition, ZP1 and ZP4 also have "Trefoil domain." Recombinant/native human zona proteins have been used to investigate their binding characteristics to the capacitated and/or acrosome-reacted spermatozoa. These investigations revealed that ZP1, ZP3, and ZP4 primarily bind to the head region of the capacitated human spermatozoa, whereas ZP2 binds to the acrosome-reacted sperm. However, using transgenic mice, N-terminal region of human ZP2 has also been shown to play an important role in binding of sperm to the egg. ZP1, ZP3, and ZP4 lead to dose-dependent increase in acrosome reaction, suggesting that in humans more than one ZP glycoprotein is responsible for induction of acrosome reaction. Glycosylation of these proteins, in particular, *N*-linked glycosylation as well as sialyl-Lewis^{x}, is essential for inducing acrosome reaction. Studies delineating downstream signaling events associated with induction of acrosome reaction reveal subtle differences between ZP3 and ZP1/ZP4 with respect to activation of Gi protein-coupled receptor and protein kinase A. The role of mutations in the zona proteins and ZP autoantibodies leading to infertility in women is suggestive and needs more rigorous experimentations for confirming their role in female infertility. The above-mentioned aspects of the human ZP glycoproteins have been discussed in this review.

1. STRUCTURE OF HUMAN ZONA PELLUCIDA

Oocytes from virtually all vertebrates are surrounded by at least one glycoprotein coat. This coat has a different name in each major vertebrate lineage: the chorion in fish, the vitelline envelope in amphibians, the perivitelline envelope in reptiles and birds, and the zona pellucida (ZP) in mammals. ZP is a highly organized dynamic structure that is essential for oogenesis, fertilization, and early embryo development. During fertilization, the ZP matrix plays a crucial role by serving as a relative taxon-specific substrate for sperm binding, an agonist for regulated exocytosis of the spermatozoon's acrosomal vesicle and helps in blocking polyspermy. In humans, ZP thickness varies from 10 to 31 μm (Bertrand, Van den Bergh, & Englert, 1995). The outer surface of the ZP of mature human oocyte has spongy appearance and is comprised of a delicate meshwork of thin interconnected filaments in a regular alternating pattern of wide (pores) and tight (compact) meshes as revealed by high-resolution scanning electron microscopy (SEM) (Familiari et al., 1992; Familiari, Relucenti, Heyn, Micara, & Correr, 2006). The pores appear larger at the outer surface of the ZP matrix than the inner surface giving it a "spongy" appearance. On the contrary, the inner surface of ZP matrix is more compact. Transmission electron microscopy revealed that filaments are 0.1–0.4 μm in length and 10–14 nm in thickness (Familiari et al., 1992).

The ultrastructure of ZP matrix of the immature or atretic human oocytes is more compact as compared to "spongy" appearance of the mature oocytes. These observations further suggest that the porous region of the human ZP is limited to approximately 25% of the external surface, while the compact region constitutes approximately 75% of the total ZP. The amorphous spongy outer surface of zona with larger pores may facilitate sperm permeability, as ZP with a more compact and smoother outer surface has been shown to be less penetrable (Familiari et al., 1992; Familiari, Nottola, Micara, Aragona, & Motta, 1988). However, in another independent study, no correlation could be established between the morphology of ZP and success of in-vitro fertilization (IVF) (Magerkurth, Töpfer-Peterson, Schwartz, & Michelmann, 1999). SEM studies further revealed that ZP of viable human blastocysts also show "spongy" ZP with numerous fenestrations formed by network of filaments, whereas that of unhealthy blastocysts show an altered ZP outer surface comparable to that of immature/atretic oocytes (Nottola et al., 2005).

2. BIOCHEMICAL CHARACTERISTICS OF THE HUMAN ZP MATRIX

Analysis of the isolated zonae from pools of unfertilized human eggs after heat solubilization and iodination by two-dimensional sodium dodecyl sulfate polyacrylamide gel electrophoresis (2-D SDS-PAGE) under reducing conditions revealed three acidic proteins with molecular weights ranging from 90 to 110 kDa (ZP1; now designated as ZP4), 64 to 78 kDa (ZP2), and 57 to 73 kDa (ZP3). However, ZP2 and ZP4 comigrated at 90–120 kDa under nonreducing conditions (Shabanowitz & O'Rand, 1988). Another independent study showed that human ZP3 has marked charged heterogeneity, which may be due to varying extent of glycosylation (Bercegeay, Jean, Lucas, & Barriere, 1995). Further characterization of human ZP by SDS-PAGE under nonreduced conditions followed by Western blotting employing polyclonal antibodies generated against the recombinant bonnet monkey (*Macaca radiata*) ZP2, ZP3, and ZP4 (previously designated as ZP1/ZPB), which have >90% sequence identity at amino acid (aa) level with their counterpart in human ZP, showed 100, 53, and 60 kDa bands corresponding to human ZP2, ZP3, and ZP4, respectively (Gupta et al., 1998). Under reduced conditions, ZP2 showed dominant band of 65 kDa and faint reactivity at 96 kDa. ZP3 localizes to 58 kDa band and antibodies against bonnet monkey ZP4 showed reactivity with 63 kDa band and faint reactivity with 53 kDa band. Further, analysis by 2-D SDS-PAGE followed by Western blot showed that human ZP2 is composed of two chains of 65 (major component) and 96 (minor component). The 65 kDa component has higher degree of charged isomers as compared to the 96 kDa component. Human ZP3 comprises a broad band with multiple isomers in the range of 58–68 kDa. Human ZP4 resolves as two bands of 58–63 kDa and 45–55 kDa, each consisting of multiple isomers. Human ZP2 is less acidic as compared to ZP3 and ZP4 (Gupta et al., 1998). In another study, using antibodies against synthetic peptides corresponding to human ZP2 and ZP3 has characterized ZP2 and ZP3 as 90–110 and 53–60 kDa glycoproteins, respectively (Bauskin, Franken, Eberspaecher, & Donner, 1999). The presence of fourth glycoprotein, ZP1, has also been demonstrated by indirect immunofluorescence employing ZP1-specific monoclonal antibodies (MAbs, devoid of reactivity in ELISA with recombinant human ZP2, ZP3, and ZP4) generated against synthetic peptide (219–258 aa residues) designed on the basis of deduced aa sequence of human ZP1 (Ganguly, Bukovsky, et al., 2010). Distinct and specific staining of the ZP matrix of human oocytes was reported by using these MAbs (Ganguly, Bukovsky, et al., 2010). It is possible that one of the chain from 58 to 63 kDa or 45 to 55 kDa previously interpreted as ZP4 (Gupta et al., 1998) may correspond to human ZP1, as polyclonal antibodies against bonnet monkey recombinant ZP4 might have crossreacted with human ZP1 due to high aa sequence identity between the two proteins (47% identity in polypeptide sequence of human ZP1 with human ZP4). These studies conclusively show that human ZP matrix is composed of four glycoproteins in contrast to mouse ZP, which is composed of three glycoproteins, *Zp4* being a pseudogene. Further, human zona proteins have been purified using immunoaffinity columns prepared by using highly specific MAbs against recombinant human ZP2, ZP3, and ZP4 (Bukovsky et al., 2008; Chiu, Wong, Lee, et al., 2008). Their characterization by SDS-PAGE revealed ~120, ~58, and ~65 kDa bands corresponding to ZP2, ZP3, and ZP4 (also contaminated with ZP1), respectively (Chiu, Wong, Lee, et al., 2008).

3. GENOMIC ORGANIZATION AND AMINO ACID SEQUENCE OF THE HUMAN ZP GLYCOPROTEINS

The genes encoding human *Zp1*, *Zp2*, *Zp3*, and *Zp4* are localized on chromosomes 11, 16, 7, and 1, respectively (Hughes & Barrat, 1999; Table 1). The human *Zp1* gene has 12 exons and encodes a polypeptide of 638 aa (Lefievre et al., 2004; Table 1). The human *Zp2* gene has 19 exons

Table 1 Characteristics of Human Zona Pellucida Glycoproteins

Gene Encoding Zona Protein	Location on Chromosome	Genome Size (kb)	Number of Exons	Length of Polypeptide (Amino Acid)	Calculated Molecular Weight (Da)
Zp1	11q 12.2	11	12	638	57
Zp2	16p 12.3-p12.2	14	19	745	82
Zp3	7q 11.23	18.3	8	424	47
Zp4	1q 43	17	13	540	59

(one more than that in mouse) transcribed into a 2235 nucleotide (nt) long mRNA transcript, which encodes a 745 aa long polypeptide (Table 1). The human ZP2 shares a sequence identity of 57% at aa level with mouse ZP2 (Liang & Dean, 1993). The human *Zp3* gene contains 8 exons spanning 18.3 kb, the transcript of which encodes a 424 aa polypeptide. The aa sequence identity shared between human ZP3 and mouse ZP3 is 67% (Chamberlin & Dean, 1990). There is a report of the presence of a second polymorphic locus for human ZP3, which, due to an extra G residue in exon 8, encodes a truncated protein of 372 aa residues beside full-length 424 aa polypeptide (van Duin et al., 1992). Further, in human ovaries, a novel bipartite RNA transcript-encoded by *POMZP3* gene which is derived from a gene homologous to rat *POM121* (encoding a nuclear pore membrane protein) and *Zp3* is also reported (Kipersztok, Osawa, Liang, Modi, & Dean, 1995). The 5′ region of *POMZP3* gene has 77% identity to the 5′ end of the coding region of *POM121*. The 3′ end of the *POMZP3* transcript has 99% identity to the *Zp3* and appears to have arisen from a duplication of its four exons (5–8). It is still not known in which gene the variation originally occurred because of failure in amplifying these genes separately by polymerase chain reaction due to high homologous intronic regions (Männikkö et al., 2005). In-situ hybridization studies revealed that both fragments are localized to chromosome 7q 11.23. Exon map for human *Zp4* (existing as a pseudogene in mouse) spans 11 kb and the transcript encodes a 540 aa long polypeptide (Table 1).

In addition to the genomic clones, the cDNA clones of human ZP1 (Ganguly, Bukovsky, et al., 2010), human ZP2 (Liang & Dean, 1993), human ZP3 (Chamberlin & Dean, 1990), and human ZP4 (Harris et al., 1994) have also been characterized. Comparative analyses of the deduced aa sequence of human zona proteins with the respective homolog from other species reveal overall conservation of backbone structure along with considerable homology at the aa sequence level throughout the course of evolution. For example, human ZP3 has 67%, 74%, and 93.9% aa sequence identity with mouse, porcine, and bonnet monkey (*M. radiata*) ZP3, respectively. In addition to the sequence homology within a given ZP protein from different species, human ZP1 has 47% identity with human ZP4, suggesting that *Zp1* and *Zp4* are paralogous genes that may have evolved from a common ancestral gene either by gene duplication or exon swapping (Swanson, Nielsen, & Yang, 2003). Further, by analysis of the deduced aa sequence, some common structural elements that are shared among all the four human ZP glycoproteins are observed, which are as follows.

3.1 Signal Peptide

All four human ZP glycoproteins have N-terminal hydrophobic signal peptide that target these to the secretory pathway through cotranslational import into the endoplasmic reticulum. The signal peptide is 25 aa long in human ZP1, 38 aa in human ZP2, 22 aa in human ZP3, and 18 aa in human ZP4 (Fig. 1). The signal peptide is ultimately cleaved-off from the mature protein by an enzyme, signal peptidase, present in the oocyte. However, it is not clear at which stage of the zona protein biosynthesis and/or assembly into the zona matrix the signal peptide of human ZP glycoproteins is removed.

3.2 ZP Domain

All four human ZP glycoproteins share a motif designated as the "ZP domain" (Fig. 1), which is also present in several other proteins such as transforming growth factor (TGF)-β receptor type III, uromodulin, pancreatic secretory granule protein GP2, α and β tectorins, DMBT-1 (deleted in brain tumor-1), Nomp A (no mechanoreceptor potential A), dumpy and cuticulin-1, *Drosophila* genes *miniature* and *dusky*, etc. (Jovine, Darie, Litscher, & Wassarman, 2005; Roch, Alonso, & Akam, 2003). The "ZP domain" consists of approximately 260 aa including 8 conserved cysteine residues and is predicted to have high β-strand content with additional conservation of hydrophobicity, polarity, and turn-forming tendency (Bork & Sander, 1992). Mutation in "ZP domain" of α-tectorin results in defective tectoral membrane assembly, underlying human hearing disorder suggesting thereby that this domain plays an important role in the polymerization of human zona proteins into a filamentous structure (Jovine, Qi, Williams, Litscher, & Wassarman, 2002). It is located at the C-terminus of the respective zona proteins, generally near the membrane-anchoring region (Bork & Sander, 1992;

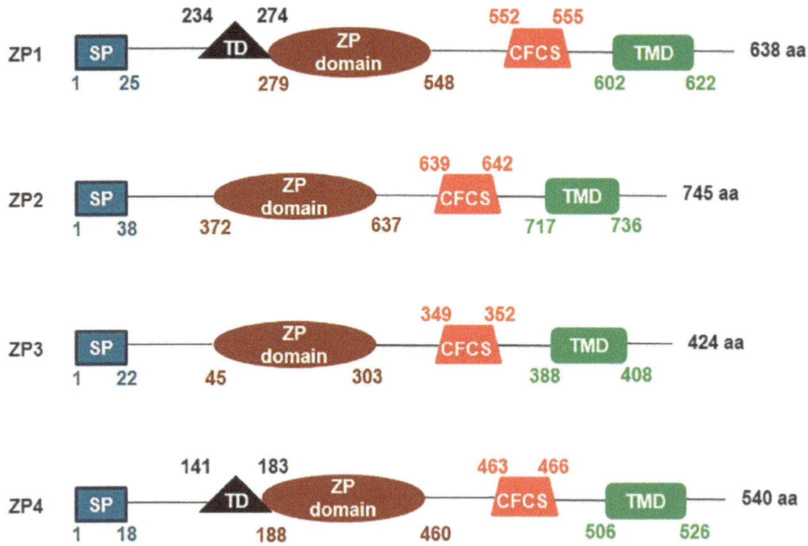

Fig. 1 Schematic representation of the structural elements of human ZP glycoproteins. Cloning and sequencing of human zona proteins revealed 638, 745, 424, and 540 amino acid long polypeptides of ZP1, ZP2, ZP3, and ZP4, respectively. The predicted domains of all four human zona proteins such as signal peptide (SP), Trefoil domain (TD), "ZP domain," consensus furin cleavage site (CFCS), and transmembrane-like domain (TMD) along with the amino acid spanning respective domain have been shown. All four human zona proteins also have short cytoplasmic tail ranging from 9 to 16 amino acids. Various domains are not drawn to scale.

Wassarman, 2008). It has a bipartite structure with ZP-N and ZP-C subdomains separated by a protease-sensitive region (Jovine et al., 2005). Two synthetic peptides predicted as "aggregation-prone" are designed based on the sequence of human ZP1-N domain, which self-assemble into amyloid-like fibrils (Louros, Iconomidou, Giannelou, & Hamodrakas, 2013). Further, six synthetic peptides corresponding to a common interface of human ZP2, ZP3, and ZP4 self-assemble into fibrils with distinct amyloid-like features, suggesting that ZP domain plays an important role in polymerization and self-assembly of ZP glycoproteins (Louros et al., 2016). Using mouse zona protein, it has also been demonstrated that ZP-N subdomain is sufficient for its polymerization (Jovine, Janssen, Litscher, & Wassarman, 2006).

3.3 Trefoil Domain

"Trefoil" or "P-domain," a 42 aa domain having a characteristic pattern of six conserved cysteines in a trefoil-like arrangement is found in a family of small polypeptides called the Trefoil family (Thim, 1989). This module has been described in several proteins that have diverse biological activities like spasmolytic polypeptide, intestinal trefoil factor, breast cancer-associated peptide (pS2), etc. (Bork, 1993). Most of the proteins containing the TD are thought to be growth factors that also contribute to mucosal defense mechanisms, particularly after injury of the gastrointestinal tract. Interestingly, human ZP1 and ZP4 also have TD, which is absent in ZP2 and ZP3 (Fig. 1). The presence of such a module in the heavily glycosylated ZP glycoproteins suggests a more general role, such as specific binding to carbohydrates. However, the significance of its presence with respect to the functions of human ZP1 and ZP4 is still not clear.

3.4 Consensus Furin Cleavage Site

Consensus furin cleavage site (CFCS) is present in all the four human ZP proteins immediately after the ZP domain. CFCS site in human ZP1, ZP2, ZP3, and ZP4 is comprised of RQRR, RHRR, RNRR, and SRRR, respectively (Fig. 1). Expression of recombinant human ZP3 in mammalian cells using either site-directed mutagenesis of the CFCS or treatment with a competitive inhibitor of all furin family members or interference with Golgi modification by Brefeldin A suggests that cleavage at CFCS is critical for its secretion. The importance of CFCS in the assembly of human ZP3 in ZP matrix is further shown by using "human ZP3 rescue" mouse model (Kiefer & Saling, 2002). It is likely that in mature human zona proteins, C-terminal propeptide is lost by the proteolytic cleavage at CFCS by proprotein convertase enzyme either in the Golgi or at the egg plasma membrane.

3.5 Transmembrane-Like Domain

Downstream of CFCS, hydrophobic transmembrane-like domain (TMD) is present in the human ZP proteins (Fig. 1). Analysis of the mouse zona proteins by 2-D thin layer chromatography reveals the absence of TMD (including short cytoplasmic tail) downstream of CFCS, suggesting that zona proteins are processed at their CFCS site prior to the secretion and incorporation into the ZP matrix (Litscher, Qi, & Wassarman, 1999). However, no such information is available with respect to human ZP glycoproteins.

3.6 Short Cytoplasmic Tail

Human ZP1, ZP2, ZP3, and ZP4 have short cytoplasmic tail comprising of 16, 9, 16, and 14 aa residues, respectively (Fig. 1). However, its functional significance with respect to human ZP proteins is not known. In mouse, cytoplasmic tail of ZP2 and ZP3 prevent their premature intracellular interactions and thereby play an important role in their incorporation into the ZP matrix (Jimenez-Movilla & Dean, 2011).

4. EXPRESSION OF ZP GLYCOPROTEINS IN HUMAN OVARIES

Expression of human ZP3 has been documented in the oocytes of fetal ovary (Törmälä et al., 2008). ZP3 mRNA has been detected from the 11th week of gestation, and maximum expression of ZP3 is observed around 20th week of gestation, at which time follicle formation is also initiated. Expression of the transcription factor—factor in the germline alpha (FIGα), which is essential for the expression of ZP genes and formation of primordial follicles, has also been observed in the human fetal and adult ovaries (Bayne, Martins da Silva, & Anderson, 2004; Huntriss et al., 2002; Törmälä et al., 2008). In contrast to mice, the precise role of FIGα in the expression of human ZP genes is not very well understood. Expression of ZP1 mRNA is low in the fetal and adult ovaries (Törmälä et al., 2008).

It is not clear whether human ZP proteins are synthesized in the oocytes, granulosa cells, or both. There are studies which suggest that ZP3 is expressed in both the oocytes and granulosa cells of most of the primordial follicles and recruited follicles (Grootenhuis, Philipsen, de Breet-Grijsbach, & van Duin, 1996). Human ZP1 and ZP2 along with ZP3 have also been detected in both the oocytes and the granulosa cells as early as in the primordial follicles. The detection of ZP proteins in the quiescent primordial follicles suggests that these proteins are present since oogenesis (Gook, Edgar, Borg, & Martic, 2008). The possibility that ZP proteins in granulosa cells may represent those selectively endocytosed into the granulosa cells after synthesis elsewhere needs to be further investigated. However, there is growing evidence that during human folliculogenesis, ZP proteins are expressed only in the oocytes (Eberspaecher, Becker, Bringmann, van der Merwe, & Donner, 2001; Huntriss et al., 2002; Lefievre et al., 2004; Törmälä et al., 2008).

Our understanding of the expression of the respective zona proteins as a function of folliculogenesis and their subsequent assembly in the ZP matrix in humans is very limited. Employing murine MAbs generated against recombinant human zona proteins that specifically react with ZP2, ZP3, and ZP4 reveal that ZP2 is not detectable in the oocytes in resting primordial follicles but show reactivity with the oocytes in growing and antral follicles.

However, antibodies also react with ZP2 of primordial follicles undergoing atresia. MAbs against human ZP2 is of interest since they do not recognize resting healthy primordial follicles with Balbiani bodies (Bukovsky et al., 2008). ZP3 is also detectable in oocytes in growing and antral follicles. However, MAbs against ZP4 react with ZP of oocytes in primordial, growing, and antral follicles (Bukovsky et al., 2008). No reactivity of these MAbs is observed with other cell types of ovary, suggesting that the expression of these proteins is specific to the oocytes. MAbs against human ZP1 do not react with oocytes in healthy primordial and primary follicles but react with oocytes of primordial/primary follicles undergoing atresia. Oocytes in secondary and antral follicles show strong surface ZP1 expression (Ganguly, Bukovsky, et al., 2010). Besides reactivity with follicular oocytes, all ZP MAbs also stain degenerating intravascular oocytes in either ovarian or extra ovarian venules (Ganguly, Bukovsky, et al., 2010).

It is of interest to understand, how the expression of various zona proteins is regulated in oocytes during folliculogenesis. Analysis of the upstream sequences of mouse and human ZP2 and ZP3 genes reveal the presence of five short conserved DNA sequences (4–12 bp) that are 60%–100% identical, upstream of the TATAA box (I, IIA, IIB, III, and IV; Millar, Lader, Liang, & Dean, 1991). Mutation analysis establishes that the 12 bp element IV, present approximately 200 bp upstream of TATAA box, is both necessary and sufficient for high-level expression of a reporter gene under ZP promoter in mouse oocytes. Oligonucleotide containing the conserved upstream regulatory elements from either ZP2 or ZP3 forms DNA–protein complexes of identical mobility in gel retardation assay, suggesting the involvement of similar regulatory processes in the expression and binding of common regulatory factors to the conserved element IV. Further, a putative transcription factor known as zona pellucida gene activating protein-1 (ZAP-1) has been implicated in coordinated and oocyte-specific expression of mouse and human ZP2 and ZP3 genes (Millar, Lader, & Dean, 1993).

5. ROLE OF HUMAN ZP GLYCOPROTEINS IN SPERM BINDING AND INDUCTION OF ACROSOME REACTION

ZP glycoproteins play a critical role in accomplishment of fertilization by acting as ligand for binding of spermatozoa to the egg followed by induction of acrosomal exocytosis and block to polyspermy. Initial studies showed that capacitated sperm treated with intact human zonae or acid-disaggregated zonae led to a significant increase in induction of acrosome reaction (Cross, Morales, Overstreet, & Hanson, 1988; Franken, Bastiaan, & Oehninger, 2000). Heat-solubilized human ZP also induces dose-dependent increase in acrosomal exocytosis in capacitated human sperm (Bhandari, Bansal, Talwar, & Gupta, 2010), suggesting that ZP is the primary physiological inducer of acrosomal exocytosis in the egg-bound spermatozoa. During fertilization in humans, it is pertinent to delineate which zona protein is involved in sperm binding and/or induction of acrosome reaction. Is the role of individual human ZP glycoproteins during sperm binding and/or induction of acrosome reaction same as that observed in mouse model or additional proteins play a role in these events? The human ZP matrix is composed of four glycoproteins as compared to three glycoproteins that constitute mouse ZP. However, the presence of ZP4 in human ZP matrix (which is absent in mouse ZP) does not explain the taxon-specific recognition of human sperm–egg interaction as transgenic mice expressing human ZP4 in mouse ZP matrix failed to bind human spermatozoa (Yauger, Boggs, & Dean, 2011). To delineate the respective role of human zona proteins in binding to the spermatozoa and induction of the acrosome reaction, various investigators have used either recombinant or purified native human zona proteins, which is briefly described below.

5.1 Human ZP1

Both *E. coli*- and baculovirus-expressed recombinant human ZP1 corresponding to 26–551 aa residues binds to mildly fixed (fixed using 0.5% paraformaldehyde for 10 min at room temperature) human capacitated spermatozoa on the acrosomal cap and equatorial region (Figs. 2 and 3). Higher percentage of binding is observed in the acrosomal cap. However, some sperm also display binding in the mid-piece and postacrosome region (Ganguly, Bukovsky, et al., 2010). In the calcium ionophore-induced acrosome-reacted spermatozoa, binding of the recombinant protein is restricted to the equatorial region of the sperm. Binding of the recombinant human ZP1 to the capacitated spermatozoa is specific as shown by competitive displacement studies as well as inhibition of its binding in the presence of ZP1-specific MAbs (Ganguly, Bukovsky, et al., 2010). *E. coli*-expressed ZP1 binds to the capacitated spermatozoa as efficiently as baculovirus-expressed ZP1. Further, *E. coli*-expressed ZP1 competitively inhibits the binding of baculovirus-expressed recombinant ZP1 to the capacitated human spermatozoa and vice versa, suggesting that the carbohydrate residues may be dispensable and the polypeptide backbone of human ZP1

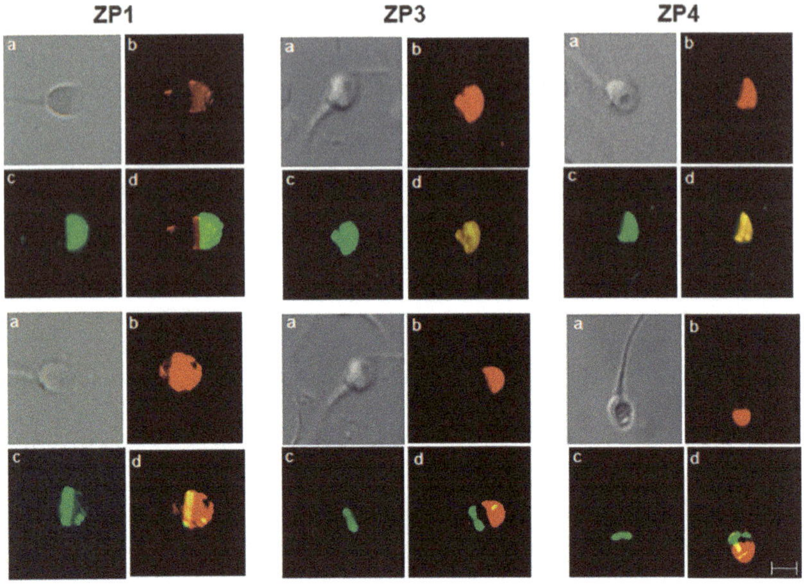

Fig. 2 Binding profile of recombinant human zona proteins with human spermatozoa. Capacitated (*upper panel*) and acrosome-reacted (*lower panel*) human sperm were incubated with fluorescein isothiocyanate (FITC)-conjugated *E. coli*-expressed recombinant human ZP1, ZP3, and ZP4. The acrosome status of the sperm was simultaneously determined by the binding of tetramethylrhodamine isothiocyanate-conjugated *Pisum sativum* agglutinin (TRITC-PSA). In each panel, the subpanels are represented as (a) phase contrast, (b) PSA-TRITC fluorescence, (c) FITC-ZP protein fluorescence, and (d) overlap of fluorescence frames. The scale bar represents 2.5 μm. *Part of the photographs are reproduced from Chakravarty, S., Kadunganattil, S., Bansal, P., Sharma, R. K., & Gupta, S. K. (2008). Relevance of glycosylation of human zona pellucida glycoproteins for their binding to capacitated human spermatozoa and subsequent induction of acrosomal exocytosis.* Molecular Reproduction and Development, 75, *75–88.*

may be sufficient for its binding to the spermatozoa. Further, baculovirus-expressed recombinant human ZP1 corresponding to 273–551 aa residues also binds primarily to the acrosomal cap of the capacitated human spermatozoa, suggesting that "ZP domain" is sufficient for binding of human ZP1 to the capacitated spermatozoa (Ganguly, Bansal, Gupta, & Gupta, 2010).

Interestingly, incubation of the capacitated human sperm with baculovirus-expressed recombinant human ZP1 (26–551 aa residues) as well as recombinant protein corresponding to "ZP domain" (273–551 aa residues) leads to a significant increase in the acrosomal exocytosis (Ganguly, Bansal, et al., 2010; Ganguly, Bukovsky, et al., 2010; Table 2). However, *E. coli*-expressed recombinant human ZP1 (26–551 aa residues), though

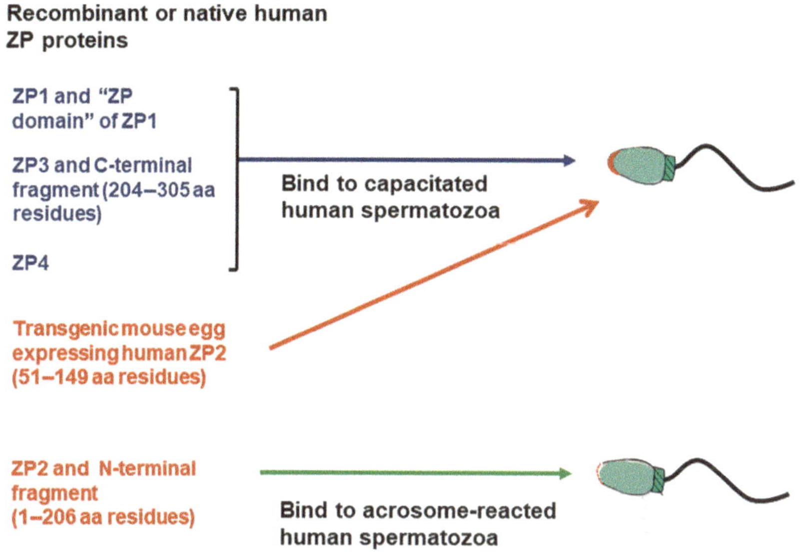

Fig. 3 Schematic diagram to illustrate the binding of human zona proteins to sperm. Using either recombinant or native human ZP proteins, it has been shown that ZP1, ZP3, and ZP4 primarily bind to the capacitated human sperm, whereas ZP2 binds to the acrosome-reacted sperm. In transgenic mouse model, ZP2 has also been shown to play a role in binding of human sperm to mouse egg expressing human ZP2. The epitopes/domains of ZP1, ZP2, and ZP3 responsible for their binding to sperm are also shown.

bind to capacitated spermatozoa but fail to induce any significant increase in the acrosomal exocytosis suggesting that glycosylation of ZP1 is critical for induction of acrosomal exocytosis.

5.2 Human ZP2

Binding studies of *E. coli*-expressed recombinant human ZP2 fragment corresponding to 1–206 aa residues with capacitated as well as acrosome-reacted human sperm reveal that it binds only to the acrosome-reacted spermatozoa (Fig. 3). The binding was observed in the acrosomal region and mid-piece (Tsubamoto et al., 1999). The fluorochrome-conjugated *E. coli*-expressed human ZP2 (38–645 aa residues) and baculovirus-expressed ZP2 (1–745 aa residues) fail to bind to the capacitated human sperm but showed binding to the equatorial region of acrosome-reacted sperm (Fig. 3). Both *E. coli*- and baculovirus-expressed recombinant ZP2 show very similar binding profiles (Chakravarty et al., 2008). Further native human ZP2 (~120 kDa) purified through MAb-based immunochromatography column

Table 2 Induction of Acrosome Reaction in Capacitated Human Sperm by Human ZP Glycoproteins

Capacitated Human Sperm Treated With	Nature of Zona Protein	Percent Induction of Acrosome Reaction[a]	References
ZP1	Baculovirus-expressed recombinant protein	~26	Ganguly, Bukovsky, et al. (2010)
"ZP domain" (273–551 aa residues) of ZP1	Baculovirus-expressed recombinant protein	~34	Ganguly, Bansal, et al. (2010)
ZP3	Baculovirus-expressed recombinant protein	~26	Chakravarty, Suraj, and Gupta (2005) and Chakravarty, Kadunganattil, Bansal, Sharma, and Gupta (2008)
ZP3	Mammalian (CHO cells)-expressed recombinant protein	~18	van Duin et al. (1994) and Bray, Son, Kumar, Harris, and Meizel (2002)
ZP3	Recombinant protein expressed in PA-1 cells	~10	Dong et al. (2001)
ZP3	Immunoaffinity-purified native protein	~31	Chiu, Wong, Chung, et al. (2008)
C-terminal fragment (214–305 aa residues) of ZP3	Baculovirus-expressed recombinant protein	~17	Bansal, Chakrabarti, and Gupta (2009)
ZP4	Baculovirus-expressed recombinant protein	~13	Chakravarty et al. (2005) and Chakravarty et al. (2008)
ZP4	Immunoaffinity-purified native protein	~24	Chiu, Wong, Chung, et al. (2008)

[a]Percent of sperm undergoing acrosome reaction in the presence of ZP protein minus spontaneous acrosome reaction.
CHO, Chinese hamster ovary cells; *PA-1*, human ovarian teratocarcinoma cell line.

also fails to bind to the capacitated human sperm and shows binding to the acrosome region only in the acrosome-reacted spermatozoa (Chiu, Wong, Lee, et al., 2008). Using transgenic mice wherein mouse oocytes expressing human ZP2, either alone or coexpressed with other human zona proteins bind human sperm, which penetrate the ZP matrix and accumulate in the perivitelline space (Baibakov, Boggs, Yauger, Baibakov, & Dean, 2012). However, human sperm are unable to fuse with mouse egg membrane expressing human ZP2. Employing recombinant peptides, the binding site of ZP2 to human sperm was localized to N-terminus of ZP2 (51–149 aa residues; Avella, Baibakov, & Dean, 2014). These experiments suggest that ZP2 domain corresponding to 51–149 aa residues is also involved and important for human gamete recognition and penetration through the ZP matrix (Avella et al., 2014; Baibakov et al., 2012; Fig. 3). However, no significant increase in the acrosome-reaction is observed when capacitated human sperm are treated with either recombinant or purified native human ZP2 (Chakravarty et al., 2008, 2005; Chiu, Wong, Chung, et al., 2008).

5.3 Human ZP3

Both *E. coli*- and baculovirus-expressed recombinant human ZP3 bind to either anterior head or the equatorial region of the capacitated human spermatozoa (Chakravarty et al., 2008; Figs. 2 and 3). Acrosome-reacted sperm fail to show any binding of the recombinant human ZP3 to the anterior head and binding is primarily observed in the equatorial region (Chakravarty et al., 2008; Fig. 2). Purified native human ZP3 shows binding to the acrosome, equatorial region and mid-piece of the capacitated spermatozoa, whereas it could only bind to mid-piece of the acrosome-reacted spermatozoa (Chiu, Wong, Lee, et al., 2008).

Recombinant human ZP3 expressed either using baculovirus or mammalian expression systems shows dose-dependent increase in the acrosomal exocytosis (Bray et al., 2002; Caballero-Campo et al., 2006; Chakravarty et al., 2008, 2005; Dong et al., 2001; Jose et al., 2010; van Duin et al., 1994; Table 2). The role of human ZP3 in the induction of acrosome reaction is further confirmed by using purified native human ZP3 (Chiu, Wong, Chung, et al., 2008; Table 2). Induction of acrosome reaction by *E. coli*-expressed recombinant human ZP3 (likely to be devoid of glycosylation) has also been reported (Chapman, Kessopoulou, Andrews, Hornby, & Barratt, 1998). However, in this study the capacitated sperm were treated with *E. coli*-expressed recombinant ZP3 protein for 18 h, which is a very long time period and does not mimic physiological conditions. Chakravarty et al. (2008, 2005) failed to observe any significant increase in the induction of acrosome reaction in the capacitated sperm when treated with *E. coli*-expressed recombinant human ZP3 for 60 min. Under similar experimental conditions, ~34%–38% capacitated sperm undergo acrosome reaction when treated with baculovirus-expressed recombinant ZP3 as compared to ~52% in the presence of 10 μM calcium ionophore (A23187), a chemical agonist of acrosome reaction, used as a positive control and ~9%–12% in case of spontaneous acrosome reaction (Chakravarty et al., 2005).

Attempts have been made to identify domain of human ZP3 that is responsible for its ability to induce acrosome reaction. Mature human ZP3 has 12 cysteine residues. Mapping of disulfide linkages by mass spectrometric analysis of recombinant human ZP3 expressed in mammalian cells reveals that first four cysteine residues form linkages between Cys^{46} and Cys^{140}, and Cys^{78} and Cys^{99} that forms a loop-within-loop motif. The next four cysteine residues form disulfide bonds between Cys^{217} and Cys^{282}, and Cys^{239} and Cys^{300} leading to formation of crossover motif. The disulfide bonds could not be mapped for four cysteine residues, namely, Cys^{319}, Cys^{321}, Cys^{322}, and Cys^{327} which lie within a tight cluster (Zhao et al., 2004). Schematic diagram of human ZP3 depicting the disulfide linkages is shown in Fig. 4. ZP3 when expressed as either N-terminal fragments (1–175 and 23–175 aa residues) or C-terminal fragments (214–348 and 214–305 aa residues) using baculovirus expression system shows significant induction of acrosome reaction with both the C-terminal fragments of

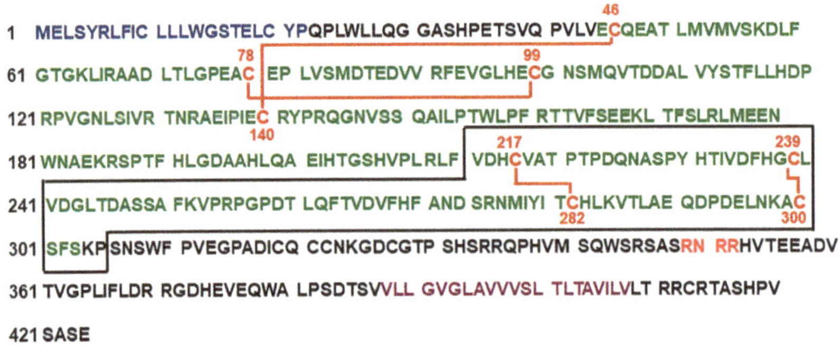

Fig. 4 Amino acid sequence of human ZP3 showing mapped disulfide bonds and minimal epitope capable of inducing acrosome reaction. The amino acid sequence of human ZP3 was obtained from Uniprot Protein Database (accession number, AAA61336.1). Various structural elements such as signal peptide (1–22 aa residues; *blue color*), ZP domain (45–303 aa residues; *green color*), consensus furin cleavage site (349–352 aa residues; *pink color*), and transmembrane-like domain (388–408 aa residues; *purple color*) are shown. The disulfide bonds mapped by LC-QTOF mass spectrometry by Zhao et al. (2004) are also shown. The N-terminal subdomain has loop-within-loop disulfide linkages, and C-terminal subdomain has crossover disulfide linkages. The minimal epitope of human ZP3 (214–305 aa residues) responsible for inducing acrosome reaction is shown in the *box*.

human ZP3, whereas N-terminal fragments fail to induce any significant increase in acrosome reaction (Bansal et al., 2009; Table 2; Fig. 4). Interestingly, both N- and C-terminal fragments are shown to bind to the anterior head of the capacitated spermatozoa (Bansal et al., 2009). Further, chemically deglycosylated C-terminal fragment (214–348 aa residues) expressed in baculovirus or that expressed in *E. coli* fails to induce any significant increase in acrosome reaction again reiterating the relevance of glycosylation during induction of acrosome reaction.

5.4 Human ZP4

In addition to human ZP1, ZP2, and ZP3, the role of human ZP4 in binding to the capacitated human spermatozoa has also been investigated. It has been shown that fluorescein isothiocyanate (FITC)-conjugated baculovirus-expressed recombinant human ZP4 binds to the acrosomal cap and equatorial region of the capacitated human sperm (Chakravarty et al., 2008). In contrast to ZP3, the major binding profile of ZP4 is observed on the acrosomal cap. Acrosome-reacted sperm do not show any binding to the acrosomal cap, and binding is restricted to only equatorial region. FITC-conjugated *E. coli*-expressed recombinant human ZP4 also shows similar binding profile as observed with the baculovirus-expressed human ZP4 (Chakravarty et al., 2008; Fig. 2). The specificity of the binding of recombinant ZP4 to capacitated spermatozoa has been further confirmed by competitive displacement studies with recombinant human ZP2 and ZP3, which fail to show any significant decrease in ZP4 binding to spermatozoa. Interestingly, triple labeling experiments reveal that recombinant human ZP3 and ZP4 can simultaneously bind to the same capacitated spermatozoon, suggesting that these glycoproteins may have distinct binding sites on spermatozoa (Chakravarty et al., 2008). Binding of baculovirus-expressed ZP4 is inhibited in the presence of *E. coli*-expressed ZP4 and vice versa, suggesting that glycosylation may not be critical for binding of ZP4 to the capacitated sperm. In another independent study, binding of sperm has been demonstrated to immobilized recombinant human ZP4 (Chirinos et al., 2011). Native immunoaffinity-purified human ZP4 has been shown to bind to the entire head region of acrosome-intact spermatozoa, which is lost after acrosome reaction (Chiu, Wong, Lee, et al., 2008).

Incubation of capacitated human sperm with baculovirus-expressed recombinant human ZP4 leads to a significant increase in acrosome reaction, whereas *E. coli*-expressed human ZP4 fails to induce any significant changes in acrosome reaction (Chakravarty et al., 2005; Table 2). The role of human ZP4 in induction of acrosome reaction has also been confirmed by independent studies employing either native or recombinant ZP4 (Caballero-Campo et al., 2006; Chiu, Wong, Chung, et al., 2008; Table 2). Interestingly, incubation of capacitated sperm with recombinant ZP4 led to considerable decrease in progressive motility, concomitant with an increase in nonprogressive sperm motility. However, there is an increase in curvilinear velocity and the amplitude of lateral head displacement, suggesting that ZP4 not only increases acrosome reaction but modifies sperm motility thereby facilitating fertilization (Caballero-Campo et al., 2006).

Induction of acrosome reaction mediated by the baculovirus-expressed recombinant human ZP4 is inhibited by MAbs (MA-1662 and MA-1671) against human ZP4, which also binds to the ZP matrix of human eggs. Epitope mapping studies reveal that MA-1671 recognizes an epitope corresponding to 126–130 aa residues (PARDR), whereas MA-1662 maps to 256–260 aa residues (ENELV), suggesting that these epitopes of human ZP4 may be relevant for induction of acrosome reaction (Xu et al., 2012).

5.5 In Humans More Than One Zona Protein Is Involved in Binding of Sperm to the Oocyte and Induction of Acrosome Reaction

From the above studies, it is quite evident that in humans sperm–egg binding, all the four human zona proteins play a role (Fig. 3). Further, glycosylation of the human zona proteins is not critical for their binding to the sperm. In contrast to mouse, where ZP3 is the primary agonist for induction of acrosome reaction, in humans, ZP1, ZP3, and ZP4 are competent to induce acrosome reaction when incubated with capacitated human sperm. There is a need to further investigate the relative importance of their ability to induce acrosome reaction when present in the ZP matrix, which will mimic physiological conditions.

6. RELEVANCE OF GLYCOSYLATION OF HUMAN ZONA PROTEINS IN FERTILIZATION

Using lectins as histochemical markers for carbohydrates on ZP, it has been shown that the most pronounced difference between the ZP of humans and other mammals is the presence of very high concentration of D-mannose residues in human ZP, reflecting a high content of asparagines-linked oligosaccharides (Maymon et al., 1994). The glycosylation profile of human ZP is different as compared to other mammalian species (Jiménez-Movilla et al., 2004). Binding studies with various lectins and antibodies also reveal the presence of sialyl-Lewis^a, sialyl-Lewis^x, Neu5Acα2-3Galβ1,4Glc-NAc, Galβ1,3GalNAc-Ser/Thr, Neu5Ac-GalNAc, fucosylated oligosaccharides, *N*-acetylgalactosamine residues, galactose residues, and *N*-acetylglucosamine residues on human ZP. Further, ultrastructural cytochemical findings reveal that glycosylation profile of external surface is different as compared to the internal surface of ZP matrix, which may be responsible for differential sperm-binding affinity between the outer and the inner regions of the ZP (Jiménez-Movilla et al., 2004). Additionally, in sugar competition assays, mannose has been implicated in human fertilization (Mori, Daitoh, Irahara, Kamada, & Aono, 1989). In addition to mannose, several oligosaccharide moieties along with complex glycoconjugates bearing selectin-like ligands have also been shown to be involved in the human sperm–egg binding (Miranda et al., 1997; Oehninger, Patankar, Seppala, & Clark, 1998). In hemizona assay, by periodate oxidation, sialic acid on human ZP has been shown to play a role in sperm–egg binding. Further treatment of human ZP with either neuraminidase or endo-beta-galactosidase leads to an increase in the sperm binding to hemizona, suggesting that there are other potential binding sites on ZP for spermatozoa that may be obscured by sialic acid and lactosaminoglycan (Ozgur, Patankar, Oehninger, & Clark, 1998). Ultrasensitive mass spectrometric analyses reveal the presence of sialyl-Lewis^x sequence [NeuAcα2-Galβ1-4(fucα1-3)GlcNAc] on the *N*- and *O*-linked glycans of human ZP (Pang et al., 2011). The sialyl-Lewis^x plays an important role in human sperm–egg binding as preincubation of sperm with either sialyl-Lewis^x or its conjugate with bovine serum albumin leads to a significant decrease in the binding of sperm to the ZP in hemizona assay. Further, antibodies to sialyl-Lewis^x also inhibit human sperm-ZP binding (Pang et al., 2011).

Lectin-binding studies reveal the presence of both *N*- and *O*-linked glycosylation in the baculovirus-expressed human ZP3 and ZP4 (Chakravarty et al., 2008). Reduced *N*-linked oligosaccharides of baculovirus-expressed human ZP3 and ZP4 synthesized in the presence of Tunicamycin result in significantly reduced ability of these recombinant proteins to induce acrosome reaction in capacitated human spermatozoa. On the other hand, reduction of *O*-linked glycosylation does not alter their acrosome reaction induction capability in capacitated human spermatozoa (Chakravarty et al., 2008). Further, removal of *N*-linked glycans by *N*-glycosidase-F of purified native human ZP3/ZP4 also results in significant decrease in acrosome reaction as compared to untreated ZP3/ZP4. Whereas removal of *O*-linked glycans by alkali hydrolysis (β-elimination) does not lead to any significant decrease in acrosome reaction mediated by native human ZP3/ZP4 further reiterating that *N*-linked glycans of ZP3/ZP4 are more relevant than *O*-linked glycans in mediating acrosome reaction (Chiu, Wong, Chung, et al., 2008).

7. SIGNALING EVENTS ASSOCIATED WITH HUMAN ZP GLYCOPROTEINS-MEDIATED ACROSOME REACTION

ZP-induced acrosomal exocytosis involves at least two different receptor-mediated signaling pathways in spermatozoa plasma membrane. One is a guanine nucleotide-binding regulatory protein (Gi protein)-coupled receptor that activates phospholipase Cβ1 (PLCβ1)-mediated signaling pathway and the other is a tyrosine kinase (TK) receptor coupled to PLCγ (Baldi, Luconi, Bonaccorsi, & Forti, 2002; Doherty, Tarchala, Radwanska, & De Jonge, 1995). Phospholipase C leads to the hydrolysis of phosphatidylinositol bisphosphate (PIP_2) to diacylglycerol and inositol triphosphate (IP_3) followed by translocation of protein kinase C (PKC) to the plasma membrane and its activation (Patrat, Serres, & Jouannet, 2000). IP_3 may further facilitate an increase in intracellular calcium ($[Ca^{2+}]_i$) release through modulation of IP_3-sensitive intracellular calcium stores by binding to its receptor (Walensky & Snyder, 1995). ZP binding to spermatozoa is also accompanied by depolarization of spermatozoa membrane potential, an increase in intracellular pH through Na^+/H^+ exchanger activation and an increase in $[Ca^{2+}]_i$ via voltage-operated calcium channels (VOCCs), leading to fusion of the spermatozoa plasma membrane with the outer acrosomal membrane resulting in acrosome reaction and release of the acrosomal contents.

The induction of acrosome reaction mediated by either acid- or heat-solubilized human ZP (SIZP) involves activation of Gi protein-coupled receptor as, in the presence of pertussis toxin (an inhibitor of Gi protein-mediated signaling pathway), a significant inhibition of SIZP-mediated acrosomal exocytosis is observed (Bastiaan, Franken, & Wranz, 1999; Bhandari et al., 2010; Lee, Check, & Kopf, 1992). Human SIZP also involves activation of TK as preincubation of capacitated human sperm with Herbimycin A (inhibitor of TK) leads to significant reduction in acrosome reaction (Bhandari et al., 2010). Selective inhibition of protein kinase A (PKA) by KT5720, PKC by Staurosporine, and phosphoinositide 3

(PI3)-kinase by either Wortmannin or LY294002 leads to significant decrease in the human SIZP-mediated acrosome reaction, suggesting their involvement in acrosome reaction (Bhandari et al., 2010; Bielfeld, Faridi, Zaneveld, & De Jonge, 1994; Fisher, Brewis, Barratt, Cooke, & Moore, 1998; Liguori, de Lamirande, Minelli, & Gagnon, 2005; Liu & Baker, 1997). Induction of acrosome reaction by human SIZP is accompanied by an increase in $[Ca^{2+}]_i$, which depends on extracellular Ca^{2+} as chelating the extracellular calcium by ethylene glycol-bis(β-aminoethyl ether)-*N*,*N*,*N*′,*N*′-tetraacetic acid (EGTA) leads to inhibition of acrosome reaction (Bhandari et al., 2010). Further increase in $[Ca^{2+}]_i$-mediated by SIZP is dependent on T-type VOCCs as pretreatment with Pimozide and Mibefradil leads to inhibition of acrosome reaction (Fig. 5). However, L-type VOCCs inhibitors such as Nifedipine and Verapamil fail to inhibit SIZP-mediated acrosome reaction (Bhandari et al., 2010; Patrat et al., 2006; Fig. 5). Chloride movement by the mammalian sperm has been reported to occur during the zona-initiated acrosome reaction. The gamma aminobutyric acid (GABA)-A receptor linked to chloride channels is also involved in human SIZP-mediated acrosome reaction (Bhandari et al., 2010; Breitbart, 2002; Patrat et al., 2006).

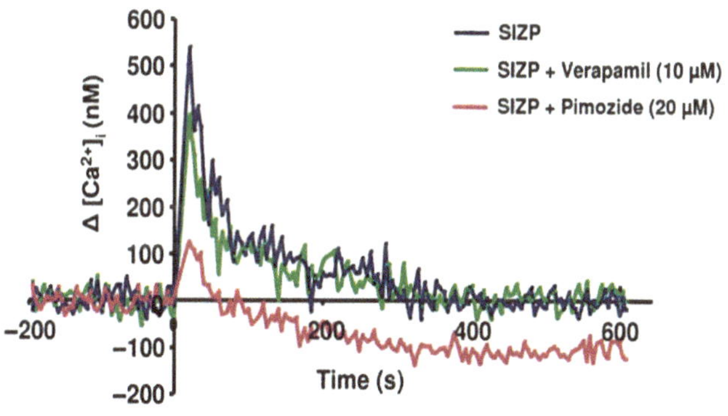

Fig. 5 Profile of intracellular calcium release in sperm treated with human heat-solubilized isolated zona pellucida (SIZP). Treatment of fluo-3 acetoxymethyl ester-labeled capacitated human sperm with SIZP led to an increase in intracellular calcium $[Ca^{2+}]_i$. Pretreatment of labeled capacitated human sperm with Pimozide, T-type VOCC inhibitor, led to a significant inhibition of $[Ca^{2+}]_i$, whereas Verapamil, L-type VOCC inhibitor, did not show any significant decrease of $[Ca^{2+}]_i$, suggesting that SIZP primarily uses T-type VOCC to increase $[Ca^{2+}]_i$. *The data are reproduced from Bhandari, B., Bansal, P., Talwar, P., & Gupta, S. K. (2010). Delineation of downstream signalling components during acrosome reaction mediated by heat solubilized human zona pellucida.* Reproductive Biology and Endocrinology, 8, *7*.

Recombinant as well as purified native human ZP proteins have also been used to delineate the signaling events associated with induction of acrosome reaction. Both baculovirus-expressed recombinant human ZP3 and immunoaffinity-purified native human ZP3 induce dose-dependent increase in acrosome reaction, which is inhibited by pretreatment of capacitated sperm with pertussis toxin, suggesting that activation of Gi protein-coupled receptor is critical for ZP3-mediated acrosome reaction (Chakravarty et al., 2005; Chiu, Wong, Chung, et al., 2008). On the other hand, failure to observe any significant decrease in the induction of acrosome reaction by baculovirus-expressed recombinant human ZP1 and ZP4 as well as native human ZP4, in the presence of pertussis toxin, suggests that induction of acrosome reaction by ZP1/ZP4 is not dependent on Gi protein-coupled receptor activation (Chakravarty et al., 2005; Chiu, Wong, Chung, et al., 2008; Ganguly, Bukovsky, et al., 2010; Schuffner et al., 2002). These observations suggest that there are subtle differences in the signaling events during acrosome reaction mediated by the respective ZP glycoproteins. Inhibitors-based approach suggests that T-type VOCCs are critical for induction of acrosome reaction by human ZP3 (Chiu, Wong, Chung, et al., 2008), whereas, inhibitors of L-type VOCCs do not significantly inhibits the ZP3-mediated induction of acrosome reaction. Baculovirus-expressed recombinant protein corresponding to C-terminal fragment of human ZP3 (aa residues 214–348) also primarily uses T-type VOCCs to induce acrosome reaction (Bansal et al., 2009). However, another independent study using baculovirus-expressed recombinant human ZP3 observes the involvement of both T- and L-type VOCCs in ZP3-mediated increase in $[Ca^{2+}]_i$ and thereby acrosome reaction (Jose et al., 2010). Recombinant human ZP1 and ZP4 (as well as native ZP4) activate both T- and L-type VOCCs for $[Ca^{2+}]_i$ influx and acrosome reaction (Chiu, Wong, Chung, et al., 2008; Ganguly, Bukovsky, et al., 2010). Various signaling events associated with induction of acrosome reaction by ZP4 are summarized in Fig. 6. Extracellular Ca^{2+} is critical for induction of acrosome reaction by ZP1, ZP3, and ZP4 as preincubation of capacitated sperm with EGTA leads to decrease in the acrosome reaction to the levels of spontaneous acrosome reaction (Bansal et al., 2009; Chiu, Wong, Chung, et al., 2008; Ganguly, Bukovsky, et al., 2010).

Human ZP3 and ZP4 also activate TK for the acrosome reaction, but the role of TK in ZP1-mediated acrosome reaction is yet to be elucidated (Chiu, Wong, Chung, et al., 2008). Among the various downstream effector molecules, the role of PKA in downstream signaling mediated by ZP3 is not critical for acrosome reaction (Chiu, Wong, Chung, et al., 2008), suggesting

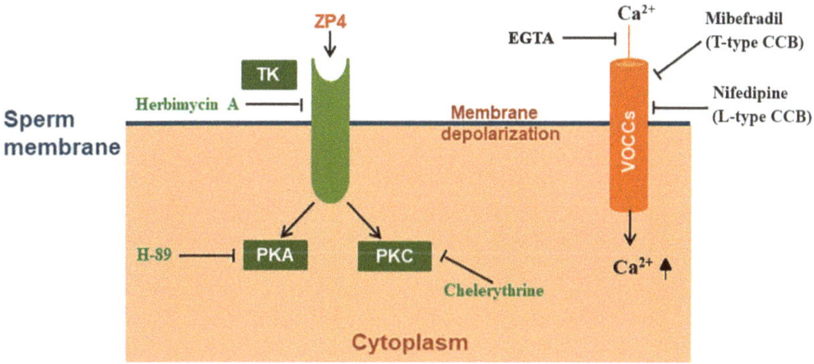

Fig. 6 Schematic representation of the signaling events associated with human ZP4-mediated acrosome reaction. Human ZP4 binds to the tyrosine kinase (TK) receptor as preincubation of the capacitated human sperm with Herbimycin A inhibits ZP4-mediated acrosome reaction. Binding of ZP4 to TK leads to activation of the protein kinase A (PKA) and protein kinase C (PKC). The relevance of PKA and PKC in ZP4-mediated acrosome reaction was confirmed by the use of their respective inhibitors, H-89 and Chelerythrine. Binding of ZP4 to spermatozoon also activates cation channels present on the spermatozoa plasma membrane leading to membrane depolarization and activation of L- and T-type VOCCs. Mibefradil, T-type calcium channel blocker (CCB), as well as Nifedipine, L-type CCB, inhibit ZP4-mediated acrosome reaction, suggesting that both T- and L-type VOCCs are important. Extracellular Ca^{2+} is also important in ZP4-mediated acrosome reaction as addition of chelating agent, EGTA, inhibits induction of acrosome reaction.

its redundancy and perhaps supplementation by parallel signaling pathways. However, activation of PKA is critical for both ZP1- and ZP4-mediated induction of acrosome reaction as its pharmacological inhibitor, H-89, specifically inhibits the ZP1-/ZP4-induced acrosome reaction (Chiu, Wong, Chung, et al., 2008; Ganguly, Bukovsky, et al., 2010; Fig. 6). These studies suggest that the downstream signaling pathways involved in the ZP1- and ZP4-mediated acrosome reaction are similar but are subtly different as compared to those activated by ZP3. By and large, similar downstream signaling events employed by ZP1 and ZP4 to induce acrosome reaction may be due to high aa sequence identity between these two proteins. Activation of PKC by ZP1, ZP3, and ZP4 is critical as pretreatment of capacitated sperm with Chelerythrine-specific inhibitor of PKC leads to a significant reduction of acrosome reaction (Chiu, Wong, Chung, et al., 2008; Ganguly, Bukovsky, et al., 2010). It is likely that the above pathways activated by three ZP glycoproteins converge at a common downstream signaling event such as membrane fusion to induce the acrosome reaction and thus work as a ZP complex during in vivo conditions.

8. RELEVANCE OF ZP MATRIX IN AVOIDANCE OF POLYSPERMY DURING FERTILIZATION

During union between the spermatozoon and the egg leading to successful fertilization, ZP glycoproteins not only play a major role in sperm–egg binding and acrosome reaction in the zona-bound spermatozoa but also eliminate binding of additional sperm subsequent to fertilization as a part of the process to block polyspermy, thereby preventing formation of nonviable polyploid embryos. The entire fertilization process is precise and involves various check points to avoid polyspermy. Our understanding of the mechanisms responsible for avoidance of polyspermy in humans is very limited due to the nonavailability of suitable laboratory models. However, various studies in mouse model have proposed two broad working models: (i) "ZP2-cleavage model" and (ii) "ZP3 glycan-release model." "ZP2-cleavage model" suggests that the cleavage of ZP2 at ^{166}LA DE^{169} by ovastacin, a metalloendoprotease, released following cortical granule exocytosis renders the ZP nonpermissive for gamete recognition. Whereas "ZP3 glycan-release model" suggests the release of *O*-glycans from ZP3 Ser^{332} and Ser^{334} residues by glycosidase subsequent to cortical granule exocytosis leading to the formation of ZP3f and thereby making ZP refractory for sperm binding (Avella, Xiong, & Dean, 2013; Gahlay, Gauthier, Baibakov, Epifano, & Dean, 2010). In humans, block to polyspermy is attributed to "oocyte membrane block," also known as "fast block," which primarily involves depolarization of the oocyte membrane after binding of the first spermatozoa and transiently prevents any subsequent sperm binding to the oocyte. The other possibility is "zona reaction" also known as "slow block," which involves exocytosis of the cortical granules leading to release of various enzymes such as hydrolase, proteinase, and peroxidase, which prevent subsequent binding/penetration of following sperm by modifying the ZP glycoproteins and also by "hardening" of the ZP. However, the mechanistic details of these two processes in humans leading to block in polyspermy are largely unclear. Time-lapse cinematography studies reveal that once a leading spermatozoon penetrates the ZP matrix and attaches to the oocyte membrane, the following sperm are arrested from further penetration into the ZP matrix within 10 s (Mio et al., 2012). "Juno" named after human goddess of fertility and marriage, present on the mouse and human egg membrane, binds with Izumo-1 (named after a Japanese marriage shrine) present on the sperm leading to the successful accomplishment of fertilization (Bianchi, Doe, Goulding, & Wright, 2014). Mating studies of Juno-deficient (Juno$^{-/-}$) female mice with male mice of proven fertility fail to produce any litters. Further, Juno is rapidly shed from the egg membrane after fertilization and is weakly detectable in zona-intact-fertilized mouse eggs at telophase II and undetectable at the pronuclear phase, suggesting that it plays an important role in avoidance of polyspermy at egg membrane (oolemma) level (Bianchi et al., 2014). Modification of the human ZP glycoproteins following fertilization has been reported (Bauskin et al., 1999; Moos, Faundes, Kopf, & Schultz, 1995). Limited proteolysis of ZP2 at the amino-terminal domain to a 60–73 kDa species from 90 to 110 kDa following sperm–egg fusion during fertilization has been documented, which may have relevance in blocking polyspermy (Bauskin et al., 1999).

9. ZP DEFECTS AND FEMALE FACTOR INFERTILITY: CLINICAL SIGNIFICANCE

ZP dysmorphology like dark ZP, large perivitelline space, oval, or irregular shaped ZP has been shown to be associated with reduced implantation and pregnancy rates following IVF as compared to normal oocytes (Sauerbrun-Cutler et al., 2015). However, multiple ZP dysmorphological features do not affect pregnancy rates by intracytoplasmic sperm injection (ICSI; Rienzi, Vajta, & Ubaldi, 2011). Predictive potential of the ZP/oocyte morphology in pregnancy outcome remains to be established. In addition to several studies pertaining to ZP morphology and its implications in fertility, few studies have also been undertaken to investigate the role of mutations in the genes encoding human zona proteins as factor(s) for female infertility. Analysis of *Zp1*, *Zp2*, *Zp3*, and *Zp4* genes in women whose eggs fail to fertilize using IVF as compared to those with successful fertilization following IVF as well as women with proven fertility show 1.5-fold increase in sequence variation in *Zp1* and *Zp3* genes (Männikkö et al., 2005). Two single-base substitutions in the sequence of the *Zp3* gene are found to exist with increased frequency as compared to the control fertile groups. One of these mapped to a conserved motif in the upstream regulatory region (element IIA) of the gene, suggesting that changes in the expression levels of the ZP genes may lead to altered matrix formation with adverse consequences for fertility. Another independent study in infertile women enrolled in the infertility clinic reveals sequence variations in genes encoding *Zp2* and *Zp3* that might be associated with most frequent morphological anomalies observed in the ZP (Pökkylä, Lakkakorpi, Nuojua-Huttunen, &

Tapanainen, 2011). Sequencing of human *Zp1*, *Zp2*, *Zp3*, *Zp4* and regulatory element for the *Zp3* gene from three infertile women with abnormal oocyte ZP appearance reveal eight synonymous and nonsynonymous previously reported polymorphisms only in *Zp1*, *Zp2*, and *Zp3*, suggesting that genetic changes may not be responsible for abnormal oocyte ZP appearance (Margalit et al., 2012). In another study, mutations in the genes encoding four ZP glycoproteins in six members of the family were studied. There was no ZP in any of the eggs collected from the index patient. A homozygous transcript deletion of 8 bp encompassing nucleotides 1169–1176 in *Zp1* in all the six members of the family is observed. It results in the production of truncated form of ZP1 (404 aa residues). Defective ZP1 protein and normal ZP3 protein are colocalized throughout the oocytes and are not expressed on the oocyte cell surface. These observations suggest that aberrant ZP1 prevents the formation of the ZP around oocyte (Huang et al., 2014). However, additional studies are needed to further delineate the role of mutations in the genes encoding human zona proteins leading to either abnormal morphology of the ZP or failure of fertilization during IVF.

10. SIGNIFICANCE OF ZP AUTOANTIBODIES IN WOMEN WITH "UNEXPLAINED INFERTILITY"

Generation of antibodies against ZP proteins has been used as one of the promising approach to develop immunocontraceptive vaccine for wildlife population management (reviewed in Gupta & Minhas, 2017). Do naturally occurring autoantibodies against ZP are one of the possible causes for "unexplained infertility" in women? Initial indirect immunofluorescence studies employing porcine oocytes show the presence of anti-ZP antibodies in 15%–68% infertile women, suggesting their role in inducing infertility (Mori et al., 1978; Nishimoto, Mori, Yamada, & Nishimura, 1980; Shivers & Dunbar, 1977). Subsequently, studies employing human ZP have documented a lower incidence of autoantibodies against ZP in infertile women (Nayudu, Freemann, & Trounson, 1982). The role of ZP autoantibodies in "unexplained infertility" and the success of IVF have been controversial as some studies showed a positive association between the presence of ZP autoantibodies and infertility (Horejsí, Martínek, Nováková, Madar, & Brandejska, 2000; Mori et al., 1978; Nishimoto et al., 1980; Papale et al., 1994; Shivers & Dunbar, 1977; Singh & Mhaskar, 1985) and others fail to find any correlation with presence of ZP autoantibodies and infertility (Nayudu et al., 1982; Sacco & Moghissi, 1979). Further, autoantibodies against ZP have been documented as one of the factors in idiopathic premature ovarian failure (Hovav et al., 1994; Kelkar, Meherji, Kadam, Gupta, & Nandedkar, 2005; Takamizawa, Shibahara, Shibayama, & Suzuki, 2007) and endometriosis (Ulcovà-Gallovà, Bouse, Svàbek, Turek, & Rokyta, 2002). It is difficult to draw any conclusions from the above conflicting findings, and hence there is a need to develop more specific diagnostic tools for documenting ZP autoantibodies for an appropriate clinical surveillance, which will also facilitate in identification of the women that may benefit from immune modulating therapy to regain fertility (Forges, Monnier-Barbarino, Faure, & Béné, 2004).

11. CONCLUSIONS

Biochemical and molecular biology tools have enabled better understanding of the structure of human ZP glycoproteins. Availability of recombinant/purified native human zona proteins has shown that more than one zona protein is involved in binding to the spermatozoa and induction of the acrosome reaction, thereby challenging the simple paradigm established in mouse model that ZP3 is the primary sperm receptor and ZP2 acts as secondary sperm receptor. Associations of abnormal morphology of ZP matrix, aberrations in genes encoding human zona proteins, and autoantibodies against ZP with infertility in women are suggestive but will need more rigorous validation. However, lack of a good experimental animal model is one of the major limitations to understand in a comprehensive manner the functional role of human ZP proteins as present in the ZP matrix of human oocytes rather than in isolation during fertilization as well as block of polyspermy.

ACKNOWLEDGMENTS

S.K.G. would like to acknowledge J.C. Bose National Fellowship (SB/S2/JCB-040/2015) by Science and Engineering Research Board, Department of Science and Technology, Government of India. S.K.G. would also like to acknowledge the help provided by Mrs. Vidisha Minhas in preparation of this chapter.

REFERENCES

Avella, M. A., Baibakov, B., & Dean, J. (2014). A single domain of the ZP2 zona pellucida protein mediates gamete recognition in mice and humans. *Journal of Cell Biology*, *205*, 801–809.

Avella, M. A., Xiong, B., & Dean, J. (2013). The molecular basis of gamete recognition in mice and humans. *Molecular Human Reproduction*, *19*, 279–289.

Baibakov, B., Boggs, N. A., Yauger, B., Baibakov, G., & Dean, J. (2012). Human sperm bind to the N-terminal domain of ZP2 in humanized zonae pellucidae in transgenic mice. *Journal of Cell Biology, 197*, 897–905.

Baldi, E., Luconi, M., Bonaccorsi, L., & Forti, G. (2002). Signal transduction pathways in human spermatozoa. *Journal of Reproductive Immunology, 53*, 121–131.

Bansal, P., Chakrabarti, K., & Gupta, S. K. (2009). Functional activity of human ZP3 primary sperm receptor resides toward its C-terminus. *Biology of Reproduction, 81*, 7–15.

Bastiaan, H., Franken, D., & Wranz, P. (1999). G-protein regulation of the solubilized human zona pellucida-mediated acrosome reaction and zona pellucida binding. *Journal of Assisted Reproduction and Genetics, 16*, 332–336.

Bauskin, A. R., Franken, D. R., Eberspaecher, U., & Donner, P. (1999). Characterization of human zona pellucida glycoproteins. *Molecular Human Reproduction, 5*, 534–540.

Bayne, R. A., Martins da Silva, S. J., & Anderson, R. A. (2004). Increased expression of the FIGα transcription factor is associated with primordial follicle formation in the human fetal ovary. *Molecular Human Reproduction, 10*, 373–381.

Bercegeay, S., Jean, M., Lucas, H., & Barriere, P. (1995). Composition of human zona pellucida as revealed by SDS-PAGE after silver staining. *Molecular Reproduction and Development, 41*, 355–359.

Bertrand, E., Van den Bergh, M., & Englert, Y. (1995). Does zona pellucida thickness influence the fertilization rate? *Human Reproduction, 10*, 1189–1193.

Bhandari, B., Bansal, P., Talwar, P., & Gupta, S. K. (2010). Delineation of downstream signalling components during acrosome reaction mediated by heat solubilized human zona pellucida. *Reproductive Biology and Endocrinology, 8*, 7.

Bianchi, E., Doe, B., Goulding, D., & Wright, G. J. (2014). Juno is the egg Izumo receptor and is essential for mammalian fertilization. *Nature, 508*, 483–487.

Bielfeld, P., Faridi, A., Zaneveld, L. J., & De Jonge, C. J. (1994). The zona pellucida-induced acrosome reaction of human spermatozoa is mediated by protein kinases. *Fertility and Sterility, 61*, 536–541.

Bork, P. (1993). A trefoil domain in the major rabbit zona pellucida protein. *Protein Science, 2*, 669–670.

Bork, P., & Sander, C. (1992). A large domain common to sperm receptors (Zp2 and Zp3) and TGF-beta type III receptor. *FEBS Letters, 300*, 237–240.

Bray, C., Son, J. H., Kumar, P., Harris, J. D., & Meizel, S. (2002). A role for the human sperm glycine receptor/Cl^- channel in the acrosome reaction initiated by recombinant ZP3. *Biology of Reproduction, 66*, 91–97.

Breitbart, H. (2002). Intracellular calcium regulation in sperm capacitation and acrosomal reaction. *Molecular and Cellular Endocrinology, 187*, 139–144.

Bukovsky, A., Gupta, S. K., Bansal, P., Chakravarty, S., Chaudhary, M., Svetlikova, M., et al. (2008). Production of monoclonal antibodies against recombinant human zona pellucida glycoproteins: Utility in immunolocalization of respective zona proteins in ovarian follicles. *Journal of Reproductive Immunology, 78*, 102–114.

Caballero-Campo, P., Chirinos, M., Fan, X. J., González-González, M. E., Galicia-Chavarria, M., Larrea, F., et al. (2006). Biological effects of recombinant human zona pellucida proteins on sperm function. *Biology of Reproduction, 74*, 760–768.

Chakravarty, S., Kadunganattil, S., Bansal, P., Sharma, R. K., & Gupta, S. K. (2008). Relevance of glycosylation of human zona pellucida glycoproteins for their binding to capacitated human spermatozoa and subsequent induction of acrosomal exocytosis. *Molecular Reproduction and Development, 75*, 75–88.

Chakravarty, S., Suraj, K., & Gupta, S. K. (2005). Baculovirus-expressed recombinant human zona pellucida glycoprotein-B induces acrosomal exocytosis in capacitated spermatozoa in addition to zona pellucida glycoprotein-C. *Molecular Human Reproduction, 11*, 365–372.

Chamberlin, M. E., & Dean, J. (1990). Human homolog of the mouse sperm receptor. *Proceedings of the National Academy of Sciences of the United States of America*, *87*, 6014–6018.

Chapman, N., Kessopoulou, E., Andrews, P., Hornby, D., & Barratt, C. R. (1998). The polypeptide backbone of recombinant human zona pellucida glycoprotein-3 initiates acrosomal exocytosis in human spermatozoa in vitro. *The Biochemical Journal*, *330*, 839–845.

Chirinos, M., Cariño, C., González-González, M. E., Arreola, E., Reveles, R., & Larrea, F. (2011). Characterization of human sperm binding to homologous recombinant zona pellucida proteins. *Reproductive Sciences*, *18*, 876–885.

Chiu, P. C., Wong, B. S., Chung, M. K., Lam, K. K., Pang, R. T., Lee, K. F., et al. (2008). Effects of native human zona pellucida glycoproteins-3 and -4 on acrosome reaction and zona pellucida binding of human spermatozoa. *Biology of Reproduction*, *79*, 869–877.

Chiu, P. C., Wong, B. S., Lee, C. L., Pang, R. T., Lee, K. F., Sumitro, S. B., et al. (2008). Native human zona pellucida glycoproteins: Purification and binding properties. *Human Reproduction*, *23*, 1385–1393.

Cross, N. L., Morales, P., Overstreet, J. W., & Hanson, F. W. (1988). Induction of acrosome reaction by the human zona pellucida. *Biology of Reproduction*, *38*, 235–244.

Doherty, C. M., Tarchala, S. M., Radwanska, E., & De Jonge, C. J. (1995). Characterization of two second messenger pathways and their interactions in eliciting the human sperm acrosome reaction. *Journal of Andrology*, *16*, 36–46.

Dong, K. W., Chi, T. F., Juan, Y. W., Chen, C. W., Lin, Z., Xiang, X. Q., et al. (2001). Characterization of the biologic activities of a recombinant human zona pellucida protein 3 expressed in human ovarian teratocarcinoma (PA-1) cells. *American Journal of Obstetrics and Gynecology*, *184*, 835–844.

Eberspaecher, U., Becker, A., Bringmann, P., van der Merwe, L., & Donner, P. (2001). Immunohistochemical localization of zona pellucida proteins ZPA, ZPB and ZPC in human, cynomolgus monkey and mouse ovaries. *Cell & Tissue Research*, *303*, 277–287.

Familiari, G., Nottola, S. A., Macchiarelli, G., Micara, G., Aragona, C., & Motta, P. M. (1992). Human zona pellucida during in vitro fertilization: An ultrastructural study using saponin, ruthenium red, and osmium-thiocarbohydrazide. *Molecular Reproduction and Development*, *32*, 51–61.

Familiari, G., Nottola, S. A., Micara, G., Aragona, C., & Motta, P. M. (1988). Is the sperm-binding capability of the zona pellucida linked to its surface structure? A scanning electron microscopic study of human in vitro fertilization. *Journal of in vitro Fertilization and Embryo Transfer*, *5*, 134–143.

Familiari, G., Relucenti, M., Heyn, R., Micara, G., & Correr, S. (2006). Three-dimensional structure of the zona pellucida at ovulation. *Microscopic Research and Technique*, *69*, 415–426.

Fisher, H. M., Brewis, I. A., Barratt, C. L., Cooke, I. D., & Moore, H. D. (1998). Phosphoinositide 3-kinase is involved in the induction of the human sperm acrosome reaction downstream of tyrosine phosphorylation. *Molecular Human Reproduction*, *4*, 849–855.

Forges, T., Monnier-Barbarino, P., Faure, G. C., & Béné, M. C. (2004). Autoimmunity and antigenic targets in ovarian pathology. *Human Reproduction Update*, *10*, 163–175.

Franken, D. R., Bastiaan, H. S., & Oehninger, S. C. (2000). Physiological induction of the acrosome reaction in human sperm: Validation of a microassay using minimal volumes of solubilized homologous zona pellucida. *Journal of Assisted Reproduction and Genetics*, *17*, 374–378.

Gahlay, G., Gauthier, L., Baibakov, B., Epifano, O., & Dean, J. (2010). Gamete recognition in mice depends on the cleavage status of an egg's zona pellucida protein. *Science*, *329*, 216–219.

Ganguly, A., Bansal, P., Gupta, T., & Gupta, S. K. (2010). "ZP domain" of human zona pellucida glycoprotein-1 binds to human spermatozoa and induces acrosomal exocytosis. *Reproductive Biology and Endocrinology*, *8*, 110.

Ganguly, A., Bukovsky, A., Sharma, R. K., Bansal, P., Bhandari, B., & Gupta, S. K. (2010). In humans, zona pellucida glycoprotein-1 binds to spermatozoa and induces acrosomal exocytosis. *Human Reproduction*, *25*, 1643–1656.

Gook, D. A., Edgar, D. H., Borg, J., & Martic, M. (2008). Detection of zona pellucida proteins during human folliculogenesis. *Human Reproduction*, *23*, 394–402.

Grootenhuis, A. J., Philipsen, H. L., de Breet-Grijsbach, J. T., & van Duin, M. (1996). Immunocytochemical localization of ZP3 in primordial follicles of rabbit, marmoset, rhesus monkey and human ovaries using antibodies against human ZP3. *Journal of Reproduction and Fertility. Supplementary*, *50*, 43–54.

Gupta, S. K., & Minhas, V. (2017). Wildlife population management: Are contraceptive vaccines a feasible proposition. *Frontiers in Bioscience (Scholar Edition)*, *9*, 357–374.

Gupta, S. K., Yurewicz, E. C., Sacco, A. G., Kaul, R., Jethanandani, P., & Govind, C. K. (1998). Human zona pellucida glycoproteins: Characterization using antibodies against recombinant non human primate ZP1, ZP2 and ZP3. *Molecular Human Reproduction*, (11), 1058–1064.

Harris, J. D., Hibler, D. W., Fontenot, G. K., Hsu, K. T., Yurewicz, E. C., & Sacco, A. G. (1994). Cloning and characterization of zona pellucida genes and cDNAs from a variety of mammalian species: The ZPA, ZPB and ZPC gene families. *DNA Sequence*, *4*, 361–393.

Horejsí, J., Martínek, J., Nováková, D., Madar, J., & Brandejska, M. (2000). Autoimmune antiovarian antibodies and their impact on the success of an IVF/ET program. *Annals of the New York Academy of Sciences*, *900*, 351–356.

Hovav, Y., Almagor, M., Benbenishti, D., Margalioth, E. J., Kafka, I., & Yaffe, H. (1994). Immunity to zona pellucida in women with low response to ovarian stimulation, in unexplained infertility and after multiple IVF attempts. *Human Reproduction*, *9*, 643–645.

Huang, H. L., Lv, C., Zhao, Y. C., Li, W., He, X. M., Li, P., et al. (2014). Mutant ZP1 in familial infertility. *New England Journal of Medicine*, *370*, 1220–1226.

Hughes, D. C., & Barrat, C. L. (1999). Identification of the true human orthologue of the mouse Zp1 gene: Evidence for greater complexity in the mammalian zona pellucida? *Biochemistry and Biophysics Acta*, *1447*, 303–306.

Huntriss, J., Gosden, R., Hinkins, M., Oliver, B., Miller, D., Rutherford, A. J., et al. (2002). Isolation, characterization and expression of the human factor in the germline alpha (FIGα) gene in ovarian follicles and oocytes. *Molecular Human Reproduction*, *8*, 1087–1095.

Jiménez-Movilla, M., Avilés, M., Gómez-Torres, M. J., Fernández-Colom, P. J., Castells, M. T., de Juan, J., et al. (2004). Carbohydrate analysis of the zona pellucida and cortical granules of human oocytes by means of ultrastructural cytochemistry. *Human Reproduction*, *19*, 1842–1855.

Jimenez-Movilla, M., & Dean, J. (2011). ZP2 and ZP3 cytoplasmic tails prevent premature interactions and ensure incorporation into the zona pellucida. *Journal of Cell Science*, *124*, 940–950.

Jose, O., Hernandez-Hernandez, O., Chirinos, M., González-González, M. E., Larrea, F., Almanza, A., et al. (2010). Recombinant human ZP3-induced sperm acrosome reaction: Evidence for the involvement of T- and L-type voltage-gated calcium channels. *Biochemical and Biophysical Research Communications*, *395*, 530–534.

Jovine, L., Darie, C. C., Litscher, E. S., & Wassarman, P. M. (2005). Zona pellucida domain proteins. *Annual Review of Biochemistry*, *74*, 83–114.

Jovine, L., Janssen, W. G., Litscher, E. S., & Wassarman, P. M. (2006). The PLAC1-homology region of the ZP domain is sufficient for protein polymerization. *BMC Biochemistry*, *7*, 11.

Jovine, L., Qi, H., Williams, Z., Litscher, E. S., & Wassarman, P. M. (2002). The ZP domain is a conserved module for polymerization of extracellular proteins. *Nature Cell Biology*, *4*, 457–461.

Kelkar, R. L., Meherji, P. K., Kadam, S. S., Gupta, S. K., & Nandedkar, T. D. (2005). Circulating auto-antibodies against the zona pellucida and thyroid microsomal antigen in women with premature ovarian failure. *Journal of Reproductive Immunology, 66*, 53–67.

Kiefer, S. M., & Saling, P. (2002). Proteolytic processing of human zona pellucida proteins. *Biology of Reproduction, 66*, 407–414.

Kipersztok, S., Osawa, G. A., Liang, L. F., Modi, W. S., & Dean, J. (1995). POMZP3, a bipartite transcript derived from human ZP3 and a POM121 homologue. *Genomics, 25*, 354–359.

Lee, M. A., Check, J. H., & Kopf, G. S. (1992). A guanine nucleotide-binding regulatory protein in human sperm mediates acrosomal exocytosis induced by the human zona pellucida. *Molecular Reproduction and Development, 31*, 78–86.

Lefievre, L., Conner, S. J., Salpekar, A., Olufowobi, O., Ashton, P., Pavlovic, B., et al. (2004). Four zona pellucida glycoproteins are expressed in the human. *Human Reproduction, 19*, 1580–1586.

Liang, L. F., & Dean, J. (1993). Conservation of mammalian secondary sperm receptor genes enables the promoter of the human gene to function in mouse oocytes. *Developmental Biology, 156*, 399–408.

Liguori, L., de Lamirande, E., Minelli, A., & Gagnon, C. (2005). Various protein kinases regulate human sperm acrosome reaction and the associated phosphorylation of Tyr residues and of the Thr-Glu-Tyr motif. *Molecular Human Reproduction, 11*, 211–221.

Litscher, E. S., Qi, H., & Wassarman, P. M. (1999). Mouse zona pellucida glycoproteins mZP2 and mZP3 undergo carboxy-terminal proteolytic processing in growing oocytes. *Biochemistry, 38*, 12280–12287.

Liu, D. Y., & Baker, H. W. (1997). Protein kinase C plays an important role in the human zona pellucida-induced acrosome reaction. *Molecular Human Reproduction, 3*, 1037–1043.

Louros, N. N., Chrysina, E. D., Baltatzis, G. E., Patsouris, E. S., Hamodrakas, S. J., & Iconomidou, V. A. (2016). A common 'aggregation-prone' interface possibly participates in the self-assembly of human zona pellucida proteins. *FEBS Letters, 590*, 619–630.

Louros, N. N., Iconomidou, V. A., Giannelou, P., & Hamodrakas, S. J. (2013). Structural analysis of peptide-analogues of human zona pellucida ZP1 protein with amyloidogenic properties: Insights into mammalian zona pellucida formation. *PLoS One, 8*(9), e73258.

Magerkurth, C., Töpfer-Peterson, E., Schwartz, P., & Michelmann, H. W. (1999). Scanning electron microscopy analysis of the human zona pellucida: Influence of maturity and fertilization on morphology and sperm binding pattern. *Human Reproduction, 14*, 1057–1066.

Männikkö, M., Törmälä, R.-M., Tuuri, T., Haltia, A., Martikainen, H., Ala-Kokko, L., et al. (2005). Association between sequence variations in genes encoding human zona pellucida glycoproteins and fertilization failure in IVF. *Human Reproduction, 20*, 1578–1585.

Margalit, M., Paz, G., Yavetz, H., Yogev, L., Amit, A., Hevlin-Schwartz, T., et al. (2012). Genetic and physiological study of morphologically abnormal human zona pellucida. *European Journal of Obstetrics, Gynecology, and Reproductive Biology, 165*, 70–76.

Maymon, B. B., Maymon, R., Ben-Nun, I., Ghetler, Y., Shalgi, R., & Skutelsky, E. (1994). Distribution of carbohydrates in the zona pellucida of human oocytes. *Journal of Reproduction and Fertility, 102*, 81–86.

Millar, S. E., Lader, E. S., & Dean, J. (1993). ZAP-1 DNA binding activity is first detected at the onset of zona pellucida gene expression in embryonic mouse oocytes. *Developmental Biology, 158*, 410–413.

Millar, S. E., Lader, E., Liang, L. F., & Dean, J. (1991). Oocyte-specific factors bind a conserved upstream sequence required for mouse zona pellucida promoter activity. *Molecular and Cellular Biology, 11*, 6197–6204.

Mio, Y., Iwata, K., Yumoto, K., Kai, Y., Sargant, H. C., Mizoguchi, C., et al. (2012). Possible mechanism of polyspermy block in human oocytes observed by time-lapse cinematography. *Journal of Assisted Reproduction and Genetics, 29*, 951–956.

Miranda, P. V., Gonzalez-Echeverria, F., Marin-Briggiler, C. I., Brandelli, A., Blaquier, J. A., & Tezon, J. G. (1997). Glycosidic residues involved in human sperm-zona pellucida binding in vitro. *Molecular Human Reproduction, 3*, 399–404.

Moos, J., Faundes, D., Kopf, G. S., & Schultz, R. M. (1995). Composition of the human zona pellucida and modifications following fertilization. *Human Reproduction, 10*, 2467–2471.

Mori, K., Daitoh, T., Irahara, M., Kamada, M., & Aono, T. (1989). Significance of D-mannose as a sperm receptor site on the zona pellucida in human fertilization. *American Journal of Obstetrics and Gynecology, 161*, 207–211.

Mori, T., Nishimoto, T., Kitagawa, M., Noda, Y., Nishimura, T., & Oikawa, T. (1978). Possible presence of autoantibodies to zona pellucida in infertile women. *Experientia, 34*, 797–799.

Nayudu, P. L., Freemann, I. E., & Trounson, A. O. (1982). Zona pellucida antibodies in human sera. *Journal of Reproduction and Fertility, 65*, 77–84.

Nishimoto, T., Mori, T., Yamada, I., & Nishimura, T. (1980). Autoantibodies to zona pellucida in infertile and aged women. *Fertility and Sterility, 34*, 552–556.

Nottola, S. A., Makabe, S., Stallone, T., Familiari, G., Correr, S., & Macchiarelli, G. (2005). Surface morphology of the zona pellucida surrounding human blastocysts obtained after in vitro fertilization. *Archieves of Histology and Cytology, 68*, 133–141.

Oehninger, S., Patankar, M., Seppala, M., & Clark, G. F. (1998). Involvement of selectin-like carbohydrate binding specificity in human gamete interaction. *Andrologia, 30*, 269–274.

Ozgur, K., Patankar, M. S., Oehninger, S., & Clark, G. F. (1998). Direct evidence for the involvement of carbohydrate sequences in human sperm-zona pellucida binding. *Molecular Human Reproduction, 4*, 318–324.

Pang, P. C., Chiu, P. C., Lee, C. L., Chang, L. Y., Panico, M., Morris, H. R., et al. (2011). Human sperm binding is mediated by the sialyl-Lewis(x) oligosaccharide on the zona pellucida. *Science, 333*, 1761–1764.

Papale, M. L., Grillo, A., Leonardi, E., Giuffrida, M., Palumbo, M., & Palumbo, G. (1994). Assessment of the relevance of zona pellucida antibodies in follicular fluid of in-vitro fertilization (IVF) patients. *Human Reproduction, 9*, 1827–1831.

Patrat, C., Auer, J., Fauque, P., Leandri, R. L., Jouannet, P., & Serres, C. (2006). Zona pellucida from fertilized human oocytes induces a voltage-dependent calcium influx and the acrosome reaction in spermatozoa, but cannot be penetrated by sperm. *BMC Developmental Biology, 6*, 59.

Patrat, C., Serres, C., & Jouannet, P. (2000). The acrosome reaction in human spermatozoa. *Biology of the Cell, 92*, 255–266.

Pökkylä, R. M., Lakkakorpi, J. T., Nuojua-Huttunen, S. H., & Tapanainen, J. S. (2011). Sequence variations in human ZP genes as potential modifiers of zona pellucida architecture. *Fertility and Sterility, 95*, 2669–2672.

Rienzi, L., Vajta, G., & Ubaldi, F. (2011). Predictive value of oocyte morphology in human IVF: A systematic review of the literature. *Human Reproduction Update, 17*, 34–45.

Roch, F., Alonso, C. R., & Akam, M. (2003). Drosophila *miniature* and *dusky* encode ZP proteins required for cytoskeletal reorganisation during wing morphogenesis. *Journal of Cell Science, 116*, 1199–1207.

Sacco, A. G., & Moghissi, K. S. (1979). Anti-zona pellucida activity in human sera. *Fertility and Sterility, 31*, 503–506.

Sauerbrun-Cutler, M.-T., Vega, M., Breborowicz, A., Gonzales, E., Stein, D., Lederman, M., et al. (2015). Oocyte zona pellucida dysmorphology is associated with diminished in-vitro fertilization success. *Journal of Ovarian Research, 8*, 5.

Schuffner, A. A., Bastiaan, H. S., Duran, H. E., Lin, Z. Y., Morshedi, M., Franken, D. R., et al. (2002). Zona pellucida-induced acrosome reaction in human sperm: Dependency on activation of pertussis toxin-sensitive G(i) protein and extracellular calcium, and priming effect of progesterone and follicular fluid. *Molecular Human Reproduction, 8*, 722–727.

Shabanowitz, R. B., & O'Rand, M. G. (1988). Characterization of the human zona pellucida from fertilized and unfertilized eggs. *Journal of Reproduction and Fertility, 82*, 151–161.

Shivers, C. A., & Dunbar, B. S. (1977). Autoantibodies to zona pellucida: A possible cause for infertility in women. *Science, 197*, 1082–1084.

Singh, J., & Mhaskar, A. M. (1985). Enzyme-linked immunosorbent determination of autoantibodies to zona pellucida as a possible cause of infertility in women. *Journal of Immunological Methods, 79*, 133–141.

Swanson, W. J., Nielsen, R., & Yang, Q. (2003). Pervasive adaptive evolution in mammalian fertilization proteins. *Molecular Biology and Evolution, 20*, 18–20.

Takamizawa, S., Shibahara, H., Shibayama, T., & Suzuki, M. (2007). Detection of antizona pellucida antibodies in the sera from premature ovarian failure patients by a highly specific test. *Fertility and Sterility, 88*, 925–932.

Thim, L. (1989). A new family of growth factor-like peptides. "Trefoil" disulphide loop structures as a common feature in breast cancer associated peptide (pS2), pancreatic spasmolytic polypeptide (PSP), and frog skin peptides (spasmolysins). *FEBS Letters, 250*, 85–90.

Törmälä, R.-M., Jääskeläinen, M., Lakkakorpi, J., Liakka, A., Tapanainen, J. S., & Vaskivuo, T. E. (2008). Zona pellucida components are present in human fetal ovary before follicle formation. *Molecular and Cellular Endocrinology, 289*, 10–15.

Tsubamoto, H., Hasegawa, A., Nakata, Y., Naito, S., Yamasaki, N., & Koyama, K. (1999). Expression of recombinant human zona pellucida protein 2 and its binding capacity to spermatozoa. *Biology of Reproduction, 61*, 1649–1654.

Ulcovà-Gallovà, Z., Bouse, V., Svàbek, I., Turek, J., & Rokyta, Z. (2002). Endometriosis in reproductive immunology. *American Journal of Reproductive Immunology, 47*, 269–274.

van Duin, M., Polman, J. E., de Breet, I. T., van Ginneken, K., Bunschoten, H., Grootenhuis, A., et al. (1994). Recombinant human zona pellucida protein ZP3 produced by Chinese hamster ovary cells induces the human sperm acrosome reaction and promotes sperm-egg fusion. *Biology of Reproduction, 51*, 607–617.

van Duin, M., Polman, J. E., Verkoelen, C. C., Bunschoten, H., Meyerink, J. H., Olijve, W., et al. (1992). Cloning and characterization of the human sperm receptor ligand ZP3: Evidence for a second polymorphic allele with a different frequency in the Caucasian and Japanese populations. *Genomics, 14*, 1064–1070.

Walensky, L. D., & Snyder, S. H. (1995). Inositol 1,4,5-triphosphate receptors selectively localized to the acrosome of mammalian sperm. *The Journal of Cell Biology, 130*, 857–869.

Wassarman, P. M. (2008). Zona pellucida glycoproteins. *Journal of Biological Chemistry, 283*, 24285–24289.

Xu, W. X., Bhandari, B., He, Y. P., Tang, H. P., Chaudhary, S., Talwar, P., et al. (2012). Mapping of epitopes relevant for induction of acrosome reaction on human zona pellucida glycoprotein-4 using monoclonal antibodies. *American Journal of Reproductive Immunology, 68*, 465–475.

Yauger, B., Boggs, N. A., & Dean, J. (2011). Human ZP4 is not sufficient for taxon-specific sperm recognition of the zona pellucida in transgenic mice. *Reproduction, 141*, 313–319.

Zhao, M., Boja, E. S., Hoodbhoy, T., Nawrocki, J., Kaufman, J. B., Kresge, N., et al. (2004). Mass spectrometry analysis of recombinant human ZP3 expressed in glycosylation-deficient CHO cells. *Biochemistry, 43*, 12090–12104.

CHAPTER THIRTEEN

Structure of Zona Pellucida Module Proteins

Marcel Bokhove, Luca Jovine[1]
Department of Biosciences and Nutrition & Center for Innovative Medicine, Karolinska Institutet, Huddinge, Sweden
[1]Corresponding author: e-mail address: luca.jovine@ki.se

Contents

Abstract

The egg coat, an extracellular matrix made up of glycoprotein filaments, plays a key role in animal fertilization by acting as a gatekeeper for sperm. Egg coat components polymerize using a common zona pellucida (ZP) "domain" module that consists of two related immunoglobulin-like domains, called ZP-N and ZP-C. The ZP module has also been recognized in a large number of other secreted proteins with different biological functions, whose mutations are linked to severe human diseases. During the last decade, tremendous progress has been made toward understanding the atomic architecture of the ZP module and the structural basis of its polymerization. Moreover, sperm-binding regions at the N-terminus of mollusk and mammalian egg coat subunits were found to consist of domain repeats that also adopt a ZP-N fold. This discovery revealed an unexpected link between invertebrate and vertebrate fertilization and led to the first structure of an egg coat–sperm protein recognition complex. In this review we summarize these exciting findings, discuss their functional implications, and outline future challenges that must be addressed in order to develop a comprehensive view of this family of biomedically important extracellular molecules.

Current Topics in Developmental Biology, Volume 130
ISSN 0070-2153
https://doi.org/10.1016/bs.ctdb.2018.02.007

ABBREVIATIONS

BG betaglycan
CCS consensus cleavage site
EGF epidermal growth factor
EHP external hydrophobic patch
ENG endoglin/CD105
Ig immunoglobulin
IHP internal hydrophobic patch
TGF transforming growth factor
UMOD uromodulin/Tamm–Horsfall protein
VE vitelline envelope
VERL vitelline envelope receptor for lysin
ZP zona pellucida

1. INTRODUCTION: THE ZP "DOMAIN" MODULE

The egg coat, called zona pellucida (ZP) in mammals and vitelline envelope (VE) in nonmammals, is a specialized extracellular matrix that provides the growing oocyte with rigidity and protection from external factors. At fertilization, it constitutes a species-restricted barrier for sperm, which needs to penetrate it in order to fuse with the plasma membrane of the oocyte. After gamete fusion, modification of the ZP/VE contributes to the postfertilization block to polyspermy. Finally, the resulting hardened structure protects the developing embryo until hatching and, in mammals, implantation (Wassarman & Litscher, 2016).

The egg coat matrix consists of an intertwined three-dimensional meshwork of filaments formed by secreted glycoprotein components whose number varies from 3–4 (mammalian ZP) (Bleil & Wassarman, 1980; Lefièvre et al., 2004) to >30 (mollusk VE) (Aagaard, Vacquier, MacCoss, & Swanson, 2010). A common element among VE/ZP subunits is the presence of a C-terminal region of approximately 260 amino acids, including eight strictly conserved cysteines, which was also recognized in other extracellular proteins and was originally called ZP domain (Fig. 1A) (Bork & Sander, 1992). During the following 25 years, the number of proteins containing this element—which we will henceforth refer to as ZP module for the reasons explained later—has significantly expanded, spanning the evolutionary tree of multicellular eukaryotes from Cnidarians to human. Notably, all these molecules share an N-terminal signal peptide that directs their precursors to the secretory pathway and, in most cases, a relatively large

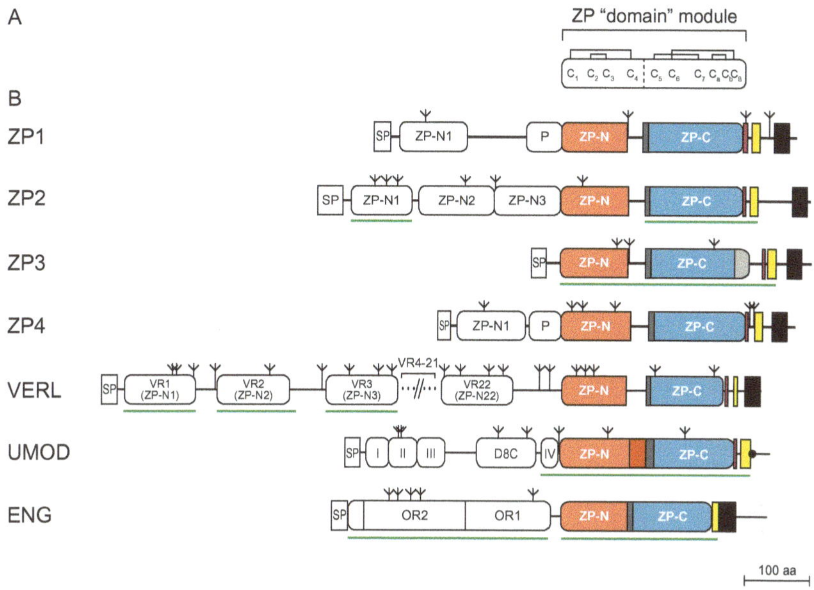

Fig. 1 Architecture of a representative set of ZP module proteins. (A) Typical disulfide connectivity of the N-terminal and C-terminal regions of the ZP module. C_a and C_b are missing in ZP3, whereas ENG also lacks C_8. For clarity, the relative spacing between cysteines is not drawn to scale. (B) ZP module protein features. *SP*, signal peptide; *ZP-N*, ZP-N domain (*salmon*); *ZP-C*, ZP-C domain (*blue*); *ZP-NX*, N-terminal isolated ZP-Ns; *P*, trefoil domain; *roman numerals*, EGF-like domains; *D8C*, domain with eight conserved cysteines; *ORX*, orphan domains. ZP module motifs are indicated by *colored bars*: *red*, structured ZP-N/ZP-C interdomain linker (as opposed to flexible linkers, depicted as *black lines*); *dark gray*, IHP; *light gray*, ZP3-specific subdomain; *magenta*, CCS; *yellow*, EHP; *black*, transmembrane helix. The *black dot* represents the GPI anchor of UMOD; N-glycans are depicted by inverted tripods. Protein regions resolved to date by X-ray crystallography are indicated by *horizontals green lines* below the architecture schemes.

number of glycosylation sites as well as a C-terminal membrane anchoring element. The latter, which can either be a single-spanning transmembrane helix or a glycosylphosphatidylinositol (GPI) anchor, is separated from the end of the ZP module by a consensus protease cleavage site (CCS; often also referred to as CFCS because it matches the recognition site for furin in several members of the family) (Fig. 1B) (Jovine, Darie, Litscher, & Wassarman, 2005; Litscher & Wassarman, 2015). In addition to egg ZP subunits, several human ZP module proteins are of particular biomedical interest. These include homopolymeric uromodulin (UMOD)/Tamm–Horsfall protein, the most abundant protein in human urine, and glycoprotein 2, the major membrane protein in the zymogen granules of the exocrine pancreas;

homo/heteropolymeric inner ear tectorial membrane components α- and β-tectorins; and nonpolymeric transforming growth factor (TGF)-β superfamily coreceptors endoglin (ENG)/CD105 and betaglycan (BG) (Jovine et al., 2005). Although these molecules are not involved in reproduction, their study has significantly contributed to our understanding of egg coat biology by yielding valuable information into the structure and biological role of ZP module proteins in general.

What is the function of the ZP module? Experiments in oocytes showed that it is responsible for mediating the incorporation of mammalian ZP2 and ZP3 and into ZP filaments (Jovine, Qi, Williams, Litscher, & Wassarman, 2002). This agrees with the observation that most ZP module proteins are found in extracellular matrices (Jovine et al., 2005) and was confirmed by parallel biochemical analyses of native UMOD (Jovine et al., 2002), as well as a subsequent report on chicken ZP1 (Sasanami et al., 2006). Further studies revealed that polymerization of ZP3 and UMOD is controlled by two conserved hydrophobic motifs, the so-called internal- and external hydrophobic patches (IHP/EHP), which flank the C-terminal half of the ZP module (Jovine, Qi, Williams, Litscher, & Wassarman, 2004; Schaeffer, Santambrogio, Perucca, Casari, & Rampoldi, 2009). Although the details of the polymerization mechanism remain to be established, the current model is that the ZP module is activated when the CCS that separates it from the EHP is cleaved by a specific protease at either the *trans*-Golgi or plasma membrane level. By ultimately resulting in dissociation of the EHP, the action of this enzyme—recently identified as serine protease hepsin in the case of UMOD (Brunati et al., 2015)—both releases the mature form of the ZP module from the plasma membrane and concomitantly exposes its IHP, triggering protein incorporation into growing polymers (Jovine et al., 2004). Consistent with their highly mosaic architecture, the diverse biological functions of ZP module proteins are thought to derive from the combination of their common filament scaffold to different types and numbers of additional domains. For example, regions N- and C-terminal to the ZP modules of ZP2 and ZP3 have been, respectively, implicated in the interaction with sperm (Avella, Baibakov, & Dean, 2014; Bleil, Greve, & Wassarman, 1988; Chen, Litscher, & Wassarman, 1998; Williams, Litscher, Jovine, & Wassarman, 2006), whereas the D8C domain preceding the ZP module of UMOD carries a high-mannose glycan that binds uropathogenic bacteria (Cavallone, Malagolini, Monti, Wu, & Serafini-Cessi, 2004).

In addition to the aforementioned hydrophobic patch duplication (Jovine et al., 2004), several independent observations raised the possibility that the ZP "domain" element actually consisted of two distinct moieties. These included protease-sensitive sites located between the first half of the element and the IHP, suggesting the presence of an interdomain linker that would also coincide with a conserved intron/exon boundary at the DNA level (Jovine et al., 2004); bipartite disulfide bond clusters derived from mass spectrometric analyses (Boja, Hoodbhoy, Fales, & Dean, 2003; Darie, Biniossek, Jovine, Litscher, & Wassarman, 2004; Kanai et al., 2008; Yonezawa & Nakano, 2003) (Fig. 1A); and the identification of a set of proteins that contains only the N-terminal half of the element (Cocchia et al., 2000; Jovine, Janssen, Litscher, & Wassarman, 2006; Yan et al., 2001). Although the C-terminal half is only found in the context of a full element (Jovine et al., 2006), these considerations collectively suggested that there is no true ZP "domain," but rather a ZP module consisting of two separate domains denominated ZP-N and ZP-C (Jovine et al., 2004, 2005, 2006). As summarized in this review, structural biology has played a major role in conclusively addressing this question, as well as bringing many additional insights into the function and assembly of ZP module proteins.

2. STRUCTURES OF THE ZP-N DOMAIN

Although secondary structure predictions suggested a predominance of β-strands, no significant tertiary structure match for the ZP-N moiety of the ZP module could be obtained using different fold recognition algorithms (Callebaut, Mornon, & Monget, 2007). At the same time, experimental investigations of the 3D structure of ZP module proteins were long hindered by their highly complex posttranslational modifications, such as intra- and intermolecular disulfide bonds as well as N- and O-linked glycosylation. Due to advances in recombinant protein expression technology, and in particular the use of fusion proteins and specialized cell lines, several structures of the ZP-N domain have been determined by X-ray crystallography during the course of the last 10 years (Table 1; Bokhove, Sadat Al Hosseini, et al., 2016; Han et al., 2010; Monné et al., 2008; Raj et al., 2017). The first to be reported was that of the ZP-N domain of murine ZP3, which was recombinantly expressed as a maltose-binding protein fusion using a highly engineered strain of *Escherichia coli* (Monné et al., 2008).

2.1 First Structure of a ZP-N Domain: Murine ZP3

The structure of the N-terminal half of mouse ZP3, determined in three different crystal forms (Table 1), showed that this part of the protein indeed

Table 1 Crystal Structures of Egg Coat Subunits and Other ZP Module Proteins

Protein(s)	Domain(s)	Species (UniProt)	Residues (Construct/Resolved)	N-/O-Glycosylation Sites	Space Group	Resolution (Å)	PDB	References
ZP3	ZP-N	*M. musculus* (P10761)	42–143 (102/102)	—	$I\,2\,2\,2$	2.90	3D4C	Monné, Han, Schwend, Burendahl, and Jovine (2008)
ZP3	ZP-N	*M. musculus* (P10761)	42–143 (102/102)	—	$P\,1$	2.30	3D4G[a]	Monné et al. (2008)
ZP3	ZP-N	*M. musculus* (P10761)	42–143 (102/102)	—	$P\,2_1\,2_1\,2$	3.10	3EF7	Monné et al. (2008)
ZP3	ZP-N, ZP-C +subdomain	*G. gallus* (P79762)	53–347, 359–382 (319/297)	1[b] (N), 1 (O)	$P\,4_1\,2_1\,2$	2.60, 2.00	3NK3, 3NK4	Han et al. (2010)
ZP2	ZP-N1	*M. musculus* (P20239)	35–138 (104/92)	1[b] (N)	$P\,2_1\,2_1\,2_1$	0.95	5II6	Raj et al. (2017)
ZP2	ZP-C	*M. musculus* (P20239)	463–664 (202/159)	—	$P\,6_1$	2.25	5BUP	Bokhove, Nishimura, et al. (2016)
VERL	VR1 + linker	*H. rufescens* (Q8WR62)	38–175 (138/109)	4[b] (N)	$P\,6_5\,2\,2$	2.00	5II4	Raj et al. (2017)
VERL	VR1	*H. rufescens* (Q8WR62)	38–151 (114/108)	3[b] (N)	$P\,6_5\,2\,2$	1.80	5II5	Raj et al. (2017)
VERL	VR2 + linker	*H. rufescens* (Q8WR62)	176–298 (123/105)	1–3[b] (O)	$C\,1\,2\,1$	2.50	5MR2	Raj et al. (2017)
VERL/lysin complex	VR2 + linker	*H. rufescens* (Q8WR62)	176–298 (123/116)	1–3[b] (O)	$P\,1\,2_1\,1$	1.80	5MR3	Raj et al. (2017)
VERL	VR3	*H. rufescens* (Q8WR62)	340–453 (114/101)	3 (N)	$P\,1\,2_1\,1$	2.90	5IIC	Raj et al. (2017)
VERL/lysin complex	VR3	*H. rufescens* (Q8WR62)	340–453 (114/105)	3 (N)	$P\,1$	1.70	5IIA	Raj et al. (2017)

VERL/lysin complex	VR3	*H. rufescens* (Q8WR62)	340–453 (114/105)	3 (N)	$P\,3_1\,2\,1$	1.64	5IIB	Raj et al. (2017)
BG	ZP-C	*R. norvegicus* (P26342)	591–763 (173/173)	1 (N)	$P\,2_1\,2_1\,2_1$	2.00	3QW9	Lin, Hu, Zhu, Woodruff, and Jardetzky (2011)
BG	ZP-C	*M. musculus* (O88393)	591–757 (167/152)	1 (N)	$P\,2_1\,2_1\,2_1$	2.70	4AJV	Diestel et al. (2013)
UMOD	EGF IV, ZP-N, ZP-C	*H. sapiens* (P07911)	295–610 (316/316)	$2+1^b$ (N)	$H\,3\,2$	3.20	4WRN	Bokhove, Nishimura, et al. (2016)
ENG	ZP-N, ZP-C	*H. sapiens* (P17813)	338–581 (244/244)	—	$P\,6_5$	2.70	5HZV	Saito et al. (2017)

[a]Structure factors and model coordinates derived from the same dataset reprocessed at 2.05 Å resolution in space group $P\,2_1\,2\,2_1$ are now also available with PDB ID 5OSQ.
[b]Site(s) was/were mutated in the construct used for crystallization.

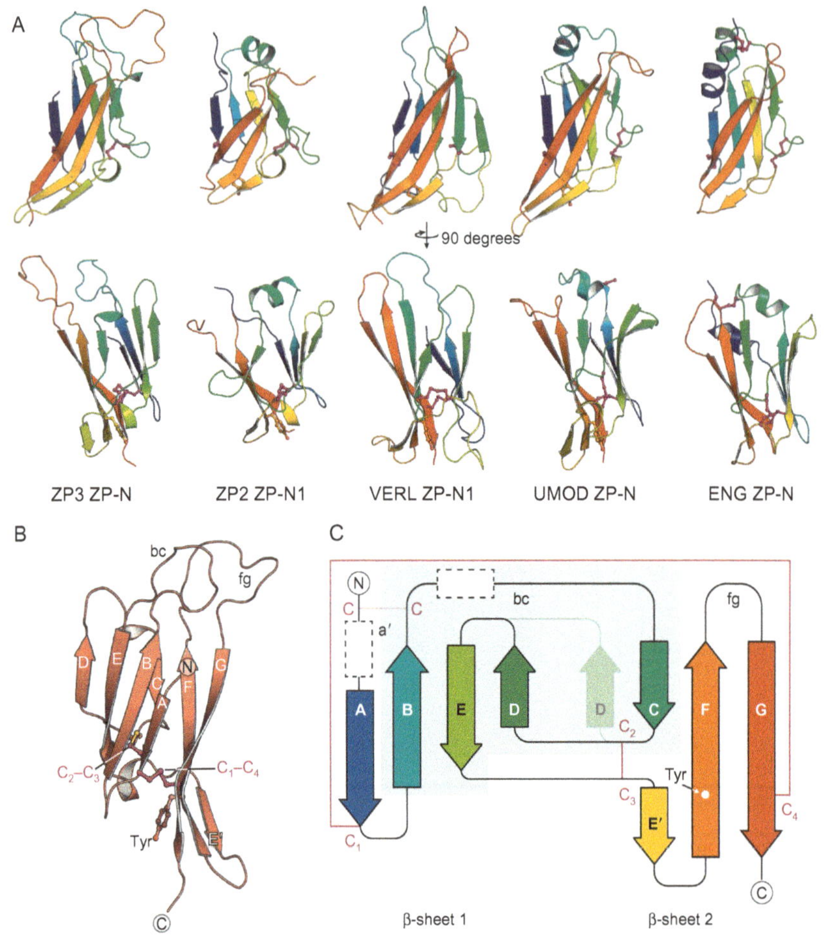

Fig. 2 Structures and features of different ZP-N domains. (A) Perpendicular side-by-side comparison of a representative set of ZP-N structures. Cartoons are rainbow colored from the N-terminus (*blue*) to the C-terminus (*red*), with the conserved tyrosine and disulfide bonds (*magenta*) shown as *ball* and *sticks*. (B) Close-up of murine ZP3 ZP-N shown as a *salmon cartoon*. N- and C-termini are encircled and β-strands are labeled A–G following the standard convention used for Ig-like domains, with the tyrosine and disulfides depicted as in panel (A). (C) Topology of the ZP-N domain with β-strands colored as in panel (A) and labeled as in panel (B). Semitransparent and dashed features indicate elements found in some but not all ZP-N domains. The transparent D strand highlights the strand-switched topology of VERL ZP-Ns, and the *light cyan* background indicates the $(3,1)_N$ Greek key motif.

folds into a compact, isolated domain (Fig. 2A–C). Remarkably, when analyzed using secondary structure matching (Krissinel & Henrick, 2004), the ZP-N structure strongly resembles immunoglobulin (Ig)-like domains despite complete absence of sequence identity (a common characteristic of Ig-like proteins). The ZP-N fold consists of two stacked antiparallel β-sheets, one containing strands A–B–E–D (β-sheet 1) and the other consisting of strands C–F–G (β-sheet 2) (Figs. 2C and 4A). This β-sandwich is held together by a hydrophobic core of buried residue side chains and two disulfide bonds with 1–4, 2–3 connectivity that link strands A and G and the cd and ef loops, respectively. The C_1–C_4 disulfide is located on the solvent-exposed edge of stacked β-strands A and G, whereas the C_2–C_3 disulfide is completely buried (Fig. 2B and C).

The C-type Ig-like structural topology of ZP-N follows a $(3,1)_N$ Greek key motif (Bork, Holm, & Sander, 1994; Hutchinson & Thornton, 1993) that results in stacking of the three- and four-stranded β-sheets (Figs. 2C and 4A). Compared to Ig-like domains, however, ZP-N contains an extra E′ strand that extends the short C strand of the C–F–G sheet. The resulting C + E′ combination matches the size of the F and G strands, which in ZP-N are much longer than those found in Ig-like domains (Fig. 4A). Moreover, the E′–F–G sheet extends from underneath the A–B–E–D sheet to form an important structural feature in conjunction with the partially exposed, hydrophobically stacked A strand. The F strand contains a tyrosine (Y111), which lies next to C_4 and is one of the few residues to be strictly conserved in addition to the cysteines. Consistent with this observation, two lines of evidence suggest that this residue plays an important role in polymerization: first, a corresponding tyrosine is missing in ENG, which does not assemble into polymers (Saito et al., 2017); second, mutation of the equivalent residue in α-tectorin results in nonsyndromal deafness by causing malformation of the tectorial membrane (Legan et al., 2005; Monné et al., 2008). Another feature of the ZP-N domain is the presence of extended bc and fg loops, which coincide with the location of the complementarity-determining regions (CDR) 1 and 3 of antibodies. Like CDRs, these regions are highly variable in length, chemical nature, and structure so that—as discussed in more detail later—they can not only maintain the structural integrity of ZP-N and adjacent domains but also mediate intermolecular interactions.

An important consequence of the structure of mouse ZP3 ZP-N was that, complementing an independent bioinformatic analysis (Callebaut et al., 2007), it supported the presence of isolated ZP-N domains in the N-terminal regions of ZP1, ZP2, and ZP4 (Fig. 1B) (Monné et al., 2008). Moreover, it suggested that—despite insignificant sequence identity—the 22 tandem repeats of VE receptor for lysin (VERL), a 2-MDa component of the mollusk VE with sperm receptor activity (Swanson & Vacquier, 1997), may also adopt a ZP-N fold (Fig. 1B) (Swanson et al., 2011).

These hypotheses were experimentally confirmed by the crystal structures of ZP2 ZP-N1 and VERL repeats (Raj et al., 2017) discussed in the following section as well as Section 6.

2.2 Other ZP-N Domain Structures

Six additional ZP-N domain structures have become available following the initial mouse ZP3 ZP-N report (Table 1). As evident when they are displayed side by side, representatives of all these structures share the same overall fold, the E′ extension, the angle between the two β-sheets, and the location of the conserved disulfides and tyrosine residue (Fig. 2A). However, already at this level interesting differences can be observed. For example, the structure of ZP2 ZP-N1 (Raj et al., 2017) is more compact and has shorter strands than the others; moreover, the bc loop of ZP2 ZP-N1 contains an α-helix, a feature that is missing in ZP3 but is also found in the ZP-N domains of UMOD and ENG (Fig. 2A and dashed element in Fig. 2C). Notably, as further discussed in Section 6, in ZP2 the bc helix is included in a ~30-amino acid region suggested to mediate sperm recognition (Avella et al., 2014; Raj et al., 2017); on the other hand, the bc helix of UMOD carries the cysteine that tethers an epidermal growth factor domain (EGF-IV) to the ZP-N (Bokhove, Nishimura, et al., 2016) and that of ENG packs against another helix (a′), while also being connected to the loop that precedes the latter by an additional disulfide bridge (Saito et al., 2017). Thus, the ZP-N domains of both UMOD and ENG display a similar combination of bc helix packing and intramolecular disulfide bonding.

Although they adopt the same overall structure of other ZP-N domains, the N-terminal repeats of VERL present significant local differences (Raj et al., 2017). In particular, as reflected by a shorter cd loop, the D strand of VERL repeats extends β-sheet 2 instead of belonging to β-sheet 1 as found in all other ZP-Ns (Fig. 2A and C). Such an A–B–E and D–C–F–G arrangement, which changes the ZP-N Greek key motif to a $(2,2)_N$ class (Hutchinson & Thornton, 1993), can also be found in S-type Ig-like domains (Bork et al., 1994). Another interesting feature that separates VERL repeats from other ZP-Ns is the extended C_2–C_3 disulfide-carrying ee′ loop, which contributes to the gamete-binding interface by becoming ordered upon interaction with lysin in the complex structures of VERL repeats 2 and 3 (see Section 6) (Raj et al., 2017). Despite these differences, all available ZP-N coordinate sets clearly belong to the same structural family.

3. STRUCTURES OF THE ZP-C DOMAIN

The finding that the ZP-N region of ZP3 folds into a distinct domain resembling the Igs immediately brought further support to the suggestion that the C-terminal half of the ZP module, i.e., the ZP-C domain, also formed an isolated domain (Jovine et al., 2004). Structural information on ZP-C became available as part of the first structure of a complete ZP module, that of chicken ZP3 (Han et al., 2010). Although this homolog of ZP3 is natively hypoglycosylated, its structural complexity nonetheless required expression in mammalian cells; moreover, proteolytic removal of a short peptide immediately preceding the CCS region was essential to obtain well-diffracting crystals. After describing the main characteristics of the ZP-C domain of avian ZP3, whose ZP-N moiety is highly similar to that of mouse, this section will discuss other ZP-C structures solved to date. In Section 4, the features of both domains will be finally compared to those of Igs.

3.1 Structure of Avian ZP3 ZP-C

Except for a ZP3-specific insertion discussed in more detail later, the ZP-C domain of avian ZP3 is closely related to ZP-N despite having a completely different sequence and disulfide bond pattern (Han et al., 2010). ZP-C has a V-type Ig-like fold whose basic B–C–D–E Greek key signature accommodates two additional strands (C′ and C″) compared to ZP-N; this in turn generates two partially overlapping Greek key-like motifs that involve β-strands B–C–C′–C″ and C″–D–E–F, respectively. Combined with other secondary structure elements, these features give rise to a β-sandwich whose sheets consist of strands A–B–E–D (β-sheet 1) and strands C″–C′–C–F–G–A″ (β-sheet 2) (Figs. 3A, left panel; B, top panel; C and 4B). Although ZP-C lacks the additional E′ strand of ZP-N, its β-sheet 2 is augmented by the aforementioned C′–C″ β-hairpin and an extra strand that runs parallel to strand G; the latter addition is referred to as strand A″ in avian ZP3 ZP-C (Han et al., 2010) (Fig. 3C) and A′ in other ZP-C domains (Bokhove, Nishimura, et al., 2016; Lin et al., 2011; Saito et al., 2017) (Fig. 4B). The reason for this different nomenclature is that avian ZP3 ZP-C contains a further A′ strand in its β-sheet 1, but whether this is a general feature of ZP3-type ZP-Cs remains to be established. As in the case of the ZP-N domain, the β-sheets in the ZP-C structure are held together by hydrophobic interactions; in addition, they are stabilized by an invariant disulfide that connects adjacent β-strands C and F (C_5–C_7).

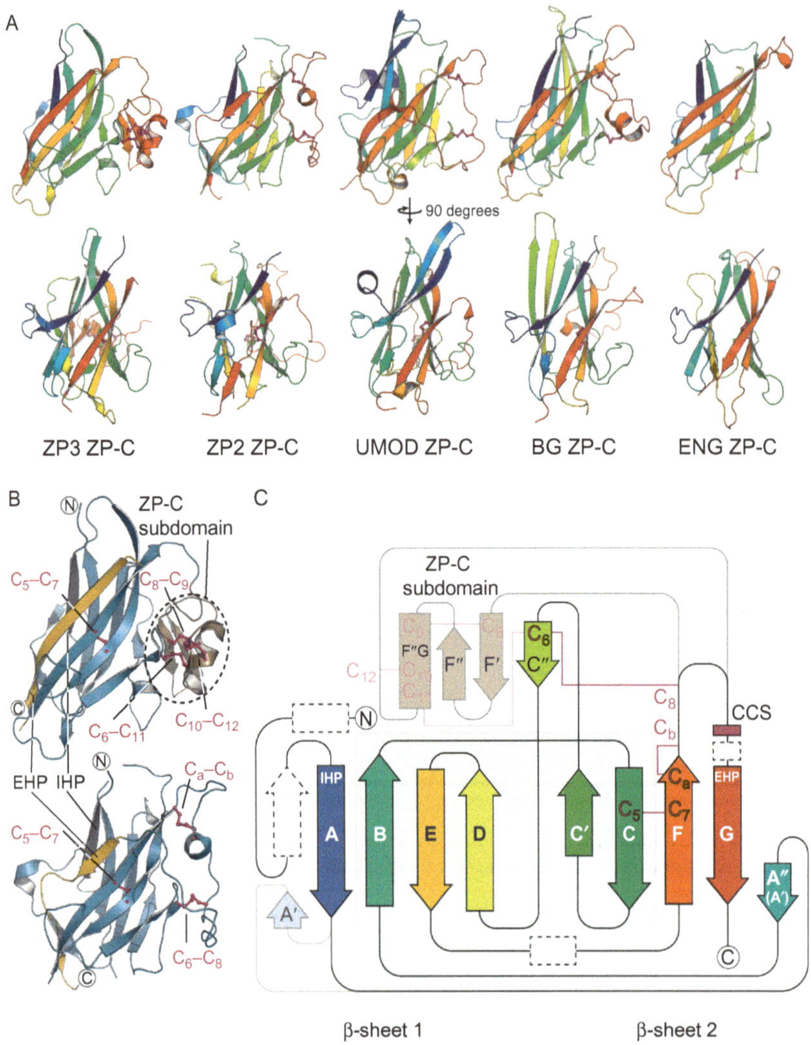

Fig. 3 Comparison and features of ZP-C domain structures. (A) Perpendicular side-by-side comparisons of all ZP-C structures determined to date, represented using the same conventions as in Fig. 2A. (B) Close-up of avian ZP3 ZP-C (*top*) and murine ZP2 ZP-C (*bottom*), depicted as in Fig. 2B. The ZP3-specific extension, IHP, and EHP are colored *light gray*, *dark gray*, and *yellow*, respectively. (C) Topology of the ZP-C domain with β-strands colored as in panel (A) and labels as in panel (B). Important ZP-C features are indicated, with semitransparent and dashed features representing elements found only in some ZP-C domains. For clarity, β-strand C″ (*light green*) has been lifted out of β-sheet 2 to highlight its interaction with the ZP3-specific extension (*light gray*).

Unlike in the other ZP-Cs discussed later, ZP3's remaining disulfides (C_6–C_{11}, C_8–C_9, and C_{10}–C_{12}) are all clustered together in a region unique to this particular protein. This ZP3-specific subdomain, a strand–strand–helix motif (F′–F″–F″G) that remotely resembles EGF-like domains, is inserted between β-strands F and G of ZP-C and expands β-sheet 2 via parallel pairing of strands F′ and C″ (Han et al., 2010) (Fig. 3B and C). The disulfide connectivity of the avian ZP3-specific subdomain is consistent with the pattern suggested for pig ZP3 on the basis of mass spectrometry measurements (Kanai et al., 2008); however, because of the clustering of its Cys residues, the fold of the subdomain could in principle also accommodate the alternative disulfide connectivity C_6–C_8, C_9–C_{11}, and C_{10}–C_{12} proposed for other homologues of ZP3 (Boja et al., 2003; Darie et al., 2004; Kanai et al., 2008; Zhao et al., 2004). As discussed in Section 5, the absolute conservation of the subdomain in all ZP3 homologues suggests that this plays a crucial role in protein–protein interactions required for egg coat assembly. At the same time, the ZP3 sequence that immediately follows the subdomain is under positive Darwinian selection in mammals (Jansa, Lundrigan, & Tucker, 2003; Swann, Cooper, & Breed, 2007; Swanson, Yang, Wolfner, & Aquadro, 2001; Turner & Hoekstra, 2006), where the C-terminal region of the protein has long been implicated in the interaction with sperm (Chen et al., 1998; Williams et al., 2006).

The elucidation of the structure of ZP3 ZP-C also resolved the IHP/EHP, two conserved regions thought to be highly important for ZP module protein polymerization (Jovine et al., 2004; Schaeffer et al., 2009). Interestingly, neither the IHP nor the EHP are independent structural elements; instead, they are an integral part of the protein fold by constituting the A and G strands of ZP-C, respectively (Fig. 3B and C). The EHP is preceded by the CCS, which is part of the extended fg loop just after the ZP3-specific subdomain (Fig. 3C). Interestingly, as in the case of the corresponding loop in ZP-N domains, the ZP3-specific subdomain, the large fg loop, and the CCS are all located in the region that corresponds to CDR3, the most variable CDR in the Ig variable domains. As further discussed later, cleavage of the CCS is believed to activate ZP module proteins for polymerization by causing ejection of the EHP (Jovine et al., 2004). However, considering that the IHP is an integral part of the ZP-C fold and makes extensive hydrophobic interactions within the β-sandwich and the EHP in particular, the mechanism by which ejection and assembly take place remains enigmatic.

3.2 Other ZP-C Domain Structures

The information obtained on ZP3 was recently complemented by a structure of the ZP-C domain of ZP2, the other major component of the mammalian ZP (Bokhove, Nishimura, et al., 2016). Furthermore, structures of ZP-C domains from nonfertilization-related proteins have also become available from parallel studies on BG (Diestel et al., 2013; Lin et al., 2011), UMOD (Bokhove, Nishimura, et al., 2016), and ENG (Saito et al., 2017) (Table 1).

The β-sandwich of ZP2 ZP-C is very similar to that of ZP3, but lacks the A′ strand in β-sheet 1 and the ZP-C subdomain. Nonetheless, ZP2 ZP-C contains the extended fg loop and, except for the conserved disulfide C_5–C_7 between β-strands C and F (Fig. 3A–C), its remaining disulfides (C_6–C_8, C_a–C_b) clamp down this loop onto the C′–C″ insert and β-sheet 2 (Fig. 3B).

Whereas ZP3 forms heteropolymers with other ZP module-containing egg coat subunits, UMOD exclusively homopolymerizes. Although the ZP-C domain of UMOD is very similar to that of ZP3 and ZP2, its N-terminus contains two additional elements, an α-helix and a β-strand, that precede the strand corresponding to the IHP of ZP3 ZP-C. Notably, the presence of these extra elements has important consequences for UMOD's proposed assembly mechanism (Section 5).

An important insight came from the observation that the structure of UMOD ZP-C contains the same C_5–C_7, C_6–C_8, and C_a–C_b disulfide bonds found in the ZP-Cs of ZP2, BG, and—allowing for the C_6–C_{11} variant and with the exception of C_a–C_b which is missing—ZP3 (Fig. 3). This is because, based on mass spectrometry studies, it was initially believed that there were two types of ZP module proteins, whose different disulfide bond patterns correlated with the respective polymerization properties (Boja et al., 2003; Darie et al., 2004; Jovine et al., 2005). The crystal structure of UMOD ZP-C, however, shows that no such distinction exists nor is the C_5–C_6, C_7–C_a, and C_b–C_8 disulfide connectivity originally proposed for the so-called type II ZP-C domains compatible with any of the elucidated ZP-C structures. In other words, there are no type I and type II ZP "domains" displaying local structural differences as a result of alternative disulfide bond connectivity. Rather, as discussed in more detail in Section 5, hetero- or homoassembly is dictated by the nature of the interdomain linker (Bokhove, Nishimura, et al., 2016) and, possibly, the ZP3-specific subdomain.

While the ZP-C domains of ZP3, ZP2, and UMOD are very similar, the ZP-C of BG is characterized by having a longer β-strand E

(Diestel et al., 2013; Lin et al., 2011) whereas that of ENG shows a much more compact, minimal fold (Saito et al., 2017) (Fig. 3A). This results from the fact that, although its β-sandwich contains the same number of secondary structure elements as ZP2 together with the invariant C_5–C_7 disulfide, ENG ZP-C completely lacks the extended fg loop and cysteines therein. By leaving cysteine C_6 in the C′–C″ insert unpaired and thus free to form an intermolecular disulfide, this arrangement contributes to the physiological homodimerization of ENG (Saito et al., 2017). Interestingly, consistent with the fact that ENG does not require EHP ejection, membrane release, and polymerization for its biological function, lack of an extended fg loop in its ZP-C domain also results in the absence of the CCS. Both of these features, which are conserved in all polymerization-competent ZP-C domains, are well defined in the X-ray map of ZP2 ZP-C (Bokhove, Nishimura, et al., 2016). This shows that the fg loop and the CCS are solvent exposed and pointing away and downward from β-sheet 2, so that the CCS can easily be recognized and cleaved by its specific protease (Fig. 3A). The crystal structure of UMOD shows a similar organization of the fg loop and orientation of the CCS (Fig. 3A; Bokhove, Nishimura, et al., 2016).

Considering that—just like ENG—BG lacks the CCS and does not polymerize, it is remarkable that its ZP-C domain is very similar to that of ZP2, with the fg loop adopting a comparable outward-facing conformation; however, in the case of BG, the loop contains a short α-helix and points upward (Diestel et al., 2013; Lin et al., 2011) (Fig. 3A). Further investigations will be required to establish whether the structural differences between ENG and BG ZP-C domains reflect a different evolutionary origin, or rather the fact that ENG binds bone morphogenetic protein 9 via its orphan domain (Saito et al., 2017), whereas BG uses the ZP-C fg loop to interact with TGF-β (Diestel et al., 2013).

4. ZP-N AND ZP-C COMPARED TO IG-LIKE DOMAINS

As introduced earlier, a β-sandwich consisting of four- and three-stranded β-sheets that follow a Greek key motif is also a characteristic feature of Ig-like domains (Figs. 2C, 3C, and 4). In particular, ZP-N domains are most similar to C-type Ig-like domains (Fig. 4A), whereas ZP-C domains resemble V-type Ig-like domains (Fig. 4B). Also in the case of Ig-like domains, the two β-sandwich sheets are held together by hydrophobic residues; however, Ig-like domains lack the E′ strand of ZP-Ns and, notwithstanding variations among superfamily members, their sheets are generally connected by a single disulfide that bridges opposite strands B and F (Bork et al., 1994) (Fig. 4).

One interesting observation that arises from the comparison of the available ZP-N structures is that the conserved tyrosine located next to C_4 is also found in the N-terminal isolated ZP-N repeats of VERL and ZP2, which are most likely not involved in polymerization (Jovine et al., 2002; Raj et al., 2017). Furthermore, this residue is also present—albeit less exposed—in several V/C/S-type Igs (Halaby, Poupon, & Mornon, 1999). In agreement with mutational studies of ZP3 Y111 (Monné et al., 2008), these observations suggest that the tyrosine is generally important for the folding of the domain. However, an additional role of this amino acid in filament assembly remains warranted in the case of ZP-Ns that belong to polymerization-competent ZP modules. This is because, whereas a deafness-associated semi-dominant mutation of the conserved tyrosine of α-tectorin (Y1870C) actively disrupts tectorial membrane filaments (presumably by interfering with the correct formation of ZP-N disulfide C_1–C_4; Monné et al., 2008), the matrix is not affected in normal-hearing animals heterozygous for a targeted deletion in the *Tecta* gene (Legan et al., 2005). Conservation of the tyrosine also suggests that, although their sequences have diverged beyond recognition, ZP-N domains might be more related to Ig-like domains than initially anticipated. Despite its absence in ZP-C domains, recurrence of the tyrosine may thus reflect a common evolutionary lineage of Ig-like and ZP-N domains, rather than a true feature of the latter.

Similar to ZP-Ns, ZP-C domains consist of a β-sandwich of two hydrophobically stacked β-sheets—β-sheet 1 with four strands and β-sheet 2 with six strands—that follow a Greek key-like motif. However, as detailed in Section 3.1, ZP-Cs lack an E′ strand and their Greek motif is broken by the presence of the C′–C″ hairpin (Fig. 4B; red); moreover, they contain an additional A′ strand in β-sheet 2 (Fig. 4B; yellow). Remarkably, the presence of a C′–C″ hairpin and an A′ β-strand are characteristics that distinguish V-type from C-type Ig superfamily members; thus, these features clearly establish ZP-Cs as V-type Ig-like molecules (Fig. 4B). At the same time, ZP-C domains differ from the latter because their invariant C_5–C_7 disulfide connects neighboring strands C and F, rather than opposite strands B and F. Notably, such an atypical linkage is also found within domain 2 of cell surface glycoprotein CD4 (Ryu et al., 1990; Wang et al., 1990).

Another interesting resemblance between Igs and ZP module proteins, which might be significant from an evolutionary point of view, is their quaternary structure. Namely, similar to an antibody light chain that consists of a

V-type and a C-type Ig-like domain, a ZP module consists of a (C-type-like) ZP-N and a (V-type-like) ZP-C (Fig. 4). Further extending this common domain organization, the N-terminal isolated ZP-N repeats of ZP2, ZP1, and ZP4 are reminiscent of the C-type repeats of antibody heavy chains. Even though the biological functions of egg ZP subunits are far removed from the way antibodies work, the parallel becomes stronger if one considers ligand binding by ENG (Saito et al., 2017). Further studies will nonetheless be required to more firmly establish the possible relationship between ZP module proteins and antibodies.

In conclusion the C_1–C_4 and C_2–C_3 disulfides, together with the E′ extension and conserved tyrosine, constitute the hallmarks of the ZP-N domain and suggest that its fold defines a new Ig superfamily subtype (Monné et al., 2008). Similarly, although its structural relation to ZP-N makes the ZP module internally symmetric (Han et al., 2010), the ZP-C domain defines a second subtype. This contains at least a C_5–C_7 disulfide and a cysteine-containing C′–C″ hairpin, as well as—in the majority of cases—an extended disulfide-rich fg loop with a CCS.

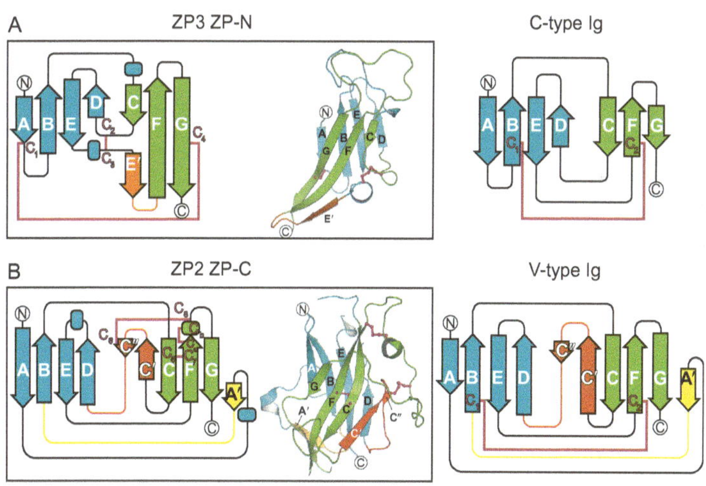

Fig. 4 Topology of ZP-N and ZP-C and their relationship with Ig-like domains. (A) Comparison of ZP3 ZP-N with C-type Ig-like domains. β-strands are labeled according to standard Ig terminology, helices are indicated by *rectangles*. Opposing β-sheets 1 and 2 are colored *blue* and *green*, respectively, with termini encircled. The additional E′ strand is *orange* and disulfides are *magenta*. Due to the fact that their D strand belongs to β-sheet 2 rather than β-sheet 1, VERL repeats resemble S-type Ig-like domains. (B) Comparison of ZP2 ZP-C with V-type Ig-like domains. Features are indicated as in panel (A), except for the additional A′ and C′, C″ strands which are colored *yellow* and *red*, respectively.

5. STRUCTURES OF COMPLETE ZP MODULES: INSIGHTS INTO POLYMERIZATION

The previously discussed ZP-N and ZP-C domains of ZP3, UMOD, and ENG (Figs. 2 and 3) were actually derived from complete ZP module crystal structures (Table 1); here we discuss them within the context of the full ZP module (Fig. 5A). Whereas as mentioned earlier ENG naturally lacks a CCS, mutation of the CCS of both ZP3 and UMOD was required to obtain soluble material for crystallographic analysis by preventing the premature aggregation or assembly of their respective ZP modules. Despite the absence of the CCS, the corresponding structures still provide valuable insights into ZP module structure and possible assembly mechanism. Remarkably, while non-polymerizing ENG crystallized as a monomer with an exposed unpaired cysteine (Saito et al., 2017), ZP3 and UMOD are both dimeric but adopt highly different conformations (Bokhove, Nishimura, et al., 2016; Han et al., 2010). These observations have significant implications for understanding the different polymerization properties of these proteins.

The structure of full-length ZP3 reveals that the ZP-N and ZP-C moieties of its ZP module are connected via a long, flexible interdomain linker that is only partly defined (Fig. 5A); notably, this region carries a conserved N-glycan, as well as an O-glycan that has been implicated in sperm binding in chicken (Han et al., 2010). ZP-N/ZP-C intramolecular interactions are mediated by the EHP through extensive interactions between ZP-N β-sheet 2 and residues in both the ef loop and the A, A′, A″, G, and F strands of ZP-C. On the other hand, the antiparallel dimer interface is mainly formed by the insertion of the ZP-N fg loop into a negatively charged pocket on the surface of ZP-C. Consistent with this interaction, which reshapes part of the fg loop into an F′ strand that interacts antiparallelely with ZP-C E′, deletion of F′ residues or disruption of one of the salt bridges that constitutes the intermolecular interface abolishes secretion. This indicates that dimer formation (Fig. 5A) is essential for biogenesis of avian ZP3 (Han et al., 2010). Considering that human ZP3 is also a homodimer in solution (Zhao et al., 2004), the dimeric state of ZP3 is likely to represent a dormant preassembly form of its heteropolymerizing ZP module (Fig. 5B).

Structural studies of the ZP module of urinary UMOD were pursued to gain insights into the homopolymeric assembly of what was initially expected to be the other type of ZP module proteins. The UMOD dimer is very different from that of ZP3 (Fig. 5A) (Bokhove, Nishimura, et al., 2016). In contrast to the latter, the ZP-N and ZP-C domains of UMOD do not

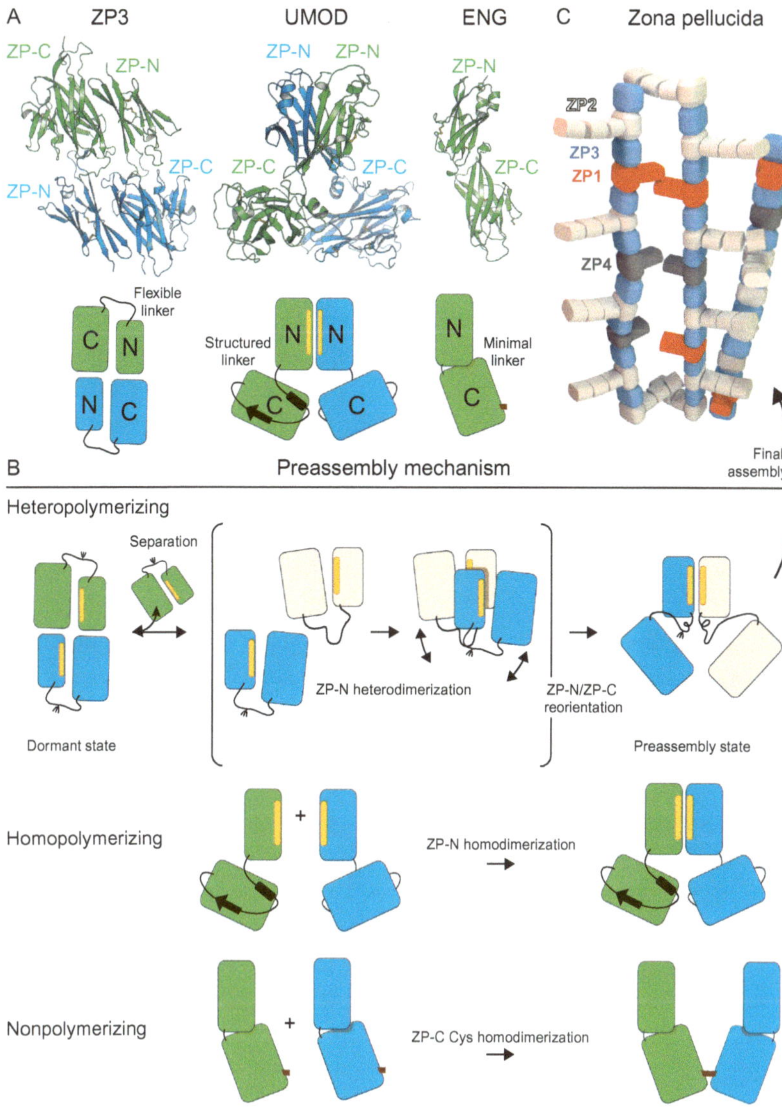

Fig. 5 Comparison of complete ZP module structures in terms of relative domain organization, quaternary structure, and proposed preassembly mechanism. (A) Dimeric ZP3 and UMOD are indicated in *blue* and *green* cartoons, while ENG is *green*; ZP-N and ZP-C are labeled. The scheme at the *bottom* shows the dormant homodimer of ZP3 with its flexible interdomain linker, the preassembly homodimer of UMOD with its structured interdomain linker and the monomer of ENG with a minimal interdomain linker and free ZP-C cysteine C_6 (*small brown bar*). (B) The dormant ZP3 homodimer exchanges one of its molecules with that of another ZP subunit (*white*). Heterodimerization requires reorienting of the ZP-N and ZP-C domains of ZP3 into an extended UMOD-like preassembly state. For both hetero- and homopolymeric ZP modules, this intermediate state is then followed by assembly into polymers via a yet-to-be-determined mechanism. In the case of the nonpolymerizing ZP module of ENG, on the contrary, the minimal interdomain linker hinders domain reorganization; however, the free ZP-C cysteine can mediate back-to-back dimerization. (C) Model of the mammalian ZP. Whereas in human this contains four subunits (ZP1–4), the ZP of other species can lack ZP1 (dog, fox, pig, bovine) or ZP4 (mouse).

interact; rather, its ZP module displays an elongated conformation with ZP-N and ZP-C diametrically opposed. This extended configuration is stabilized by a highly structured, rigid interdomain linker that—as mentioned in Section 3.2—consists of an α-helix and a β-strand that keeps ZP-N and ZP-C separate (Figs. 3A and 5A). As a consequence of this arrangement, the ZP-N domains of two UMOD molecules can interact with each other laterally via hydrophobic interaction of their A/G strand faces (Fig. 5A). Notably, the resulting configuration would not be compatible with the homodimeric interface required for secretion of ZP3, because it would cause clashes between the ZP-C domains interacting intermolecularly with the fg loops of such a ZP-N/ZP-N unit. Furthermore, unlike in the case of ZP3, dimerization of UMOD is not required for secretion; however, its ZP-N/ZP-N interaction is essential for polymerization (Bokhove, Nishimura, et al., 2016) and agrees with the finding that a basolaterally secreted isoform of UMOD that is truncated shortly after ZP-N forms homodimers in vivo (Micanovic et al., 2018). Taken together, these observations suggest that the crystal structure of homodimeric UMOD represents a preassembly state, whose activation depends on CCS cleavage by hepsin (Brunati et al., 2015) and whose relevance may also extend to heteropolymeric ZP module proteins (Fig. 5B). For the reasons outlined earlier, the adoption of this state by ZP3 would, however, require the disassembly of its closed, dormant form into elongated monomers. This event would follow ZP3 secretion and proteolytic cleavage at the CFCS (Litscher, Qi, & Wassarman, 1999), which would cause disengagement of ZP-N (and the ZP-C IHP) from the severed C-terminal fragment including the EHP (Han et al., 2010; Jovine et al., 2004). Consistent with such a scenario, ZP3 and ZP2 are known to traffic independently inside the oocyte (Hoodbhoy et al., 2006) and ZP4 ZP-N is sufficient for interaction with ZP3 (Suzuki et al., 2015); finally, isolated ZP3 ZP-Ns have a tendency to self-interact and form filamentous aggregates in vitro (Jovine et al., 2006), although the resulting material must necessarily only mimic the in vivo situation due to the absence of other ZP subunits. In this regard, ZP-N/ZP-N interactions involving ZP3 and ZP1/2/4 would result into heterodimeric variants of the homodimeric preassembly state of UMOD (Fig. 5B). Although the mechanism regulating the assembly of the these building blocks into their final polymeric form remains to be elucidated, the absolute conservation of the ZP3-specific subdomain in all vertebrate egg coats (including the fish VE, a protective matrix with no sperm-binding activity) suggests that it may be essential for interaction with other ZP/VE components. In relation to this point it is interesting to notice that, although both ZP3 and ZP2 are required for ZP formation in the mouse, the structural function of ZP2 is carried out by ZP1-like subunits in fish and can be artificially replaced by ZP4 in transgenic mice (Avella et al., 2014). By combining early biochemical and electron microscopy data on mouse ZP filaments (Bleil & Wassarman, 1980; Greve & Wassarman, 1985) with the phenotype of knockout mice for the genes encoding ZP1–3 (Liu et al., 1996; Rankin et al., 1996, 2001; Rankin, Talbot, Lee, & Dean, 1999), the current knowledge on the domain architecture of ZP subunits (Fig. 1B), and the structural information discussed earlier, a generic model of the supramolecular structure of the mammalian ZP can be suggested (Fig. 5C). In this model, μm-long filaments contain a structural repeat of ~14 nm formed by alternation of ZP3 and either ZP2 or (if present) ZP4. In the species that also express ZP1, such as mouse and human, this less abundant subunit would be occasionally incorporated instead of ZP2/4 and stabilize the ZP by introducing intermolecular cross-links between filaments.

Although this section was focused on the possible mechanism of ZP module-mediated polymerization, the elongated structure of the ZP region of nonpolymerization-competent ENG also substantiates the idea that a rigid linker between ZP-N and ZP-C domains enforces an extended conformation of the module (Fig. 5A) (Saito et al., 2017). However, since the ENG ZP-N A/G strand face is covered by the fg loop of the ZP-C domain, the ENG ZP module does not form UMOD-like homodimers. On the other hand, ENG forms yet another kind of homodimer in vivo (Gougos & Letarte, 1988). This depends on two intermolecular disulfide bonds, one of which is mediated by C_6 in the C′–C″ insert (Figs. 3A and 5B) (Saito et al., 2017) and the other by a cysteine that immediately follows ZP-C (Guerrero-Esteo, Sanchez-Elsner, Letamendia, & Bernabeu, 2002). However, together with the absence of a ZP-N/ZP-N dimerization interface and EHP release, this configuration prevents any higher-order assembly of the ENG ZP module.

6. HOW LIFE BEGINS: EGG ZP-N DOMAIN RECOGNITION BY SPERM

As described in Section 2, the structural similarity between the N-terminal repeats of mammalian ZP2 and mollusk VERL revealed that, despite being separated by 600 million years of divergent evolution, these egg coat proteins use a common ZP-N domain framework to interact with sperm. This finding had major functional implications that led to the first structure determination of an egg coat–sperm protein recognition complex

(Raj et al., 2017). This is because, whereas a binding partner of ZP2 has yet to be identified, VERL has long been known to interact with lysin, a 16 kDa protein released from sperm upon the acrosome reaction (Lewis, Talbot, & Vacquier, 1982). A highly amphipatic molecule that adopts a five-helical bundle fold (Shaw, McRee, Vacquier, & Stout, 1993), lysin dissolves the VE in a species-specific, nonenzymatic way (Lewis et al., 1982; Swanson & Vacquier, 1997). As generally observed in the case of reproductive proteins—including ZP2 and ZP3—(Swanson & Vacquier, 2002), lysin and the first two repeats of VERL (VR1–2) evolve rapidly under positive Darwinian selection; on the contrary the remaining 20 repeats of VERL (VR3–22) are homogenized by concerted evolution (Galindo, Moy, Swanson, & Vacquier, 2002; Galindo, Vacquier, & Swanson, 2003).

To understand how sequence variation affects gamete recognition in the marine gastropod mollusk abalone (Lyon & Vacquier, 1999), VERL repeat–lysin complexes were characterized biochemically and their affinities were quantified (Raj et al., 2017). This showed that, whereas lysin does not bind highly sequence-divergent VR1, it interacts weakly but species-specifically with moderately sequence-divergent VR2 (Kd ~0.5 μM); on the other hand, lysin and conserved repeat VR3 form a nonspecies-specific complex with nanomolar affinity. Atomic-resolution structures of the VR2–lysin and VR3–lysin complexes revealed that VR2, which contains two additional cysteine residues compared to other VERL repeats, forms an intermolecularly disulfide-bonded antiparallel homodimer that binds hydrophobically two copies of lysin on the opposite faces of the dimer interface (Fig. 6A); VR3 forms a similar 1:1 complex with lysin, but—consistent with a higher-affinity interaction—this involves a larger number of contacts. Except for the aforementioned VERL loop ee′, which orders upon binding, the two proteins essentially interact as rigid bodies, forming an extensive interface largely mediated by VERL β-strands B, D, E, and loops de and ee′, and lysin α-helices 2 and 4–5. Together with the analysis of mutants designed on the basis of the structures, these studies suggested that divergence of the N-terminal sequence of VERL inactivated VR1 and lowered the binding affinity of VR2 compared to VR3–22, thus generating species specificity by amplifying the effect of positive selection on lysin (Raj et al., 2017).

A recurrent feature of our crystal structures is the presence of VERL repeat homodimers stabilized by intermolecular contacts between conserved residues in β-strand A and the e′f loop of the ZP-N fold. This suggests a model whereby two intertwined VERL molecules generate a filament branch that exposes VR1 repeats on the surface of the VE, followed by a thin layer corresponding to the covalently bound antiparallel VR2

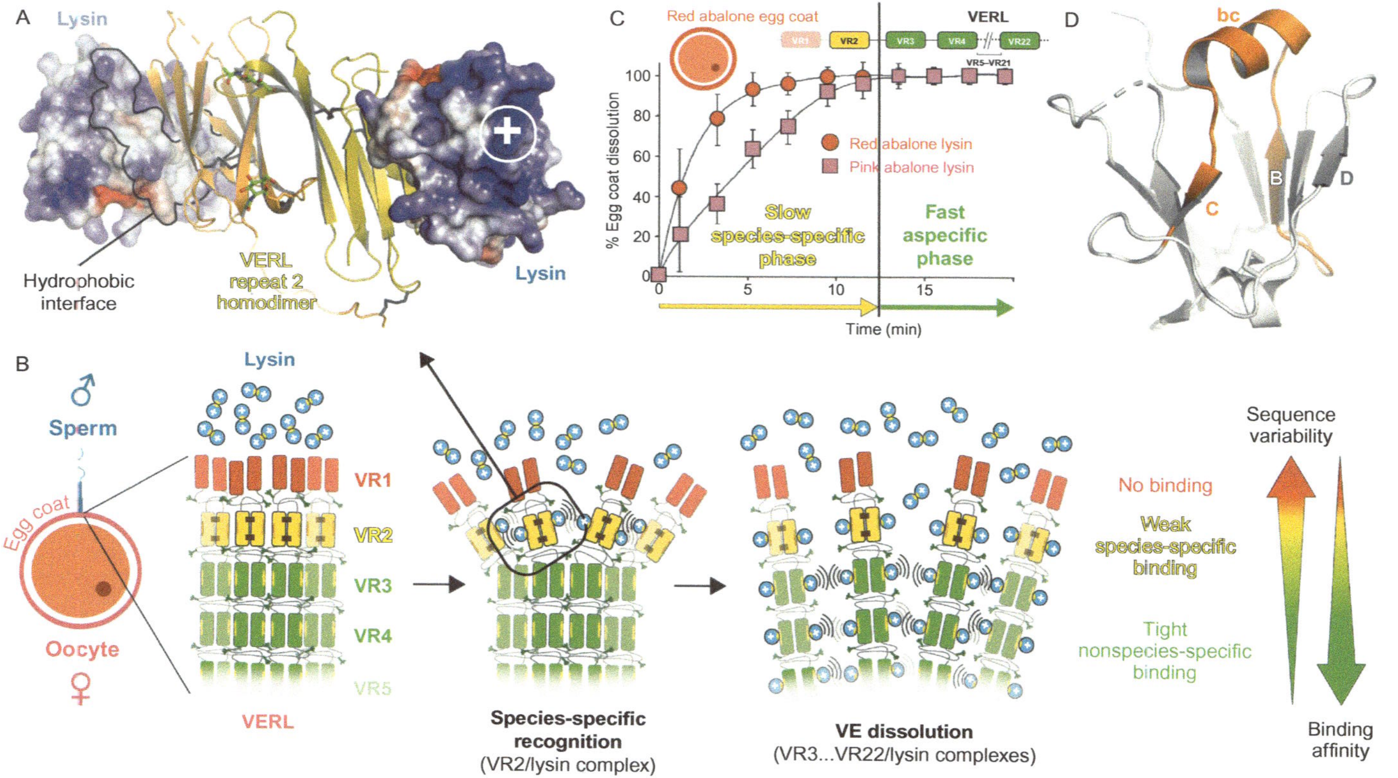

Fig. 6 Structural basis of gamete interaction. (A) Crystal structure of the species-specific complex between egg VERL VR2 and sperm lysin. VR2 repeats are in cartoon representation (*light* and *dark yellow*), with disulfide bonds and N-glycans represented by *gray* and *green sticks*, respectively. Lysin is shown as a surface colored by electrostatic potential (positive, *blue*; negative, *red*). (B) Model of abalone VE architecture and its dissolution by lysin at fertilization. (C) The two phases of VE dissolution by lysin correlate with lysin's initial low-affinity, species-specific interaction with VR2 and its subsequent high-affinity, nonspecies-specific binding to VR3–22, respectively. (D) Crystal structure of ZP2 ZP-N1 (*gray*), with the region suggested to regulate human sperm interaction highlighted in *orange*. *Panels (A) and (B) were adapted with permission from Raj, I., Sadat Al Hosseini, H., Dioguardi, E., Nishimura, K., Han, L., Villa, A., et al. (2017). Structural basis of egg coat-sperm recognition at fertilization.* Cell, 169, *1315–1326.e17; panel (C) was adapted with permission from Lyon, J. D., & Vacquier, V. D. (1999). Interspecies chimeric sperm lysins identify regions mediating species-specific recognition of the abalone egg vitelline envelope.* Developmental Biology, 214, *151–159.*

homodimer and a thick layer made up of stacked, noncovalently paired repeats 3–22 (Fig. 6B, left panel). In the absence of lysin, adjacent VERL branches are held together by lateral contacts between their hydrophobic patches—an interaction also supported by crystal packing of VR3; similarly, lysin is also released under the form of a loosely attached homodimer (Kresge, Vacquier, & Stout, 2000; Raj et al., 2017). Upon its low-affinity interaction with VR2, which acts as a species-specific checkpoint for sperm attachment, lysin's hydrophobic interface starts unraveling the VE by replacing the lateral interaction between VERL branches (Fig. 6B, middle panel). This process is accelerated when it extends to repeats VR3–22, which bind lysin much tighter in a nonspecies-specific way. As supported by molecular dynamics simulations in a seawater-like environment, the juxtaposition of many copies of the highly positively charged surface of lysin on the repeats of adjacent VERL branches would push the latter apart by electrostatic repulsion, ultimately generating a hole for sperm penetration and fusion (Fig. 6B, right panel).

Notably, this model suggests that the series and timing of events in the VE dissolution process are linearly encoded by VERL's domain structure and primary sequence (Fig. 6C). Moreover, the VERL–lysin complex structures show that the species specificity of gamete recognition is regulated in a much complex way than anticipated. This is because it does not simply involve binary changes that affect directly interacting residues on counterpart egg and sperm molecules; rather, recognition is determined by a subtle interplay between the overall affinity of the binding surfaces of different VERL repeats and the variation of lysin sequences (Raj et al., 2017). Finally, it is interesting to notice that not only ZP2 shares the same fold as VERL, but—as mentioned in Section 2.2—it is also thought to contain a sperm-interacting region (Avella et al., 2014) that partially overlaps in space with the lysin-binding surface of VERL repeats (Fig. 6D). The exact functional implications of this similarity, which is complicated by the different D strand location within the ZP-N domains of VERL and ZP2 (Fig. 2C), remain to be elucidated.

7. CONCLUDING REMARKS AND FUTURE DIRECTIONS

During the course of the last decade, we have progressed from a situation where no single structure of an egg coat component or ZP module protein in general was available, to a remarkable understanding of what ZP/VE building blocks look like at the molecular level and how they may interact to form polymers. Most importantly, as described in the previous section, a first atomic-resolution view of how the egg coat is recognized by sperm at the beginning of fertilization was also recently obtained (Raj et al., 2017). This revealed that not only the polymeric core of the egg coat, but also its sperm-interacting region is structurally conserved from mollusk to human. By creating an unexpected link between egg–sperm interaction in vertebrates and invertebrates and suggesting a detailed mechanism for egg coat penetration by sperm, the implications of this finding clearly extend well beyond the realm of protein chemistry.

Despite these major advances, many important questions remain open. Because all the structures of polymeric ZP module proteins so far determined describe the soluble precursor form of the corresponding molecules, we lack detailed information on which conformational changes occur upon C-terminal cleavage of the precursors or what the mature proteins exactly look like in their polymeric state. Similarly, the molecular basis of egg coat cross-linking by ZP1 remains unknown, and so is the relative arrangement of the isolated ZP-N domains constituting the N-terminal domain of ZP2. Concerning the latter, three functionally crucial aspects that are yet to be addressed are whether there is a counterpart of ZP2 on sperm, how postfertilization cleavage of ZP2 regulates the interaction of the ZP with sperm and what molecular mechanisms underlie ZP hardening. Additionally, it remains unclear if the observed effect of zinc sparks on ZP compaction (Que et al., 2017) is due to nonspecific interaction with the matrix or mediated by defined binding sites within one or more ZP subunits. Finally, structural biology has already brought precious insights into the molecular basis of kidney and vascular diseases caused by mutations in *UMOD* and *ENG*, respectively (Bokhove, Nishimura, et al., 2016; Saito et al., 2017). Although reports of pathogenic mutations affecting human ZP genes remain more rare because of the infertility issues associated with such variants, a number of cases were recently described (Barbaux, El Khattabi, & Ziyyat, 2017; Chen et al., 2017; Huang et al., 2014; Liu et al., 2017; Yang et al., 2017). It is our hope that, as additional structural information on the corresponding proteins becomes progressively available, this will not only help to fully elucidate a truly fundamental biological problem such as fertilization but also contribute to the reproductive medicine of the future.

ACKNOWLEDGMENTS

We thank all current and past members of our laboratory for their contributions to our understanding of egg coat structure. We are also very grateful to Tsukasa Matsuda (Nagoya University), Luca Rampoldi (San Raffaele Scientific Institute, Milan), and Daniele de Sanctis (ESRF, Grenoble) for many discussions throughout the years. This work was supported by Karolinska Institutet; the Center for Biosciences (CB) and the Center for Innovative Medicine (CIMED); Swedish Research Council Grants 2012-5093 and 2016-03999; the Göran Gustafsson Foundation for Research in Natural Sciences and Medicine; the Sven and Ebba-Christina Hagberg foundation; an EMBO Young Investigator award; and the European Research Council under the European Union's Seventh Framework Programme (FP7/2007-2013)/ERC Grant agreement 260759.

REFERENCES

Aagaard, J. E., Vacquier, V. D., MacCoss, M. J., & Swanson, W. J. (2010). ZP domain proteins in the abalone egg coat include a paralog of VERL under positive selection that binds lysin and 18-kDa sperm proteins. *Molecular Biology and Evolution*, *27*, 193–203.

Avella, M. A., Baibakov, B., & Dean, J. (2014). A single domain of the ZP2 zona pellucida protein mediates gamete recognition in mice and humans. *The Journal of Cell Biology*, *205*, 801–809.

Barbaux, S., El Khattabi, L., & Ziyyat, A. (2017). ZP2 heterozygous mutation in an infertile woman. *Human Genetics*, *136*, 1489–1491.

Bleil, J. D., Greve, J. M., & Wassarman, P. M. (1988). Identification of a secondary sperm receptor in the mouse egg zona pellucida: Role in maintenance of binding of acrosome-reacted sperm to eggs. *Developmental Biology*, *128*, 376–385.

Bleil, J. D., & Wassarman, P. M. (1980). Structure and function of the zona pellucida: Identification and characterization of the proteins of the mouse oocyte's zona pellucida. *Developmental Biology*, *76*, 185–202.

Boja, E. S., Hoodbhoy, T., Fales, H. M., & Dean, J. (2003). Structural characterization of native mouse zona pellucida proteins using mass spectrometry. *The Journal of Biological Chemistry*, *278*, 34189–34202.

Bokhove, M., Nishimura, K., Brunati, M., Han, L., de Sanctis, D., Rampoldi, L., et al. (2016). A structured interdomain linker directs self-polymerization of human uromodulin. *Proceedings of the National Academy of Sciences of the United States of America*, *113*, 1552–1557.

Bokhove, M., Sadat Al Hosseini, H., Saito, T., Dioguardi, E., Gegenschatz-Schmid, K., Nishimura, K., et al. (2016). Easy mammalian expression and crystallography of maltose-binding protein-fused human proteins. *Journal of Structural Biology*, *194*, 1–7.

Bork, P., Holm, L., & Sander, C. (1994). The immunoglobulin fold. Structural classification, sequence patterns and common core. *Journal of Molecular Biology*, *242*, 309–320.

Bork, P., & Sander, C. (1992). A large domain common to sperm receptors (Zp2 and Zp3) and TGF-beta type III receptor. *FEBS Letters*, *300*, 237–240.

Brunati, M., Perucca, S., Han, L., Cattaneo, A., Consolato, F., Andolfo, A., et al. (2015). The serine protease hepsin mediates urinary secretion and polymerisation of zona pellucida domain protein uromodulin. *eLife*, *4*, e08887.

Callebaut, I., Mornon, J.-P., & Monget, P. (2007). Isolated ZP-N domains constitute the N-terminal extensions of zona pellucida proteins. *Bioinformatics*, *23*, 1871–1874.

Cavallone, D., Malagolini, N., Monti, A., Wu, X.-R., & Serafini-Cessi, F. (2004). Variation of high mannose chains of tamm-horsfall glycoprotein confers differential binding to type 1-fimbriated *Escherichia coli*. *The Journal of Biological Chemistry*, *279*, 216–222.

Chen, T., Bian, Y., Liu, X., Zhao, S., Wu, K., Yan, L., et al. (2017). A recurrent missense mutation in ZP3 causes empty follicle syndrome and female infertility. *American Journal of Human Genetics*, *101*, 459–465.

Chen, J., Litscher, E. S., & Wassarman, P. M. (1998). Inactivation of the mouse sperm receptor, mZP3, by site-directed mutagenesis of individual serine residues located at the combining site for sperm. *Proceedings of the National Academy of Sciences of the United States of America*, *95*, 6193–6197.

Cocchia, M., Huber, R., Pantano, S., Chen, E. Y., Ma, P., Forabosco, A., et al. (2000). PLAC1, an Xq26 gene with placenta-specific expression. *Genomics*, *68*, 305–312.

Darie, C. C., Biniossek, M. L., Jovine, L., Litscher, E. S., & Wassarman, P. M. (2004). Structural characterization of fish egg vitelline envelope proteins by mass spectrometry. *Biochemistry*, *43*, 7459–7478.

Diestel, U., Resch, M., Meinhardt, K., Weiler, S., Hellmann, T. V., Mueller, T. D., et al. (2013). Identification of a novel TGF-β-binding site in the zona pellucida C-terminal (ZP-C) domain of TGF-β-Receptor-3 (TGFR-3). *PLoS One*, *8*, e67214.

Galindo, B. E., Moy, G. W., Swanson, W. J., & Vacquier, V. D. (2002). Full-length sequence of VERL, the egg vitelline envelope receptor for abalone sperm lysin. *Gene*, *288*, 111–117.

Galindo, B. E., Vacquier, V. D., & Swanson, W. J. (2003). Positive selection in the egg receptor for abalone sperm lysin. *Proceedings of the National Academy of Sciences of the United States of America*, *100*, 4639–4643.

Gougos, A., & Letarte, M. (1988). Identification of a human endothelial cell antigen with monoclonal antibody 44G4 produced against a pre-B leukemic cell line. *Journal of Immunology*, *141*, 1925–1933.

Greve, J. M., & Wassarman, P. M. (1985). Mouse egg extracellular coat is a matrix of interconnected filaments possessing a structural repeat. *Journal of Molecular Biology*, *181*, 253–264.

Guerrero-Esteo, M., Sanchez-Elsner, T., Letamendia, A., & Bernabeu, C. (2002). Extracellular and cytoplasmic domains of endoglin interact with the transforming growth factor-β receptors I and II. *The Journal of Biological Chemistry*, *277*, 29197–29209.

Halaby, D. M., Poupon, A., & Mornon, J. (1999). The immunoglobulin fold family: Sequence analysis and 3D structure comparisons. *Protein Engineering*, *12*, 563–571.

Han, L., Monné, M., Okumura, H., Schwend, T., Cherry, A. L., Flot, D., et al. (2010). Insights into egg coat assembly and egg-sperm interaction from the X-ray structure of full-length ZP3. *Cell*, *143*, 404–415.

Hoodbhoy, T., Avilés, M., Baibakov, B., Epifano, O., Jiménez-Movilla, M., Gauthier, L., et al. (2006). ZP2 and ZP3 traffic independently within oocytes prior to assembly into the extracellular zona pellucida. *Molecular and Cellular Biology*, *26*, 7991–7998.

Huang, H.-L., Lv, C., Zhao, Y.-C., Li, W., He, X.-M., Li, P., et al. (2014). Mutant ZP1 in familial infertility. *The New England Journal of Medicine*, *370*, 1220–1226.

Hutchinson, E. G., & Thornton, J. M. (1993). The Greek key motif: Extraction, classification and analysis. *Protein Engineering*, *6*, 233–245.

Jansa, S. A., Lundrigan, B. L., & Tucker, P. K. (2003). Tests for positive selection on immune and reproductive genes in closely related species of the murine genus *Mus*. *Journal of Molecular Evolution*, *56*, 294–307.

Jovine, L., Darie, C. C., Litscher, E. S., & Wassarman, P. M. (2005). Zona pellucida domain proteins. *Annual Review of Biochemistry*, *74*, 83–114.

Jovine, L., Janssen, W. G., Litscher, E. S., & Wassarman, P. M. (2006). The PLAC1-homology region of the ZP domain is sufficient for protein polymerisation. *BMC Biochemistry*, *7*, 11.

Jovine, L., Qi, H., Williams, Z., Litscher, E., & Wassarman, P. M. (2002). The ZP domain is a conserved module for polymerization of extracellular proteins. *Nature Cell Biology*, *4*, 457–461.

Jovine, L., Qi, H., Williams, Z., Litscher, E. S., & Wassarman, P. M. (2004). A duplicated motif controls assembly of zona pellucida domain proteins. *Proceedings of the National Academy of Sciences of the United States of America, 101*, 5922–5927.

Kanai, S., Kitayama, T., Yonezawa, N., Sawano, Y., Tanokura, M., & Nakano, M. (2008). Disulfide linkage patterns of pig zona pellucida glycoproteins ZP3 and ZP4. *Molecular Reproduction and Development, 75*, 847–856.

Kresge, N., Vacquier, V. D., & Stout, C. D. (2000). 1.35 and 2.07 Å resolution structures of the red abalone sperm lysin monomer and dimer reveal features involved in receptor binding. *Acta Crystallographica. Section D, Biological Crystallography, 56*, 34–41.

Krissinel, E., & Henrick, K. (2004). Secondary-structure matching (SSM), a new tool for fast protein structure alignment in three dimensions. *Acta Crystallographica. Section D, Biological Crystallography, 60*, 2256–2268.

Lefièvre, L., Conner, S. J., Salpekar, A., Olufowobi, O., Ashton, P., Pavlovic, B., et al. (2004). Four zona pellucida glycoproteins are expressed in the human. *Human Reproduction, 19*, 1580–1586.

Legan, P. K., Lukashkina, V. A., Goodyear, R. J., Lukashkin, A. N., Verhoeven, K., Van Camp, G., et al. (2005). A deafness mutation isolates a second role for the tectorial membrane in hearing. *Nature Neuroscience, 8*, 1035–1042.

Lewis, C. A., Talbot, C. F., & Vacquier, V. D. (1982). A protein from abalone sperm dissolves the egg vitelline layer by a nonenzymatic mechanism. *Developmental Biology, 92*, 227–239.

Lin, S. J., Hu, Y., Zhu, J., Woodruff, T. K., & Jardetzky, T. S. (2011). Structure of betaglycan zona pellucida (ZP)-C domain provides insights into ZP-mediated protein polymerization and TGF-β binding. *Proceedings of the National Academy of Sciences of the United States of America, 108*, 5232–5236.

Litscher, E. S., Qi, H., & Wassarman, P. M. (1999). Mouse zona pellucida glycoproteins mZP2 and mZP3 undergo carboxy-terminal proteolytic processing in growing oocytes. *Biochemistry, 38*, 12280–12287.

Litscher, E. S., & Wassarman, P. M. (2015). *A guide to zona pellucida domain proteins, Wiley series on protein and peptide science*. Hoboken, New Jersey: John Wiley & Sons, Inc.

Liu, W., Li, K., Bai, D., Yin, J., Tang, Y., Chi, F., et al. (2017). Dosage effects of ZP2 and ZP3 heterozygous mutations cause human infertility. *Human Genetics, 136*, 975–985.

Liu, C., Litscher, E. S., Mortillo, S., Sakai, Y., Kinloch, R. A., Stewart, C. L., et al. (1996). Targeted disruption of the *mZP3* gene results in production of eggs lacking a zona pellucida and infertility in female mice. *Proceedings of the National Academy of Sciences of the United States of America, 93*, 5431–5436.

Lyon, J. D., & Vacquier, V. D. (1999). Interspecies chimeric sperm lysins identify regions mediating species-specific recognition of the abalone egg vitelline envelope. *Developmental Biology, 214*, 151–159.

Micanovic, R., Khan, S., Janosevic, D., Lee, M. E., Hato, T., Srour, E. F., et al. (2018). Tamm-Horsfall protein regulates mononuclear phagocytes in the kidney. *The Journal of the American Society of Nephrology, 29*, 841–856.

Monné, M., Han, L., Schwend, T., Burendahl, S., & Jovine, L. (2008). Crystal structure of the ZP-N domain of ZP3 reveals the core fold of animal egg coats. *Nature, 456*, 653–657.

Que, E. L., Duncan, F. E., Bayer, A. R., Philips, S. J., Roth, E. W., Bleher, R., et al. (2017). Zinc sparks induce physiochemical changes in the egg zona pellucida that prevent polyspermy. *Integrative Biology, 9*, 135–144.

Raj, I., Sadat Al Hosseini, H., Dioguardi, E., Nishimura, K., Han, L., Villa, A., et al. (2017). Structural basis of egg coat-sperm recognition at fertilization. *Cell, 169*, 1315–1326.e17.

Rankin, T., Familari, M., Lee, E., Ginsberg, A., Dwyer, N., Blanchette-Mackie, J., et al. (1996). Mice homozygous for an insertional mutation in the *Zp3* gene lack a zona pellucida and are infertile. *Development*, *122*, 2903–2910.

Rankin, T. L., O'Brien, M., Lee, E., Wigglesworth, K., Eppig, J., & Dean, J. (2001). Defective zonae pellucidae in *Zp2*-null mice disrupt folliculogenesis, fertility and development. *Development*, *128*, 1119–1126.

Rankin, T., Talbot, P., Lee, E., & Dean, J. (1999). Abnormal zonae pellucidae in mice lacking ZP1 result in early embryonic loss. *Development*, *126*, 3847–3855.

Ryu, S. E., Kwong, P. D., Truneh, A., Porter, T. G., Arthos, J., Rosenberg, M., et al. (1990). Crystal structure of an HIV-binding recombinant fragment of human CD4. *Nature*, *348*, 419–426.

Saito, T., Bokhove, M., Croci, R., Zamora-Caballero, S., Han, L., Letarte, M., et al. (2017). Structural basis of the human endoglin-BMP9 interaction: Insights into BMP signaling and HHT1. *Cell Reports*, *19*, 1917–1928.

Sasanami, T., Ohtsuki, M., Ishiguro, T., Matsushima, K., Hiyama, G., Kansaku, N., et al. (2006). Zona pellucida domain of ZPB1 controls specific binding of ZPB1 and ZPC in Japanese quail (*Coturnix japonica*). *Cells, Tissues, Organs*, *183*, 41–52.

Schaeffer, C., Santambrogio, S., Perucca, S., Casari, G., & Rampoldi, L. (2009). Analysis of uromodulin polymerization provides new insights into the mechanisms regulating ZP domain-mediated protein assembly. *Molecular Biology of the Cell*, *20*, 589–599.

Shaw, A., McRee, D. E., Vacquier, V. D., & Stout, C. D. (1993). The crystal structure of lysin, a fertilization protein. *Science*, *262*, 1864–1867.

Suzuki, K., Tatebe, N., Kojima, S., Hamano, A., Orita, M., & Yonezawa, N. (2015). The hinge region of bovine zona pellucida glycoprotein ZP3 is involved in the formation of the sperm-binding active ZP3/ZP4 complex. *Biomolecules*, *5*, 3339–3353.

Swann, C. A., Cooper, S. J. B., & Breed, W. G. (2007). Molecular evolution of the carboxy terminal region of the zona pellucida 3 glycoprotein in murine rodents. *Reproduction*, *133*, 697–708.

Swanson, W. J., Aagaard, J. E., Vacquier, V. D., Monné, M., Sadat Al Hosseini, H., & Jovine, L. (2011). The molecular basis of sex: Linking yeast to human. *Molecular Biology and Evolution*, *28*, 1963–1966.

Swanson, W. J., & Vacquier, V. D. (1997). The abalone egg vitelline envelope receptor for sperm lysin is a giant multivalent molecule. *Proceedings of the National Academy of Sciences of the United States of America*, *94*, 6724–6729.

Swanson, W. J., & Vacquier, V. D. (2002). The rapid evolution of reproductive proteins. *Nature Reviews. Genetics*, *3*, 137–144.

Swanson, W. J., Yang, Z., Wolfner, M. F., & Aquadro, C. F. (2001). Positive Darwinian selection drives the evolution of several female reproductive proteins in mammals. *Proceedings of the National Academy of Sciences of the United States of America*, *98*, 2509–2514.

Turner, L. M., & Hoekstra, H. E. (2006). Adaptive evolution of fertilization proteins within a genus: Variation in ZP2 and ZP3 in deer mice (*Peromyscus*). *Molecular Biology and Evolution*, *23*, 1656–1669.

Wang, J. H., Yan, Y. W., Garrett, T. P., Liu, J. H., Rodgers, D. W., Garlick, R. L., et al. (1990). Atomic structure of a fragment of human CD4 containing two immunoglobulin-like domains. *Nature*, *348*, 411–418.

Wassarman, P. M., & Litscher, E. S. (2016). A bespoke coat for eggs: Getting ready for fertilization. *Current Topics in Developmental Biology*, *117*, 539–552.

Williams, Z., Litscher, E. S., Jovine, L., & Wassarman, P. M. (2006). Polypeptide encoded by mouse *ZP3* exon-7 is necessary and sufficient for binding of mouse sperm in vitro. *Journal of Cellular Physiology*, *207*, 30–39.

Yan, C., Pendola, F. L., Jacob, R., Lau, A. L., Eppig, J. J., & Matzuk, M. M. (2001). *Oosp1* encodes a novel mouse oocyte-secreted protein. *Genesis*, *31*, 105–110.

Yang, P., Luan, X., Peng, Y., Chen, T., Su, S., Zhang, C., et al. (2017). Novel zona pellucida gene variants identified in patients with oocyte anomalies. *Fertility and Sterility*, *107*, 1364–1369.

Yonezawa, N., & Nakano, M. (2003). Identification of the carboxyl termini of porcine zona pellucida glycoproteins ZPB and ZPC. *Biochemical and Biophysical Research Communications*, *307*, 877–882.

Zhao, M., Boja, E. S., Hoodbhoy, T., Nawrocki, J., Kaufman, J. B., Kresge, N., et al. (2004). Mass spectrometry analysis of recombinant human ZP3 expressed in glycosylation-deficient CHO cells. *Biochemistry*, *43*, 12090–12104.

CHAPTER FOURTEEN

Egg Coat Proteins Across Metazoan Evolution

Emily E. Killingbeck[1], Willie J. Swanson[1]
Department of Genome Sciences, University of Washington, Seattle, WA, United States
[1]Corresponding authors: e-mail address: emilyek@u.washington.edu; wjs18@u.washington.edu

Contents

Abstract

All animal oocytes are surrounded by a glycoproteinaceous egg coat, a specialized extracellular matrix that serves both structural and species-specific roles during fertilization. Egg coat glycoproteins polymerize into the extracellular matrix of the egg coat using a conserved protein–protein interaction module—the zona pellucida (ZP) domain—common to both vertebrates and invertebrates, suggesting that the basic structural

Current Topics in Developmental Biology, Volume 130
ISSN 0070-2153
https://doi.org/10.1016/bs.ctdb.2018.03.005

features of egg coats have been conserved across hundreds of millions of years of evolution. Egg coat proteins, as with other proteins involved in reproduction, are frequently found to be rapidly evolving. Given that gamete compatibility must be maintained for the fitness of sexually reproducing organisms, this finding is somewhat paradoxical and suggests a role for adaptive diversification in reproductive protein evolution. Here we review the structure and function of metazoan egg coat proteins, with an emphasis on the potential role their evolution has played in the creation and maintenance of species boundaries.

1. INTRODUCTION

Fertilization, the union of a single sperm and an egg, is essential to metazoan reproduction. The first contact in fertilization is between sperm and the extracellular matrix (ECM) of the egg coat, a maternally derived glycoprotein envelope present in all sexually reproducing animals as well as many asexual metazoans (Conner, Lefievre, Hughes, & Barratt, 2005; Shu, Suter, & Räsänen, 2015; Wong & Wessel, 2006a). Egg coats vary in size from a few tens of microns to over 15 cm (Lombardi, 1998) and are called different names in each major vertebrate lineage: the chorion in fish, the vitelline envelope in amphibians, the perivitelline envelope in reptiles and birds, and the zona pellucida (ZP) in mammals (Goudet, Mugnier, Callebaut, & Monget, 2008; Shu et al., 2015). For simplicity, however, we will collectively refer to these terms as the "egg coat" throughout this review.

Despite their varied nomenclature, the overall structure and function of the egg coat are conserved across vertebrates and invertebrates (Goudet et al., 2008; Han et al., 2010). Egg coats mediate fertilization via sperm recognition and binding, establish blocks to polyspermy, and protect the embryo from biotic (e.g., pathogens, predators) and abiotic (e.g., dehydration, UV radiation, salinity, pollutants) threats (Shu et al., 2015). Egg coats affect embryonic performance by providing a dispersal and attachment medium in aquatic taxa, and in viviparous species they protect the developing embryo until the egg coat hatches and implants in the wall of the uterus (Menkhorst & Selwood, 2008; Monne, Han, & Jovine, 2006; Shu et al., 2015). In addition to their conserved function, egg coat ultrastructure is also conserved across metazoans, consisting of a fibrous matrix of conserved components with common protein domains (Goudet et al., 2008; Monne et al., 2006; Monne & Jovine, 2011; Wong & Wessel, 2006a). Notably, the domain composition of egg coat proteins and the number of genes encoding them is more variable in invertebrates (Shu et al., 2015; Wong & Wessel, 2006a) than in vertebrates (Jovine, Darie, Litscher, & Wassarman, 2005; Litscher & Wassarman, 2007; Monne et al., 2006; Wong & Wessel, 2006a).

While the basic structure of the egg coat has been conserved for more than 600 million years (Han et al., 2010; Litscher & Wassarman, 2007; Monne et al., 2006; Monne & Jovine, 2011), the proteins that make up the egg coat, as with many proteins involved in reproduction, are frequently found to be rapidly evolving (Aagaard, Yi, MacCoss, & Swanson, 2006; Findlay & Swanson, 2010; Palumbi, 2009; Turner & Hoekstra, 2008; Vacquier & Swanson, 2011). This rapid evolution of reproductive proteins is somewhat paradoxical: given the fundamentality of fertilization to species propagation, sperm and egg proteins might be expected to be highly conserved to maintain compatibility. However, this rapid evolution suggests a role for positive Darwinian evolution in creating and maintaining species specificity during sperm–egg interaction (Claw & Swanson, 2012; Meslin et al., 2012; Palumbi, 2009; Turner & Hoekstra, 2008).

In this review we will discuss the composition and evolutionary history of metazoan egg coat proteins, with an emphasis on the role these factors play in the structure, function, and evolution of the metazoan egg coat in fertilization.

2. EGG COAT PROTEINS

Animal egg coat proteins share a common polymerization module called the ZP domain (Jovine, Qi, Williams, Litscher, & Wassarman, 2004; Jovine et al., 2005; Litscher & Wassarman, 2014; Wassarman, 2008; Wilburn & Swanson, 2017; Wong & Wessel, 2006a). ZP domain-containing proteins (ZP proteins) are found in the egg coats of all vertebrate taxa (Litscher, Williams, & Wassarman, 2009; Wong & Wessel, 2006a) and some invertebrate species, including gastropod mollusks (Aagaard, Vacquier, MacCoss, & Swanson, 2010; Aagaard et al., 2006; Monne et al., 2006), cephalochordates (Putnam et al., 2008; Xu et al., 2012), and urochordates (Sawada et al., 2002; Yamada, Saito, Taniguchi, Sawada, & Harada, 2009).

Phylogenetic analyses of ZP proteins suggest that the last common ancestor of vertebrates had at least one ancestral ZP gene, with all major ZP gene subfamilies emerging before the divergence of fish and amphibians ~360 million years ago (Smith, Paton, Hughes, & Burt, 2005; Spargo & Hope, 2003). ZP glycoprotein subfamilies have a historically complicated nomenclature, but recent consensus defines six subfamilies that evolved through gene duplication and pseudogenization: ZP1, ZP2/ZPA, ZP3/ZPC, ZP4/ZPB, ZPAX, and ZPD (Fig. 1) (Claw & Swanson, 2012; Conner et al., 2005; Goudet et al., 2008; Shu et al., 2015; Wong & Wessel, 2006a). ZP4, for instance, is a pseudogene in mouse, ZP1 is a pseudogene in dog, pig, cat, and cow, and ZPD and ZPAX have been pseudogenized or lost in all mammals (Goudet et al., 2008; Meslin et al., 2012; Shu et al., 2015). Such findings suggest that the evolution of ZP genes occurs mainly by gene death, with the accumulation of stop codons and/or insertions/deletions that disrupt reading frame and cause the loss of protein-coding ability (Goudet et al., 2008).

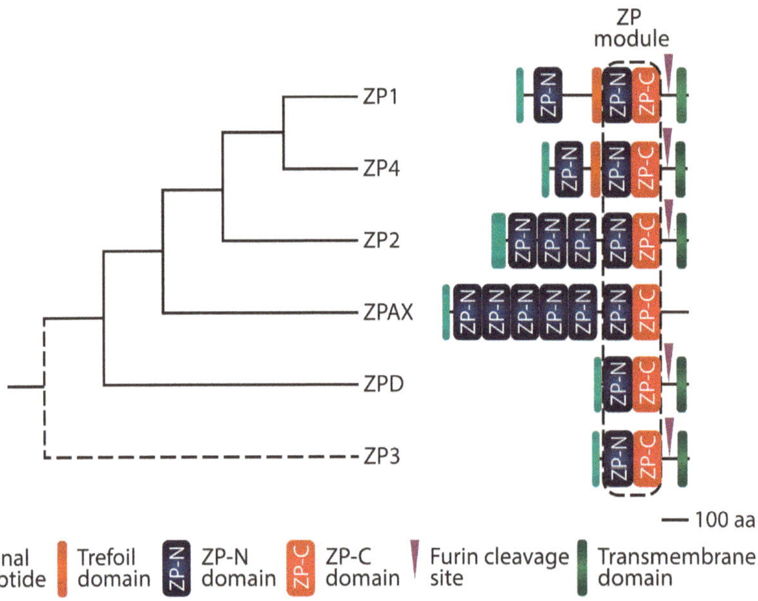

Fig. 1 Phylogeny and domain structure of zona pellucida (ZP) glycoproteins. ZP3 is thought to be the ancestral ZP gene, but its position in the tree is unknown (as indicated by the *dashed line*). *aa*, amino acid. Schematics for ZP1, ZP2, ZP3, and ZP4 are based on the human homologs, and ZPD and ZPAX are based on the homologs from *Xenopus tropicalis. Adapted from Callebaut, I., Mornon, J. P., & Monget, P. (2007). Isolated ZP-N domains constitute the N-terminal extensions of zona pellucida proteins.* Bioinformatics (Oxford, England), 23*(15), 1871–1874. btm265 [pii]; Claw, K. G., & Swanson, W. J. (2012). Evolution of the egg: New findings and challenges.* Annual Review of Genomics and Human Genetics, 13, *109–125. doi: 10.1146/annurev-genom-090711-163745; Goudet, G., Mugnier, S., Callebaut, I., & Monget, P. (2008). Phylogenetic analysis and identification of pseudogenes reveal a progressive loss of zona pellucida genes during evolution of vertebrates.* Biology of Reproduction, 78*(5), 796–806. biolreprod.107.064568 [pii]; Wilburn, D. B., & Swanson, W. J. (2018). Egg, comparative vertebrate. In M. A. Skinner (Ed.),* Encyclopedia of reproduction *(2nd ed.). Academic Press.*

Vertebrate taxa differ in the number and type of ZP proteins incorporated in their egg coat matrix, with anywhere from zero to many copies of each ZP subfamily represented in each lineage (Shu et al., 2015). ZP3, however, is the only universal ZP gene in vertebrates, suggesting that it may be the ancestral gene to all other ZP gene families, in agreement with its more minimal architecture (see Fig. 1) (Goudet et al., 2008; Litscher & Wassarman, 2014; Shu et al., 2015; Wassarman & Litscher, 2016). ZP3 likely duplicated several times hundreds of millions of years ago, giving rise to three to four ZP genes in fish (ZP1, ZP3, ZPAX, variants of ZP1 and ZP3), four to five ZP genes in amphibians (ZP2–4, ZPD, ZPAX), six ZP genes in birds (ZP1–4, ZPD, ZPAX), and three to four ZP genes (ZP1–3, sometimes ZP4) in mammals (Fig. 2) (Wassarman & Litscher, 2016). ZP4 shares a common ancestral gene with ZP1 and is present in rats as well as in humans and other

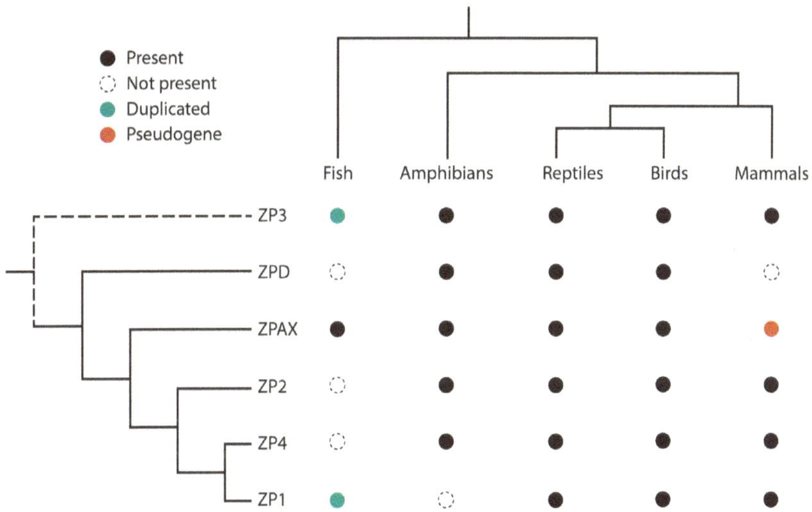

Fig. 2 Patterns of ZP gene duplication and loss are highly variable among the major vertebrate lineages; the fish clade comprises teleost fish. *Black circle*: gene present in genome; *dashed circle*: gene not present in genome; *teal circle*: gene duplicated in genome; *red circle*: gene pseudogenized in genome (Goudet et al., 2008; Meslin et al., 2012). *Adapted from Goudet, G., Mugnier, S., Callebaut, I., & Monget, P. (2008). Phylogenetic analysis and identification of pseudogenes reveal a progressive loss of zona pellucida genes during evolution of vertebrates.* Biology of Reproduction, 78*(5), 796–806. biolreprod.107.064568 [pii]; Shu, L., Suter, M. J., & Räsänen, K. (2015). Evolution of egg coats: Linking molecular biology and ecology.* Molecular Ecology, 24*(16), 4052–4073. 10.1111/mec.13283; Spargo, S. C., & Hope, R. M. (2003). Evolution and nomenclature of the zona pellucida gene family.* Biology of Reproduction, 68*(2), 358–362; Wong, J. L., & Wessel, G. M. (2006). Defending the zygote: Search for the ancestral animal block to polyspermy.* Current Topics in Developmental Biology, 72, *1–151. S0070-2153(05)72001-9 [pii].*

primates but is pseudogenized in the mouse genome (Conner et al., 2005; Goudet et al., 2008; Monne et al., 2006; Wassarman & Litscher, 2016; Wilburn & Swanson, 2016). In summary, ZP1–4 are found in mammals and other vertebrates, ZPD only in amphibians and birds, and ZPAX only in fish, amphibians, and birds (Wassarman & Litscher, 2016).

The ZP domain, despite being named for its abundance in mammalian egg coats, is not found solely in reproductive proteins (Jovine et al., 2005; Litscher & Wassarman, 2007, 2014). ZP subfamilies share an ancestral gene with the CUZD1/DMBT1 gene subfamily (CUB and ZP-like domains 1/Deleted in Malignant Brain Tumors 1), which includes proteins that incorporate two domains, the CUB domain and the ZP domain (Goudet et al., 2008). The CUB domain is found almost exclusively in extracellular and plasma membrane-associated proteins, many of which are involved in developmental processes such as embryogenesis and organogenesis (Bork, 1991; Bork & Beckmann, 1993). The CUB domain and ZP domain are also present in two families of proteins involved in sperm–egg recognition: the CUB domain in male spermadhesins, and the ZP domain in female ZP proteins (Goudet et al., 2008; Monne et al., 2006; Topfer-Petersen & Calvete, 1996). CUZD1/DMBT1 proteins are known to be expressed in the female reproductive tract, consistent with a role in fertilization (Goudet et al., 2008).

2.1 ZP Gene Losses Among Vertebrates

Given the diversity of ZP pseudogenes across the vertebrate phylogeny, ZP genes are thought to have evolved mainly through lineage-specific gene losses (Aagaard et al., 2010; Goudet et al., 2008; Meslin et al., 2012). For instance, the presence of both ZP1 and ZP4 in chicken, rat, chimpanzee, and human implies that the gene duplication that permitted the divergence of the ZP1 and ZP4 occurred early in the vertebrate lineage, before the separation of birds and mammals (~310 Mya), but after the divergence of fish (Conner et al., 2005; Goudet et al., 2008). ZP4 is a pseudogene in mouse, indicating that the loss of ZP4 occurred after the divergence of mouse and rat (Goudet et al., 2008). In mammals, only primates and rodents have a ZP1 gene, although ZP1 is present as a pseudogene in cow and dog, suggesting that the death of ZP1 in those species happened after the divergence between primate and rodent groups and other mammals (Goudet et al., 2008). The persistence of both ZP1 and ZP4 across the higher vertebrates suggests that there is a functional importance to these paralogs, as both have been retained (Conner et al., 2005).

After the divergence of birds and mammals, the ZPAX and ZPD genes seem to have been lost in mammals but not in birds (Goudet et al., 2008). Loss of ZPAX in mammals is predicted to have occurred before the divergence of humans and monkeys, as similar mutations were observed in human and chimpanzee ZPAX pseudogenes (Goudet et al., 2008).

In fish, the phylogeny of ZP genes is less well known due to both genome and gene duplications, particularly of ZP3: for instance, there are four copies of ZP3 in *Oryzias latipes* and three in *Danio rerio* (Goudet et al., 2008; Meslin et al., 2012; Sano et al., 2013; Smith et al., 2005).

The persistence of the ZP2 and ZP3 subfamilies across vertebrate lineages suggests that both genes are functionally significant (Goudet et al., 2008). In fact, the egg coats of all mammals contain ZP2 and ZP3 proteins, along with one or both of the ZP1 and ZP4 proteins (Goudet et al., 2008). These findings may imply that sperm–egg interactions in mammals requires the presence of ZP2 and ZP3 as well as one or both of ZP1 and ZP4 (Goudet et al., 2008).

It has been proposed that the pervasive loss of ZP genes in mammals could be due to taxon-specific differences in selective environments (Goudet et al., 2008; Wong & Wessel, 2006a). ZP genes may be lost in mammals because they no longer play a role in egg coat matrix formation or sperm–egg interactions (Goudet et al., 2008; Meslin et al., 2012; Shu et al., 2015). However, another possibility is that in animals with internal fertilization, embryos no longer develop in external environments and are thus no longer subject to the ecological aspects of natural selection that embryos of animals with external fertilization and/or external development (e.g., fish, amphibians, birds) are subject to (Goudet et al., 2008; Shu et al., 2015; Wong & Wessel, 2006a). This loss of selection on embryonic performance could result in the genes encoding additional structures or functions of egg coats being pseudogenized (Goudet et al., 2008; Shu et al., 2015; Wong & Wessel, 2006a).

2.2 Structure of ZP Proteins

ZP proteins share a common structural organization with four main features: (1) an N-terminal secretory signal peptide (SP) that marks them as secreted proteins; (2) the ZP domain, a conserved sequence of ~260 amino acids including 8 or 10 invariant cysteine residues that adopt two alternative disulfide bond connectivities; (3) a recognition site for members of the proprotein convertase family of proteolytic enzymes called a consensus furin cleavage site (CFCS); and (4) a C-terminal propeptide (CTP) that includes a single-spanning transmembrane (TM) domain (Jovine et al., 2005; Monne et al., 2006; Wilburn & Swanson, 2017). These elements play crucial roles in the secretion and assembly of ZP subunits (Monne et al., 2006).

3. THE ZP MODULE

All animal egg coat proteins share a common molecular basis, the ZP domain (Jovine et al., 2005; Shu et al., 2015; Wassarman & Litscher, 2016; Wong & Wessel, 2006a). Egg coat subunits polymerize using this conserved structural motif, suggesting that the basic architecture of animal egg coats has been conserved over hundreds of millions of years of evolution (Han et al., 2010; Litscher & Wassarman, 2007; Monne et al., 2006; Monne & Jovine, 2011).

The ZP domain, the structural element that gives ZP proteins their name, was first identified in ZP2 and ZP3 by pattern-based sequence analysis (Bork & Sander, 1992; Wassarman & Litscher, 2016). ZP domains are conserved protein–protein interaction modules comprised of two related immunoglobulin-like domains, ZP-N and ZP-C, that each contain characteristic disulfide bonding patterns (Jovine, Janssen, Litscher, & Wassarman, 2006; Monne & Jovine, 2011). ZP-N (~120 amino acids) and ZP-C (~130 amino acids) both have four conserved cysteine residues present as intramolecular disulfide bonds (Jovine et al., 2005; Wilburn & Swanson, 2016, 2017).

Biochemical data indicate that only ZP-N is required for protein polymerization (Jovine et al., 2006), and many ZP proteins contain tandem arrays of ZP-N repeats that have evolved independently of each other and from their associated ZP-C motifs (Callebaut, Mornon, & Monget, 2007; Swanson et al., 2011; Wilburn & Swanson, 2016, 2017). While the combined ZP-N/ZP-C pair has classically been referred to as the "ZP domain," ZP-N should be considered a domain of its own independent of ZP-C, and we will use the term "ZP module" as put forth by Bokhove et al. to refer to the combined ZP-N/ZP-C unit (Fig. 3; see also Fig. 1) (Bokhove et al., 2016; Callebaut et al., 2007; Monne & Jovine, 2011; Wilburn & Swanson, 2017).

The ZP module consists of two β sheets whose strands enclose a hydrophobic core comprising 8 or 10 cysteine residues. In Type I ZP modules the hydrophobic core contains eight invariant cysteines, a structure homologous to ZP3. In Type II ZP modules, the hydrophobic core contains 10 invariant

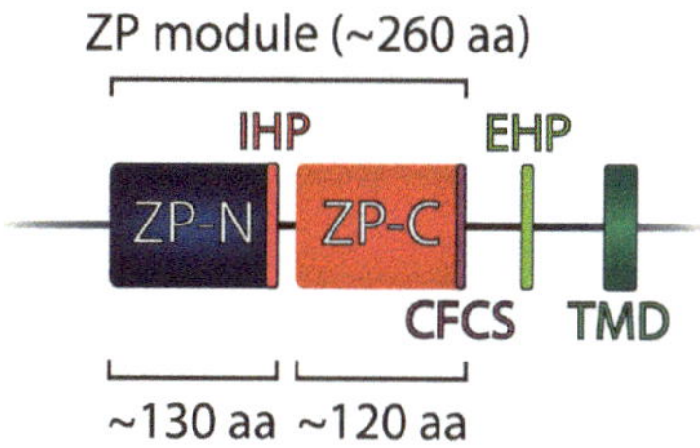

Fig. 3 Schematic of the ZP module, which comprises adjacent ZP-N and ZP-C structural domains. The location of other common ZP protein structural features is indicated on the schematic, including the consensus furin cleavage site (CFCS), the transmembrane domain (TMD) if present, the external hydrophobic patch (EHP), and the internal hydrophobic patch (IHP). The EHP and IHP are involved in ZP protein polymerization. *aa*, amino acid. *Adapted from Jovine, L., Darie, C. C., Litscher, E. S., & Wassarman, P. M. (2005). Zona pellucida domain proteins.* Annual Review of Biochemistry, *74, 83–114. doi: 10.1146/annurev.biochem.74.082803.133039; Wassarman, P. M. (2008). Zona pellucida glycoproteins.* The Journal of Biological Chemistry, 283*(36), 24285–24289. doi: 10.1074/jbc.R800027200.*

cysteines and is homologous to ZP1/ZP2/ZP4 (Bork & Sander, 1992; Claw & Swanson, 2012; Jovine et al., 2005; Monne & Jovine, 2011).

Additional ZP-N domains are present in single or multiple copies at the N-terminus of ZP1, ZP2, ZP4, and ZPAX (see Fig. 1) (Callebaut et al., 2007; Goudet et al., 2008; Jovine, Qi, Williams, Litscher, & Wassarman, 2002; Jovine et al., 2005, 2006; Monne, Han, Schwend, Burendahl, & Jovine, 2008; Wassarman & Litscher, 2016; Wilburn & Swanson, 2017). After the crystal structure of mouse ZP3 ZP-N was solved, the sequences of the other ZP proteins N-terminal to the ZP module were threaded through this three-dimensional structure (Monne et al., 2008). By this analysis, it was determined that the N-terminal regions of ZP1 and ZP4 each contain an additional copy of the ZP-N domain, and the N-terminal region of ZP2 has three additional ZP-N domain repeats in tandem, connected by short linkers (Monne et al., 2008; Monne & Jovine, 2011). ZPAX, too, has several additional ZP-N repeats in its N-terminus—for instance, there are five additional copies in *Xenopus tropicalis* (Callebaut et al., 2007; Goudet et al., 2008).

3.1 Other Roles for ZP Module-Containing Proteins

The presence of the ZP module is not limited to egg coat proteins and is found in hundreds of extracellular proteins with diverse functions in mammals, amphibians, birds, fish, flies, worms, mollusks, and urochordates (Jovine et al., 2005; Litscher & Wassarman, 2007, 2014). Proteins containing

ZP modules are structural components of animal tissues, serve as receptors, mechanotransducers, and antimicrobials, and are involved in cell signaling, differentiation, and morphogenesis (Bork & Sander, 1992; Chung, Zhu, Han, & Kernan, 2001; Fernandes et al., 2010; Heiman & Shaham, 2009; Jovine et al., 2005; Litscher & Wassarman, 2007; Plaza, Chanut-Delalande, Fernandes, Wassarman, & Payre, 2010; Wassarman & Litscher, 2016). ZP module-containing proteins organize and shape highly specialized apical structures in epithelial cells and are involved in the functioning of taste and smell (Wassarman & Litscher, 2016). Examples of ZP module-containing proteins include TGF-β receptor III (betaglycan), uromodulin, tectorin-α and -β, endoglin, vomeroglandin, hensin, cuticlins, oikosins, and mesoglein (Litscher & Wassarman, 2014; Wassarman & Litscher, 2016).

The presence of ZP modules in hundreds of polymeric extracellular proteins in eukaryotes, from jellyfish to humans, suggests that the structure has been conserved through at least 600 million years of evolution (Monne et al., 2006; Wassarman & Litscher, 2016). In fact, a *Saccharomyces cerevisiae* mating protein called α-agglutinin/Sag1p Ig III adopts a three-dimensional fold similar to ZP-N, so it is possible that the ZP module has been conserved for closer to 1 billion years of evolution (Swanson et al., 2011; Wassarman & Litscher, 2016).

Mutations in ZP module-containing proteins cause severe human pathologies, including deafness, vascular disease, renal disease, cancer, and potentially infertility (Wassarman & Litscher, 2016).

3.2 Not All Egg Coat Proteins Are ZP Proteins

While ZP proteins appear to be the core building blocks of egg coats in vertebrates, the genes encoding egg coat proteins in invertebrates are not as conserved across taxa (Shu et al., 2015; Wong & Wessel, 2006a). Although the egg coats of some marine invertebrates do contain ZP modules, several different egg coat genes are found in other invertebrates. Examples include egg bindin receptor 1 (EBR1) and rendezvin in sea urchins (Vacquier & Swanson, 2011; Wong & Wessel, 2006a); OBi1 in sea stars (Hart, Sunday, Popovic, Learning, & Konrad, 2014); chorion genes in *Drosophila* (Jagadeeshan & Singh, 2007; Papantonis, Swevers, & Iatrou, 2015), silk moths (Papantonis et al., 2015), other lepidopterans (Carter et al., 2013), and mosquitoes (Marinotti et al., 2014); and the Brownie and Citrus genes in the cockroach *Blattella germanica* (Irles, Belles, & Piulachs, 2009; Irles & Piulachs, 2011). These genes function as structural components, facilitate sperm–egg interactions, and protect embryos (Shu et al., 2015).

4. SYNTHESIS AND POLYMERIZATION OF ZP PROTEINS

As has been previously noted, the basic molecular structure of the egg coat has been conserved through hundreds of millions of years of evolution. For instance, recombinant mouse egg coat subunits can incorporate into the egg coats of *Xenopus* oocytes due to their common polymerization domain, the ZP module (Doren et al., 1999; Monne et al., 2006). How do ZP proteins polymerize to form the ECM of the egg coat?

After cleavage of the SP, ZP protein precursors are transported through the endoplasmic reticulum (ER) and the Golgi, remaining bound to the membrane of these organelles by their TM domain (Monne et al., 2006). In the ER/Golgi the ZP proteins form disulfide bonds and are modified with N- and O-linked oligosaccharides (Monne et al., 2006). The membrane-anchored proteins are then packaged into large vesicles (~2 μm in diameter), which fuse with the plasma membrane of the oocyte (Monne et al., 2006). After membrane fusion, or potentially prior to fusion within the *trans*-Golgi, ZP precursors are cleaved at their CFCS (Monne et al., 2006). This C-terminal processing is dependent on the TM domain and releases mature ZP proteins into the perivitelline space where they incorporate into the innermost layer of the growing egg coat via their ZP module (Monne et al., 2006).

ZP modules have been shown to interact with each other directly in the polymerization of ZP proteins (Jovine et al., 2002). The protofilaments formed by ZP proteins are organized in a right-handed double helix with frequent branching, creating a reticular network (Wong & Wessel, 2006a). ZP proteins can interact heterospecifically, permitting a diverse assembly of proteins within the reticular network of protofilaments (Wong & Wessel, 2006a). For instance, both urinary and cochlear ZP proteins can incorporate into the mouse egg coat if the whole ZP module and adjacent C-terminus are intact (Jovine et al., 2002). The autoaggregation and polymerization of ZP proteins are advantageous to ECM formation as additional motifs associated with the ZP module can be incorporated without interfering with matrix assembly (Wong & Wessel, 2006a).

Other structural features of ZP proteins appear to relate to the specific functions of each subunit: cross-links between ZP filaments are thought to be established by the N-terminal region of ZP1 through its trefoil domain, and regions N- and C-terminal to the ZP module of ZP2 and ZP3, respectively, are thought to be involved in sperm–egg interaction (Monne et al., 2006; Monne & Jovine, 2011).

ZP proteins vary across species in their sites of synthesis. In mammals and amphibians, ZP proteins are synthesized solely in the ovary by oocytes and/or follicle cells, whereas in fish and birds ZP proteins are synthesized in the ovary and/or liver in response to estrogen and transported via the bloodstream to the ovary, where they self-assemble around eggs (Jovine et al., 2005; Litscher & Wassarman, 2014; Sano et al., 2013; Wassarman & Litscher, 2016). Consequently, TM domains are not present in fish ZP proteins synthesized by the liver (Hyllner, Westerlund, Olsson, & Schopen, 2001; Sugiyama, Murata, Iuchi, Nomura, & Yamagami, 1999; Wong & Wessel, 2006a). The timing of ZP gene expression in teleosts likely coincides with vitellogenesis, such that soluble ZP proteins lacking a TM domain can be transported to the ovarian follicles along with vitellogenic proteins, limiting ZP protein precipitation in circulation and ensuring the movement of proteins essential to oogenesis (Callard, Riley, & Perez, 1990a, 1990b; Polzonetti-Magni, Mosconi, Soverchia, Kikuyama, & Carnevali, 2004; Schneider, 1996).

In terms of the sites of ZP protein synthesis, it has been proposed that simpler egg coats may contain only oocyte-derived proteins, whereas more elaborate egg coat matrices may require additional contributions from somatic tissue (Wong & Wessel, 2006a). More complex egg coats are often associated with mechanically protective roles, such as resistance to environmental hazards like osmotic shock and desiccation (Wong & Wessel, 2006a). In support of this hypothesis, in animals whose egg coats are very robust, such as fish and birds, ZP proteins are often synthesized in the liver (Bausek, Waclawek, Schneider, & Wohlrab, 2000; Chang, Lu, Lai, Kou, & Huang, 1999; Hyllner et al., 2001; Okumura et al., 2004; Wong & Wessel, 2006a).

Proteins that assemble in the extracellular space must have mechanisms to avoid premature association within the cell as they are synthesized (Monne et al., 2006). ZP protein precursors are stabilized in a soluble, nonpolymerization-competent conformation by two short, conserved motifs: an external hydrophobic patch (EHP) in the C-terminus between the CFCS and the TM domain, if present, and an internal hydrophobic patch (IHP) between the ZP-N and ZP-C of the ZP module (see Fig. 3) (Jovine et al., 2004; Monne et al., 2006). Cleavage of ZP precursors at the CFCS, an event required for secretion of mammalian ZP proteins and incorporation of both fish and mammalian subunits into the inner layer of the growing egg coat, dissociates mature polypeptides from the EHP and activates them for polymerization (Jovine et al., 2004; Litscher, Qi, & Wassarman, 1999; Qi, Williams, & Wassarman, 2002; Sugiyama et al., 1999). Thus, despite fish and mammalian egg coat proteins differing in their sites of synthesis and

C-terminal architecture, they share a common assembly mechanism (Monne et al., 2006). Because this mechanism relies on highly conserved elements such as the presence of the EHP/IHP and cleavage at the CFCS, it is likely common to all ZP proteins (Jovine et al., 2005, 2004; Monne et al., 2006).

Notably, in experiments where ZP proteins are truncated just upstream of the TM domain, proteins lacking a TM domain are secreted normally but are not cleaved at the CFCS or incorporated into the egg coat (Jovine et al., 2002, 2004). This suggests that TM domains in ZP proteins are not involved in specific interactions, but help with proper localization and/or topological orientation of nascent proteins for proteolytic processing and assembly (Litscher & Wassarman, 2007). Whereas mammalian ZP precursors lacking a TM domain are not assembled into the egg coat, ZP precursors from fish or birds that lack a TM domain endogenously undergo cleavage at the CFCS and assemble into the egg coat normally upon reaching the ovary (Darie et al., 2005; Sugiyama et al., 1999).

Ultrastructural analyses of egg coats suggest that they consist of filaments of similar dimensions across organisms, but how egg coat subunits are organized into these polymers is less clear (Iconomidou et al., 2000; Jovine et al., 2005; Monne et al., 2006; Nara et al., 2006). Biochemical, electron microscopy, and gene knockout studies are most consistent with a model in which the filaments are a linear repetition of ZP2/ZP3 heterodimers, with the interface between ZP2 and ZP3 running perpendicular to the axis of the filaments (Dean, 2004; Wassarman, 1988; Wassarman & Mortillo, 1991). Filament formation is therefore dependent on the interaction between Type I (ZP3) and Type II (ZP1/ZP2/ZP4) ZP modules (Monne et al., 2006).

ZP1 is thought to be responsible for cross-linking these filaments into a three-dimensional matrix, mediated by its N-terminal trefoil domain (see Fig. 1) (Greve & Wassarman, 1985; Monne et al., 2006; Rankin, Talbot, Lee, & Dean, 1999). In the mammalian egg coat, ZP1 is expressed at much lower levels than the other subunits, so ZP1 (and ZP4, if present) would be incorporated only rarely in place of ZP2 (Monne et al., 2006). By contrast, in fish, the presence of two or more ZP1 homologs (at least one of which is highly expressed (Brivio, Bassi, & Cotelli, 1991)), as well as the lack of a ZP2 homolog, predicts an egg coat with a much higher number of cross-links (Monne et al., 2006). This is in keeping with the significant resistance to mechanical and chemical stress that fish egg coats display, even prior to hardening (Monne et al., 2006). Thus the composition of egg coats in terms of the number of ZP1 homologs they contain may suggest physical properties, by giving an estimate of the number of cross-links (Jovine et al., 2005; Monne et al., 2006). Notably, the formation of additional intra- and intermolecular disulfide bonds has been implicated in the hardening of the mammalian egg coat (Iwamoto et al., 1999; Monne et al., 2006).

5. EGG COAT STRUCTURE

5.1 Mammals

Mouse egg coats are the best understood of all vertebrates, and the most well studied of mammals specifically (Monne et al., 2006; Shu et al., 2015). Mouse egg coats are comprised of homologs of ZP1 (~200 kDa), ZP2 (~120 kDa), and ZP3 (~83 kDa) (Goudet et al., 2008; Litscher & Wassarman, 2007; Wassarman, 2008). These proteins assemble into ~2–3 μm long filaments, with a structural repeat of ~150 Å, and are cross-linked into a highly porous, 6.5 μm thick elastic network (Monne et al., 2006). The egg coats of other mammals, including humans, are thought to have a similar structure with the inclusion of an additional ZP1-like subunit, ZP4 (Goudet et al., 2008; Monne & Jovine, 2011).

By electron microscopy, it was suggested that mouse ZP2 and ZP3 assemble into micron-long polymers which are cross-linked into a three-dimensional matrix by disulfide-bonded ZP1 homodimers (Greve & Wassarman, 1985; Monne & Jovine, 2011; Wassarman & Mortillo, 1991). This model is confirmed by the phenotypes of mice lacking the genes for the individual ZP subunits (Monne & Jovine, 2011). Homozygous ZP1 knockout mice produce an egg coat, but it is loose and insufficiently cross-linked and mutant females are less fertile than wild type, suggesting that ZP1 interconnects ZP fibrils and that a structurally intact egg coat is integral to fertilization (Claw & Swanson, 2012; Rankin et al., 1999; Wassarman & Litscher, 2013). Homozygous knockout female mice lacking either ZP2 or ZP3 fail to construct an egg coat and are infertile, indicating that ZP2 and ZP3 depend on each other for incorporation into the egg coat (Liu et al., 1996; Rankin et al., 1996, 2001; Wassarman & Litscher, 2013, 2016). Mutant female mice with a single ZP3 allele assemble an egg coat and reproduce normally, but their egg coat is less than half the thickness of wild type (Wassarman & Litscher, 2013, 2016; Wassarman, Qi, & Litscher, 1997).

5.2 Birds

The egg coats of birds consist of two layers separated by a thin continuous membrane, with the inner or perivitelline layer representing a 1- to 3.5-μm thick network of fibers analogous to the mammalian egg coat as it mediates the species-specific binding of sperm (Monne et al., 2006). After fertilization, oocytes acquire the continuous membrane (0.1–0.5 μm thick) and the outer layer (3–8 μm thick), which are thought to be involved in blocks to polyspermy (Ichikawa, Matsuzaki, Hiyama, Mizushima, & Sasanami, 2016; Monne et al., 2006).

As with the mammalian egg coat, the avian perivitelline layer contains several glycoproteins: homologs to ZP1, ZP2, ZP3, ZP4, and ZPD have been found in quail, and genes for ZP1, ZP2, ZP3, ZP4, ZPD, and ZPAX are present in the chicken genome (Goudet et al., 2008; Ichikawa et al., 2016; Meslin et al., 2012; Monne et al., 2006; Serizawa et al., 2011; Smith et al., 2005).

Avian ZP3 is ~32–42 kDa, and avian ZP1 exists as both a ~97 kDa monomer and a homodimer held together by intermolecular disulfide bonds (Monne et al., 2006). ZP1 is secreted by the liver in response to estrogens and is characterized by a proline/glutamine-rich repeat region N-terminal to the trefoil domain and a short CTP lacking a TM domain, as would be predicted by its liver synthesis (Bausek et al., 2000; Monne et al., 2006; Sasanami, Pan, & Mori, 2003). Chicken ZPD (~42 kDa) loosely associates with the perivitelline layer. Together with dimeric ZP1, ZPD has been implicated in sperm activation (Ichikawa et al., 2016; Monne et al., 2006; Okumura et al., 2004). ZP3 is thought to be responsible for sperm binding in the perivitelline layer of both chicken and quail (Ichikawa et al., 2016). Upon binding of sperm, avian ZP1 is degraded and hole-like structures appear in the perivitelline layer (Monne et al., 2006; Takeuchi et al., 2001).

5.3 Amphibians

Amphibians include anurans (frogs), which reproduce by external fertilization in water, and urodeles (salamanders), which reproduce by internal fertilization in the female cloaca (Monne et al., 2006; Watanabe & Onitake, 2002). The ~1 μm thick egg coat of *Xenopus laevis* is the best studied of the anurans and consists of five major ZP glycoproteins that are synthesized by the oocyte: ZP2 (gp69/64), ZP3 (gp43/41), ZP4 (gp37), ZPAX (gp120/112), and ZPD (gp80) (Goudet et al., 2008; Hedrick, 2008; Kubo et al., 1997, 2000, 1999; Lindsay, Wallace, & Hedrick, 2001; Lindsay, Yang, & Hedrick, 2002; Monne et al., 2006; Tian, Gong, & Lennarz, 1999).

Xenopus ZPAX contains a ~600 amino acid N-terminal region and a short CTP lacking a predicted TM domain and is related to ZP2 with its Type II ZP module and absence of a trefoil domain (Monne et al., 2006).

In agreement with homology to ZP2, *X. tropicalis* ZPAX has five additional ZP-N repeats in its N-terminus (Callebaut et al., 2007).

Xenopus ZPD has a simple architecture, consisting of an SP, a ZP module, a CFCS, and a TM domain, and appears to represent a subfamily of its own: although its ZP module sequence is most similar to a Type II ZP module, it lacks 2 of the 10 conserved cysteines (Lindsay et al., 2002; Monne et al., 2006).

Fertilization in urodels has not been as well studied, with the exception of the newt *Cynops pyrrhogaster. C. pyrrhogaster* eggs can undergo polyspermy, as no fertilization envelope forms after fertilization (Watanabe & Onitake, 2002). A transcriptome assembly from ovary, testis, and oviduct found homologs to all six ZP subfamilies—ZP1, ZP2, ZP3, ZP4, ZPAX, and ZPD—expressed in the ovaries of *C. pyrrhogaster* (Watanabe & Takayama-Watanabe, 2014). Notably, the authors found six distinct paralogs of ZP3 (Watanabe & Takayama-Watanabe, 2014).

5.4 Teleost Fish

Teleost fish are highly diverse, constituting almost half of the total number of vertebrates (Monne et al., 2006). This diversity is reflected in the architecture of teleost egg coats, which vary in thickness, structure, and number of layers both between and within species (Monne et al., 2006). A single layer of follicle cells surrounds oocytes as they grow (Monne et al., 2006). In response to external signals, follicle cells produce 17β-estradiol to induce synthesis of both egg yolk (vitellogenin) and ZP protein precursors (Monne et al., 2006). In most teleosts, soluble ZP protein precursors are secreted by hepatocytes and travel in the blood to the oocyte for incorporation into the egg coat (Arukwe & Goksoyr, 2003; Sugiyama & Iuchi, 2000). ZP precursors are deposited in the perivitelline space, at the base of long microvilli that stretch from the plasma membrane of the oocyte to the follicle cells (Monne et al., 2006). Egg coat assembly proceeds as the oocyte accumulates egg yolk, resulting in a radially striated structure of helicoidal glycoprotein bundles separated by extended microvilli (Monne et al., 2006).

The egg coats of rainbow trout (*Oncorhynchus mykiss*) have a thin outer layer and a ~50 μm thick inner layer of three major subunits: VEα (~58 kDa), VEβ (~52 kDa), and VEγ (~47 kDa) (Darie, Litscher, & Wassarman, 2008; Litscher & Wassarman, 2007; Monne et al., 2006). Like avian ZP1, these subunits have an N-terminal proline/glutamine-rich repeat region and a short CTP lacking a predicted TM domain (Brivio et al., 1991; Monne et al., 2006). VEα and

VEβ are very similar in sequence and contain a trefoil domain immediately before a Type II ZP module, suggesting homology to mammalian ZP1/ZP2/ZP4, whereas VEγ contains a sperm combining site-like sequence C-terminal to its Type I ZP module, suggesting homology to ZP3 (Brivio et al., 1991; Darie, Biniossek, Jovine, Litscher, & Wassarman, 2004; Litscher & Wassarman, 2007; Monne et al., 2006). Rainbow trout egg coats consist of VEα/γ and VEβ/γ heterodimers (Litscher & Wassarman, 2007).

The egg coats of other teleost fish have similar compositions to rainbow trout, although in species such as carp, goldfish, and zebrafish, ZP genes are synthesized in the ovaries and thus contain a predicted TM domain in their CTP (Chang, Hsu, Wang, Tsao, & Huang, 1997; Litscher & Wassarman, 2007; Mold et al., 2001). In medaka, ZP subunits are synthesized by both the liver and oocytes (Kanamori, Naruse, Mitani, Shima, & Hori, 2003). Additionally, ZP3 genes are often duplicated within teleosts: for instance, there are four ZP3 genes in medaka and three in zebrafish (Goudet et al., 2008).

No ZP2 orthologs have been identified in teleosts, despite the classification of some fish egg coat genes as ZP2 (Hughes, 2007). These genes are in fact ZP1 homologs, as evidenced by the presence of a trefoil domain in these proteins (see Fig. 1) (Berois, Arezo, & Papa, 2011; Hughes, 2007). In agreement with the absence of the ZP2 subfamily in teleosts, there are no N-terminal ZP-N repeat regions in teleost ZP proteins, except for some homologs of ZPAX (Hedrick, 2008).

Teleost sperm lack an acrosome, a secretory vesicle that facilitates sperm–egg contact and egg coat dissolution. Instead, sperm reach the egg plasma membrane through the micropyle, a funnel-shaped, narrow channel through the egg coat (Hart & Donovan, 1983; Iwamatsu, Onitake, Matsuyama, Satoh, & Yukawa, 1997; Litscher & Wassarman, 2007; Monne et al., 2006; Wong & Wessel, 2006a). The micropyle attracts sperm by chemotaxis, and its precise diameter prevents polyspermy by permitting one sperm to pass at a time (Amanze & Iyengar, 1990; Lombardi, 1998; Yanagimachi et al., 2013). In contrast to mammals, birds, amphibians, mollusks, echinoderms, and urochordates, teleost sperm do not need to dissolve the egg coat to reach the egg and fuse with its plasma membrane (Hart, 1990; Lombardi, 1998). Therefore, it has been argued that fish egg coat proteins play a purely structural role in fertilization (Hart, 1990; Litscher & Wassarman, 2007; Monne et al., 2006).

The structure of the micropyle has convergently evolved in at least two animal orders with different modes of reproduction, the dipterans and the teleosts (Wong & Wessel, 2006a). In fish the micropyle is formed after the deposition of the egg coat, by the retraction of the cytoplasmic process of a specialized follicle cell called the micropylar cell that extends to the oocyte surface (Hart, 1990; Lombardi, 1998). In zebrafish it has been shown that mutants in the gene bucky ball have an excessive number of follicle cells that develop as micropylar cells, leading to multiple functional micropyles and polyspermic fertilization (Marlow & Mullins, 2008). Micropyle architecture differs across teleost species, from simple channels traversing the egg coat to more elaborate structures with outer sperm catchment areas that funnel sperm into the micropyle (Amanze & Iyengar, 1990; Cherr & Clark, 1986; Hart, 1990; Hart, Pietri, & Donovan, 1984; Yamagami, Hamazaki, Yasumasu, Masuda, & Iuchi, 1992).

After fertilization, teleost egg coats undergo a cortical reaction and secrete a transglutaminase, which hardens the egg coat by introducing cross-links between egg coat subunits, likely via their proline/glutamine-rich repeat regions (Monne et al., 2006; Sugiyama & Iuchi, 2000). The hardened egg coat has a different morphology and protects the developing embryo against environmental hazards and pathogens. Notably, mammalian ZP proteins lack an N-terminal proline/glutamine-rich repeat region, and covalent linkages between ZP subunits have not been detected in mammalian eggs or embryos (Litscher & Wassarman, 2007; Wassarman, 1988).

5.5 Mollusks

Marine invertebrates are some of the first model systems in the study of fertilization, facilitated by their numerous, accessible gametes as an externally fertilizing group (Turner & Hoekstra, 2008). Much of the work has centered on the gastropod abalone (genus *Haliotis*). The abalone egg coat is known to mediate species specificity in gamete interactions and triggers the sperm acrosome reaction (Lyon & Vacquier, 1999; Monne et al., 2006). This exocytic event releases lysin, a dimeric 16 kDa protein that binds to the vitelline envelope receptor for lysin (VERL), a highly repetitive ~1 MDa egg coat glycoprotein, ~90% of which is 22 repeats of ~153 residues homologous to the vertebrate ZP-N polymerization domain (Galindo, Moy, Swanson, & Vacquier, 2002; Swanson & Vacquier, 1997; Wilburn & Swanson, 2016). These ZP-N repeats are thought to generate an interlocking network of hydrogen bonds, stabilizing the egg coat supramolecular structure (Wilburn & Swanson, 2016). The VERL protein consists of an SP, the array of 22 tandem ZP-N repeats, a Type II ZP module, a CFCS, and a TM domain (Galindo et al., 2002; Monne et al., 2006).

Lysin dissolves the abalone egg coat through nonenzymatic means, likely competing for hydrogen bonds between VERL repeats and splaying the fibers of the egg coat, allowing sperm to pass (Swanson & Vacquier, 1997; Wilburn & Swanson, 2016). Lysin dissolves conspecific egg coats faster than heterospecific egg coats, suggesting species specificity in its activity (Swanson & Vacquier, 1997; Wilburn & Swanson, 2016). In keeping with this pattern of species specificity, the sequence of lysin homologs from different species of abalone is extremely divergent as a result of strong adaptive evolution (Monne et al., 2006; Yang, Swanson, & Vacquier, 2000). VERL ZP-N repeats 1 and 2 are also rapidly evolving, whereas repeats 3–22 evolve neutrally and are homogenized (>98% identical within a given species) by unequal crossing over and concerted evolution (Galindo, Vacquier, & Swanson, 2003; Monne et al., 2006; Swanson, Aquadro, & Vacquier, 2001; Swanson & Vacquier, 1998; Wilburn & Swanson, 2016). Lysin and VERL experience correlated rates of evolution and display intergenic linkage disequilibrium despite not being physically linked (Clark et al., 2009). This coevolution between a male protein and a female receptor suggests that identification of coevolving proteins in the two sexes could potentially find other interacting fertilization proteins—for instance, mammalian ZP interactors (Clark et al., 2009; Hart et al., 2018).

There are other ZP module-containing proteins in the abalone egg coat in addition to VERL; in fact, at least 30 additional ZP proteins were identified in abalone egg coats by expressed sequence tag sequencing and shotgun proteomics (Aagaard et al., 2010). These additional ZP proteins included a paralog of VERL called VEZP14, which also contains a putative lysin-binding motif and is rapidly evolving (Aagaard et al., 2010).

The gastropod mollusk *Tegula*, a genus of free-spawning marine snails, also possesses an ortholog to VERL that binds *Tegula* lysin (Hellberg, Dennis, Arbour-Reily, Aagaard, & Swanson, 2012). Whereas abalone VERL has 22 tandem ZP-N repeats, *Tegula* VERL contains only one (Hellberg et al., 2012). Of note, however, is the presence of a homogenized array of three to four residues, mainly serine–proline–threonine or serine–proline–threonine–threonine, that is repeated 70 times between *Tegula* VERL's ZP-N repeat and its ZP module (Hellberg et al., 2012).

In the basal mollusk chiton, follicle cells surround a gelatinous protective layer around the egg coat called the jelly hull (Buckland-Nicks, Koss, & Chia, 1988; Litscher & Wassarman, 2014; Wong & Wessel, 2006a). These follicle cells shrink on contact with seawater, leaving behind pores in the hull that are almost continuous with pores in the egg coat (Alliegro & Wright, 1983; Buckland-Nicks et al., 1988; Mozingo, Vacquier, & Chandler, 1995;

Wong & Wessel, 2006a). These tunnels focus sperm and facilitate access to the egg plasma membrane, somewhat akin to micropyles (Buckland-Nicks, 1993; Wong & Wessel, 2006a).

In the bivalve *Unio*, a freshwater mussel, the egg coat is attached to the egg plasma membrane solely at the vegetal pole (Focarelli, Renieri, & Rosati, 1988). As the only site of sperm binding and fusion, this region is the functional equivalent of a micropyle (Wong & Wessel, 2006a). The attachment point is a crater on the egg coat comprised exclusively of the sperm-receptive gp273 glycoprotein, marking the only fusogenic region of the egg; the remainder of the egg coat consists of the structural glycoprotein gp180 (Focarelli & Rosati, 1995).

5.6 Sea Urchin

In sea urchin, the egg coat is estimated to contain 25 major glycoproteins (Gache, Niman, & Vacquier, 1983; Longo, 1981; Niman et al., 1984). Two of these have been shown to play a structural role: p160 and rendezvin^{VL} (Wong & Wessel, 2006a). p160 is a 160-kDa protein composed of five CUB domains and a putative TM domain and is predicted to be an integral membrane protein with protein–protein interaction motifs facing the ECM of the egg (Haley & Wessel, 2004). p160 is found at the tips of microvilli, suggesting that it links the egg plasma membrane to the egg coat until fertilization, at which point it is cleaved to separate the two and establish one block to polyspermy (Haley & Wessel, 2004).

Rendezvin^{VL} is a splice variant of the oocyte-specific rendezvin gene that is trafficked to the vitelline layer, where it reunites at fertilization with another rendezvin splice variant that associates with cortical granules: specialized organelles that are exocytosed into the perivitelline space between the egg coat and the egg that participate in the slow block to polyspermy (Claw & Swanson, 2012; Wong & Wessel, 2006b). The two splice variants create the fertilization envelope—modifications to the egg coat that form the block to polyspermy in sea urchins—likely via heterologous interactions between their CUB domains (Wong & Wessel, 2006b).

Analogous to lysin and VERL in the abalone and *Tegula* systems, sea urchins encode a pair of interacting fertilization proteins called bindin and EBR1 (Wilburn & Swanson, 2016). Bindin is the sole protein in sea urchin sperm acrosomes and mediates the species-specific binding of sperm to the egg coat via its receptor, EBR1 (Kamei & Glabe, 2003; Zigler, McCartney, Levitan, & Lessios, 2005). EBR1 has a similar architecture to

VERL in that it contains ~19 tandem EBR repeats consisting of alternating CUB and thrombospondin type 1 (TSP-1) domains in *Strongylocentrotus franciscanus*, with the last 10 EBR repeats being highly conserved and species-specific (Kamei & Glabe, 2003). *Strongylocentrotus purpuratus* EBR1 has a slightly modified architecture of ~8 EBR repeats that share 88% identity with the *S. franciscanus* EBR repeat, but an entirely differently species-specific domain consisting of 11 hyalin-like repeats (Kamei & Glabe, 2003). These species-specific domains in *S. purpuratus* and *S. franciscanus* EBR1 are known to function in species-specific sperm adhesion (Kamei & Glabe, 2003).

5.7 Insects

The egg coats of flies (dipterans) are composed of two layers: an inner vitelline layer and an outer chorion layer (Degrugillier & Leopold, 1976; Mouzaki, Zarani, & Margaritis, 1991; Pascucci, Perrino, Mahowald, & Waring, 1996; Turner & Mahowald, 1976). The chorion is synthesized by the surrounding follicle cells and protects the egg from desiccation, mechanical stress, and pathogens after it is laid (Papantonis et al., 2015; Wong & Wessel, 2006a).

Dipteran eggs are fertilized internally as the egg travels down the oviduct (Wong & Wessel, 2006a). Sperm are released in a controlled manner from the spermatheca, so the presence of a micropyle in dipterans, as previously noted, is somewhat unexpected (Degrugillier & Leopold, 1976; Mouzaki et al., 1991; Neubaum & Wolfner, 1999; Turner & Mahowald, 1976). It is likely that the importance of the egg coat in preventing desiccation takes precedence, making the micropyle less of a structure whose primary role is to block polyspermy and more one that supports gamete interactions while allowing gas exchange during embryogenesis (Li, Hodgeman, & Christensen, 1996; Wong & Wessel, 2006a).

The *Drosophila* chorion is composed of ~20 structural proteins synthesized by ovarian follicle cells and contains six major chorion proteins (cp): cp15, cp16, cp18, cp19, cp36, and cp38 (Papantonis et al., 2015; Parks, Wakimoto, & Spradling, 1986). cp36 and cp38 are expressed early in chorion formation (Parks et al., 1986), and cp15, cp16, cp18, and cp19 are expressed late (Griffin-Shea, Thireos, & Kafatos, 1982; Spradling & Mahowald, 1980; Spradling, 1981). These chorion proteins are stabilized in the final egg coat structure by peroxidase-mediated tyrosine cross-links (Konstandi et al., 2005).

Several additional genes in *Drosophila* are required for the assembly of the chorion, including fs(2)QJ42 and defective chorion 1 (dec-1) (Pascucci et al., 1996). The protein products of these genes are found throughout the egg coat rather than localized at the micropyle, suggesting that they play a role in structural integrity rather than sperm–egg interactions; this hypothesis is supported by the loss-of-function phenotypes associated with these loci (Pascucci et al., 1996; Wong & Wessel, 2006a). Females homozygous for fs(2)QJ42 produce chorions with altered permeability properties, and females who fail to synthesize dec-1 display morphological abnormalities in the chorion layers (Pascucci et al., 1996).

In silk moths, such as the cultivated *Bombyx mori* and the wild oak silk moth *Antheraea polyphemus*, the chorion contains more than 100 distinct components by two-dimensional gel electrophoresis (Kafatos et al., 1987). Silk moths possess an additional egg coat structure called the lamellar chorion, which in *Bombyx* contains unusual cysteine-rich proteins termed high-cysteine (Hc) proteins whose presence is unique to *Bombyx* (Hamodrakas, Kamitsos, & Papanikolaou, 1984; Nadel & Kafatos, 1980). These Hc proteins give the egg coat enhanced hardness and reduced permeability, likely as an adaptation to prolonged winter diapause (Nadel & Kafatos, 1980). Cysteine-rich proteins have also been found in the chorions of the mosquitoes *Aedes aegypti* and *Anopheles gambiae* (Amenya et al., 2010; Marinotti et al., 2014).

In the cockroach *B. germanica*, two highly abundant chorion genes, Brownie and Citrus, have been recently described (Irles et al., 2009; Irles & Piulachs, 2011). Brownie is expressed in follicle cells, forming a structure called the sponge-like body in a cavity left behind by those cells in late choriogenesis, when Brownie expression is at its highest (Irles et al., 2009). The sponge-like body is a complex structure that combines the micropyle and the aeropyle, a feature of insect eggs that functions in gas exchange (Irles et al., 2009).

Citrus is involved in chorion formation, as females treated with RNAi to Citrus laid fragile eggs showing discontinuous deposition of chorion proteins, resulting in increased permeability (Irles & Piulachs, 2011). Citrus has an SP and is rich in glycine, tyrosine, and proline, as with other insect chorion proteins where these abundances serve structural roles (such as tyrosine cross-linking) (Irles & Piulachs, 2011). The protein contains 33 repeats of the same motif: 30–40 residues in length, rich in glutamic acid at the N-terminus and glycine, tyrosine, and proline repetitions at the C-terminus (Irles & Piulachs, 2011). The first four N-terminal repeats are the least conserved, and database searches found no homologous proteins or motifs (Irles & Piulachs, 2011). Both Brownie and Citrus show an absence of cysteines and a high concentration of tyrosines, suggesting that they could be cross-linked as has been described in the chorions of dipterans (Irles & Piulachs, 2011; Konstandi et al., 2005).

5.8 Cephalochordates and Urochordates

Cephalochordates, urochordates, and vertebrates represent the three extant groups of chordate animals (Delsuc, Brinkmann, Chourrout, & Philippe, 2006). Ascidians (sea squirts) are mostly hermaphroditic urochordates that produce self-sterile gametes due to the presence of a self-incompatibility system (Monne et al., 2006; Sawada et al., 2004). The ascidian egg coat plays a role in species-specific binding of sperm and egg, participates in the slow block to polyspermy, and prevents self-fertilization in self-sterile species (Lambert & Goode, 1992; Monne et al., 2006).

Ascidian egg coats contain multiple homologs to ZP proteins, most of which possess epidermal growth factor (EGF)-like repeats (Yamada et al., 2009). For instance, the major component of the egg coat in *Halocynthia roretzi* is HrVC120 (precursor to HrVC70), which consists of an SP, 13 EGF-like repeats, a Type II-like ZP module, a CFCS, and TM domain (Ban, Harada, Yokosawa, & Sawada, 2005; Kurn, Sommer, Bosch, & Khalturin, 2007; Monne et al., 2006; Sawada et al., 2002). HrVC70, the product of proteolytic cleavage of HrVC120 within the last EGF repeat, contains 12 EGF repeats and shows sperm receptor activity (Kurn et al., 2007; Sawada et al., 2004). Because of its sperm-binding capabilities and polymorphic nature, HrVC70 has been proposed as a potential allorecognition molecule mediating self-sterility (Sawada et al., 2004). In *Halocynthia aurantium*, a homolog to HrVC120 called HaVC130 shows high similarity to HrVC120 but has 14 EGF domain repeats, suggesting that the number of EGF repeats could play a role in species specificity of sperm–egg interactions (Ban et al., 2005).

Ciona intestinalis, another urochordate, has four homologs of ZP proteins called Vc16, Vc20, Vc182, and Vc569 (Kurn et al., 2007). The ZP proteins consist of an SP, a ZP module, and a TM domain; both Vc182 and Vc16 have a CFCS, and all four have varying numbers of EGF repeats: six for Vc569, two for Vc20, one for Vc16, and none for Vc182 (Kurn et al., 2007). All four ZP proteins are expressed in the developing oocyte (Kurn et al., 2007). A more recent proteomic characterization of *C. intestinalis* egg coats found an additional 7 ZP proteins, bringing the total to 11; all 7 contain single or multiple EGF-like domains (Yamada et al., 2009).

ZP proteins are present in cephalochordate egg coats as well: five were identified in *Branchiostoma belcheri* by mass spectrometry, termed BbZP1–5 (Xu et al., 2012). Each BbZP has a C-terminal ZP module, and the majority contain a low-density lipoprotein receptor domain and a von Willebrand factor type A domain (Xu et al., 2012). However, none have the EGF-like domain frequently observed in the ZP proteins of urochordates (Xu et al., 2012). Only BbZP1 has a TM domain, and BbZP1, 3, and 4 have CFCSs (Xu et al., 2012). The five BbZPs are synthesized primarily by the developing oocytes (Xu et al., 2012).

Cephalochordate ZP proteins are evolutionary homologs of the ZP1, ZP2, and ZPAX subfamilies, whereas urochordate ZP proteins appear to be more closely related to ZP3 (Xu et al., 2012). Regardless, analyses of cephalochordate and urochordate ZP homologs suggest that vertebrate ZP proteins have an invertebrate origin, or at least arose at the base of chordate evolution (Xu et al., 2012).

6. ZP PROTEINS IN FERTILIZATION

In species-specific mating, gamete recognition ensures that a single sperm fertilizes the egg while preventing polyspermy that can lead to embryo death. The first contact in gamete recognition is mediated by the ECM of the egg coat surrounding ovulated eggs (Baibakov, Boggs, Yauger, Baibakov, & Dean, 2012; Vacquier, 1998). As with the ECM in other tissues, the egg coat ECM is a critical intermediary in cell–cell communication (Wong & Wessel, 2006a). Initial contact with the egg coat triggers a cascade of reactions in sperm, increasing metabolism and motility and initiating the acrosome reaction, releasing the contents of the sperm secretory vesicle into the local environment by exocytosis (Neill & Vacquier, 2004; Okamura & Nishiyama, 1978; Tulsiani, Abou-Haila, Loeser, & Pereira, 1998; Wassarman, 1999). In most animals, these events signify the beginning of successful gamete recognition (Wong & Wessel, 2006a).

The molecules that elicit sperm activation vary significantly among animals, often requiring overlapping receptor–ligand interactions that enhance species specificity (Wong & Wessel, 2006a). Sperm activation results in a shift to chemotactic motility toward the egg coat and initiates the acrosome reaction, exposing additional sperm–egg-binding partners in the form of membrane-associated proteins in the luminal face of the acrosome and soluble proteins released into the local environment (Neill & Vacquier, 2004; Wong & Wessel, 2006a). Some of these soluble proteins aid sperm progression through the egg coat ECM (Wong & Wessel, 2006a). The acrosome reaction thus signifies sperm activation and the successful achievement of initial gamete contact, except in teleost fish whose sperm lack an acrosome (Hart, 1990; Wong & Wessel, 2006a).

It is likely that both ZP2 and ZP3 play important roles in sperm binding. Early experiments in the 1970s demonstrated that solubilized ZP glycoproteins from unfertilized hamster and mouse egg coats could inhibit binding of hamster sperm to ovulated eggs in vitro, suggesting the presence of receptors in the solubilized ZPs that could bind sperm and prevent their interaction with ovulated eggs (Gwatkin & Williams, 1977; Wassarman & Litscher, 2016). In mice, it was found that solubilized ZP3 inhibited sperm binding to ovulated mouse eggs in vitro, whereas solubilized ZP1 and ZP2 did not, suggesting that in mice the egg coat receptor for sperm is ZP3 (Bleil & Wassarman, 1980). ZP3 is able to trigger the acrosome reaction, and acrosome-reacted sperm bind to ZP2, making ZP3 the primary receptor for sperm and ZP2 the secondary receptor (Arnoult, Zeng, & Florman, 1996; Bleil, Greve, & Wassarman, 1988; Bleil & Wassarman, 1983, 1986; Jungnickel, Sutton, Wang, & Florman, 2007; Wassarman & Litscher, 2016).

Over time, it was shown that O-glycans on serine-332 and -334 in the C-terminus of ZP3, located in a so-called sperm combining site, are essential for gamete recognition, and that their postfertilization cleavage could account for another observation in mice: the inability of mouse sperm to bind to solubilized ZPs from two-cell embryos (Baibakov et al., 2012; Chen, Litscher, & Wassarman, 1998; Florman & Wassarman, 1985). However, elimination of glycosyltransferases required for the candidate O-glycans did not affect fertility, and mass spectrometry on native mouse egg coats found no evidence of glycosylation at the two serine residues (Baibakov et al., 2012; Boja, Hoodbhoy, Fales, & Dean, 2003; Williams, Xia, Cummings, McEver, & Stanley, 2007). Furthermore, mutating serine-332 and -334 to preclude modification with O-glycans did not alter sperm recognition or fertility, even if endogenous ZP3 was removed, suggesting that these sites cannot be intrinsically involved in sperm binding (Gahlay, Gauthier, Baibakov, Epifano, & Dean, 2010; Liu, Litscher, & Wassarman, 1995).

Such observations suggest that ZP proteins other than ZP3 may be involved in sperm–egg recognition. Mice lacking ZP1 are fertile, despite reduced fecundity, ruling out an essential role for ZP1 in gamete interactions (Rankin et al., 1999). ZP4 is a pseudogene in mice, and human sperm do not bind to transgenic mouse eggs expressing recombinant human ZP4 (Goudet et al., 2008; Yauger, Boggs, & Dean, 2011). This leaves ZP2 as a candidate.

Although human sperm can bind to the egg coats of *Homo sapiens*, *Gorilla gorilla*, and *Hylobates lar* (gibbon), they do not bind to the egg coats of baboon and rhesus macaque or to other mammals, including mice (Bedford, 1977; Lanzendorf, Holmgren, Johnson, Scobey, & Jeyendran, 1992). Notably, however, replacing mouse ZP2 with human ZP2 is sufficient to permit human sperm to bind to the recombinant mouse egg coat (Baibakov et al., 2012). In this gain-of-function assay, the site of gamete recognition was localized to a domain of ~115 amino acids in the N-terminus of ZP2 (Baibakov et al., 2012).

Following fertilization, the proteolytic cleavage of ZP2 into two fragments of approximately 23 and 90 kDa, which remain disulfide bonded, prevents sperm from binding to two-cell mouse embryos (Bleil, Beall, & Wassarman, 1981). The cleavage of ZP2 is catalyzed by ovastacin, an oocyte-specific member of the astacin family of metalloendoproteases that is released into the perivitelline space upon fusion of the egg cortical granules with the plasma membrane (Bleil & Wassarman, 1980; Burkart, Xiong, Baibakov, Jimenez-Movilla, & Dean, 2012; Moller & Wassarman, 1989; Quesada, Sanchez, Alvarez, & Lopez-Otin, 2004). Mutating the cleavage site of ZP2 or eliminating the gene encoding ovastacin leaves ZP2 intact following fertilization, allowing sperm to bind to the egg coat despite cortical granule exocytosis (Burkart et al., 2012; Gahlay et al., 2010).

Taken together, these observations suggest a molecular basis of gamete recognition in which sperm bind to an N-terminal domain of ZP2, followed by egg coat penetration and gamete fusion (Avella, Xiong, & Dean, 2013). After fertilization, ovastacin is released from egg cortical granules, cleaving extracellular ZP2 and eliminating the sperm-binding domain, accounting for the inability of sperm to bind to two-cell embryos (Burkart et al., 2012; Gahlay et al., 2010).

While ZP2 has been historically considered a secondary egg coat receptor that binds acrosome-reacted sperm, there is precedence for ZP2 acting in primary gamete recognition in *X. laevis*, where its ZP2 homolog gp69/64 inhibits sperm binding to eggs in vitro (Tian, Gong, Thomsen, & Lennarz, 1997). Following fertilization, *X. laevis* ZP2 is cleaved by a zinc metalloprotease, suggesting that postfertilization proteolysis of ZP2 could apply more generally across vertebrates as a block to polyspermy (Lindsay & Hedrick, 2004).

6.1 ZP-N Repeats in Sperm Binding

In addition to their C-terminal ZP module, some ZP proteins have N-terminal extensions that are thought to be involved in species-specific gamete recognition (Callebaut et al., 2007). These extensions are made up of single or multiple copies of the ZP-N domain, which is related to the N-terminal region of the ZP module. The presence of these ZP-N repeats in ZP1, ZP2, ZP4, and ZPAX suggests that ZP proteins may have evolved around a common ZP-N architecture (Callebaut et al., 2007; Han et al., 2010).

With the exception of some homologs of the ZPAX subfamily, ZP-N repeats are not found in the egg coat proteins of fish, whose ZP proteins play a structural role like their counterparts in other metazoans but are likely not involved in sperm binding due to the presence of the micropyle (Litscher & Wassarman, 2007; Monne & Jovine, 2011). Further, although the ZP module is conserved in hundreds of extracellular proteins unrelated to fertilization, none of these have been shown to possess ZP-N repeats (Jovine et al., 2005). These considerations suggest that ZP-N repeats could have functional roles specific to fertilization, in agreement with what is known about the ZP-N architecture and sperm-binding properties of ZP2 (Avella, Baibakov, & Dean, 2014; Baibakov et al., 2012; Bleil et al., 1988; Callebaut et al., 2007; Gahlay et al., 2010; Monne & Jovine, 2011).

Notably, the mollusk and ascidian egg coat proteins VERL and HrVC70 both contain C-terminal ZP modules preceded by repeats that have been implicated in sperm binding (Ban et al., 2005; Galindo et al., 2002; Sawada et al., 2004; Swanson & Vacquier, 1997). While the 12 repeats of HrVC70 are EGF-like domains, the 22 tandem repeats of abalone VERL are known to adopt a ZP-N fold (Sawada et al., 2004; Swanson et al., 2011; Wilburn & Swanson, 2016).

Furthermore, a ZP-N domain in the N-terminal region of VEZP14, another abalone egg coat protein thought to be involved in sperm binding, facilitated the discovery of a ZP-N-like fold in the Ig III domain of the yeast protein alpha-agglutinin/Sag1p by fold recognition analysis (Aagaard et al., 2010; Swanson et al., 2011). This Sag1p Ig III ZP-N is directly involved in haploid yeast cell interactions during mating, a process that mirrors sperm–egg interactions in higher eukaryotes (Dranginis, Rauceo, Coronado, & Lipke, 2007; Monne & Jovine, 2011). Despite a separation of almost 1 billion years of evolution, gamete recognition in metazoans and mating in yeast appear to share common structural features, potentially mediated by the ZP-N domain (Monne & Jovine, 2011; Swanson et al., 2011).

7. EVOLUTION OF EGG COAT PROTEINS

The rapid evolution of reproductive proteins is a recurring observation among natural populations of animals (Turner & Hoekstra, 2008; Wilburn & Swanson, 2016). Because reproductive proteins regulate an essential process—fertilization—that fundamentally impacts fitness, the rapid divergence characteristic of reproductive proteins is striking and suggests that they may evolve under adaptive evolution (Clark, Aagaard, & Swanson, 2006).

The first evidence for rapid evolution of reproductive proteins came from free-spawning marine invertebrates such as abalone and sea urchins (Kresge, Vacquier, & Stout, 2001; Zigler et al., 2005). Whereas the evolution of gamete interactions in internally fertilizing species can be confounded by physiological and behavioral aspects of mating that may also be under selection, in taxa that release millions of gametes into the external environment for fertilization, gamete interactions can be readily observed and the targets of selection are clear (Clark et al., 2006; Kosman & Levitan, 2014; Turner & Hoekstra, 2008; Vacquier & Swanson, 2011). External fertilization is also thought to be the ancestral mating strategy, providing insight into how gametes in general and gametic compatibility in particular have evolved (Kosman & Levitan, 2014; Ruppert, Fox, & Barnes, 2004).

Since their inception in marine invertebrates, studies on reproductive protein evolution have accumulated in diverse taxa. Evolutionary sequence analyses from insects to vertebrates have found evidence for the rapid, adaptive evolution of reproductive proteins with functionally diverse roles (Turner & Hoekstra, 2008). This rapid evolution of reproductive proteins can contribute to reproductive isolation between diverging taxa, and the evolution of reproductive isolation is integral to the process of speciation (Coyne & Orr, 2004; Hart et al., 2018; Meslin et al., 2012; Turner & Hoekstra, 2008). For instance, rapid evolution may drive changes to amino acids important in sperm–egg interaction, creating or reinforcing reproductive barriers (Meslin et al., 2012).

Rapid evolution can be tested for by comparing the frequency of nucleotide substitutions at codons within genes, usually between species (Wilburn & Swanson, 2016). In the absence of selection, most nucleotide (and thus amino acid) substitutions will occur at a rate that reflects the basal mutation rate. The frequency of synonymous substitutions (d_S) provides an estimate of this rate, and under neutral evolution nonsynonymous substitutions (d_N) should occur at a similar frequency, leading to a d_N/d_S ratio of approximately 1. However, since most nonsynonymous substitutions are likely to alter protein structure and negatively impact function, nonsynonymous substitutions should occur less frequently and d_N/d_S will be less than 1. Sites where nonsynonymous substitutions are disfavored are considered under negative or purifying selection (Goldman & Yang, 1994; Wong, Yang, Goldman, & Nielsen, 2004;

Yang, 1998, 2007). Alternatively, in situations where rapid change may be adaptive, nonsynonymous substitutions can occur more frequently than the mutation rate and d_N/d_S will be greater than 1, reflecting positive Darwinian selection (Claw & Swanson, 2012; Goldman & Yang, 1994; Wilburn & Swanson, 2016).

Numerous evolutionary forces have been implicated in the rapid evolution of reproductive proteins, including pathogen resistance, sperm competition, cryptic female choice, sexual conflict, reinforcement, and avoidance of heterospecific fertilization (Clark et al., 2006; Howard, 1999; Kosman & Levitan, 2014; Swanson & Vacquier, 2002a, 2002b; Turner & Hoekstra, 2008). These mechanisms reflect the potential for selection to act on reproductive proteins at various stages throughout fertilization (Turner & Hoekstra, 2008).

To accomplish fertilization, sperm and egg must encounter foreign molecules in the form of the gametes of the opposite sex (Turner & Hoekstra, 2008). Gametes are additionally exposed to novel environments in the process of fertilization, whether as sperm traveling through the female reproductive tract, or egg and sperm released into the surrounding environment in externally fertilizing species. These exposures render gametes vulnerable to microbial pathogens, making pathogen resistance an exogenous force that can drive the divergence of reproductive proteins (Clark et al., 2006). For instance, microbial attack may impose a constant selective pressure for diversification of gamete surface proteins, necessitating continual adaptation of reproductive proteins to this changing recognition surface (Clark et al., 2006; Vacquier, Swanson, & Lee, 1997).

Interactions between gametes represent another source of selection. In species where females mate with multiple males, females may exhibit cryptic female choice in choosing between sperm of different males (Turner & Hoekstra, 2008). In males, competition among sperm can lead to strong selection to rapidly penetrate and fertilize the egg (Turner & Hoekstra, 2008). Females, however, favor a lower fertilization rate because the larger energy investment of their gametes means polyspermy is more detrimental to female fitness. Consequently, female gamete proteins will evolve to lower the fertilization rate, and sperm proteins will evolve to increase it (Clark et al., 2006). Such opposing optimal fertilization rates can lead to sexual conflict at the gamete level, maintaining polymorphism and leading to a coevolutionary arms race between sperm and egg proteins (Clark et al., 2006; Gavrilets, 2000; Palumbi, 1999; Rice & Holland, 1997; Turner & Hoekstra, 2008).

Heterospecific interactions are also subject to selection. For instance, reinforcement is a selective force driven by the negative consequences of hybridization: if gametes from different species meet and the resulting hybrids have reduced fitness, changes to protein interactions that reduce heterospecific fertilization will be favored, leading to the evolution of reproductive barriers (Kosman & Levitan, 2014; Turner & Hoekstra, 2008). Species that overlap geographically may show only partial gametic incompatibility, so reinforcement is predicted to be strongest in sympatric populations, as opposed to allopatric populations where no heterospecific sperm is present (Dobzhansky, 1940; Kosman & Levitan, 2014; Lessios, 2007).

These and other selective forces acting on reproductive proteins can be challenging to separate and may occur simultaneously in a single species (Coyne & Orr, 2004; Turner & Hoekstra, 2008).

7.1 Reproductive Proteins as Species Barriers

Sperm–egg interaction is a species-specific event mediated by the recognition and binding of complementary molecules on the surface of the egg coat and on the sperm plasma membrane (Tulsiani et al., 1998; Vacquier, 1998; Vieira & Miller, 2006). The importance of the egg coat in maintaining species boundaries is evidenced by the fact that its removal permits heterospecific sperm binding to the egg plasma membrane; for instance, guinea pig, mouse, rat, human, rabbit, goat, dolphin, cattle, horse, and pig sperm can all successfully penetrate hamster oocytes lacking an egg coat (Vieira & Miller, 2006; Yanagimachi & Phillips, 1984).

As the first step in gamete recognition, mutations affecting the interface between sperm and the egg coat can create barriers to fertilization that may ultimately lead to speciation—particularly because sperm and egg coat proteins from ascidians to mollusks to humans are known to diverge rapidly as a result of adaptive evolution (Ban et al., 2005; Galindo et al., 2003; Grayson & Civetta, 2012; Meslin et al., 2012; Metz, Robles-Sikisaka, & Vacquier, 1998; Monne et al., 2006; Smith et al., 2005; Swanson, Nielsen, & Yang, 2003; Swanson & Vacquier, 2002a, 2002b; Vieira & Miller, 2006). Experimental studies have demonstrated that adaptive evolution acting on functional domains within reproductive proteins is sufficient to cause reproductive isolation among closely related species, with implications for the mechanisms of speciation (Clark et al., 2006; Coyne & Orr, 2004; Lyon & Vacquier, 1999; Sainudiin et al., 2005; Turner & Hoekstra, 2008).

In support of this hypothesis, variation in gamete recognition proteins within a species influences reproductive compatibility; protein divergence predicts the likelihood of hybrid fertilization better than neutral genetic markers, and polymorphism in sperm–egg recognition proteins determines reproductive success among individuals within a population (Kosman & Levitan, 2014; Levitan, 2012; Levitan & Ferrell, 2006; Levitan & Stapper, 2010; Palumbi, 1999; Zigler et al., 2005). Gametic compatibility, therefore, influences the effectiveness of reproductive isolation across species boundaries (Evans & Marshall, 2005; Kosman & Levitan, 2014; Palumbi, 1999; Vieira & Miller, 2006).

Furthermore, a comparatively small number of nonsynonymous mutations can alter reproductive protein interactions, with as few as 10 amino acid changes in sea urchins leading to incompatibility (Zigler et al., 2005). Nonsynonymous substitutions occur frequently at sperm–egg-binding sites in many species, indicating that these substitutions may influence gamete interactions (Clark, Findlay, Yi, MacCoss, & Swanson, 2007; Hellberg et al., 2012; Kosman & Levitan, 2014; Lyon & Vacquier, 1999; Springer & Crespi, 2007; Torgerson, Kulathinal, & Singh, 2002).

The species-specific reproduction of externally fertilizing marine invertebrates led to the hypothesis that divergence in gamete recognition proteins could lead to speciation (Coyne & Orr, 2004; Vacquier & Swanson, 2011). In sea urchins, sympatric species with overlapping habitats show the highest rates of sperm bindin evolution and the highest sequence polymorphism, with strong blocks to hybrid fertilization caused by the failure of heterospecific sperm bindin to bind to egg coats (Lessios, 2007; Metz, Kane, Yanagimachi, & Palumbi, 1994; Palumbi & Lessios, 2005; Palumbi & Metz, 1991; Vacquier, 1998; Vacquier & Swanson, 2011). Notably, in the three sea urchin genera that contain allopatric species, bindin has been found to be evolving slowly (Vacquier & Swanson, 2011). These observations provide support for the hypothesis of reinforcement, where the rapid evolution of bindin in sympatric species may be a mechanism to prevent hybridization and reinforce species boundaries (Lessios, 2007; Palumbi & Lessios, 2005; Vacquier & Swanson, 2011; Wilburn & Swanson, 2016). Conversely, in allopatric populations with no overlapping habitats, interspecific hybrids cannot occur, eliminating selective pressure to reinforce differences in gamete recognition proteins and thus slowing the evolutionary rate of bindin (Vacquier & Swanson, 2011).

Gamete recognition shows species specificity in abalone as well (Lyon & Vacquier, 1999; Turner & Hoekstra, 2008). In in vitro egg coat dissolution experiments, significantly more lysin is required to dissolve egg coats in heterospecific pairings than in conspecific pairings, indicating that the interaction of lysin and VERL is the species-specific step in abalone fertilization (Swanson & Vacquier, 1997; Turner & Hoekstra, 2008). Swapping lysin amino acid sequences between two species of abalone to create chimeric proteins determined that the N- and C-termini of lysin—in particular the hypervariable N-terminus, which is always species-unique—mediated species-specific recognition of the abalone egg coat (Lyon & Vacquier, 1999). Notably, a recent NMR solution structure of lysin found that its N- and C-termini are close together in physical space, forming a nexus that also contains most of lysin's positively selected residues (Wilburn, Tuttle, Klevit, & Swanson, 2018). These regions are thus good candidates for determining which parts of lysin structurally interact with VERL to maintain species boundaries (Turner & Hoekstra, 2008; Wilburn & Swanson, 2016).

8. FINAL COMMENTS

The egg coats surrounding all metazoan eggs are closely related and share common structural features, such as the ZP module. Furthermore, the genes encoding egg coat proteins likely share a common ancestral gene, potentially ZP3 (Goudet et al., 2008; Litscher & Wassarman, 2014; Shu et al., 2015; Wassarman & Litscher, 2016). Despite the deep homology of the egg's extracellular barrier, however, a frequent observation among egg coat proteins is their rapid evolution.

In humans, approximately 10% of in vitro fertilization (IVF) attempts fail, with no known cause (Liu, Clarke, Martic, Garrett, & Baker, 2001). Such cases of unexplained infertility may be due to incompatibility between sperm–egg recognition proteins, driven by the rapid evolution of reproductive loci (Clark et al., 2006). Correlating this sequence divergence with protein function may help to elucidate the molecular basis of reproductive incompatibilities, with implications for infertility, reproductive isolation, and speciation (Clark et al., 2006; Turner & Hoekstra, 2008; Wilburn & Swanson, 2016).

Anomalies in the structure and thickness of the egg coat are known to affect reproductive fitness, with IVF success being one particularly well-studied aspect; notably, a high degree of sequence variation has been observed in all four human ZP glycoproteins (ZP1–4) in infertile women presenting to IVF clinics (Mannikko et al., 2005; Nayernia et al., 2002; Oehninger et al., 1996; Pokkyla, Lakkakorpi, Nuojua-Huttunen, & Tapanainen, 2011; Sun, Xu, Cao, Su, & Guo, 2005).

On the other hand, ZP glycoproteins are also potential targets for contraception (Paterson, Jennings, Wilson, & Aitken, 2002). Recently, it was shown that the mouse sperm-binding region of ZP2, when conjugated to agarose beads inserted into the uterus of fertile mice, provided a long-acting method of contraception by decoying and binding sperm, preventing their access to the egg (Avella, Baibakov, Jimenez-Movilla, Sadusky, & Dean, 2016). This same construct, when paired with the human ZP2 sperm-binding region, could select for sperm that are better able to bind and penetrate the egg coat, suggesting the suitability of this method for assisted reproductive technologies as well (Avella et al., 2016).

Continued research into the structure, function, and evolution of metazoan egg coat proteins has great potential to contribute to both human health and the field of evolutionary biology.

ACKNOWLEDGMENTS

We would like to thank Dr. Damien Wilburn for his insightful comments on the manuscript. This work was supported by NIH grant R01HD076862 (to W.J.S.) and NIH fellowships T32HG000035-22 and F31HD093441 (to E.E.K.).

REFERENCES

Aagaard, J. E., Vacquier, V. D., MacCoss, M. J., & Swanson, W. J. (2010). ZP domain proteins in the abalone egg coat include a paralog of VERL under positive selection that binds lysin and 18-kDa sperm proteins. *Molecular Biology and Evolution*, *27*(1), 193–203. https://doi.org/10.1093/molbev/msp221.

Aagaard, J. E., Yi, X., MacCoss, M. J., & Swanson, W. J. (2006). Rapidly evolving zona pellucida domain proteins are a major component of the vitelline envelope of abalone eggs. *Proceedings of the National Academy of Sciences of the United States of America*, *103*(46), 17302–17307. 0603125103 [pii].

Alliegro, M. C., & Wright, D. A. (1983). Polyspermy inhibition in the oyster, Crassostrea virginica. *The Journal of Experimental Zoology*, *227*(1), 127–137. https://doi.org/10.1002/jez.1402270117.

Amanze, D., & Iyengar, A. (1990). The micropyle: A sperm guidance system in teleost fertilization. *Development (Cambridge, England)*, *109*(2), 495–500.

Amenya, D. A., Chou, W., Li, J., Yan, G., Gershon, P. D., James, A. A., et al. (2010). Proteomics reveals novel components of the Anopheles gambiae eggshell. *Journal of Insect Physiology*, *56*(10), 1414–1419. https://doi.org/10.1016/j.jinsphys.2010.04.013.

Arnoult, C., Zeng, Y., & Florman, H. M. (1996). ZP3-dependent activation of sperm cation channels regulates acrosomal secretion during mammalian fertilization. *The Journal of Cell Biology*, *134*(3), 637–645.

Arukwe, A., & Goksoyr, A. (2003). Eggshell and egg yolk proteins in fish: Hepatic proteins for the next generation: Oogenetic, population, and evolutionary implications of endocrine disruption. *Comparative Hepatology*, *2*(1), 4.

Avella, M. A., Baibakov, B., & Dean, J. (2014). A single domain of the ZP2 zona pellucida protein mediates gamete recognition in mice and humans. *The Journal of Cell Biology*, *205*(6), 801–809. https://doi.org/10.1083/jcb.201404025.

Avella, M. A., Baibakov, B. A., Jimenez-Movilla, M., Sadusky, A. B., & Dean, J. (2016). ZP2 peptide beads select human sperm in vitro, decoy mouse sperm in vivo, and provide reversible contraception. *Science Translational Medicine*, *8*(336), 336ra60. https://doi.org/10.1126/scitranslmed.aad9946.

Avella, M. A., Xiong, B., & Dean, J. (2013). The molecular basis of gamete recognition in mice and humans. *Molecular Human Reproduction*, *19*(5), 279–289. https://doi.org/10.1093/molehr/gat004.

Baibakov, B., Boggs, N. A., Yauger, B., Baibakov, G., & Dean, J. (2012). Human sperm bind to the N-terminal domain of ZP2 in humanized zonae pellucidae in transgenic mice. *The Journal of Cell Biology*, *197*(7), 897–905. https://doi.org/10.1083/jcb.201203062.

Ban, S., Harada, Y., Yokosawa, H., & Sawada, H. (2005). Highly polymorphic vitelline-coat protein HaVC80 from the ascidian, Halocynthia aurantium: Structural analysis and involvement in self/nonself recognition during fertilization. *Developmental Biology*, *286*(2), 440–451. S0012-1606(05)00523-3 [pii].

Bausek, N., Waclawek, M., Schneider, W. J., & Wohlrab, F. (2000). The major chicken egg envelope protein ZP1 is different from ZPB and is synthesized in the liver. *The Journal of Biological Chemistry*, *275*(37), 28866–28872. 275/37/28866 [pii].

Bedford, J. M. (1977). Sperm/egg interaction: The specificity of human spermatozoa. *The Anatomical Record*, *188*(4), 477–487. https://doi.org/10.1002/ar.1091880407.

Berois, N., Arezo, M. J., & Papa, N. G. (2011). Gamete interactions in teleost fish: The egg envelope. Basic studies and perspectives as environmental biomonitor. *Biological Research*, *44*(2), 119–124. doi:/S0716-97602011000200.

Bleil, J. D., Beall, C. F., & Wassarman, P. M. (1981). Mammalian sperm-egg interaction: Fertilization of mouse eggs triggers modification of the major zona pellucida glycoprotein, ZP2. *Developmental Biology*, *86*(1), 189–197. 0012-1606(81)90329-8 [pii].

Bleil, J. D., Greve, J. M., & Wassarman, P. M. (1988). Identification of a secondary sperm receptor in the mouse egg zona pellucida: Role in maintenance of binding of acrosome-reacted sperm to eggs. *Developmental Biology*, *128*(2), 376–385. 0012-1606(88)90299-0 [pii].

Bleil, J. D., & Wassarman, P. M. (1980). Mammalian sperm-egg interaction: Identification of a glycoprotein in mouse egg zonae pellucidae possessing receptor activity for sperm. *Cell*, *20*(3), 873–882. 0092-8674(80)90334-7 [pii].

Bleil, J. D., & Wassarman, P. M. (1983). Sperm-egg interactions in the mouse: Sequence of events and induction of the acrosome reaction by a zona pellucida glycoprotein. *Developmental Biology*, *95*(2), 317–324. 0012-1606(83)90032-5 [pii].

Bleil, J. D., & Wassarman, P. M. (1986). Autoradiographic visualization of the mouse egg's sperm receptor bound to sperm. *The Journal of Cell Biology*, *102*(4), 1363–1371.

Boja, E. S., Hoodbhoy, T., Fales, H. M., & Dean, J. (2003). Structural characterization of native mouse zona pellucida proteins using mass spectrometry. *The Journal of Biological Chemistry*, *278*(36), 34189–34202. https://doi.org/10.1074/jbc.M304026200.

Bokhove, M., Nishimura, K., Brunati, M., Han, L., de Sanctis, D., Rampoldi, L., et al. (2016). A structured interdomain linker directs self-polymerization of human uromodulin. *Proceedings of the National Academy of Sciences of the United States of America*, *113*(6), 1552–1557. https://doi.org/10.1073/pnas.1519803113.

Bork, P. (1991). Complement components C1r/C1s, bone morphogenic protein 1 and Xenopus laevis developmentally regulated protein UVS.2 share common repeats. *FEBS Letters*, *282*(1), 9–12. 0014-5793(91)80433-4 [pii].

Bork, P., & Beckmann, G. (1993). The CUB domain: A widespread module in developmentally regulated proteins. *Journal of Molecular Biology*, *231*(2), 539–545.

Bork, P., & Sander, C. (1992). A large domain common to sperm receptors (Zp2 and Zp3) and TGF-beta type III receptor. *FEBS Letters*, *300*(3), 237–240. 0014-5793(92)80853-9 [pii].

Brivio, M. F., Bassi, R., & Cotelli, F. (1991). Identification and characterization of the major components of the Oncorhynchus mykiss egg chorion. *Molecular Reproduction and Development*, *28*(1), 85–93. https://doi.org/10.1002/mrd.1080280114.

Buckland-Nicks, J. (1993). Hull cupules of chiton eggs: Parachute structures and sperm focusing devices? *The Biological Bulletin*, *184*(3), 269–276. https://doi.org/10.2307/1542445.

Buckland-Nicks, J., Koss, R., & Chia, F. S. (1988). Fertilization in a chiton: Acrosome-mediated sperm-egg fusion. *Gamete Research*, *21*(3), 199–212. https://doi.org/10.1002/mrd.1120210302.

Burkart, A. D., Xiong, B., Baibakov, B., Jimenez-Movilla, M., & Dean, J. (2012). Ovastacin, a cortical granule protease, cleaves ZP2 in the zona pellucida to prevent polyspermy. *The Journal of Cell Biology*, *197*(1), 37–44. https://doi.org/10.1083/jcb.201112094.

Callard, I. P., Riley, D., & Perez, L. (1990a). Vertebrate vitellogenesis: Molecular model for multihormonal control of gene regulation. *Progress in Clinical and Biological Research*, *342*, 343–348.

Callard, I. P., Riley, D., & Perez, L. (1990b). Vitellogenesis in reptiles as a model for mammalian sex-differentiated hepatic protein synthesis. *The Journal of Experimental Zoology. Supplement*, *4*, 106–111.

Callebaut, I., Mornon, J. P., & Monget, P. (2007). Isolated ZP-N domains constitute the N-terminal extensions of zona pellucida proteins. *Bioinformatics (Oxford, England)*, *23*(15), 1871–1874 [btm265 [pii]].

Carter, J. M., Baker, S. C., Pink, R., Carter, D. R., Collins, A., Tomlin, J., et al. (2013). Unscrambling butterfly oogenesis. *BMC Genomics*, *14*. 283-2164-14-283. https://doi.org/10.1186/1471-2164-14-283.

Chang, Y. S., Hsu, C. C., Wang, S. C., Tsao, C. C., & Huang, F. L. (1997). Molecular cloning, structural analysis, and expression of carp ZP2 gene. *Molecular Reproduction and Development*, *46*(3), 258–267. https://doi.org/10.1002/(SICI)1098-2795(199703)46:33.0.CO;2-O.

Chang, Y. S., Lu, L. F., Lai, C. Y., Kou, Y. H., & Huang, F. L. (1999). Purification, characterization, and molecular cloning of an outer layer protein of carp fertilization envelope. *Molecular Reproduction and Development*, *54*(2), 186–193. https://doi.org/10.1002/(SICI)1098-2795(199910)54:23.0.CO;2-G.

Chen, J., Litscher, E. S., & Wassarman, P. M. (1998). Inactivation of the mouse sperm receptor, mZP3, by site-directed mutagenesis of individual serine residues located at the combining site for sperm. *Proceedings of the National Academy of Sciences of the United States of America*, *95*(11), 6193–6197.

Cherr, G. N., & Clark, W. H., Jr. (1986). Induction of the acrosomal reaction in sperm from the white sturgeon, Acipenser transmontanus. *Advances in Experimental Medicine and Biology*, *207*, 235–249.

Chung, Y. D., Zhu, J., Han, Y., & Kernan, M. J. (2001). nompA encodes a PNS-specific, ZP domain protein required to connect mechanosensory dendrites to sensory structures. *Neuron*, *29*(2), 415–428. S0896-6273(01)00215-X [pii].

Clark, N. L., Aagaard, J. E., & Swanson, W. J. (2006). Evolution of reproductive proteins from animals and plants. *Reproduction (Cambridge, England)*, *131*(1), 11–22. 131/1/11 [pii].

Clark, N. L., Findlay, G. D., Yi, X., MacCoss, M. J., & Swanson, W. J. (2007). Duplication and selection on abalone sperm lysin in an allopatric population. *Molecular Biology and Evolution*, *24*(9), 2081–2090. msm137 [pii].

Clark, N. L., Gasper, J., Sekino, M., Springer, S. A., Aquadro, C. F., & Swanson, W. J. (2009). Coevolution of interacting fertilization proteins. *PLoS Genetics*, *5*(7), e1000570. https://doi.org/10.1371/journal.pgen.1000570.

Claw, K. G., & Swanson, W. J. (2012). Evolution of the egg: New findings and challenges. *Annual Review of Genomics and Human Genetics*, *13*, 109–125. https://doi.org/10.1146/annurev-genom-090711-163745.

Conner, S. J., Lefievre, L., Hughes, D. C., & Barratt, C. L. (2005). Cracking the egg: Increased complexity in the zona pellucida. *Human Reproduction (Oxford, England), 20*(5), 1148–1152. deh835 [pii].

Coyne, J. A., & Orr, H. A. (2004). *Speciation*. Sunderland, MA: Sinauer Associates, Inc.

Darie, C. C., Biniossek, M. L., Gawinowicz, M. A., Milgrom, Y., Thumfart, J. O., Jovine, L., et al. (2005). Mass spectrometric evidence that proteolytic processing of rainbow trout egg vitelline envelope proteins takes place on the egg. *The Journal of Biological Chemistry, 280*(45), 37585–37598. M506709200 [pii].

Darie, C. C., Biniossek, M. L., Jovine, L., Litscher, E. S., & Wassarman, P. M. (2004). Structural characterization of fish egg vitelline envelope proteins by mass spectrometry. *Biochemistry, 43*(23), 7459–7478. https://doi.org/10.1021/bi0495937.

Darie, C. C., Litscher, E. S., & Wassarman, P. M. (2008). Structure, processing, and polymerization of rainbow trout egg vitelline envelope proteins. In C. Popescu, A. D. Zamfir, & N. Dinca (Eds.), *Applications of mass spectrometry in life safety* (pp. 23–36). Dordrecht, The Netherlands: Springer. https://doi.org/10.1007/978-1-4020-8811-7_2.

Dean, J. (2004). Reassessing the molecular biology of sperm-egg recognition with mouse genetics. *BioEssays: News and Reviews in Molecular, Cellular and Developmental Biology, 26*(1), 29–38. https://doi.org/10.1002/bies.10412.

Degrugillier, M. E., & Leopold, R. A. (1976). Ultrastructure of sperm penetration of house fly eggs. *Journal of Ultrastructure Research, 56*(3), 312–325.

Delsuc, F., Brinkmann, H., Chourrout, D., & Philippe, H. (2006). Tunicates and not cephalochordates are the closest living relatives of vertebrates. *Nature, 439*, 965.

Dobzhansky, T. (1940). Speciation as a stage in evolutionary divergence. *The American Naturalist, 74*(753), 312–321. https://doi.org/10.1086/280899.

Doren, S., Landsberger, N., Dwyer, N., Gold, L., Blanchette-Mackie, J., & Dean, J. (1999). Incorporation of mouse zona pellucida proteins into the envelope of Xenopus laevis oocytes. *Development Genes and Evolution, 209*(6), 330–339. 92090330.427 [pii].

Dranginis, A. M., Rauceo, J. M., Coronado, J. E., & Lipke, P. N. (2007). A biochemical guide to yeast adhesins: Glycoproteins for social and antisocial occasions. *Microbiology and Molecular Biology Reviews: MMBR, 71*(2), 282–294. 71/2/282 [pii].

Evans, J. P., & Marshall, D. J. (2005). Male-by-female interactions influence fertilization success and mediate the benefits of polyandry in the sea urchin Heliocidaris erythrogramma. *Evolution; International Journal of Organic Evolution, 59*(1), 106–112.

Fernandes, I., Chanut-Delalande, H., Ferrer, P., Latapie, Y., Waltzer, L., Affolter, M., et al. (2010). Zona pellucida domain proteins remodel the apical compartment for localized cell shape changes. *Developmental Cell, 18*(1), 64–76. https://doi.org/10.1016/j.devcel.2009.11.009.

Findlay, G. D., & Swanson, W. J. (2010). Proteomics enhances evolutionary and functional analysis of reproductive proteins. *BioEssays: News and Reviews in Molecular, Cellular and Developmental Biology, 32*(1), 26–36. https://doi.org/10.1002/bies.200900127.

Florman, H. M., & Wassarman, P. M. (1985). O-linked oligosaccharides of mouse egg ZP3 account for its sperm receptor activity. *Cell, 41*(1), 313–324. 0092-8674(85)90084-4 [pii].

Focarelli, R., Renieri, T., & Rosati, F. (1988). Polarized site of sperm entrance in the egg of a freshwater bivalve, Unio elongatulus. *Developmental Biology, 127*(2), 443–451.

Focarelli, R., & Rosati, F. (1995). The 220-kDa vitelline coat glycoprotein mediates sperm binding in the polarized egg of Unio elongatulus through O-linked oligosaccharides. *Developmental Biology, 171*(2), 606–614.

Gache, C., Niman, H. L., & Vacquier, V. D. (1983). Monoclonal antibodies to the sea urchin egg vitelline layer inhibit fertilization by blocking sperm adhesion. *Experimental Cell Research, 147*(1), 75–84.

Gahlay, G., Gauthier, L., Baibakov, B., Epifano, O., & Dean, J. (2010). Gamete recognition in mice depends on the cleavage status of an egg's zona pellucida protein. *Science (New York, NY), 329*(5988), 216–219. https://doi.org/10.1126/science.1188178.

Galindo, B. E., Moy, G. W., Swanson, W. J., & Vacquier, V. D. (2002). Full-length sequence of VERL, the egg vitelline envelope receptor for abalone sperm lysin. *Gene, 288*(1–2), 111–117. S0378111902004596 [pii].

Galindo, B. E., Vacquier, V. D., & Swanson, W. J. (2003). Positive selection in the egg receptor for abalone sperm lysin. *Proceedings of the National Academy of Sciences of the United States of America, 100*(8), 4639–4643. https://doi.org/10.1073/pnas.0830022100.

Gavrilets, S. (2000). Rapid evolution of reproductive barriers driven by sexual conflict. *Nature, 403*(6772), 886–889. https://doi.org/10.1038/35002564.

Goldman, N., & Yang, Z. (1994). A codon-based model of nucleotide substitution for protein-coding DNA sequences. *Molecular Biology and Evolution, 11*(5), 725–736.

Goudet, G., Mugnier, S., Callebaut, I., & Monget, P. (2008). Phylogenetic analysis and identification of pseudogenes reveal a progressive loss of zona pellucida genes during evolution of vertebrates. *Biology of Reproduction, 78*(5), 796–806. biolreprod.107.064568 [pii].

Grayson, P., & Civetta, A. (2012). Positive selection and the evolution of izumo genes in mammals. *International Journal of Evolutionary Biology, 2012*. 958164.

Greve, J. M., & Wassarman, P. M. (1985). Mouse egg extracellular coat is a matrix of interconnected filaments possessing a structural repeat. *Journal of Molecular Biology, 181*(2), 253–264. 0022-2836(85)90089-0 [pii].

Griffin-Shea, R., Thireos, G., & Kafatos, F. C. (1982). Organization of a cluster of four chorion genes in Drosophila and its relationship to developmental expression and amplification. *Developmental Biology, 91*(2), 325–336. 0012-1606(82)90039-2 [pii].

Gwatkin, R. B., & Williams, D. T. (1977). Receptor activity of the hamster and mouse solubilized zona pellucida before and after the zona reaction. *Journal of Reproduction and Fertility, 49*(1), 55–59.

Haley, S. A., & Wessel, G. M. (2004). Proteolytic cleavage of the cell surface protein p160 is required for detachment of the fertilization envelope in the sea urchin. *Developmental Biology, 272*(1), 191–202.

Hamodrakas, S. J., Kamitsos, E. I., & Papanikolaou, A. (1984). Laser-Raman spectroscopic studies of the eggshell (chorion) of Bombyx mori. *International Journal of Biological Macromolecules, 6*(6), 333–336.

Han, L., Monne, M., Okumura, H., Schwend, T., Cherry, A. L., Flot, D., et al. (2010). Insights into egg coat assembly and egg-sperm interaction from the X-ray structure of full-length ZP3. *Cell, 143*(3), 404–415. https://doi.org/10.1016/j.cell.2010.09.041.

Hart, N. H. (1990). Fertilization in teleost fishes: Mechanisms of sperm-egg interactions. *International Review of Cytology, 121*, 1–66.

Hart, N. H., & Donovan, M. (1983). Fine structure of the chorion and site of sperm entry in the egg of brachydanio. *Journal of Experimental Zoology, 227*(2), 277–296. https://doi.org/10.1002/jez.1402270212.

Hart, N. H., Pietri, R., & Donovan, M. (1984). The structure of the chorion and associated surface filaments in oryzias—Evidence for the presence of extracellular tubules. *The Journal of Experimental Zoology, 230*(2), 273–296. https://doi.org/10.1002/jez.1402300213.

Hart, M. W., Stover, D. A., Guerra, V., Mozaffari, S. V., Ober, C., Mugal, C. F., et al. (2018). Positive selection on human gamete-recognition genes. *PeerJ, 6*. e4259. https://doi.org/10.7717/peerj.4259.

Hart, M. W., Sunday, J. M., Popovic, I., Learning, K. J., & Konrad, C. M. (2014). Incipient speciation of sea star populations by adaptive gamete recognition coevolution. *Evolution; International Journal of Organic Evolution, 68*(5), 1294–1305. https://doi.org/10.1111/evo.12352.

Hedrick, J. L. (2008). Anuran and pig egg zona pellucida glycoproteins in fertilization and early development. *The International Journal of Developmental Biology*, *52*(5–6), 683–701. https://doi.org/10.1387/ijdb.082580jh.

Heiman, M. G., & Shaham, S. (2009). DEX-1 and DYF-7 establish sensory dendrite length by anchoring dendritic tips during cell migration. *Cell*, *137*(2), 344–355. https://doi.org/10.1016/j.cell.2009.01.057.

Hellberg, M. E., Dennis, A. B., Arbour-Reily, P., Aagaard, J. E., & Swanson, W. J. (2012). The tegula tango: A coevolutionary dance of interacting, positively selected sperm and egg proteins. *Evolution; International Journal of Organic Evolution*, *66*(6), 1681–1694. https://doi.org/10.1111/j.1558-5646.2011.01530.x.

Howard, D. J. (1999). Conspecific sperm and pollen precedence and speciation. *Annual Review of Ecology and Systematics*, *30*(1), 109–132. https://doi.org/10.1146/annurev.ecolsys.30.1.109.

Hughes, D. C. (2007). ZP genes in avian species illustrate the dynamic evolution of the vertebrate egg envelope. *Cytogenetic and Genome Research*, *117*(1–4), 86–91. 000103168 [pii].

Hyllner, S. J., Westerlund, L., Olsson, P. E., & Schopen, A. (2001). Cloning of rainbow trout egg envelope proteins: Members of a unique group of structural proteins. *Biology of Reproduction*, *64*(3), 805–811.

Ichikawa, Y., Matsuzaki, M., Hiyama, G., Mizushima, S., & Sasanami, T. (2016). Sperm-egg interaction during fertilization in birds. *The Journal of Poultry Science*, *53*(3), 173–180. https://doi.org/10.2141/jpsa.0150183.

Iconomidou, V. A., Chryssikos, D. G., Gionis, V., Pavlidis, M. A., Paipetis, A., & Hamodrakas, S. J. (2000). Secondary structure of chorion proteins of the teleostean fish Dentex dentex by ATR FT-IR and FT-Raman spectroscopy. *Journal of Structural Biology*, *132*(2), 112–122. https://doi.org/10.1006/jsbi.2000.4307.

Irles, P., Belles, X., & Piulachs, M. D. (2009). Brownie, a gene involved in building complex respiratory devices in insect eggshells. *PLoS One*, *4*(12), e8353. https://doi.org/10.1371/journal.pone.0008353.

Irles, P., & Piulachs, M. D. (2011). Citrus, a key insect eggshell protein. *Insect Biochemistry and Molecular Biology*, *41*(2), 101–108. https://doi.org/10.1016/j.ibmb.2010.11.001.

Iwamatsu, T., Onitake, K., Matsuyama, K., Satoh, M., & Yukawa, S. (1997). Effect of micropylar morphology and size on rapid sperm entry into the eggs of the medaka. *Zoological Science*, *14*(4), 623–628. https://doi.org/10.2108/zsj.14.623.

Iwamoto, K., Ikeda, K., Yonezawa, N., Noguchi, S., Kudo, K., Hamano, S., et al. (1999). Disulfide formation in bovine zona pellucida glycoproteins during fertilization: Evidence for the involvement of cystine cross-linkages in hardening of the zona pellucida. *Journal of Reproduction and Fertility*, *117*(2), 395–402.

Jagadeeshan, S., & Singh, R. S. (2007). Rapid evolution of outer egg membrane proteins in the Drosophila melanogaster subgroup: A case of ecologically driven evolution of female reproductive traits. *Molecular Biology and Evolution*, *24*(4), 929–938. msm009 [pii].

Jovine, L., Darie, C. C., Litscher, E. S., & Wassarman, P. M. (2005). Zona pellucida domain proteins. *Annual Review of Biochemistry*, *74*, 83–114. https://doi.org/10.1146/annurev.biochem.74.082803.133039.

Jovine, L., Janssen, W. G., Litscher, E. S., & Wassarman, P. M. (2006). The PLAC1-homology region of the ZP domain is sufficient for protein polymerisation. *BMC Biochemistry*, 7. 11-2091-7-11. 1471-2091-7-11 [pii].

Jovine, L., Qi, H., Williams, Z., Litscher, E., & Wassarman, P. M. (2002). The ZP domain is a conserved module for polymerization of extracellular proteins. *Nature Cell Biology*, *4*(6), 457–461. https://doi.org/10.1038/ncb802.

Jovine, L., Qi, H., Williams, Z., Litscher, E. S., & Wassarman, P. M. (2004). A duplicated motif controls assembly of zona pellucida domain proteins. *Proceedings of the National Academy of Sciences of the United States of America*, *101*(16), 5922–5927. https://doi.org/10.1073/pnas.0401600101.

Jungnickel, M. K., Sutton, K. A., Wang, Y., & Florman, H. M. (2007). Phosphoinositide-dependent pathways in mouse sperm are regulated by egg ZP3 and drive the acrosome reaction. *Developmental Biology, 304*(1), 116–126. S0012-1606(06)01458-8 [pii].

Kafatos, F. C., Spoerel, N., Mitsialis, S. A., Nguyen, H. T., Romano, C., Lingappa, J. R., et al. (1987). Developmental control and evolution in the chorion gene families of insects. In J. G. Scandalios & E. W. Caspari (Eds.), *Molecular genetics of development, Advances in genetics: Vol. 24* (pp. 223–242). Academic Press.

Kamei, N., & Glabe, C. G. (2003). The species-specific egg receptor for sea urchin sperm adhesion is EBR1,a novel ADAMTS protein. *Genes & Development, 17*(20), 2502–2507. https://doi.org/10.1101/gad.1133003.

Kanamori, A., Naruse, K., Mitani, H., Shima, A., & Hori, H. (2003). Genomic organization of ZP domain containing egg envelope genes in medaka (Oryzias latipes). *Gene, 305*(1), 35–45. S0378111902012118 [pii].

Konstandi, O. A., Papassideri, I. S., Stravopodis, D. J., Kenoutis, C. A., Hasan, Z., Katsorchis, T., et al. (2005). The enzymatic component of Drosophila melanogaster chorion is the Pxd peroxidase. *Insect Biochemistry and Molecular Biology, 35*(9), 1043–1057.

Kosman, E. T., & Levitan, D. R. (2014). Sperm competition and the evolution of gametic compatibility in externally fertilizing taxa. *MHR: Basic Science of Reproductive Medicine, 20*(12), 1190–1197.

Kresge, N., Vacquier, V. D., & Stout, C. D. (2001). Abalone lysin: The dissolving and evolving sperm protein. *BioEssays: News and Reviews in Molecular, Cellular and Developmental Biology, 23*(1), 95–103. https://doi.org/10.1002/1521-1878(200101)23:13.0.CO;2-C.

Kubo, H., Kawano, T., Tsubuki, S., Kawashima, S., Katagiri, C., & Suzuki, A. (1997). A major glycoprotein of Xenopus egg vitelline envelope, gp41, is a frog homolog of mammalian ZP3. *Development, Growth & Differentiation, 39*(4), 405–417.

Kubo, H., Kawano, T., Tsubuki, S., Kotani, M., Kawasaki, H., & Kawashima, S. (2000). Egg envelope glycoprotein gp37 as a Xenopus homolog of mammalian ZP1, based on cDNA cloning. *Development, Growth & Differentiation, 42*(4), 419–427.

Kubo, H., Matsushita, M., Kotani, M., Kawasaki, H., Saido, T. C., Kawashima, S., et al. (1999). Molecular basis for oviductin-mediated processing from gp43 to gp41, the predominant glycoproteins of Xenopus egg envelopes. *Developmental Genetics, 25*(2), 123–129. https://doi.org/10.1002/(SICI)1520-6408(1999)25:23.0.CO;2-3.

Kurn, U., Sommer, F., Bosch, T. C., & Khalturin, K. (2007). In the urochordate Ciona intestinalis zona pellucida domain proteins vary among individuals. *Developmental and Comparative Immunology, 31*(12), 1242–1254. S0145-305X(07)00063-8 [pii].

Lambert, C. C., & Goode, C. A. (1992). Glycolipid linkage of a polyspermy blocking glycosidase to the ascidian egg surface. *Developmental Biology, 154*(1), 95–100. 0012-1606(92)90051-H [pii].

Lanzendorf, S. E., Holmgren, W. J., Johnson, D. E., Scobey, M. J., & Jeyendran, R. S. (1992). Hemizona assay for measuring zona binding in the lowland gorilla. *Molecular Reproduction and Development, 31*(4), 264–267. https://doi.org/10.1002/mrd.1080310407.

Lessios, H. A. (2007). Reproductive isolation between species of sea urchins. *Bulletin of Marine Science, 81*(2), 191–208.

Levitan, D. R. (2012). Contemporary evolution of sea urchin gamete-recognition proteins: Experimental evidence of density-dependent gamete performance predicts shifts in allele frequencies over time. *Evolution; International Journal of Organic Evolution, 66*(6), 1722–1736. https://doi.org/10.1111/j.1558-5646.2012.01608.x.

Levitan, D. R., & Ferrell, D. L. (2006). Selection on gamete recognition proteins depends on sex, density, and genotype frequency. *Science (New York, NY), 312*(5771), 267–269. 312/5771/267 (pii).

Levitan, D. R., & Stapper, A. P. (2010). Simultaneous positive and negative frequency-dependent selection on sperm bindin, a gamete recognition protein in the sea urchin

Strongylocentrotus purpuratus. *Evolution; International Journal of Organic Evolution*, *64*(3), 785–797. https://doi.org/10.1111/j.1558-5646.2009.00850.x.

Li, J., Hodgeman, B. A., & Christensen, B. M. (1996). Involvement of peroxidase in chorion hardening in Aedes aegypti. *Insect Biochemistry and Molecular Biology*, *26*(3), 309–317.

Lindsay, L. L., & Hedrick, J. L. (2004). Proteolysis of Xenopus laevis egg envelope ZPA triggers envelope hardening. *Biochemical and Biophysical Research Communications*, *324*(2), 648–654. S0006-291X(04)02164-3 (pii).

Lindsay, L. L., Wallace, M. A., & Hedrick, J. L. (2001). A hatching enzyme substrate in the Xenopus laevis egg envelope is a high molecular weight ZPA homolog. *Development, Growth & Differentiation*, *43*(3), 305–313. dgd577 [pii].

Lindsay, L. L., Yang, J. C., & Hedrick, J. L. (2002). Identification and characterization of a unique Xenopus laevis egg envelope component, ZPD. *Development, Growth & Differentiation*, *44*(3), 205–212. 635 [pii].

Litscher, E. S., Qi, H., & Wassarman, P. M. (1999). Mouse zona pellucida glycoproteins mZP2 and mZP3 undergo carboxy-terminal proteolytic processing in growing oocytes. *Biochemistry*, *38*(38), 12280–12287. bi991154y [pii].

Litscher, E. S., & Wassarman, P. M. (2007). Egg extracellular coat proteins: From fish to mammals. *Histology and Histopathology*, *22*(3), 337–347.

Litscher, E. S., & Wassarman, P. M. (2014). Evolution, structure, and synthesis of vertebrate egg-coat proteins. *Trends in Developmental Biology*, *8*, 65–76.

Litscher, E. S., Williams, Z., & Wassarman, P. M. (2009). Zona pellucida glycoprotein ZP3 and fertilization in mammals. *Molecular Reproduction and Development*, *76*(10), 933–941. https://doi.org/10.1002/mrd.21046.

Liu, D. Y., Clarke, G. N., Martic, M., Garrett, C., & Baker, H. W. (2001). Frequency of disordered zona pellucida (ZP)-induced acrosome reaction in infertile men with normal semen analysis and normal spermatozoa-ZP binding. *Human Reproduction (Oxford, England)*, *16*(6), 1185–1190.

Liu, C., Litscher, E. S., Mortillo, S., Sakai, Y., Kinloch, R. A., Stewart, C. L., et al. (1996). Targeted disruption of the mZP3 gene results in production of eggs lacking a zona pellucida and infertility in female mice. *Proceedings of the National Academy of Sciences of the United States of America*, *93*(11), 5431–5436.

Liu, C., Litscher, E. S., & Wassarman, P. M. (1995). Transgenic mice with reduced numbers of functional sperm receptors on their eggs reproduce normally. *Molecular Biology of the Cell*, *6*(5), 577–585.

Lombardi, J. (1998). *Comparative vertebrate reproduction*. Boston: Kluwer Academic Publishers.

Longo, F. J. (1981). Effects of concanavalin A on the fertilization of sea urchin eggs. *Developmental Biology*, *82*(1), 197–202.

Lyon, J. D., & Vacquier, V. D. (1999). Interspecies chimeric sperm lysins identify regions mediating species-specific recognition of the abalone egg vitelline envelope. *Developmental Biology*, *214*(1), 151–159.

Mannikko, M., Tormala, R. M., Tuuri, T., Haltia, A., Martikainen, H., Ala-Kokko, L., et al. (2005). Association between sequence variations in genes encoding human zona pellucida glycoproteins and fertilization failure in IVF. *Human Reproduction (Oxford, England)*, *20*(6), 1578–1585. deh837 [pii].

Marinotti, O., Ngo, T., Kojin, B. B., Chou, S. P., Nguyen, B., Juhn, J., et al. (2014). Integrated proteomic and transcriptomic analysis of the Aedes aegypti eggshell. *BMC Developmental Biology*, *14*. https://doi.org/10.1186/1471-213X-14-15.

Marlow, F. L., & Mullins, M. C. (2008). Bucky ball functions in Balbiani body assembly and animal-vegetal polarity in the oocyte and follicle cell layer in zebrafish. *Developmental Biology*, *321*(1), 40–50. https://doi.org/10.1016/j.ydbio.2008.05.557.

Menkhorst, E., & Selwood, L. (2008). Vertebrate extracellular preovulatory and postovulatory egg coats. *Biology of Reproduction*, *79*(5), 790–797. https://doi.org/10.1095/biolreprod.108.068551.

Meslin, C., Mugnier, S., Callebaut, I., Laurin, M., Pascal, G., Poupon, A., et al. (2012). Evolution of genes involved in gamete interaction: Evidence for positive selection, duplications and losses in vertebrates. *PLoS One*, 7(9). e44548. https://doi.org/10.1371/journal.pone.0044548.

Metz, E. C., Kane, R. E., Yanagimachi, H., & Palumbi, S. R. (1994). Fertilization between closely related sea urchins is blocked by incompatibilities during sperm-egg attachment and early stages of fusion. *The Biological Bulletin*, *187*(1), 23–34. https://doi.org/10.2307/1542162.

Metz, E. C., Robles-Sikisaka, R., & Vacquier, V. D. (1998). Nonsynonymous substitution in abalone sperm fertilization genes exceeds substitution in introns and mitochondrial DNA. *Proceedings of the National Academy of Sciences of the United States of America*, *95*(18), 10676–10681.

Mold, D. E., Kim, I. F., Tsai, C. M., Lee, D., Chang, C. Y., & Huang, R. C. (2001). Cluster of genes encoding the major egg envelope protein of zebrafish. *Molecular Reproduction and Development*, *58*(1), 4–14. https://doi.org/10.1002/1098-2795(200101)58:13.0.CO;2-P.

Moller, C. C., & Wassarman, P. M. (1989). Characterization of a proteinase that cleaves zona pellucida glycoprotein ZP2 following activation of mouse eggs. *Developmental Biology*, *132*(1), 103–112. 0012-1606(89)90209-1 [pii].

Monne, M., Han, L., & Jovine, L. (2006). Tracking down the ZP domain: From the mammalian zona pellucida to the molluscan vitelline envelope. *Seminars in Reproductive Medicine*, *24*(4), 204–216. https://doi.org/10.1055/s-2006-948550.

Monne, M., Han, L., Schwend, T., Burendahl, S., & Jovine, L. (2008). Crystal structure of the ZP-N domain of ZP3 reveals the core fold of animal egg coats. *Nature*, *456*(7222), 653–657. https://doi.org/10.1038/nature07599.

Monne, M., & Jovine, L. (2011). A structural view of egg coat architecture and function in fertilization. *Biology of Reproduction*, *85*(4), 661–669. https://doi.org/10.1095/biolreprod.111.092098.

Mouzaki, D. G., Zarani, F. E., & Margaritis, L. H. (1991). Structure and morphogenesis of the eggshell and micropylar apparatus in the olive fly, Dacus oleae (diptera: Tephritidae). *Journal of Morphology*, *209*(1), 39–52. https://doi.org/10.1002/jmor.1052090105.

Mozingo, N. M., Vacquier, V. D., & Chandler, D. E. (1995). Structural features of the abalone egg extracellular matrix and its role in gamete interaction during fertilization. *Molecular Reproduction and Development*, *41*(4), 493–502. https://doi.org/10.1002/mrd.1080410412.

Nadel, M. R., & Kafatos, F. C. (1980). Specific protein synthesis in cellular differentiation: IV. The chorion proteins of Bombyx mori and their program of synthesis. *Developmental Biology*, *75*(1), 26–40.

Nara, M., Yonezawa, N., Shimada, T., Takahashi, K., Tanokura, M., Yumoto, F., et al. (2006). Fourier transform infrared spectroscopic analysis of the intact zona pellucida of the mammalian egg: Changes in the secondary structure of bovine zona pellucida proteins during fertilization. *Experimental Biology and Medicine (Maywood, NJ)*, *231*(2), 166–171. 231/2/166 [pii].

Nayernia, K., Adham, I. M., Shamsadin, R., Muller, C., Sancken, U., & Engel, W. (2002). Proacrosin-deficient mice and zona pellucida modifications in an experimental model of multifactorial infertility. *Molecular Human Reproduction*, *8*(5), 434–440.

Neill, A. T., & Vacquier, V. D. (2004). Ligands and receptors mediating signal transduction in sea urchin spermatozoa. *Reproduction (Cambridge, England)*, *127*(2), 141–149. https://doi.org/10.1530/rep.1.00085.

Neubaum, D. M., & Wolfner, M. F. (1999). Wise, winsome, or weird? Mechanisms of sperm storage in female animals. *Current Topics in Developmental Biology*, *41*, 67–97. S0070-2153(08)60270-7 [pii].

Niman, H. L., Hough-Evans, B. R., Vacquier, V. D., Britten, R. J., Lerner, R. A., & Davidson, E. H. (1984). Proteins of the sea urchin egg vitelline layer. *Developmental Biology*, *102*(2), 390–401.

Oehninger, S., Hinsch, E., Pfisterer, S., Veeck, L. L., Kolm, P., Schill, W. B., et al. (1996). Use of a specific zona pellucida (ZP) protein 3 antiserum as a clinical marker for human ZP integrity and function. *Fertility and Sterility*, *65*(1), 139–145. S0015-0282(16)58041-8 [pii].

Okamura, F., & Nishiyama, H. (1978). The passage of spermatozoa through the vitelline membrane in the domestic fowl, Gallus gallus. *Cell and Tissue Research*, *188*(3), 497–508.

Okumura, H., Kohno, Y., Iwata, Y., Mori, H., Aoki, N., Sato, C., et al. (2004). A newly identified zona pellucida glycoprotein, ZPD, and dimeric ZP1 of chicken egg envelope are involved in sperm activation on sperm-egg interaction. *The Biochemical Journal*, *384*(Pt 1), 191–199. https://doi.org/10.1042/BJ20040299.

Palumbi, S. R. (1999). All males are not created equal: Fertility differences depend on gamete recognition polymorphisms in sea urchins. *Proceedings of the National Academy of Sciences of the United States of America*, *96*(22), 12632–12637. https://doi.org/10.1073/pnas.96.22.12632.

Palumbi, S. R. (2009). Speciation and the evolution of gamete recognition genes: Pattern and process. *Heredity*, *102*(1), 66–76. https://doi.org/10.1038/hdy.2008.104.

Palumbi, S. R., & Lessios, H. A. (2005). Evolutionary animation: How do molecular phylogenies compare to Mayr's reconstruction of speciation patterns in the sea? *Proceedings of the National Academy of Sciences of the United States of America*, *102*(Suppl 1), 6566–6572. 0501806102 [pii].

Palumbi, S. R., & Metz, E. C. (1991). Strong reproductive isolation between closely related tropical sea urchins (genus Echinometra). *Molecular Biology and Evolution*, *8*(2), 227–239.

Papantonis, A., Swevers, L., & Iatrou, K. (2015). Chorion genes: A landscape of their evolution, structure, and regulation. *Annual Review of Entomology*, *60*, 177–194. https://doi.org/10.1146/annurev-ento-010814-020810.

Parks, S., Wakimoto, B., & Spradling, A. (1986). Replication and expression of an X-linked cluster of Drosophila chorion genes. *Developmental Biology*, *117*(1), 294–305. 0012-1606(86)90372-6 [pii].

Pascucci, T., Perrino, J., Mahowald, A. P., & Waring, G. L. (1996). Eggshell assembly in Drosophila: Processing and localization of vitelline membrane and chorion proteins. *Developmental Biology*, *177*(2), 590–598.

Paterson, M., Jennings, Z. A., Wilson, M. R., & Aitken, R. J. (2002). The contraceptive potential of ZP3 and ZP3 peptides in a primate model. *Journal of Reproductive Immunology*, *53*(1–2), 99–107. S016503780100105X [pii].

Plaza, S., Chanut-Delalande, H., Fernandes, I., Wassarman, P. M., & Payre, F. (2010). From A to Z: Apical structures and zona pellucida-domain proteins. *Trends in Cell Biology*, *20*(9), 524–532. https://doi.org/10.1016/j.tcb.2010.06.002.

Pokkyla, R. M., Lakkakorpi, J. T., Nuojua-Huttunen, S. H., & Tapanainen, J. S. (2011). Sequence variations in human ZP genes as potential modifiers of zona pellucida architecture. *Fertility and Sterility*, *95*(8), 2669–2672. https://doi.org/10.1016/j.fertnstert.2011.01.168.

Polzonetti-Magni, A. M., Mosconi, G., Soverchia, L., Kikuyama, S., & Carnevali, O. (2004). *Multihormonal control of vitellogenesis in lower vertebrates*. Academic Press.

Putnam, N. H., Butts, T., Ferrier, D. E., Furlong, R. F., Hellsten, U., Kawashima, T., et al. (2008). The amphioxus genome and the evolution of the chordate karyotype. *Nature*, *453*(7198), 1064–1071. https://doi.org/10.1038/nature06967.

Qi, H., Williams, Z., & Wassarman, P. M. (2002). Secretion and assembly of zona pellucida glycoproteins by growing mouse oocytes microinjected with epitope-tagged cDNAs for mZP2 and mZP3. *Molecular Biology of the Cell*, *13*(2), 530–541. https://doi.org/10.1091/mbc.01-09-0440.

Quesada, V., Sanchez, L. M., Alvarez, J., & Lopez-Otin, C. (2004). Identification and characterization of human and mouse ovastacin: A novel metalloproteinase similar to hatching enzymes from arthropods, birds, amphibians, and fish. *The Journal of Biological Chemistry*, *279*(25), 26627–26634. https://doi.org/10.1074/jbc.M401588200.

Rankin, T., Familari, M., Lee, E., Ginsberg, A., Dwyer, N., Blanchette-Mackie, J., et al. (1996). Mice homozygous for an insertional mutation in the Zp3 gene lack a zona pellucida and are infertile. *Development (Cambridge, England)*, *122*(9), 2903–2910.

Rankin, T. L., O'Brien, M., Lee, E., Wigglesworth, K., Eppig, J., & Dean, J. (2001). Defective zonae pellucidae in Zp2-null mice disrupt folliculogenesis, fertility and development. *Development (Cambridge, England)*, *128*(7), 1119–1126.

Rankin, T., Talbot, P., Lee, E., & Dean, J. (1999). Abnormal zonae pellucidae in mice lacking ZP1 result in early embryonic loss. *Development (Cambridge, England)*, *126*(17), 3847–3855.

Rice, W. R., & Holland, B. (1997). The enemies within: Intergenomic conflict, interlocus contest evolution (ICE), and the intraspecific red queen. *Behavioral Ecology and Sociobiology*, *41*(1), 1–10. https://doi.org/10.1007/s002650050357.

Ruppert, E., Fox, R., & Barnes, R. (2004). *Invertebrate zoology: A functional evolutionary approach* (7th ed.). Brooks Cole.

Sainudiin, R., Wong, W. S. W., Yogeeswaran, K., Nasrallah, J. B., Yang, Z., & Nielsen, R. (2005). Detecting site-specific physicochemical selective pressures: Applications to the class I HLA of the human major histocompatibility complex and the SRK of the plant sporophytic self-incompatibility system. *Journal of Molecular Evolution*, *60*(3), 315–326. https://doi.org/10.1007/s00239-004-0153-1.

Sano, K., Kawaguchi, M., Watanabe, S., Nagakura, Y., Hiraki, T., & Yasumasu, S. (2013). Inferring the evolution of teleostean zp genes based on their sites of expression. *Journal of Experimental Zoology Part B: Molecular and Developmental Evolution*, *320*(5), 332–343. https://doi.org/10.1002/jez.b.22507.

Sasanami, T., Pan, J., & Mori, M. (2003). Expression of perivitelline membrane glycoprotein ZP1 in the liver of Japanese quail (Coturnix japonica) after in vivo treatment with diethylstilbestrol. *The Journal of Steroid Biochemistry and Molecular Biology*, *84*(1), 109–116. S0960076003000086 [pii].

Sawada, H., Sakai, N., Abe, Y., Tanaka, E., Takahashi, Y., Fujino, J., et al. (2002). Extracellular ubiquitination and proteasome-mediated degradation of the ascidian sperm receptor. *Proceedings of the National Academy of Sciences of the United States of America*, *99*(3), 1223–1228. https://doi.org/10.1073/pnas.032389499.

Sawada, H., Tanaka, E., Ban, S., Yamasaki, C., Fujino, J., Ooura, K., et al. (2004). Self/nonself recognition in ascidian fertilization: Vitelline coat protein HrVC70 is a candidate allorecognition molecule. *Proceedings of the National Academy of Sciences of the United States of America*, *101*(44), 15615–15620. 0401928101 [pii].

Schneider, W. J. (1996). Vitellogenin receptors: Oocyte-specific members of the low-density lipoprotein receptor supergene family. In K. W. Jeon (Ed.), *International review of cytology: Vol. 166* (pp. 103–137). Academic Press.

Serizawa, M., Kinoshita, M., Rodler, D., Tsukada, A., Ono, H., Yoshimura, T., et al. (2011). Oocytic expression of zona pellucida protein ZP4 in Japanese quail (Coturnix japonica). *Animal Science Journal = Nihon Chikusan Gakkaiho*, *82*(2), 227–235. https://doi.org/10.1111/j.1740-0929.2010.00830.x.

Shu, L., Suter, M. J., & Räsänen, K. (2015). Evolution of egg coats: Linking molecular biology and ecology. *Molecular Ecology*, *24*(16), 4052–4073. https://doi.org/10.1111/mec.13283.

Smith, J., Paton, I. R., Hughes, D. C., & Burt, D. W. (2005). Isolation and mapping the chicken zona pellucida genes: An insight into the evolution of orthologous genes in different species. *Molecular Reproduction and Development*, *70*(2), 133–145. https://doi.org/10.1002/mrd.20197.

Spargo, S. C., & Hope, R. M. (2003). Evolution and nomenclature of the zona pellucida gene family. *Biology of Reproduction*, *68*(2), 358–362.

Spradling, A. C. (1981). The organization and amplification of two chromosomal domains containing Drosophila chorion genes. *Cell*, *27*(1 Pt 2), 193–201. 0092-8674(81)90373-1 [pii].
Spradling, A. C., & Mahowald, A. P. (1980). Amplification of genes for chorion proteins during oogenesis in Drosophila melanogaster. *Proceedings of the National Academy of Sciences of the United States of America*, 77(2), 1096–1100.
Springer, S. A., & Crespi, B. J. (2007). Adaptive gamete-recognition divergence in a hybridizing mytilus population. *Evolution; International Journal of Organic Evolution*, *61*(4), 772–783. EVO073 [pii].
Sugiyama, H., & Iuchi, I. (2000). Molecular structure and hardening of egg envelope in fish. *Recent Research Developments in Comparative Biochemistry & Physiology*, *1*(1), 139–161.
Sugiyama, H., Murata, K., Iuchi, I., Nomura, K., & Yamagami, K. (1999). Formation of mature egg envelope subunit proteins from their precursors (choriogenins) in the fish, Oryzias latipes: Loss of partial C-terminal sequences of the choriogenins. *Journal of Biochemistry*, *125*(3), 469–475.
Sun, Y. P., Xu, Y., Cao, T., Su, Y. C., & Guo, Y. H. (2005). Zona pellucida thickness and clinical pregnancy outcome following in vitro fertilization. *International Journal of Gynaecology and Obstetrics: The Official Organ of the International Federation of Gynaecology and Obstetrics*, *89*(3), 258–262. S0020-7292(05)00157-8 [pii].
Swanson, W. J., Aagaard, J. E., Vacquier, V. D., Monne, M., Al, S., Hosseini, H., et al. (2011). The molecular basis of sex: Linking yeast to human. *Molecular Biology and Evolution*, *28*(7), 1963–1966. https://doi.org/10.1093/molbev/msr026.
Swanson, W. J., Aquadro, C. F., & Vacquier, V. D. (2001). Polymorphism in abalone fertilization proteins is consistent with the neutral evolution of the egg's receptor for lysin (VERL) and positive Darwinian selection of sperm lysin. *Molecular Biology and Evolution*, *18*(3), 376–383.
Swanson, W. J., Nielsen, R., & Yang, Q. (2003). Pervasive adaptive evolution in mammalian fertilization proteins. *Molecular Biology and Evolution*, *20*(1), 18–20.
Swanson, W. J., & Vacquier, V. D. (1997). The abalone egg vitelline envelope receptor for sperm lysin is a giant multivalent molecule. *Proceedings of the National Academy of Sciences of the United States of America*, *94*(13), 6724–6729.
Swanson, W. J., & Vacquier, V. D. (1998). Concerted evolution in an egg receptor for a rapidly evolving abalone sperm protein. *Science (New York, NY)*, *281*(5377), 710–712.
Swanson, W. J., & Vacquier, V. D. (2002a). The rapid evolution of reproductive proteins. *Nature Reviews. Genetics*, *3*(2), 137–144. https://doi.org/10.1038/nrg733.
Swanson, W., & Vacquier, V. (2002b). Reproductive protein evolution. *Annual Review of Ecology and Systematics*, *33*, 161–179. https://doi.org/10.1146/annurev.ecolsys.33.010802.150439.
Takeuchi, Y., Cho, R., Iwata, Y., Nishimura, K., Kato, T., Aoki, N., et al. (2001). Morphological and biochemical changes of isolated chicken egg-envelope during sperm penetration: Degradation of the 97-kilodalton glycoprotein is involved in sperm-driven hole formation on the egg-envelope. *Biology of Reproduction*, *64*(3), 822–830.
Tian, J., Gong, H., & Lennarz, W. J. (1999). Xenopus laevis sperm receptor gp69/64 glycoprotein is a homolog of the mammalian sperm receptor ZP2. *Proceedings of the National Academy of Sciences of the United States of America*, *96*(3), 829–834.
Tian, J., Gong, H., Thomsen, G. H., & Lennarz, W. J. (1997). Gamete interactions in Xenopus laevis: Identification of sperm binding glycoproteins in the egg vitelline envelope. *The Journal of Cell Biology*, *136*(5), 1099–1108.
Topfer-Petersen, E., & Calvete, J. J. (1996). Sperm-associated protein candidates for primary zona pellucida-binding molecules: Structure-function correlations of boar spermadhesins. *Journal of Reproduction and Fertility. Supplement*, *50*, 55–61.
Torgerson, D. G., Kulathinal, R. J., & Singh, R. S. (2002). Mammalian sperm proteins are rapidly evolving: Evidence of positive selection in functionally diverse genes. *Molecular Biology and Evolution*, *19*(11), 1973–1980.

Tulsiani, D. R., Abou-Haila, A., Loeser, C. R., & Pereira, B. M. (1998). The biological and functional significance of the sperm acrosome and acrosomal enzymes in mammalian fertilization. *Experimental Cell Research*, *240*(2), 151–164. S0014-4827(98)93943-X [pii].

Turner, L. M., & Hoekstra, H. E. (2008). Causes and consequences of the evolution of reproductive proteins. *The International Journal of Developmental Biology*, *52*(5–6), 769–780. https://doi.org/10.1387/ijdb.082577lt.

Turner, F. R., & Mahowald, A. P. (1976). Scanning electron microscopy of Drosophila embryogenesis: 1. The structure of the egg envelopes and the formation of the cellular blastoderm. *Developmental Biology*, *50*(1), 95–108.

Vacquier, V. D. (1998). Evolution of gamete recognition proteins. *Science (New York, NY)*, *281*(5385), 1995–1998.

Vacquier, V. D., & Swanson, W. J. (2011). Selection in the rapid evolution of gamete recognition proteins in marine invertebrates. *Cold Spring Harbor Perspectives in Biology*, *3*(11). a002931. https://doi.org/10.1101/cshperspect.a002931.

Vacquier, V. D., Swanson, W. J., & Lee, Y. H. (1997). Positive darwinian selection on two homologous fertilization proteins: What is the selective pressure driving their divergence? *Journal of Molecular Evolution*, *44*(Suppl 1), S15–22.

Vieira, A., & Miller, D. J. (2006). Gamete interaction: Is it species-specific? *Molecular Reproduction and Development*, *73*(11), 1422–1429. https://doi.org/10.1002/mrd.20542.

Wassarman, P. M. (1988). Zona pellucida glycoproteins. *Annual Review of Biochemistry*, *57*, 415–442. https://doi.org/10.1146/annurev.bi.57.070188.002215.

Wassarman, P. M. (1999). Mammalian fertilization: Molecular aspects of gamete adhesion, exocytosis, and fusion. *Cell*, *96*(2), 175–183. S0092-8674(00)80558-9 [pii].

Wassarman, P. M. (2008). Zona pellucida glycoproteins. *The Journal of Biological Chemistry*, *283*(36), 24285–24289. https://doi.org/10.1074/jbc.R800027200.

Wassarman, P. M., & Litscher, E. S. (2013). Biogenesis of the mouse egg's extracellular coat, the zona pellucida. *Current Topics in Developmental Biology*, *102*, 243–266. https://doi.org/10.1016/B978-0-12-416024-8.00009-X.

Wassarman, P. M., & Litscher, E. S. (2016). Chapter thirty-one—A bespoke coat for eggs: Getting ready for fertilization. In P. M. Wassarman (Ed.), *Current topics in developmental biology: Vol. 117, Essays on developmental biology, Part B* (pp. 539–552). Academic Press.

Wassarman, P. M., & Mortillo, S. (1991). Structure of the mouse egg extracellular coat, the zona pellucida. *International Review of Cytology*, *130*, 85–110.

Wassarman, P. M., Qi, H., & Litscher, E. S. (1997). Mutant female mice carrying a single mZP3 allele produce eggs with a thin zona pellucida, but reproduce normally. *Proceedings of the Biological Sciences*, *264*(1380), 323–328. https://doi.org/10.1098/rspb.1997.0046.

Watanabe, A., & Onitake, K. (2002). The urodele egg-coat as the apparatus adapted for the internal fertilization. *Zoological Science*, *19*(12), 1341–1347. https://doi.org/10.2108/zsj.19.1341.

Watanabe, A., & Takayama-Watanabe, E. (2014). In silico identification of the genes for sperm-egg interaction in the internal fertilization of the newt Cynops pyrrhogaster. *The International Journal of Developmental Biology*, *58*(10 − 12), 873–879. https://doi.org/10.1387/ijdb.140193aw.

Wilburn, D. B., & Swanson, W. J. (2016). From molecules to mating: Rapid evolution and biochemical studies of reproductive proteins. *Journal of Proteomics*, *135*, 12–25. https://doi.org/10.1016/j.jprot.2015.06.007.

Wilburn, D. B., & Swanson, W. J. (2017). The "ZP domain" is not one, but likely two independent domains. *Molecular Reproduction and Development*, *84*(4), 284–285. https://doi.org/10.1002/mrd.22781.

Wilburn, D. B., Tuttle, L. M., Klevit, R. E., & Swanson, W. J. (2018). Solution structure of sperm lysin yields novel insights into molecular dynamics of rapid protein evolution. *Proceedings of the National Academy of Sciences of the United States of America*, *115*(6), 1310–1315.

Williams, S. A., Xia, L., Cummings, R. D., McEver, R. P., & Stanley, P. (2007). Fertilization in mouse does not require terminal galactose or N-acetylglucosamine on the zona pellucida glycans. *Journal of Cell Science*, *120*(Pt 8), 1341–1349. jcs.004291 [pii].

Wong, J. L., & Wessel, G. M. (2006a). Defending the zygote: Search for the ancestral animal block to polyspermy. *Current Topics in Developmental Biology*, *72*, 1–151. S0070-2153(05)72001-9 [pii].

Wong, J. L., & Wessel, G. M. (2006b). Rendezvin: An essential gene encoding independent, differentially secreted egg proteins that organize the fertilization envelope proteome after self-association. *Molecular Biology of the Cell*, *17*(12), 5241–5252. https://doi.org/10.1091/mbc.E06-07-0634.

Wong, W. S., Yang, Z., Goldman, N., & Nielsen, R. (2004). Accuracy and power of statistical methods for detecting adaptive evolution in protein coding sequences and for identifying positively selected sites. *Genetics*, *168*(2), 1041–1051. 168/2/1041 [pii].

Xu, Q., Li, G., Cao, L., Wang, Z., Ye, H., Chen, X., et al. (2012). Proteomic characterization and evolutionary analyses of zona pellucida domain-containing proteins in the egg coat of the cephalochordate, Branchiostoma belcheri. *BMC Evolutionary Biology*, *12*. 239-2148-12-239 https://doi.org/10.1186/1471-2148-12-239.

Yamada, L., Saito, T., Taniguchi, H., Sawada, H., & Harada, Y. (2009). Comprehensive egg coat proteome of the ascidian Ciona intestinalis reveals gamete recognition molecules involved in self-sterility. *The Journal of Biological Chemistry*, *284*(14), 9402–9410. https://doi.org/10.1074/jbc.M809672200.

Yamagami, K., Hamazaki, T. S., Yasumasu, S., Masuda, K., & Iuchi, I. (1992). Molecular and cellular basis of formation, hardening, and breakdown of the egg envelope in fish. *International Review of Cytology*, *136*, 51–92.

Yanagimachi, R., Cherr, G., Matsubara, T., Andoh, T., Harumi, T., Vines, C., et al. (2013). Sperm attractant in the micropyle region of fish and insect eggs. *Biology of Reproduction*, *88*(2), 47. https://doi.org/10.1095/biolreprod.112.105072.

Yanagimachi, R., & Phillips, D. M. (1984). The status of acrosomal caps of hamster spermatozoa immediately before fertilization in vivo. *Gamete Research*, *9*(1), 1–19. https://doi.org/10.1002/mrd.1120090102.

Yang, Z. (1998). Likelihood ratio tests for detecting positive selection and application to primate lysozyme evolution. *Molecular Biology and Evolution*, *15*(5), 568–573.

Yang, Z. (2007). PAML 4: Phylogenetic analysis by maximum likelihood. *Molecular Biology and Evolution*, *24*(8), 1586–1591. msm088 [pii].

Yang, Z., Swanson, W. J., & Vacquier, V. D. (2000). Maximum-likelihood analysis of molecular adaptation in abalone sperm lysin reveals variable selective pressures among lineages and sites. *Molecular Biology and Evolution*, *17*(10), 1446–1455.

Yauger, B., Boggs, N. A., & Dean, J. (2011). Human ZP4 is not sufficient for taxon-specific sperm recognition of the zona pellucida in transgenic mice. *Reproduction (Cambridge, England)*, *141*(3), 313–319. https://doi.org/10.1530/REP-10-0241.

Zigler, K. S., McCartney, M. A., Levitan, D. R., & Lessios, H. A. (2005). Sea urchin bindin divergence predicts gamete compatibility. *Evolution; International Journal of Organic Evolution*, *59*(11), 2399–2404.

FSC
www.fsc.org
MIX
Papier | Fördert gute Waldnutzung
FSC® C083411

Kolibri 360 GmbH
Ferdinand-Jühlke-Straße 7
99095 Erfurt, Deutschland
produktsicherheit@kolibri360.de